Recent Advances in Organometallic Chemistry
Synthetic, Mechanistic and Medicinal Perspective

Recent Advances in Organometallic Chemistry
Synthetic, Mechanistic and Medicinal Perspective

Edited by

Azaj Ansari
Department of Chemistry, Central University of Haryana, Mahendergarh, Haryana, India

Vinod Kumar
Department of Chemistry, Central University of Haryana, Mahendergarh, Haryana, India

ELSEVIER

Elsevier
Radarweg 29, PO Box 211, 1000 AE Amsterdam, Netherlands
125 London Wall, London EC2Y 5AS, United Kingdom
50 Hampshire Street, 5th Floor, Cambridge, MA 02139, United States

Copyright © 2024 Elsevier Inc. All rights are reserved, including those for text and data mining, AI training, and similar technologies.

Publisher's note: Elsevier takes a neutral position with respect to territorial disputes or jurisdictional claims in its published content, including in maps and institutional affiliations.

No part of this publication may be reproduced or transmitted in any form or by any means, electronic or mechanical, including photocopying, recording, or any information storage and retrieval system, without permission in writing from the publisher. Details on how to seek permission, further information about the Publisher's permissions policies and our arrangements with organizations such as the Copyright Clearance Center and the Copyright Licensing Agency, can be found at our website: www.elsevier.com/permissions.

This book and the individual contributions contained in it are protected under copyright by the Publisher (other than as may be noted herein).

Notices
Knowledge and best practice in this field are constantly changing. As new research and experience broaden our understanding, changes in research methods, professional practices, or medical treatment may become necessary.

Practitioners and researchers must always rely on their own experience and knowledge in evaluating and using any information, methods, compounds, or experiments described herein. In using such information or methods they should be mindful of their own safety and the safety of others, including parties for whom they have a professional responsibility.

To the fullest extent of the law, neither the Publisher nor the authors, contributors, or editors, assume any liability for any injury and/or damage to persons or property as a matter of products liability, negligence or otherwise, or from any use or operation of any methods, products, instructions, or ideas contained in the material herein.

ISBN: 978-0-323-90596-1

For Information on all Elsevier publications
visit our website at https://www.elsevier.com/books-and-journals

Publisher: Candice Janco
Acquisitions Editor: Gabriela Capille
Editorial Project Manager: Dan Egan
Production Project Manager: Rashmi Manoharan
Cover Designer: Matthew Limbert

Typeset by MPS Limited, Chennai, India

Contents

List of contributors xv
About the editors xix

1. **An overview of organometallic compounds: important terms and terminologies** 1
   Manjeet Kumar, Deepak Yadav, Vinod Kumar and Azaj Ansari

   1.1 Introduction 1
   1.2 Important terms and terminologies 7
   1.3 Classifications 10
   1.4 Common methods of preparation 13
   1.5 General mechanisms 15
   1.6 Applications in emerging fields 17
   1.7 Future prospective and scope 20
   1.8 Problems with solutions 20
   1.9 Objective type questions 22
   References 23

2. **Organometallic compounds: bonding and spectral characteristics** 31
   Kirandeep, Kushal Arya, Richa, Vinod Kumar and Ramesh Kataria

   2.1 Introduction 31
   2.2 Bonding in organometallic compounds 32
       2.2.1 Ionic bond 32
       2.2.2 Covalent bond 32
       2.2.3 Electron-deficient bond (polycentric localized bond) 34
       2.2.4 Delocalized bond in polynuclear systems 35
       2.2.5 Dative bonds 36
       2.2.6 Hydrogen bonding in organometallics 36
   2.3 Organometallic compounds containing metal−metal bond 38
   2.4 Spectral characteristics of organometallics 39
   2.5 Nuclear magnetic resonance spectroscopy 39
       2.5.1 Decoupling difference nuclear magnetic resonance spectroscopy 40
       2.5.2 Nuclear overhauser enhancement difference spectroscopy 40
       2.5.3 Heteronuclear experiments 41

vi Contents

2.6	Electron paramagnetic resonance	44
2.7	Electron nuclear double resonance	47
2.8	Electron spin echo envelope modulation	49
2.9	Infrared spectroscopy	50
2.10	Ultraviolet–visible spectroscopy	51
2.11	Raman spectroscopy	52
2.12	X-ray absorption spectroscopy	53
2.13	Summary	55
References		56

3. Frontier organometallic catalysis: traditional methodologies and future implications — 61
Payal Rani, Mohit Saroha, Jayant Sindhu and Ramesh Kataria

3.1	Introduction	61
3.2	Fundamental principles	62
3.3	Common methods of preparation	62
	3.3.1 Reaction of metals with organic halides	62
	3.3.2 Reaction of metals with existing chemical compounds	63
3.4	Mechanism Involved	65
	3.4.1 Oxidative addition	65
3.5	Recent developments with examples	67
3.6	Applications in emerging fields	78
	3.6.1 Photoredox catalysis in medicinal chemistry	78
	3.6.2 Organometallic bio-probes for cellular imaging	79
	3.6.3 Bio-organometallic chemistry	81
	3.6.4 Enzyme inhibition using organometallic compounds	82
	3.6.5 Organometallic compounds in solar cells	83
3.7	Future prospective and scope	86
3.8	Summary	86
3.9	Problems with solutions	87
3.10	Objective type questions	87
References		88

4. Coordination cages and clusters as functional materials — 93
Mohd Zeeshan, Farhat Vakil and M. Shahid

4.1	Introduction	93
	4.1.1 Fundamental principles	97
4.2	Common methods of preparation	97
	4.2.1 Synthesis of $[Dy_6(\mu_6\text{-}O)(\mu_3\text{-}OH)_8(H_2O)_{12}(NO_3)_6](NO_3)_2(H_2O)_2$	98
4.3	Mechanism involved	100
4.4	Recent developments with examples	101
	4.4.1 Applications in emerging fields	103
4.5	Future prospective and scope	108
4.6	Summary	109

4.7	Problems with solution	110
4.8	Objective type questions	111
	References	112

5. C–H activation: A sustainable approach for the synthesis of functionalized heterocycles 115

Indu Sharma, Mukesh Kumari, Gargi Poonia, Sunil Kumar and Ramesh Kataria

5.1	Introduction	115
5.2	Principle of C–H activation	116
5.3	Method of preparation and mechanisms involved in C–H activation: recent applications in organic synthesis	117
	5.3.1 Ruthenium-catalyzed C–H activation	117
	5.3.2 Iron-catalyzed C–H activation	118
	5.3.3 Rhodium-catalyzed C–H activation	119
5.4	Future prospective and scope	133
5.5	Summary	134
	References	136

6. Organotransition metal chemistry in asymmetric catalysis mediated by different transition metal N-heterocyclic carbene complexes 139

Dakoju Ravi Kishore, Pankaj Kalita and Naushad Ahmed

6.1	Introduction	139
6.2	Fundamental principles	141
6.3	Common methods of preparation	142
6.4	Mechanism involved	143
6.5	Recent developments with examples	144
6.6	Applications in emerging fields	148
6.7	Future prospective and scope	150
6.8	Summary	150
6.9	Problems with solutions	150
6.10	Objective type questions	151
	References	152

7. Organophosphorus compounds: recent developments and future perspective 155

Vitthalrao Swamirao Kashid and Azaj Ansari

7.1	Introduction	155
7.2	Fundamental principles	156
	7.2.1 Phosphorus-based ligands	156
	7.2.2 Classification of phosphorus-based ligands	156

	7.3	Common methods of preparation	157
		7.3.1 Pincer ligands	157
		7.3.2 Synthesis of pincer complexes	157
		7.3.3 Oxidative addition to aryl halide bond	159
		7.3.4 *Trans*-metallation	159
		7.3.5 *Trans*-cyclometallation	160
	7.4	Classification of pincer complexes	160
		7.4.1 Transition metal chemistry of pincer ligands	161
		7.4.2 Application of tridentate chelating pincer complexes in catalysis	169
		7.4.3 Chiral tridentate chelating pincer complexes	172
		7.4.4 Aminophosphines	174
		7.4.5 Phosphacyclophanes	179
	7.5	Summary	181
	7.6	Problems with solutions	181
	7.7	Objective-type questions	182
	References	183

8. Synthesis of silicon and germanium organometallic compounds and their applications in catalysis   195

Manoj Kumar Pradhan, Ranjan Kumar Mohapatra, Mohammad Azam, Snehasish Mishra and Azaj Ansari

8.1	Introduction	195
8.2	Synthesizing of silicon and germanium organometallic compounds	199
	8.2.1 Direct synthesis	200
	8.2.2 Hydrosilylation	200
	8.2.3 Silenes	200
	8.2.4 Silicones, silyl ethers, silanols, siloxanes, and siloxides	201
	8.2.5 Bonding through carbon	202
8.3	Catalysis	205
8.4	Future prospective and scope	210
8.5	Summary	211
8.6	Problems with solutions	212
8.7	Objective type questions	213
References	214

9. Sensing application of organometallic compounds   217

Durga Prasad Mishra and Ashish Kumar Sarangi

9.1	Introduction	217
9.2	Fundamental principles	218
9.3	Common methods of preparation	219
9.4	Mechanism involved	219
	9.4.1 Ligand substitution	219
	9.4.2 Oxidative addition	219
	9.4.3 Concerted track based	222

9.5	Recent developments with examples		223
9.6	Applications in emerging fields		225
	9.6.1	Chemical sensor	225
	9.6.2	Gas sensor	228
	9.6.3	Optical sensor	229
	9.6.4	Biochemical sensor	230
9.7	Future prospective and scope		231
9.8	Summary		232
9.9	Problems with solutions		232
9.10	Objective type questions		234
References			236

10. Bioorganometallic chemistry: a new horizon on organometallic landscape   239

Mudasir Ahmad Hafiz, Moniza Qayoom, Tabee Jan, Mohd Mustafa, Tabasum Maqbool and Masood Ahmad Rizvi

10.1	Introduction		239
10.2	Shift from platinum metal based therapeutics to other bio-compatible less toxic nonplatinum complexes		241
	10.2.1	Anticancerous properties of iron complexes	241
	10.2.2	Anticancer properties of copper complexes	243
	10.2.3	Anticancer properties of ruthenium complexes	245
	10.2.4	Anticancer properties of gold complexes	248
	10.2.5	Anticancer properties of rhodium and iridium complexes	248
10.3	Organoselenium compounds as potent chemotherapeutic agents		250
	10.3.1	Inorganic selenium compounds	251
	10.3.2	Organic selenium compounds	252
	10.3.3	Selenocysteine	253
	10.3.4	Selol	253
	10.3.5	Seleninic acids	254
	10.3.6	Selenophene-based derivatives	255
	10.3.7	1,2-Benzisoselenazole-3[2H]-one derivatives	255
	10.3.8	Analysis of anticancer property of selenium-bearing 4-anilinoquinazoline compounds	256
	10.3.9	Evaluation of indole chalcone and diarylketone derivatives containing selenium as tubulin polymerization inhibitory agents	257
	10.3.10	Chemopreventive applications of isoselenocyanate compounds	258
10.4	Other biological activities of organoselenium compounds		259
	10.4.1	GPx activity	259
	10.4.2	Anti-Alzheimer activity	262
	10.4.3	Cytoprotection	264
	10.4.4	Insecticidal activity	266

10.5	Different plausible mechanisms of action for bioorganometallic complexes		268
	10.5.1	Activation by means of hydrolysis	269
	10.5.2	Redox activation	271
	10.5.3	Photoactivation	275
	10.5.4	Ionizing radiation, sonodynamic and thermal activation	276
	10.5.5	Catalytic metallodrugs	278
10.6	Summary		280
10.7	Problems with solutions		281
10.8	Objective type questions		283
References			285

11. Azole-based organometallic compounds as bioactive agents — 287

Krishna, Deepak Yadav, Sunil Kumar, Meenakshi, Aman Kumar and Vinod Kumar

11.1	Introduction		287
11.2	Azoles as antifungal drugs		288
11.3	Mechanistic approach of azole-based organometallic complexes for infection		291
11.4	Azole based metal-complexes and their use in medicinal chemistry		292
11.5	Recent advances of azole-based organometallic compounds		292
	11.5.1	Azole derivatives-based ruthenium organometallic complexes	293
	11.5.2	Manganese based organometallic complexes with azole ligands	295
	11.5.3	Copper-, gold-, and platinum-based organometallic complexes with azole derivatives	296
	11.5.4	Copper- and zinc-based organometallic complexes with azole derivatives	298
	11.5.5	Metallocenyl derivatives with azole-based ligands	298
11.6	Summary		301
11.7	Objective type questions		302
References			303

12. Hybrid organometallic compounds as potent antimalarial agents — 309

Preeti Singh, Yadav Preeti, Badri Parshad, Deepak Yadav, Sushmita and Manjeet Kumar

12.1	Introduction	309
12.2	Life cycle of malaria	311
12.3	Some common examples of antimalarial drugs	312

12.4	An approach to hybrid antimalarial drugs	314
12.5	Hybrid organometallic antimalarial compounds	315
12.6	Recent developments with examples: some representative hybrid organometallic antimalarial agents	317
	12.6.1 Trioxaferroquine containing hybrid organometallic drug	317
	12.6.2 Ferrocene-quinoline containing hybrid organometallic compounds	318
	12.6.3 Organoruthenium aminoquinoline–trioxane hybrid	318
	12.6.4 Organoiridium based antimalarial hybrid complexes	319
	12.6.5 Organorhodium-based antimalarial complexes	320
	12.6.6 Organoosmium-based antimalarial complexes	321
	12.6.7 Organoplatinum-based antimalarial complexes	322
12.7	Future prospective and scope	323
12.8	Summary	324
12.9	Objective type questions	324
References		326

13. Therapeutic approach of polynuclear organometallic complexes   331

Ashish Kumar Sarangi

13.1	Introduction	331
13.2	Fundamental principles	332
13.3	Common methods of preparation	332
	13.3.1 Substitution of ligand on metal carbonyls	333
	13.3.2 Template method of preparation	333
	13.3.3 Metal fragments condensation	334
	13.3.4 Oxidative addition reactions	334
13.4	Mechanism involved	335
	13.4.1 Ligand substitution	335
	13.4.2 Oxidative addition	336
	13.4.3 Reductive elimination	338
	13.4.4 Migratory insertion	339
	13.4.5 Hydrometalation / β-hydride elimination	339
	13.4.6 Transmetalation	340
	13.4.7 Recent developments with examples	340
13.5	Applications in emerging fields	343
	13.5.1 In the field of medicine	343
	13.5.2 In the field of dye sensitized solar cell	344
13.6	Future prospective and scope	345
13.7	Summary	346
Problems with solutions		346
Objective type questions		348
References		350

14. Role of organometallic compounds in neglected tropical diseases 353

Deepak Yadav, Sushmita, Shramila Yadav, Sunil Kumar, Manjeet Kumar and Vinod Kumar

14.1	Introduction	353
14.2	Diseases	354
	14.2.1 Chagas disease	354
	14.2.2 Human African trypanosomiasis	359
	14.2.3 Leishmaniasis	362
	14.2.4 Echinococcosis	365
	14.2.5 Schistosomiasis	367
	14.2.6 Lymphatic filariasis	368
	14.2.7 Dengue	369
14.3	Summary	371
14.4	Problems with solutions	371
14.5	Objective type questions	372
References		373

15. A computational approach toward the role of biomimetic complexes in hydroxylation reactions 379

Monika, Oval Yadav, Manjeet Kumar, Ranjan Kumar Mohapatra, Vitthalrao Swamirao Kashid and Azaj Ansari

15.1	Introduction	379
15.2	Computational details	384
15.3	Recent developments with examples	384
15.4	Mechanism involved	386
	15.4.1 C—H bond activation followed by oxygen rebound mechanism	386
	15.4.2 Hydrogen abstraction followed by oxygen nonrebound mechanism	388
	15.4.3 Direct attack of oxygen atom	388
	15.4.4 Nonradical mechanism	391
	15.4.5 Regioselectivity of aliphatic versus aromatic hydroxylation	392
	15.4.6 Tunneling of reaction pathway from epoxidation to hydroxylation using cyclohexene	392
	15.4.7 Sigma pathway and pie pathway	393
15.5	Factors affecting hydroxylation	395
	15.5.1 Oxidation state (spin state) of metal ion	395
	15.5.2 Nature of ligand	396
	15.5.3 Ligand design	396
	15.5.4 Axial and equatorial ligands	397
	15.5.5 Single-state reactivity (SSR) versus two-state reactivity (TSR)	398

15.6	Applications in emerging fields	399
15.7	Future prospective and scope	401
15.8	Summary	402
Problems with solutions		402
Objective type questions		404
References		406

Index 413

List of contributors

Naushad Ahmed Department of Chemistry, Indian Institute of Technology Hyderabad, Kandi, Sangareddy, Telangana, India

Azaj Ansari Department of Chemistry, Central University of Haryana, Mahendergarh, Haryana, India

Kushal Arya Department of Chemistry and Centre for Advanced Studies in Chemistry, Panjab University, Chandigarh, India

Mohammad Azam Department of Chemistry, College of Science, King Saud University, Riyadh, Saudi Arabia

Mudasir Ahmad Hafiz Department of Chemistry, University of Kashmir, Hazratbal, Jammu and Kashmir, India

Tabee Jan Department of Chemistry, University of Kashmir, Hazratbal, Jammu and Kashmir, India

Pankaj Kalita Department of Chemistry, Nowgong Girls' College, Nagaon, Assam, India

Vitthalrao Swamirao Kashid Department of Chemistry, Gaya College of Engineering, Gaya, Bihar, India; Department of Humanities and Science (Chemistry), Malla Reddy Engineering College for Women, Hyderabad, Telangana, India

Ramesh Kataria Department of Chemistry and Centre for Advanced Studies in Chemistry, Panjab University, Chandigarh, India

Kirandeep Department of Chemistry and Centre for Advanced Studies in Chemistry, Panjab University, Chandigarh, India

Dakoju Ravi Kishore Department of Chemistry, Indian Institute of Technology Hyderabad, Kandi, Sangareddy, Telangana, India

Krishna Department of Chemistry, University of Delhi, New Delhi, Delhi, India

Aman Kumar Department of Chemistry, Central University of Haryana, Mahendergarh, Haryana, India

Manjeet Kumar Department of Chemistry, Central University of Haryana, Mahendergarh, Haryana, India

Sunil Kumar Department of Chemistry, Government P.G. College, Hisar, Haryana, India

Sunil Kumar Department of Chemistry, J. C. Bose University of Science and Technology, YMCA, Faridabad, India

Vinod Kumar Department of Chemistry, Central University of Haryana, Mahendergarh, Haryana, India

Ashish Kumar Sarangi Department of Chemistry, School of Applied Sciences, Centurion University of Technology and Management, Balangir, Odisha, India

Mukesh Kumari K. M. Government College, Narwana, Jind, India; Department of Chemistry (SBAS), Maharaja Agrasen University, Baddi, Solan, India

Tabasum Maqbool Department of Chemistry, University of Kashmir, Hazratbal, Jammu and Kashmir, India

Meenakshi Department of Chemistry, Government P.G. College, Hisar, Haryana, India

Durga Prasad Mishra Department of Pharmaceutical Chemistry, School of Pharmacy, Centurion University of Technology and Management, Balangir, Odisha, India

Snehasish Mishra School of Biotechnology, KIIT Deemed-to-be University, Bhubaneswar, Odisha, India

Ranjan Kumar Mohapatra Department of Chemistry, Government College of Engineering, Keonjhar, Odisha, India

Monika Department of Chemistry, Central University of Haryana, Mahendergarh, Haryana, India

Mohd Mustafa Department of Chemistry, University of Kashmir, Hazratbal, Jammu and Kashmir, India

Badri Parshad Wellman Center for Photomedicine, Massachusetts General Hospital, Harvard Medical School, Boston, MA, United States

Gargi Poonia Department of Chemistry and Centre for Advanced Studies in Chemistry, Panjab University, Chandigarh, India

Manoj Kumar Pradhan Department of Chemistry, Government College of Engineering, Keonjhar, Odisha, India

Yadav Preeti Department of Chemistry, Maitreyi College, University of Delhi, New Delhi, Delhi, India

Moniza Qayoom Department of Chemistry, University of Kashmir, Hazratbal, Jammu and Kashmir, India

Payal Rani Department of Chemistry, COBS&H, CCS Haryana Agricultural University, Hisar, India

Richa Department of Chemistry and Centre for Advanced Studies in Chemistry, Panjab University, Chandigarh, India

Masood Ahmad Rizvi Department of Chemistry, University of Kashmir, Hazratbal, Jammu and Kashmir, India

Mohit Saroha Department of Chemistry, University of Delhi, New Delhi, India

M. Shahid Functional Inorganic Materials Lab (FIML), Department of Chemistry, Aligarh Muslim University, Aligarh, Uttar Pradesh, India

Indu Sharma Department of Chemistry and Centre for Advanced Studies in Chemistry, Panjab University, Chandigarh, India

Jayant Sindhu Department of Chemistry, COBS&H, CCS Haryana Agricultural University, Hisar, India

Preeti Singh Department of Chemistry, Swami Vivekanand Subharti University, Meerut, Uttar Pradesh, India

Sushmita Department of Chemistry, Netaji Subhas University of Technology, Dwarka, Delhi, India

Farhat Vakil Functional Inorganic Materials Lab (FIML), Department of Chemistry, Aligarh Muslim University, Aligarh, Uttar Pradesh, India

Deepak Yadav Department of Chemistry, Gurugram University, Gurugram, Haryana, India

Oval Yadav Department of Chemistry, Central University of Haryana, Mahendergarh, Haryana, India

Shramila Yadav Department of Chemistry, Rajdhani College, University of Delhi, New Delhi, Delhi, India

Mohd Zeeshan Functional Inorganic Materials Lab (FIML), Department of Chemistry, Aligarh Muslim University, Aligarh, Uttar Pradesh, India

About the editors

Dr. Azaj Ansari is working in the Department of Chemistry at Central University of Haryana, Mahendergarh, India since 2016. He completed postgraduation in chemistry from Ranchi University, Ranchi, India, and qualified CSIR-JRF (a highly competitive test for lectureship in India). He completed PhD degree under the supervision of Prof. Gopalan Rajaraman from the Department of Chemistry, Indian Institute of Technology Bombay, Mumbai, India. His research interests focus to understand electronic structures and spectroscopic properties of metal complexes and mechanistic study of chemical reactions using computational approach. He has published more than 50 research articles in journals of national/international repute. He has supervised several PhD and master's students. He has received Best Research Award for the year 2022 and 2023 by Central University of Haryana, Mahendergarh, Haryana, India.

Prof. Vinod Kumar is working in the Department of Chemistry at Central University of Haryana, Mahendergarh, India. He completed postgraduation in chemistry and qualified the NET Examination (a highly competitive test for lectureship in India) in 2001 and obtained PhD degree in 2006 under the supervision of Prof. S. P. Singh and Prof. Ranjana Aggarwal from Kurukshetra University, Kurukshetra. For displaying a high degree of innovation, creativity, and scientific professionalism by making exemplary contribution in the field of chemical sciences, Prof. Kumar received "ISCA-Young Scientist Award" of the Science Congress in 2007 by the then President Dr. A. P. J. Abdul Kalam, "Prof R.C. Shah Memorial Lecture Award" of the Indian Science Congress in 2017, and "Haryana Yuva Vigyan Ratna Award" by the then Governor, Chief Minister, and Minister of Science and Technology, Haryana in 2017. He was invited to visit Korea Atomic Energy Research Institute (KAERI), South Korea, as a foreign expert in 2012. His research interests focus on the structural reinvestigations, greener approaches, and synthetic, mechanistic, and spectral aspects of heterocyclic compounds, mainly azoles of medicinal interests besides synthesis of Schiff bases. He has published more than 90 research papers in journals of international repute and coauthored a book, *Pericyclic Reactions* published by

Academic Press, London, and coedited a book, *Chemical Drug Design* published by De Gruyter, Berlin, Germany. Five Indian patents have been granted in addition to five published patents to Dr. Kumar. He has supervised several students of MSc, MPhil, and PhD for their dissertation or thesis work. He has been serving as an editor, guest editor, and reviewer for many national and international journals.

Chapter 1

An overview of organometallic compounds: important terms and terminologies

Manjeet Kumar[1], Deepak Yadav[2], Vinod Kumar[1] and Azaj Ansari[1]
[1]*Department of Chemistry, Central University of Haryana, Mahendergarh, Haryana, India,*
[2]*Department of Chemistry, Gurugram University, Gurugram, Haryana, India*

1.1 Introduction

Organometallic compounds have a profound preface in the field of inorganic and organic chemistry including catalysis, polymerization and many biological processes [1–4]. These compounds are generally considered for those molecules consisting of organic groups and metal centers, and there should be direct bonding between the metal center and organic groups present in the molecule. The bonding fashion may be furnished either by sigma (σ) or pi (π) bond [5–7]. Presence of organic groups are of different types in organometallic complexes, such as $-CH_3$ and $-C_6H_5$, and these groups have good bond formation ability with both the main group metals (having s and p orbitals) and transition group metals (having d and f orbitals). Also, stable molecules like carbon monoxide (CO), ethane (C_2H_4), or benzene (C_6H_6) can be bonded with transition metal atoms [8–10]. One other dimension of organometallic compounds is that main group/transition metals can form stable bonding with those molecules whose independent existence is incompetent, such as cyclobutadiene (C_4H_4) and trimethaylenemethane (C_4H_6) [11–13].

Prof. Stone remarked organometallic compounds as the growth point of chemical sciences [14]. Organometallic chemistry coalesce both the inorganic and organic chemistry area of research. The importance of this field has been also recognized through world class academic publishing houses, such as Elsevier, releases a special peer-reviewed journal (*Journal of Organometallic Chemistry*) in 1964; which is a monthly journal. Nearly after two decades of this; American Chemical Society also started publishing a special journal (*Organometallics*) from 1982; which is a biweekly journal.

Moreover, just after 5 years in 1985 John Wiley & Sons also started printing a special journal (*Applied Organometallic Chemistry*); which is a peer-reviewed monthly journal. These journals show the exciting research in organometallic chemistry that is being discovered at the interface of material science, biology, medicine, catalysis and energy transformations.

Over the last many decades, organometallic reagents have played important role in the total synthesis of generous organometallic molecules; among them, many compounds are biologically active [15−17]. In organometallic chemistry, the first organometallic compound was synthesized in 1760 by Paris military pharmacy by Louis Claude Cadet de Gassicourt [18]. It was the first main group metal-based organometallic compound; also, this compound played a significant role in the development of organometallic chemistry [19]. For the synthesis of this compound, they were used arsenic metal containing cobalt ores ($CoAs_2$ and $CoAsS_2$). Cadet named this compound as cadet's fuming arsenical liquid. It has an unpleasant odor and is toxic.

$$As_2O_3 + 4\ CH_3COOK \rightarrow [(CH_3)_2As]_2O + 2\ K_2CO_3 + 2\ CO_2$$
$$\text{dimethylarsinous anhydride}$$

Later, the first transition metal-based organometallic compound; $Na[Pt(Cl)_3(C_2H_4)]$ (Zeise's salt) was synthesized in 1827 by William Christopher Zeise [20,21]. In this, alkene bonded to the platinum (Pt) with three chloride ions. After the invention of Zeise's salt, Sir Edward Frankland (one of the originators of organometallic chemistry) also carried out some experiments in searching to isolate the radical of ethyl in 1849. He was taken ethyl iodide and zinc metal in a test tube and accidentally designed a new organometallic compound diethyl zinc (a pyrophoric liquid) [22].

$$3C_2H_5I + 3Zn \rightarrow (C_2H_5)_2Zn + C_2H_5ZnI + ZnI_2$$
$$\text{dimethylzinc}$$

After the discovery of dimethylmercury, Sir Edward Frankland also designed some more organometallic compounds like CH_3HgI in 1852, $(CH_3)_3B$ in 1860, etc. [23−27] Frankland was also given the concept of valency (explain the atoms are bonded in equal ratios in a chemical compound) and pioneer term "**Organometallic.**" In 1852 C. J. Löwig and M. E. Schweizer also contributed to the enhancement of organometallic chemistry. Both worked on the same ligand (i.e., ethyl) with different metal ions. They synthesized tetraethyl lead $(C_2H_5)_4Pb$ (used as petro-fuel additive) and in the subsequent period and they also designed $(C_2H_5)_3Bi$ and $(C_2H_5)_3Sb$ by the same manner [19,28,29]. Both A. Schafarik and W. Hallwachs contributed in developing the chemistry of alkyl aluminum iodides [30]. In 1863 C. Friedel and J. M. Crafts synthesized a new class of organometallic compounds, that is, organochlorosilanes.

$$SiCl_4 + x/2\ ZnR_2 \rightarrow R_xSiCl_{4-x} + x/2\ ZnCl_2$$

Halide free alkyl magnesium compound was optimized by treating diethyl mercury with magnesium metal by J.A. Wanklyn in 1866 [31,32].

$$Hg(C_2H_5)_2 + Mg \rightarrow Mg(C_2H_5)_2 + Hg$$

After the invention of metal carbonyl compounds, organometallic chemistry got a new effervescent path toward more valued complexes from the application point of view. The first carbonyl organometallic complex ([Pt(CO)Cl$_2$]$_2$) was developed by M. P. Schützenberger in 1868 and started a new era of carbonyl chemistry; in 1890 Prof. Ludwig Mond synthesized [Ni(CO)$_4$] by treating nickel (Ni) metal and carbonyl (CO) directly [33,34]. This was pioneering the enhancement in efforts of new transition metal-based carbonyl compounds. Also, it was used for the purification of nickel metal by the end of the 19th century in industries [35]. Also, by the end of the 19th century; another important organometallic compound *Grignard reagent* was synthesized by V. Grignard, who also won the Nobel Prize with P. Sabatier in 1912. This work was initiated by P. Barbier (by replacing Zn with Mg metal in treatment with alkyl iodides) and exceeded by his student Grignard in 1899 [36,37].

At the start of the 20th century (1901), L.F.S. Kipping innovated silicon-based organometallic compound (C$_6$H$_5$)$_2$SiO (diphenyl silicone) [38]. The first σ-organotransition-metal complex [(CH$_3$)$_3$PtI] was prepared by W. J. Pope in 1909 [39]. In the same year P. Ehrlich pioneered arsenic-based organo-compound (arsphenamine) for the effective treatment of syphilis and this compound is also known as the first magic bullet for the curing of syphilis [40].

The first half of the 20th century was a noticeable remark for the embossing of catalysis research in organometallic chemistry. The difference between heterogeneous and homogeneous catalysis was represented by a French researcher Paul Sabatier. Sabatier's work on the hydrogenation of olefins by finely disintegrated metals whereas Victor Grignard's work for the development of the so-called Grignard reagent [37]. In 1917 W. Schlenk synthesized alkyl lithium reagents through *trans*-alkylation approach [41]. In 1919 F. Hein synthesized the chromium (Cr) metal-based sandwich complexes from CrCl$_3$ and PhMgBr [42]. The first systematic study of metal carbonyl complexes was pioneered by W. Hieber in 1928 as well as he also developed the first metal-hydride Fe(CO)$_4$H$_2$ in 1931 [43,44].

$$Fe(CO)_5 + (H_2NCH_2)_2 \rightarrow (H_2NCH_2)_2Fe(CO)_3 + 2\ CO$$
$$Fe(CO)_5 + X_2 \rightarrow Fe(CO)_4X_2 + CO$$

In 1938 German chemist, Otto Roelen discovered the well-known hydroformylation of olefins by CO and H$_2$ by using dicobalt octacarbonyl

[Co$_2$(CO)$_8$] [45]. This is also known as the *oxo*-process, one of the successful industrial catalytic processes in catalytical research. From 1939 to the late 1940s Reppe worked on the catalysis of the transformation of alkynes (tetramerization to cyclo-octatetraene, 1948). This work has been recognized as alkyne cyclotrimerization [46,47]. In 1943, for example, Rochow synthesized silicon metal-based organometallic complexes by using Cu-catalyst at 300°C temperature [48].

$$3CH_3Cl + Si \rightarrow (CH)_3SiCl + Cl_2$$

This work was been initiated by R. Müller but an impediment created by World War II stopped this reaction at that time. This synthesis allows the large-scale manufacturing of organometallic compounds by the direct use of silicones. In 1951, two important breakthroughs occurred in the research field of organometallics, one was the development of a bonding model (band theory) for metal-alkene complexes which was later extended by J. Chatt and L. A. Duncanson in 1953, and the second was the isolation of first organometallic sandwich complex ferrocene [Fe(C$_5$H$_5$)$_2$] by P. Pauson and S. A. Miller [49–52]. The establishment of novel organocuprates, such as LiCu(CH$_3$)$_2$, was pioneered by H. Gilman in 1952 [53]. Further, G. Wittig introduced a new class of synthesis of alkenes from the treatment of phosphonium ylides and ketone/aldehyde in 1953, afterward, he shared the Nobel Prize with Herbert C. Brown (who has worked on organoboranes) in chemistry (1979) [54,55]. In 1955, K. Ziegler and G. Natta derived a new process of synthesizing polymers of ethylene/propylene using mixed metal (transition-metal halide/AlR$_3$) catalysts later both were also shared the Nobel Prize in Chemistry (1963) [56]. Primarily, the π-allyl transition metal compounds were introduced in 1959 by J. Smidt and W. Hafner using Palladium (Pd) metal [57]. In 1955, H. C. Longuet-Higgins suggested the concept of boron hydrides especially icosahedron close-borane dianions [B$_{12}$H$_{12}$]$^{2-}$ which was proved after five years by M. F. Hawthorne in 1960 [58]. In the same year, D. Crowfoot Hodgkins deduced the structure of vitamin B$_{12}$ by using X-ray crystallography and won the Nobel Prize in 1964, she successfully demonstrated the existence of a bond between cobalt and carbon in vitamin B$_{12}$ coenzyme [59]. L. Vaska designed an iridium (Ir) containing organometallic compound in 1963 known as Vaska's complex which has the ability to bind O$_2$ reversibly [60]. In the next year 1964, E.O. Fischer reported the first carbene complex in which transition metal features a low oxidation state while carbene carbon has an electrophilic ability [61]. In 1965 G. Wilkinson and R. S. Coffey designed a rhodium (Rh) containing homogeneous catalysis for the hydrogenation of alkenes [62]. These innovations in organometallics by Fischer and Wilkinson results in the winning and sharing of 1973s Nobel prize in chemistry. After three years in 1976 William Lipscomb also won the Nobel prize for extensive research in the field of NMR and chemical shift, boranes chemistry, the nature of chemical bonding and the atomic structure of large biochemical molecules, such as proteins and the functions of enzymes [63–65]. Also, the research of boron and metal-amide chemistry (main group element dimetallenes) was paced up by Prof. Michael Lappert, he framed several reactions and was recognized as the giant contributor to the development of organometallic chemistry in the 20th century [66,67]. First, stable disilenes having a double bond between Si–Si atoms; tetramesityldisilene ((Mes)$_2$Si = Si(Mes)$_2$) was synthesized by Robert West in 1981 [68]. This compound was synthesized by the irradiation of 2,2-bis (2,4,6-trimethylphenyl)hexamethyltrisilane with hydrocarbons in an inert environment at room temperature. In the same year (1981), the first stable carbon–phosphorus triple bonded compound (2,2-dimethylpropylidynephosphine) was designed by Prof. Gerd Becker which opened a new aisle for the phosphorus containing organometallic compounds [69]. Also, in this year (1981), R. Hoffmann and K. Fukui were jointly awarded the Nobel Prize for chemistry for the establishment of semiempirical molecular orbital concepts. Their models on molecular bonding help to understand and predict structure, reaction rates and electronic configuration in chemical reactions. Next year (1982), Prof Robert G. Bergman introduced a new idea for the development of C–H bond activation by soluble organotransition metal compounds [70]. These inventions help to improve the understanding of researchers to explore the mechanism of many organic synthesis problems furnished through C–H bond activations by organotransition metal compounds. In 1989, the development of mononuclear aluminum and gallium organometallic chemistry was pioneered by Hansgeorg Schnöckel. Later he also worked on halides of Al and Ga as well as cluster compounds of these posttransition metals [71,72]. David Milstein also introduced a new approach in organometallic chemistry by the insertion of Rh metal in simple C–C bonds inhomogeneous compounds [73]. In 1994 the first lightest metallocene [Li(C$_5$H$_5$)$_2$]$^-$ was synthesized by Prof. Sjoerd Harder by treatment of [LiCp] with PPh$_4$Cl in tetrahydrofuran taken as a solvent system. This metallocene possesses staggered conformation with D$_{5d}$ symmetry [74]. The mechanism of C–H bond activation got more assessment after the invention of the first σ-complex of silane (see Scheme 1.1) by Greg Kubas in 1995.

These observations boosted the concept of tautomerism in the organometallic compound as well as develop a better understanding of the C–H bond activation reactions [75]. The first triply bonded Mo–Ge compound conception was introduced in the organometallic compound by P. P. Power in 1996,

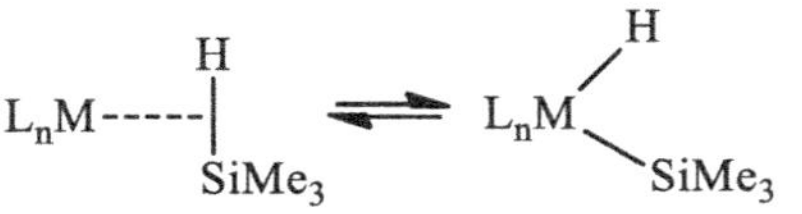

SCHEME 1.1 Tautomeric equilibrium of transition metal and first σ-complex of silane.

later, gallium—gallium triply bonded diaryldigallyne anion was also isolated by G. M. Robinson which was the first example of the steric protection of a labile structural metal atom [76,77]. The introduction of the chirality concept (stereochemistry) in organometallic compounds opened the neoteric doors for a new height in stereo-selective organic synthesis by William S. Knowles, Ryoji Noyori and K. Barry Sharpless (also jointly won the Nobel Prize 2001 in Chemistry). Their arguments established that the hydrogenation and oxidation reactions can be catalyzed by a chiral ligand coordinated metal complex [78]. In 2004, Prof. Ernesto Carmona isolated and characterized an unprecedented Zn(I)-based decamethyldizincocene, $Zn_2(\eta^5\text{-}C_5Me_5)_2$ compound and this was the first example of a stable $(Zn-Zn)^{2+}$ core compound in organometallic molecules having zinc in +1 oxidation state [79]. Y. Chauvin, R. R. Schrock, and R. H. Grubbs established a new understanding of the perspective of mechanism and application oriented explorations on molecular catalysts in alkenes (olefin) metathesis (they also shared the Nobel Prize 2005 in Chemistry) [80].

In 2004, organometallic chemistry got a new advancement in triple, quadruple and quintuple metal—metal bonds bearing compounds which pushed the limits of chemical bonding. A. Sekiguchi group has synthesized and fully characterized an interesting compound containing Si—Si triple bond, further, in 2005, Philip P. Power reported a chromium (Cr) aryls compound containing formal quintuple bonds [81,82]. Further, in the same year, the research reported by Carmona's group on pioneering the reactivity of molybdenum compounds with higher multiplicity (quadruple) Mo—Mo bonds. Furthermore, in 2013, they also demonstrated the fascinating interconversion of quadruple coordinated and quintuple coordinated Mo_2 cores by bimetallic oxidative addition of H_2 and reductive elimination of H_2 across quintuple Mo—Mo bond (see Scheme 1.2) [83].

Carmona's group still actively working toward uncovering the facts of stability and chemical reactivity of multiply bonded Mo—Mo organometallic compounds [84—88].

Organometallic chemistry is developing steadily to find solutions for challenging issues of practical applications of organometallic compounds in the realm of catalysis, medicine, materials, climate change concerns, green and sustainable chemistry, alternative energy resources, sensors, etc.

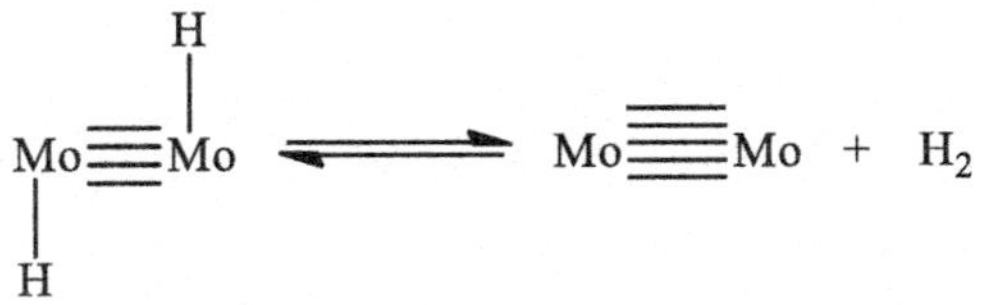

SCHEME 1.2 Depiction of an interconversion of quadruple coordinated and quintuple coordinated Mo_2 cores.

1.2 Important terms and terminologies

There are certain terms and terminologies defined in organometallic chemistry. These terms have great significance for the enhancement of this branch. Some of them are discussed below

1. *Effective atomic number (EAN) rule* [39]:
    The EAN count for an organometallic compound is the sum of electrons on metal and electrons donated by the coordinated ligands along with electrons shared by metal-metal bonds (due to presence of multinuclear metal complex). EAN rule obeys if the sum of the electrons is equal to the next noble gas configuration.

    EAN count = *electrons on metal + electrons donated by ligands + electrons shared by metal-metal bonds - oxidation state*

    Each Os is coordinated by four CO ligands along with two metal-metal bonds with the other two Os centres. EAN of Os = 76 (electrons on Os) + 8 (4CO) + 2 (2 metal-metal bonds) = 86 i.e. Rn (noble gas). Os in $Os_3(CO)_{12}$ obeys the EAN rule.

    $Mo(CO)_6$ $\qquad\qquad$ $Mo = 42e^-$
    $\qquad\qquad\qquad\qquad$ $CO \times 6 = 12e^-$
    $\qquad\qquad\qquad\qquad$ $EAN = 54e^-$ [Xe]
    $Mo(CO)_6$ *obeys EAN rule*
    $Mn(CO)_5$ $\qquad\qquad$ $Mn = 25e^-$
    $\qquad\qquad\qquad\qquad$ $CO \times 5 = 10e^-$
    $\qquad\qquad\qquad\qquad$ $EAN = 35e^-$

    The next noble gas is Kr (z = 36), so $Mn(CO)_5$ does not obey EAN rule

2. *18 electron's rule* [39]: This rule works excellent for the small, higher-field, monodentate ligands, such as H and CO. In this rule, the organometallic compounds tend to follow the 18 electron rule. For example, V$(CO)_6$ ($17e^-$) easily reduced to the $V(CO)_6^-$ anion ($18e^-$). This rule is also known as the noble-gas rule because the metals in an organometallic compound attain the noble-gas configuration. For example, $Fe(CO)_5$ follows the $18e^-$ rule. Also, $Mn(CO)_5$ has $17e^-$ while the dimerization of this compound ($Mn_2(CO)_{10}$) follows the $18e^-$ rule and achieves the noble-gas configuration. Those organometallic compounds which follow this rule are considered as stable compounds.
3. *Hapticity* [30]: This term is used to describe the coordination of ligands through multiple atoms simultaneously with the metal atom. The number of coordination of atoms to the metal atom is denoted by the prefix η (Greek letter "eta") followed by a superscript denotation indicating the

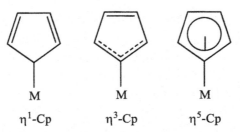

SCHEME 1.3 Representation of different hapticity modes of Cp for the coordination with metal.

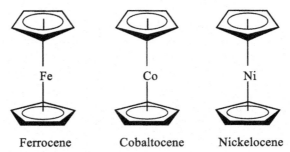

SCHEME 1.4 Examples of metallocenes.

number of ligated atoms (as shown in Scheme 1.3). Generally, cyclic, conjugated systems are found to be good for coordination with the metal atoms through various hapticity modes. For example, cyclopentadienyl anion $(C_5H_5)^-$ acts as monohapto (η^1), trihapto (η^3), or pentahapto (η^5) ligand.

4. **Metallocenes** [30]: Ferrocene (Fe(C_5H_5)$_2$) is the best example of *metallocenes* (see Scheme 1.4) [50]. This organometallic compound follows the 18 electron rule and has a stable configuration. Other examples of metallocenes have similar structures but do not essentially obey the 18 electron rule. For example, Co(C_5H_5)$_2$ (cobaltocene) and Ni(C_5H_5)$_2$ (nickelocene); both have a structure similar to ferrocene but have 19 and 20 electrons. These extra electrons caused different chemical and physical consequences, such as M—C bond length, bond enthalpy, and color of compounds.

5. **Sandwich complexes** [89–91]: The organometallic complexes containing metal atom/ion between two planar polyhapto rings are called sandwich complexes, for example, ferrocene. These complexes are also known as metallocenes. These complexes are mainly categorized into half, bent and triple decker sandwich complexes as shown in Scheme 1.5.

Generally, sandwich complexes do not follow the 18 electron's rule consistently as followed by the carbonyl and nitrosyl complexes. Also, the

Overview of organometallic compounds Chapter | 1   9

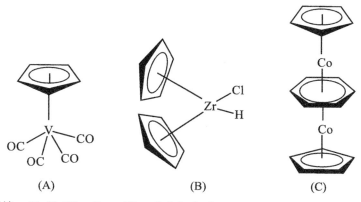

(A) = *Half*, (B) = *Bent*, (C) = *Triple decker*

SCHEME 1.5 Different types of sandwich complexes.

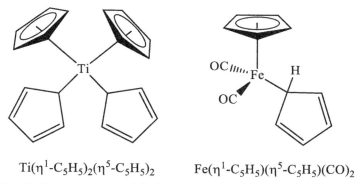

SCHEME 1.6 Common examples of fluxional organometallic compounds.

sandwich complexes show exceptional thermal stability as well as not readily oxidized in air.

6. *Fluxionality* [92,93]: On the modification in temperature, when a stereochemically nonrigid molecule shows two or more nonequivalent conformations is known as tautomer and the process is called tautomerization while if a molecule shows equivalent conformations is known as a fluxional molecule and the process of change in conformation is called fluxionality. The process of fluxionality of a compound becomes faster at high temperature whereas slower at low temperature. Such behavior is common for $Ti(\eta^1\text{-}C_5H_5)_2(\eta^5\text{-}C_5H_5)_2$ and $Fe(\eta^1\text{-}C_5H_5)(\eta^5\text{-}C_5H_5)(CO)_2$ (see Scheme 1.6).

7. *Coupling reactions* [30]: Several reactions are studied in organometallic chemistry which discussed about the bond formation between

carbon—carbon and carbon—heteroatom, such as Negishi, Suzuki, Mizoroki-Heck, Sonogashira, Kumada, Hiyama, Tsuji—Trost, and Stille reaction. These reactions are usually known as coupling reactions. In these reactions, two organic fragments are coupled with the help of an organometallic compound which acts as a catalyst. Generally, coupling reactions are validated in two categories:

a. *Homocoupling reactions*: in which two identical organic fragments are coupled together.

$$2R - X \rightarrow R - R + X_2$$

b. *Cross-coupling reactions*: in which two different organic fragments are coupled together.

$$RMgX + R'X \rightarrow R - R' + MgX_2$$

1.3 Classifications

The organometallic compounds can be classified on various criteria, but here we have categorized them mainly into two categories; one is based on bonding pattern and the second is on the presence of metal atoms in a particular organometallic compound (see Fig. 1.1) [30]. Here, we have discussed these two categories in brief:

1. Based on the bonding pattern

Organometallic compounds may appropriately be classified by the types of metal—carbon bonding. Carbon is a fairly electronegative element (2.5 on the Pauling scale) and hence might be expected to form ionic bonds only with the most electropositive elements, but also to form electron-pair covalent bonds with other elements. Also, the periodic table may be divided into very approximate regions into which the various types of organometallic compounds predominantly fall. It will be noticed that these regions are very similar to those observed in the classification of hydrides into (a) ionic, (b) volatile covalent, (c) hydrogen bridged, and (d) metal like.

The organometallic compounds of the d-block transition elements often involve not only σ but also π- or δ-bonding which is not commonly found among compounds of the main group elements (see Figs. 1.2 and 1.3). In a detailed discussion of their chemistry, therefore, d-block transition metal organometallic compounds are better taken separately from those of the main group elements. Moreover, the chemistry of organic derivatives of the transition elements may be dominated more by the ligand (especially when it occupies in several coordination positions) rather than by the periodic group.

Organometallic compounds are classified based on the bonding nature between the metal atom and the carbon atom.

Overview of organometallic compounds Chapter | 1 11

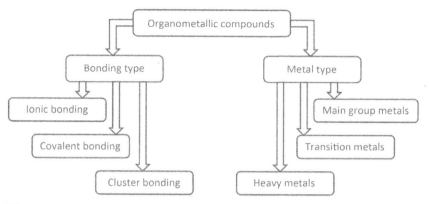

FIGURE 1.1 Classifications of the organometallic compounds.

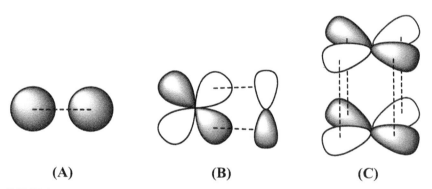

FIGURE 1.2 Pictorial depictions of (A) σ-, (B) π-, and (C) δ-bonding present in organometallic compounds.

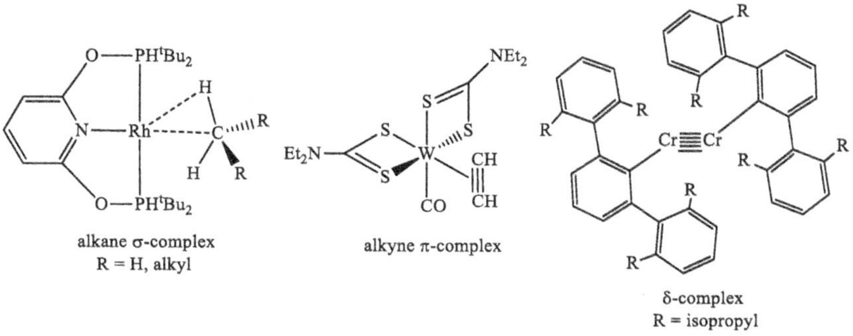

FIGURE 1.3 Some examples of σ-, π-, and δ-bonding consisting compounds.

a. *Ionic organometallic compounds*: The organometallic complexes of alkali, alkaline earth metals, lanthanides and actinides are predominantly formed ionic compounds, such as Na^+Cp^-, $Na^+C^-Ph_3$, and Cs^+Me^-.
b. *Covalent organometallic compounds* [94,95]: there are two different types of bonding possible between carbon and the metal center
   I. *σ-bonding*: when there is 2e−2c (2 electrons and 2 centers) covalent bonding between the carbon and the metal atom, such as [Be(C$_2$H$_5$)$_2$], [Sn(CH$_3$)$_4$], [Fe(CO)$_5$], [Ni(CO)$_4$], and MgCH$_3$Br.
   II. *π-bonding*: when the π-orbitals of alkene, alkyne or some other carbon group interact with the metal orbitals, such as [(η^5-C$_5$H$_5$)$_2$Fe] and [PtCl$_3$(η^2-C$_2$H$_2$)]$^-$.
c. *Complex (multicenter) organometallic compounds* [96]: The multicenter organometallic compounds contain three or more metal−metal bonding in a closed array. For example, [Fe$_3$(CO)$_9$], [Co$_4$(CO)$_{12}$], [Ru$_3$(CO)$_{12}$], and [Ir$_4$(CO)$_{12}$].

2. Based on metal type [30,39]

   Organometallic compounds can be classified into the following three types:
   a. *Based on main group*: In this category, only those organometallic compounds which contain s- or p-block metals (elements) as shown in Scheme 1.7. [(CH$_3$)$_2$As]$_2$O (cacodyl oxide) was the first main group-based organometallic compound. Other examples include SiMe$_4$, CH$_3$MgBr, etc.
   b. *Based on transition metal*: d-block metals are present in these organometallic compounds. Following are the main examples of transition metal-based organometallic complexes (see Scheme 1.8): Zeise's salt

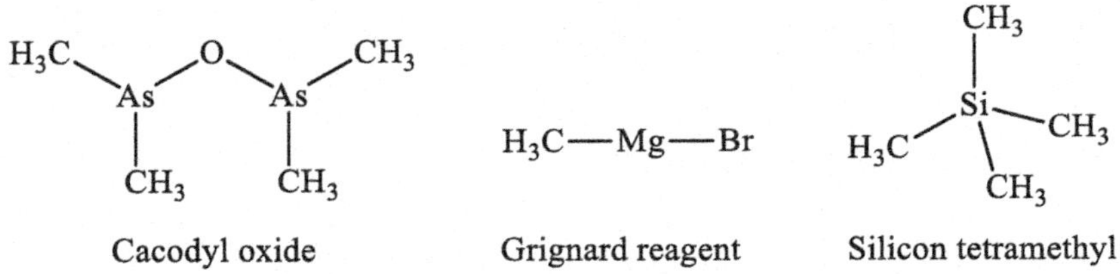

Cacodyl oxide   Grignard reagent   Silicon tetramethyl

SCHEME 1.7 Main group element-based organometallic complexes.

Zeise's salt   Monsanto's catalyst   Collmann's reagent   Vaska's complex

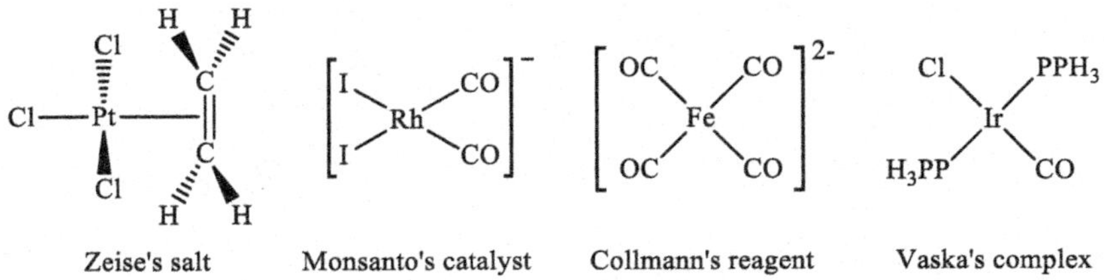

SCHEME 1.8 Transition metal-based organometallic complexes.

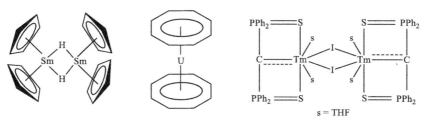

SCHEME 1.9 Heavy metal (f-block element)-based organometallic complexes.

($[PtCl_3(C_2H_4)]^-$), Gillmann's reagent (R_2CuLi), Collmann's reagent ($[Fe(CO)_4]^{2-}$), Vaska's complex ($[Ir(PPh_3)_2(CO)Cl]$), Monsanto's catalyst $[(Rh(CO)_2I_2)^-]$, palladium catalyst ($Pd(PPh_3)_4$), etc. [30].

c. *Heavy metal (Lanthanides and Actinides) based*: These organometallic compounds consisting f-block elements, such as lanthanide and actinide. Some examples are shown in Scheme 1.9, such as Uranocene ($[U(C_8H_8)_2]$) and ($[SmH(Cp)_2]_2$) [97–100].

1.4 Common methods of preparation

Usually, organometallic compounds are synthesized by two methods: (1) by using elemental metal and (2) synthesis by reactions of already formed compounds [30,39].

1. Synthesis by using elemental metal
   The following methods are used for the synthesis of organometallic compounds:
   a. *Preparation of Grignard reagent*: when the reaction of an electropositive metal atom and a halogen substituted hydrocarbon (alkyl, ethyl) are reacting together.

   $$\text{C}_6\text{H}_{11}\text{Cl} \xrightarrow{\text{Mg}} \text{C}_6\text{H}_{11}\text{MgCl}$$

   $$\text{H}_3\text{C}-\overset{\overset{\text{Br}}{|}}{\text{CH}_2} + \text{Mg} \xrightarrow{\text{ethoxyethane}} \text{H}_3\text{C}-\text{CH}_3\text{MgBr}$$

   b. *Transmetallation*: when one metal atom replaced the other metal, it is known as transmetallation. It is favorable when the displaced metal is lower than the displacing metal in electrochemical series.

$$3H_3C-Hg-CH_3 + 2\,Ga \rightarrow 2Ga(CH_3)_3 + 3Hg$$

c. *Preparation of metal carbonyl*: the direct mixing of CO with the elemental metals produced metal carbonyls.

$$Fe + 5CO \xrightarrow{200°C,\ 10\ atm} Fe(CO)_5$$
$$Ni + 4CO \xrightarrow{R.T.,\ 100\ atm} Ni(CO)_4$$
$$2Co + 8CO \xrightarrow{200°C,\ 100\ atm} Co_2(CO)_8$$

d. *Preparation of cyclopentadiene (Cp) compounds*: sodium cyclopentadiene is prepared by treating Cp with Na. Sodium cyclopentadienide is further used for the synthesis of ester derivatives.

$$2Na + 2C_5H_6 \rightarrow 2Na(C_5H_5) + H_2$$
$$Na(C_5H_5) + O=C(OEt)_2 \rightarrow NaC_5H_4CO_2Et + NaOEt$$

2. Synthesis by already present compounds

These procedures are more commonly used to synthesize organometallic compounds. The following are the typical methods:

a. *Metathesis*: in this, exchange reactions take place between the metal halides and alkylating reagents.

$$MR + M'X \rightarrow RM' + MX$$
For example: $\quad 2C_2H_5Li + CuI \rightarrow (C_2H_5)_2CuLi + LiI$

b. *Exchange of halogen and hydrogen*: in this, metal-halogen or metal-hydrogen exchange occurred.

$$RM + R'X\ or\ R'H \longrightarrow R'M + RX\ or\ RH$$

For example:

$$\text{NC-C}_6H_4\text{-Br} + CH_3(CH_2)_3-Li \xrightarrow{-100\ °C} \text{NC-C}_6H_4\text{-Li} + CH_3(CH_2)_3-Br$$

c. *Addition reactions*: In these reactions generally insertion takes place, such as carbonylation, hydrometallation, oxymetallation, and carbometallation.

$$RM + Y\ (=CO, CH_2=CH_2,\ etc) \longrightarrow R-Y-M$$

For example:

$$(CO)_5Mn-CH_3 + CO \longrightarrow (CO)_5Mn-C(CH_3)=O$$

$$MCH_2CH_3 \longrightarrow MH(CH_2CH_2)$$

Examples:

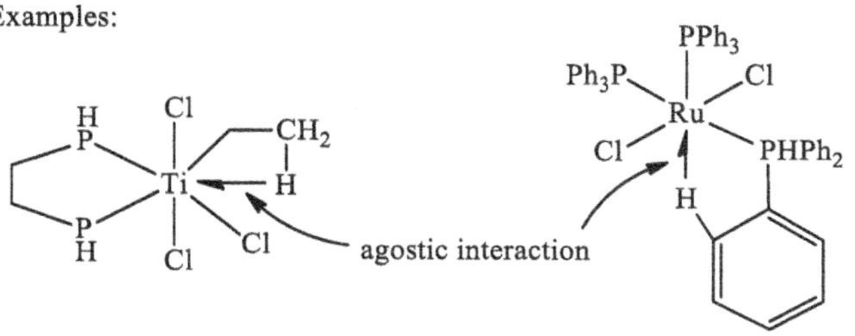

FIGURE 1.4 Examples of structurally characterized β-agostic-metal-alkyl compounds [101].

d. *Oxidative additions*: addition of alkyl halides and esters.

$$MLn + AB (=RX, ester, etc) \longrightarrow M(A)(B)Ln$$

For example:

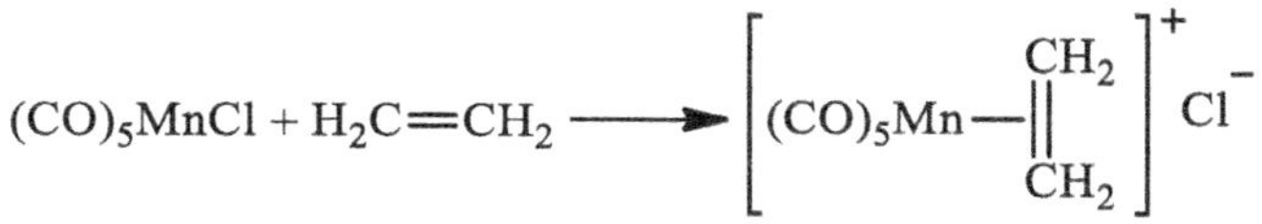

$$(CO)_5MnCl + H_2C=CH_2 \longrightarrow \left[(CO)_5Mn-\overset{CH_2}{\underset{CH_2}{||}}\right]^+ Cl^-$$

e. *Decarbonylation*: The removal of CO occurred in these reactions. Wilkinson's catalysis reactions are common example for decarbonylation reactions.

$$RCOM \longrightarrow RM + CO$$

For example:

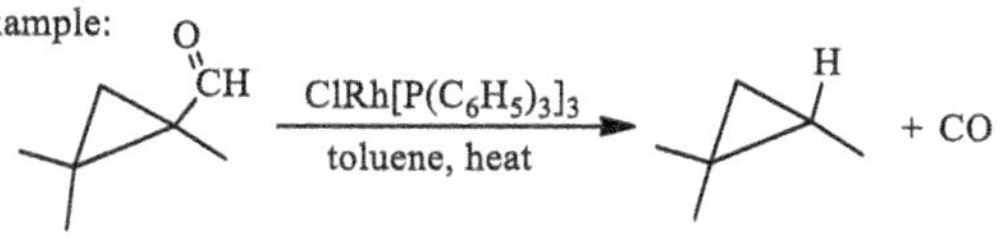

$$Rh(L)_3Cl \xrightarrow{RCHO} Rh(L)_3(RCO)(H)Cl \xrightarrow{-L} Rh(L)_2(R)(CO)(H)Cl \longrightarrow Rh(L)_2(CO)Cl + R-H$$
$L = P(C_6H_5)_3$

f. *β-H elimination reactions*: in these reactions, the β-H elimination takes place after an agostic interaction of β-hydrogen to the metal atom (see Fig. 1.4). It is also known as 3-center-2-electron interaction.

g. *Reductive carbonylation of metal oxides*: In these reactions, CO reacts with the metal oxides and CO_2 is released as a by-product in such types of reactions. An example of such a reaction (shown in Fig. 1.5) demonstrates the nitrosoarene complex to aniline in the presence of CO and an acid.

1.5 General mechanisms

Many of the common steps are involved to complete a chemical reaction by using organometallic compounds. These steps are, such as oxidative addition,

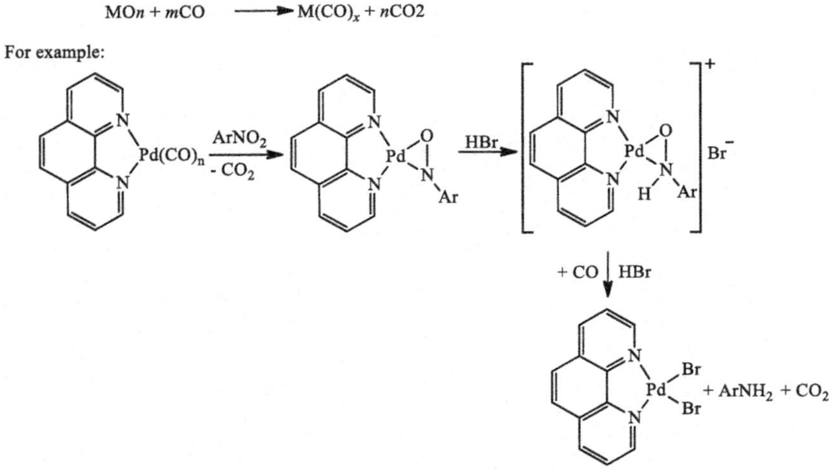

16 Recent Advances in Organometallic Chemistry

$$MOn + mCO \longrightarrow M(CO)_x + nCO2$$

For example:

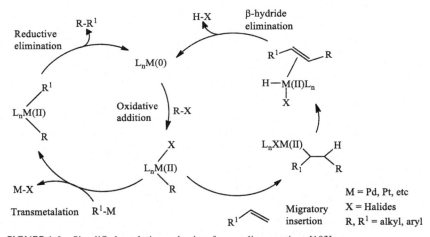

FIGURE 1.5 A simplified reaction demonstrating the reductive carbonylation of metal oxides treating with CO [102].

FIGURE 1.6 Simplified catalytic mechanism for coupling reactions [103].

transmetalation, reductive elimination, migratory insertion, and β-hydride elimination. Fig. 1.6 shows a simplified catalytic mechanism for coupling reactions [103]. The importance of coupling reactions was recognized in the field of organometallic chemistry in 2010 after the award of the Nobel Prize to Prof. Richard F. Heck, Prof. Ei-ichi Negishi, and Prof. Akira Suzuki for their determinations in Pd-catalyzed reactions in organic synthesis [104].

Primarily oxidative addition is followed by organometallic substitution (transmetalation) and then reductive elimination takes place for the formation of the desired product. In this process, the catalyst is regenerated successfully. Each step has been the subject of studies to understand the catalytic mechanism of the chemical reaction. Worldwide, several research groups, such as Denmark, Hartwig, and Lloyd-Jones, independently working to explore the transmetalation step in the Suzuki-Miyaura coupling reactions [105–107]. Furthermore, the Sonogashira coupling reaction in which C(sp) –C(sp^2) bond formation occurred due to copper acting as a cocatalyst [108]. For the cocatalyst, many refinements have been carried out to this reaction, such as careful choice of Pd catalyst as well as reaction conditions [109–111].

In the Heck reaction, the mechanism consisting β-hydride elimination step is a crucial step for the formation of the product which is different from the other named cross-coupling reactions. In the Heck mechanism, the Ni-based catalyst can also be used for the formation of the final product. This alteration brings advantages, such as higher activity under certain conditions [39].

1.6 Applications in emerging fields

There has been tremendous interest in the applications of organometallic compounds. The interest in these compounds stems from their remarkable properties. Organometallic compounds are generally used as catalysts, biomedical agents in medical sector, sensors and as notable materials in energy production and storage, etc.

1. **As catalysis:** the first biological organometallic compound recognized in nature is vitamin B_{12} which is also known as coenzyme B_{12}. It contains cobalt as a central metal in the corrin ring and furnishes many catalytic reactions [112]. Organometallic compounds are being used regularly as catalysts for the synthesis of many organic compounds. Some of them are listed below:
    a. [Rh(PPh$_3$)$_3$Cl]—used for the hydrogenation of alkenes. This compound is also known as Wilkinson's catalyst (see Fig. 1.7A).
    b. [IrI$_2$(CO)$_2$]$^-$—this compound is used for the production of acetic acid and the process of manufacturing is known as Cativa process (see Fig. 1.7B).
    c. [Rh(CO)$_2$I$_2$]$^-$—this complex is used for the production of glacial acetic acid and the process is known as Monsanto acetic acid process (see Fig. 1.7C).
    d. [Co$_2$(CO)$_8$]—this complex is used for the hydroformylation of alkenes (see Fig. 1.7D).
2. **In medicine:** Cp$_2$TiCl$_2$ is the first synthesized organometallic compound that was clinically trialed for antitumor activities [113]. After that, a number of organometallic compounds are designed and implemented in various medicinal applications, for example, ferroquine acts as an

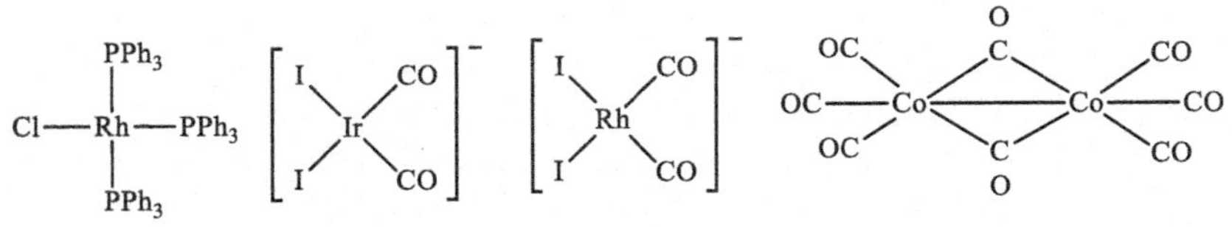

18 Recent Advances in Organometallic Chemistry

FIGURE 1.7 Different types of organometallic complexes such as (A) [Rh(PPh$_3$)$_3$Cl], (B) [IrI$_2$(CO)$_2$]$^-$, (C) [Rh(CO)$_2$I$_2$]$^-$ and (D) [Co$_2$(CO)$_8$] used in catalysis used in catalysis.

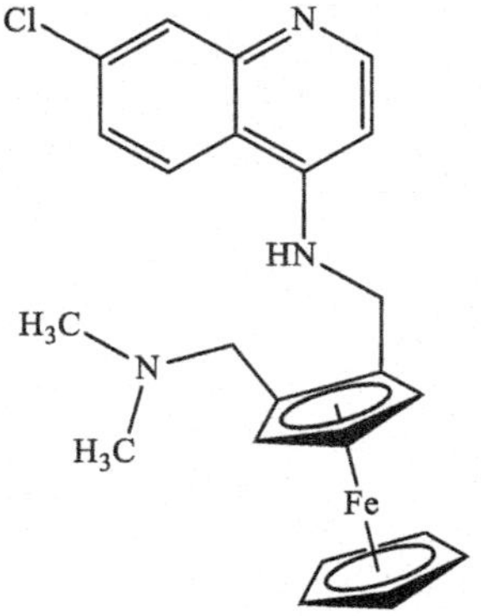

Ferroquine (7-chloro-N-[[2-[(dimethylamino)methyl]cyclopenta-1,4-dien-1-yl]methyl]quinolin-4-amine;cyclopenta-1,3-diene;iron(2+))

FIGURE 1.8 Illustration of the structure of ferroquine.

antimalarial drug (see Fig. 1.8). This compound also shows a potent antitumor activity [114]. The bent metallocene dihalides are also found promising candidates for biological activities. The titanium containing organometallic compounds is found stimulatory effect on cancer cells. Although, titanocene dichloride has similarities with cisplatin but there has never been any conclusive proof found for a common method of action for anticancer activity. A computational report on a benzyl-substituted titanocene (titanocene Y) (see Fig. 1.9) demonstrates that the Cp(R)$_2$Ti^{2+} dication coordinated with a DNA phosphate group, with additional interactions stabilizing the binding to the DNA [115,116].

3. **As reagent:** The alkali and alkaline earth metal consisting of organometallic compounds are being routinely used as reagents, such as Grignard reagent (RMgX) and Gilman's reagent (R$_2$CuLi) [39].
4. **Materials:** The silicon-based organometallic compounds have remarkable properties, such as water repellent, high heat stability and chemical attack resistance. These compounds are furnished with enormous applications, such as in oils, greases, rubber-like materials, hydraulic fluids, electrical insulators, and moisture repellent agents in fabrics. The polymers, such as polysiloxanes, which are also known as silicone and are

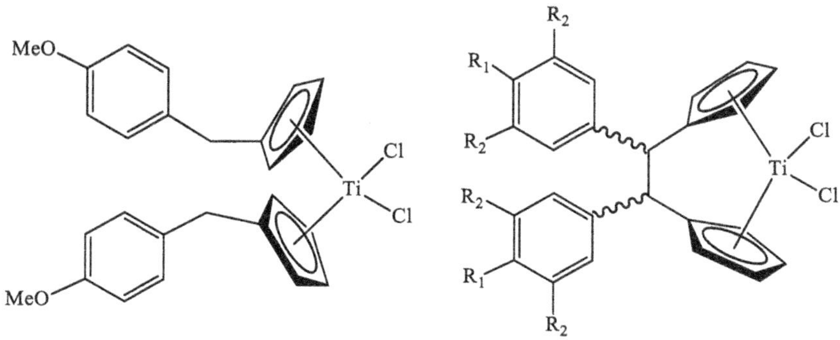

FIGURE 1.9 Titanocene Y and the *ansa*-bridged derivatives titanocene X and Z compounds.

made up of siloxane, are widely used in a large number of different fields ranging from cookware to construction materials, toys, etc. [30].

5. **Sensors:** Many organometallic polymers consisting Fe, Os, Co, and Ru show electrochemical sensing applications [117]. Recently it has been demonstrated that a redox-responsive ferrocene-based polymer shows on-off switchable pores. The vesicular membranes containing switchable pores are designed by seeded reversible addition-fragmentation chain-transfer (RAFT) polymerization approach. The porous multicompartment vesicles of ferrocene-containing polymer are redox-responsive and the membrane pores can be on-off switched through redox triggering. It is an interesting research outbreak in metallocene containing electrochemical sensing materials [118].

6. **Industrial process considerations**: On a large scale, organometallic compounds are regularly synthesized and utilized as stoichiometric catalysts or reagents for a variety of synthetic processes [30]. The organometallic compounds are used for the three foremost types of applications in areas of polymers, electronics and organic synthesis at the industrial level. The well-known heterogeneous Ziegler-Natta polymerization process is implemented for the production of polyethylene and polypropylene [39]. Also, for the production of electronic materials, aluminum alkyls are widely used via the chemical vapor deposition (CVD) technique at the industrial scale.

7. **Energy outlook:** Energy expectations are being increased rapidly as human resources increase in over the world. Hence, to meet these expectations organometallic chemistry can also play an important role. Over the last three to four decades, alternative energy resources have gained high intensity in relation to environment friendly. Recently electrocatalysis has drawn attention to the reduction of CO_2 invaluable energy and fuel transformations, such as alcohol (MeOH) and ether (MeOMe).

So far, a rhenium-based organometallic complex ([Re(bipy-tBu)(CO)$_3$Cl]) is found considerably good in CO$_2$ reduction [119].

More recently along with electro-catalytic CO$_2$ reduction, research on photocatalytic CO$_2$ reduction also got more momentum. In which, the organometallic complexes employ earth-abundant metals, such as Mn, Fe, Co, Ni, Cu, and Zn, for the advancement of CO$_2$ reduction to create energy benign products [120,121].

1.7 Future prospective and scope

Although only a few organometallic compounds are widely used for the specific purpose (for example, Rh(PPh$_3$)$_3$Cl, bistributyltinoxide, silicon polymers, ferroquine, tetraethyl leads), their value in chemical reactions either as catalysts or as intermediates are great, and it is expected that it will be increased in near future. Both inorganic and organic chemistry are incorporated by organometallic chemistry which helps to unify chemistry. But still, many problems remain to be solved. Previously insufficient attention has been given to the mechanistic studies of organometallic compound derived reactions. Our knowledge of bonding in organometallic compounds, particularly those involving transition metals, remains at the best imprecise. Although some noteworthy advancement has been carried out to design homogeneous catalysts but the knowledge about the exact composition of the catalysts under the working conditions is limited. The explorations of the synthetic and structural nature of polynuclear metal compounds containing hydrogen and organic ligands may eventually allow a better understanding of heterogeneous transformations, predominantly the reactions of hydrocarbons on metal and metal oxide surfaces. It is tough to predict the future of organometallic chemistry, but it certainly has more momentum than other areas of chemistry.

Although much progress has been accomplished toward the development of organometallic compounds but still an evaluative synthetic methodological relationship between the "environmentally benign operations" and "specific synthetic models" is still yet to be attained.

1.8 Problems with solutions

Q1. What do you understand by 18-electron rule?

Answer: The 18e$^-$ rule is also known as the noble-gas rule because the metal center in an organometallic compound accomplished the noble-gas configuration. Those organometallic compounds which follow this rule are considered as stable compounds. For example, Fe(CO)$_5$ follows the 18e$^-$ rule. Whereas Mn(CO)$_5$ has 17e$^-$ while the dimerization of this compound (Mn$_2$(CO)$_{10}$) follows the 18e$^-$ rule and achieves the stable configuration.

Overview of organometallic compounds Chapter | 1   21

Q2. How can Grignard reagent be prepared?
Answer: For the preparation of Grignard reagent, the reaction of an electropositive metal atom (mainly alkali earth metals) and a halogen-substituted hydrocarbon (alkyl, ethyl) are reacting together. For example

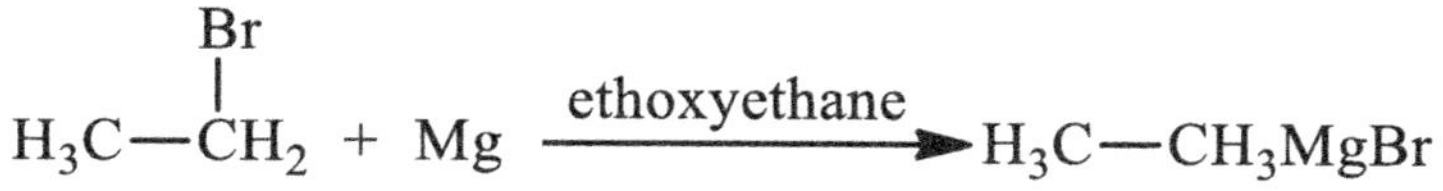

$$H_3C-\underset{\underset{Br}{|}}{CH_2} + Mg \xrightarrow{\text{ethoxyethane}} H_3C-CH_3MgBr$$

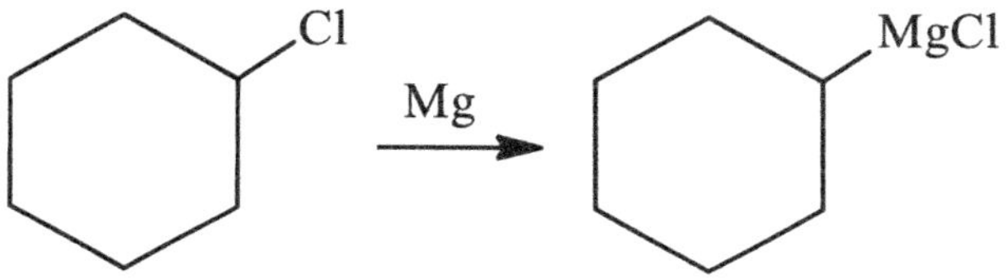

[cyclohexyl-Cl] + Mg → [cyclohexyl-MgCl]

Q3. Write a short note on agostic interaction.
Answer: In organometallic chemistry, the term "agostic interaction" describes the reaction that occurs when a transition metal center interact with a C−H bond in a ligand attached to metal. And the two electrons of C−H bond interact with empty d-orbitals of the metal center results a three-center two−electron bonding.

Examples:

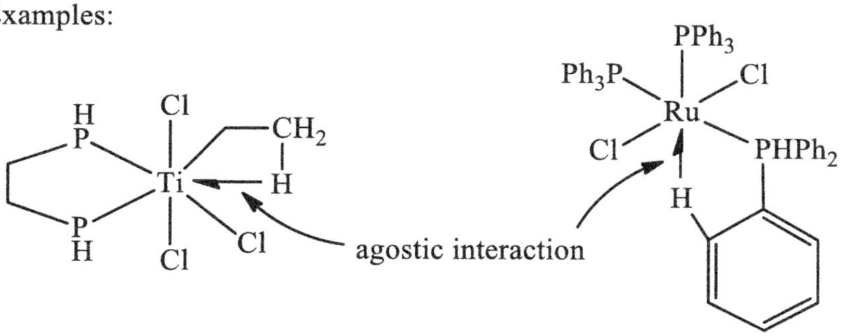

[Ti complex with H, Cl, CH₂, P-H ligands showing agostic interaction] — agostic interaction — [Ru complex with PPh₃, Ph₃P, Cl, PHPh₂, H]

Q4. Write the formulae of three organometallic compounds having multicentered bonding?
Answer: When a pair of electrons participate in a bond between three or more atoms, or when a pair of electrons occupy a bonding molecular orbital that encompasses the atomic orbitals of three or more atoms, this delocalization results in multicenter bonding.
For examples; [Fe$_3$(CO)$_9$], [Co$_4$(CO)$_{12}$], [Ru$_3$(CO)$_{12}$], and [Ir$_4$(CO)$_{12}$] show multicentered bonding.

Q5. Discuss the transmetallation with a suitable example.
Answer: In the transmetallation reaction one metal atom replaced the other metal center in an organometallic compound. It is favorable when the

displaced metal is lower than the displacing metal in electrochemical series. For example, in below reaction, mercury is replaced by the gallium metal.

$$3H_3C-Hg-CH_3 + 2Ga \rightarrow 2Ga(CH_3)_3 + 3Hg$$

vi. Write a short note on the Heck coupling reaction.

Answer: In the Heck reaction, an unsaturated halide reacts with an alkene in the presence of a Pd catalyst and a base (such as triethylamine, sodium acetate, potassium carbonate). This reaction yields a substituted alkene as a product. It is the first example among all coupling reactions which formed a carbon−carbon bond followed by a Pd(0)/Pd(II) cycle. The reaction starts with the oxidative addition of the unsaturated halide to the Pd metal. Then, a migratory insertion of the alkene to the Pd occurs. Further, a β-hydride elimination of alkene occurs as a crucial step and a subsequent addition of the base to the Pd catalyst for the regeneration of the starting catalyst and to end the cycle.

1.9 Objective type questions

i. Identify the molecule which follow the 18-electron.
   (a) $[Ni(H_2O)_6]^{2+}$
   (b) $Fe(CO)_5$
   (c) $V(CO)_6$
   (d) (d) $(\eta^6-C_6H_6)Ru$
   Answer: (b)

ii. The hapticity of cyclohelpta-1,3,5-triene in $(C_7H_8)Fe(CO)_3$ complex is:
   (a) 2
   (b) 3
   (c) 4
   (d) 5
   Answer: (c)

iii. The organometallic compound $Fe(C_5H_5)_2(CO)_2$ follows the 18-electron rule. The hapticities of the two cyclopentadienyl groups are:
   (a) 1 and 5
   (b) 3 and 3
   (c) 3 and 5
   (d) 5 and 5
   Answer: (a)

iv. Select the transition metal which present in the Gilman's reagent:
   (a) Fe
   (b) Co
   (c) Ni
   (d) Cu
   Answer: (d)

v. Which organometallic compound shows the fluxional behavior?
   (a) Ti(η^1-C$_5$H$_5$)$_2$(η^5-C$_5$H$_5$)$_2$
   (b) Cr$_2$(C$_5$H$_5$)$_2$(NO)$_4$
   (c) Ru(PPh$_3$)$_2$(NO)$_2$Cl
   (d) Mn(C$_5$H$_5$)(CO)$_3$
   Answer: (a)

vi. The complex which doesn't obey the 18-electron rule is:
   (a) Ni(CO)$_3$(PPh$_3$)
   (b) Cr(C$_5$H$_5$)$_2$
   (c) Co$_2$(CO)$_8$
   (d) Mn(CO)$_5$Cl
   Answer: (b)

vii. Identify the Wilkinson's catalyst:
   (a) [RhCl(PPh$_3$)$_3$]
   (b) [Rh(CO)$_2$I$_2$]$^-$
   (c) [HCo(CO)$_4$]
   (d) [Rh(CO)(PPh$_3$)$_2$Cl]
   Answer: (a)

viii. The organometallic compound used as fuel additive:
   (a) Mg(C$_2$H$_5$)$_2$
   (b) (C$_2$H$_5$)$_4$Pb
   (c) (CH)$_3$SiCl
   (d) Ti(C$_5$H$_5$)$_2$Cl$_2$
   Answer: (b)

ix. The organometallic compound [Co$_4$(CO)$_{12}$] have metal—metal bond:
   (a) Two
   (b) Four
   (c) Five
   (d) Six

Answer: (d)

References

[1] Frühauf H-W. Metal-assisted cycloaddition reactions in organotransition metal chemistry. Chem Rev 1997;97(3):523–96.
[2] Frenking G, Fröhlich N. The nature of the bonding in transition-metal compounds. Chem Rev 2000;100(2):717–74.
[3] Ziegler K, Holzkamp E, Breil H, Martin H. Das mülheimer normaldruck-polyäthylen-verfahren. Angew Chem 1955;67(19–20):541–7.
[4] Bertini I, Gray HB, Lippard SJ, Valentine JS. Bioinorganic chemistry. Mill Valley, CA: University Science Books; 1994. p. 1–33.
[5] Ziegler T, Tschinke V, Becke A. Theoretical study on the relative strengths of the metal-hydrogen and metal-methyl bonds in complexes of middle to late transition metals. J Am Chem Soc 1987;109(5):1351–8.

[6] Vyboishchikov SF, Frenking G. Structure and bonding of low-valent (Fischer-type) and high-valent (Schrock-type) transition metal carbene complexes. Chem -Eur J 1998;4(8):1428–38.
[7] Ushio J, Nakatsuji H, Yonezawa T. Electronic structures and reactivities of metal-carbon multiple bonds; Schrock-type metal-carbene and metal-carbyne complexes. J Am Chem Soc 1984;106(20):5892–901.
[8] Chen Y, Hartmann M, Frenking G. Ligand site preference in iron tetracarbonyl complexes Fe(CO)$_4$L (L = CO, CS, N2, NO$^+$, CN$^-$, NC$^-$, η^2-C$_2$H$_4$, η^2-C$_2$H$_2$, CCH$_2$, CH$_2$, CF$_2$, NH$_3$, NF$_3$, PH$_3$, PF$_3$, η^2-H$_2$). Z Anorg Allg Chem 2001;627(5):985–98.
[9] Kurikawa T, Takeda H, Hirano M, Judai K, Arita T, Nagao S. Electronic properties of organometallic metal − benzene complexes [M$_n$(benzene)$_m$(M = Sc − Cu)]. Organometallics 1999;18(8):1430–8.
[10] Astruc D. Organo-iron complexes of aromatic compounds. Applications in synthesis. Tetrahedron 1983;39(24):4027–95.
[11] Cram DJ, Tanner ME, Thomas R. The taming of cyclobutadiene. Angew Chem Int Ed 1991;30(8):1024–7.
[12] Kollmar H, Staemmler V. A theoretical study of the structure of cyclobutadiene. J Am Chem Soc 1977;99(11):3583–7.
[13] Slipchenko LV, Krylov AI. Electronic structure of the trimethylenemethane diradical in its ground and electronically excited states: bonding, equilibrium geometries, and vibrational frequencies. J Chem Phys 2003;118(15):6874–83.
[14] Stone FGA. Perspectives in organometallic chemistry. Nature 1971;232(5312):534–9.
[15] Zhang P, Sadler PJ. Redox-active metal complexes for anticancer therapy. Eur J Inorg Chem 2017;2017(12):1541–8.
[16] Soldevila-Barreda JJ, Metzler-Nolte N. Intracellular catalysis with selected metal complexes and metallic nanoparticles: advances toward the development of catalytic metallodrugs. Chem Rev 2019;119(2):829–69.
[17] Kotha S, Meshram M. Application of organometallics in organic synthesis. J Organomet Chem 2018;874:13–25.
[18] Seyferth D. Cadet's fuming arsenical liquid and the cacodyl compounds of Bunsen. Organometallics 2001;20(8):1488–98.
[19] Seyferth D. Zinc alkyls, Edward Frankland, and the beginnings of main-group organometallic chemistry. Organometallics 2001;20(14):2940–55.
[20] Zeise WC. Von der wirkung zwischen platinchlorid und alkohol, und von den dabei entstehenden neuen Substanzen. Ann Phys Chem 1831;97(4):497–541.
[21] P.B. Chock, J. Halpern, F.E. Paulik, S.I. Shupack, T.P. Deangelis Potassium trichloro(ethene)platinate(II) (Zeise's Salt). In: R.J. Angelici, (Eds.). *Inorganic syntheses*. 1990. p. 349–351.
[22] Frankland E. XXVII.—on the isolation of the organic radicals. J Chem Soc 1850;2(3):263–96.
[23] Larock RC. Reactivity and structure concepts in organic chemistry. Organomercury compounds in organic synthesis. Berlin/Heidelberg: Springer; 1985. p. 1–3.
[24] Meić Z. Vibrational spectra and force constants of CH$_3$HgI and CD$_3$HgI. J Mol Struct 1974;23(1):131–9.
[25] Davies AG, Smith PJ. In: Wilkinson G, Stone FGA, Abel EW, editors. Comprehensive organometallic chemistry, Vol. 1. Oxford: Pergamon Press; 1982. p. 519–627.
[26] Ingham RK, Rosenberg SD, Gilman H. Organotin compounds. Chem Rev 1960;60(5):459–539.

[27] Frankland E. About a new series of organic compounds containing boron. Ann der Chem und Pharm 1862;124(1):129−57.
[28] Tetra-ethyl lead as an addition to petrol. Br Med J 1928;1(3504):366−7.
[29] Kovarik W. Ethyl-leaded gasoline: how a classic occupational disease became an international public health disaster. Int J Occup Env Health 2005;11(4):384−97.
[30] Elschenbroich C. Organometallics. 3rd ed. Weinheim, Germany: Wiley-VCH; 2006.
[31] Rösch L, John P, Reitmeier R. Organic silicon compounds. Ullmann's *encyclopedia of industrial chemistry*. Weinheim, Germany: Wiley-VCH Verlag GmbH & Co. KGaA; 2000.
[32] Friedel C, Crafts JM. Ueber einige neue organische verbindungen des siliciums und das atomgewicht dieses elementes. Ann Chem Pharm 1863;127(1):28−32.
[33] von Ahsen B, Wartchow R, Willner H, Jonas V, Aubke F. Bis(carbonyl)platinum(II) derivatives: molecular structure of cis-Pt(CO)$_2$(SO$_3$F)$_2$, complete vibrational analysis of cis-Pt(CO)$_2$Cl$_2$, and attempted synthesis of cis-Pt(CO)$_2$F$_2$. Inorg Chem 2000;39(20):4424−32.
[34] Mond L, Langer C, Quincke F. Action of carbon monoxide on nickel. J Chem Soc Trans 1890;57(0):749−53.
[35] The extraction of nickel from its ores by the Mond process. Nature 1898;59(1516):63−4.
[36] Seyferth D. The Grignard reagents. Organometallics 2009;28(6):1598−605.
[37] The Nobel Prize in Chemistry 1912. Nobel Media AB 2020. 2020 June 5. https://www.nobelprize.org/prizes/chemistry/1912/summary/.
[38] Kipping FS. LXIX.—organic derivatives of silicon. Part XXIV. dl-derivatives of silicoethane. J Chem Soc Trans 1921;119(0):647−53.
[39] Cotton FA, Wilkinson G, Gaus PL. Basic inorganic chemistry. 3rd ed. John Wiley; 2007.
[40] Williams K. The introduction of 'chemotherapy' using arsphenamine − The first magic bullet. J R Soc Med 2009;102(8):343−8.
[41] Schlenk W, Holtz J. Über die einfachsten metallorganischen alkaliverbindungen. Ber Dtsch Chem Ges 1917;50(1):262−74.
[42] Seyferth D. Bis(benzene)chromium. 1. Franz Hein at the university of Leipzig and Harold Zeiss and Minoru Tsutsui at Yale. Organometallics 2002;21(8):1520−30.
[43] Hieber W, Bader G. Reaktionen und Derivate des Eisencarbonyls, II.: Neuartige kohlenoxyd-verbindungen von eisenhalogeniden. Ber Dtsch Chem Ges 1928;61(8):1717−22.
[44] Hieber W, Bader G. Über Metallcarbonyle. VI. Neuartige kohlenoxydverbindungen von eisenhalogeniden und ihre chemische charakterisierung. Z Anorg Allg Chem 1930;190 (1):193−214.
[45] Franke R, Selent D, Börner A. Applied hydroformylation. Chem Rev 2012;112 (11):5675−732.
[46] Reppe W, Schlichting O, Klager K, Toepel T. Cyclisierende polymerisation von acetylen I Über cyclooctatetraen. Justus Liebigs Ann Chem 1948;560(1):1−92.
[47] Reppe W, Schweckendiek W. Cyclisierende polymerisation von acetylen. III Benzol, benzolderivate und hydroaromatische verbindungen. Justus Liebigs Ann Chem 1948;560 (1):104−16.
[48] Rochow EG. The direct synthesis of organosilicon compounds. J Am Chem Soc 1945;67 (6):963−5.
[49] Mingos DMP. A historical perspective on Dewar's landmark contribution to organometallic chemistry. J Organomet Chem 2001;635(1−2):1−8.
[50] Kealy TJ, Pauson PL. A new type of organo-iron compound. Nature 1951;168 (4285):1039−40.

[51] Miller SA, Tebboth JA, Tremaine JF. 114. Dicyclopentadienyliron. J Chem Soc 1952;632–5.
[52] Werner H. At least 60 years of ferrocene: the discovery and rediscovery of the sandwich complexes. Angew Chem Int Ed 2012;51(25):6052–8.
[53] Gilman H, Jones RG, Woods LA. The preparation of methylcopper and some observations on the decomposition of organocopper compounds. J Org Chem 1952;17(12):1630–4.
[54] Wittig G, Geissler G. Zur reaktionsweise des pentaphenyl-phosphors und einiger derivate. Justus Liebigs Ann Chem 1953;580(1):44–57.
[55] Wittig G, Schöllkopf U. Über triphenyl-phosphin-methylene als olefinbildende reagenzien (I. Mittei[1]). Chem Ber 1954;87(9):1318–30.
[56] Cecchin G, Morini G, Piemontesi F. Ziegler-Natta catalysts. Kirk-Othmer encyclopedia of chemical technology. Hoboken, NJ: John Wiley & Sons, Inc; 2003.
[57] Smidt J, Hafner W, Jira R, Sedlmeier J, Sieber R, Rüttinger R, et al. Katalytische Umsetzungen von Olefinen an Platinmetall-Verbindungen Das Consortium-Verfahren zur Herstellung von Acetaldehyd. Angew Chem 1959;71(5):176–82.
[58] Pitochelli AR, Hawthorne FM. The isolation of the Icosahedral $B_{12}H_{12}^{-2}$ ion. J Am Chem Soc 1960;82(12):3228–9.
[59] Hodgkin DC, Pickworth J, Robertson JH, Trueblood KN, Prosen RJ, White JG. Structure of vitamin B_{12}: the crystal structure of the hexacarboxylic acid derived from B_{12} and the molecular structure of the vitamin. Nature 1955;176(4477):325–8.
[60] Vaska L, DiLuzio JW. Carbonyl and hydrido-carbonyl complexes of iridium by reaction with alcohols. Hydrido complexes by reaction with acid. J Am Chem Soc 1961;83(12):2784–5.
[61] Fischer EO, Maasböl A. On the existence of a tungsten carbonyl carbene complex. Angew Chem Int Ed 1964;3(8):580–1.
[62] Young JF, Osborn JA, Jardine FH, Wilkinson G. Hydride intermediates in homogeneous hydrogenation reactions of olefins and acetylenes using rhodium catalysts. Chem Commun 1965;7:131–2.
[63] Lipscomb WN. The chemical shift and other second-order magnetic and electric properties of small molecules. In: Waugh J, editor. Advances in nuclear magnetic resonance. Academic Press; 1966. p. 137–76.
[64] Eaton GR, Lipscomb WN. NMR studies of boron hydrides and related compounds. 1st ed. New York: W. A. Benjamin; 1969.
[65] Lipscomb WN, Hartsuck JA, Reeke GN, Quiocho FA, Bethge PH, Ludwig ML. The structure of carboxypeptidase A. VII. The 2.0-angstrom resolution studies of the enzyme and of its complex with glycyltyrosine, and mechanistic deductions. Brookhaven Symp Biol 1968;21(1):24–90.
[66] Gerrard W, Lappert MF. Interaction of boron trichloride with optically active alcohols and ethers. J Chem Soc 1951;1020–4.
[67] George TA, Jones K, Lappert MF. Amino-derivatives of metals and metalloids. Part II. Aminostannylation of unsaturated substrates, and the infrared spectra and structures of carbamato- and dithiocarbamato-trimethylstannanes and related compounds. J Chem Soc 1965;2157–65.
[68] West R, Fink MJ, Michl J. Tetramesityldisilene, a stable compound containing a silicon-silicon double bond. Science 1981;214(4527):1343–4.
[69] Becker G, Gresser G, Uhl W. 2,2 dimethylpropylidynephosphine, a stable compound with a phosphorus atom of coordination number 1 (acyl- and alkylidenephosphines; 15). Z Naturforsch B 1981;36:16–19.

[70] Janowicz AH, Bergman RG. Carbon-hydrogen activation in completely saturated hydrocarbons: direct observation of M + R-H → M(R)(H). J Am Chem Soc 1982;104(1):352–4.

[71] Ahlrichs R, Häser M, Schnöckel H, Tacke M. Aluminum η^2-olefin bonds in dimeric 1,4-dichloro-1,4-dialumina-2,5-cyclohexadiene. Chem Phys Lett 1989;154(2):104–10.

[72] Dohmeier C, Robl C, Tacke M, Schnöckel H. The tetrameric aluminum(I) compound[{Al(η^5-C$_5$Me$_5$)}$_4$]. Angew Chem Int Ed 1991;30(5):564–5.

[73] Gozin M, Weisman A, Ben-David Y, Milstein D. Activation of a carbon–carbon bond in solution by transition-metal insertion. Nature 1993;364(6439):699–701.

[74] Harder S, Prosenc MH. The simplest metallocene sandwich: the lithocene anion. Angew Chem Int Ed 1994;33(17):1744–6.

[75] Luo X-L, Kubas GJ, Burns CJ, Bryan JC, Unkefer CJ. Synthesis of the first examples of transition metal η^2-SiH$_4$ complexes, cis-Mo(η^2-SiH$_4$)(CO)(R$_2$PC$_2$H$_4$PR$_2$)$_2$, and evidence for an unprecedented tautomeric equilibrium between an η^2-SiH$_4$ complex and a hydridosilyl species: a model for methane coordination and activation. J Am Chem Soc 1995;117(3):1159–60.

[76] Simons RS, Power PP. (η^5-C$_5$H$_5$)(CO)$_2$MoGeC$_6$H$_3$-2,6-Mes$_2$: a transition-metal germylyne complex. J Am Chem Soc 1996;118(47):11966–7.

[77] Su J, Li X-W, Crittendon RC, Robinson GH. How short is a -Ga≡Ga- triple bond? synthesis and molecular structure of Na$_2$[Mes*$_2$C$_6$H$_3$-Ga≡Ga-C$_6$H$_3$Mes*$_2$] (Mes* = 2,4,6-i-Pr$_3$C$_6$H$_2$): the first gallyne. J Am Chem Soc 1997;119(23):5471–2.

[78] Knowles WS, Noyori R. Pioneering perspectives on asymmetric hydrogenation. Acc Chem Res 2007;40(12):1238–9.

[79] Resa I, Carmona E, Gutierrez-Puebla E, Monge A. Decamethyldizincocene, a stable compound of Zn(I) with a Zn-Zn bond. Science 2004;305(5687):1136–8.

[80] Astruc D. The metathesis reactions: from a historical perspective to recent developments. N J Chem 2005;29(1):42–56.

[81] Sekiguchi A, Kinjo R, Ichinohe M. A stable compound containing a silicon-silicon triple bond. Science 2004;305(5691):1755–7.

[82] Nguyen T, Sutton AD, Brynda M, Fettinger JC, Long GJ, Power PP. Synthesis of a stable compound with fivefold bonding between two chromium(I) centers. Science 2005;310(5749):844–7.

[83] Carrasco M, Curado N, Maya C, Peloso R, Rodríguez A, Ruiz E. Interconversion of quadruply and quintuply bonded molybdenum complexes by reductive elimination and oxidative addition of dihydrogen. Angew Chem Int Ed 2013;52(11):3227–31.

[84] Carrasco M, Mendoza I, Faust M, López-Serrano J, Peloso R, Rodríguez A. Terphenyl complexes of molybdenum and tungsten with quadruple metal–metal bonds and bridging carboxylate ligands. J Am Chem Soc 2014;136(25):9173–80.

[85] Curado N, Carrasco M, Álvarez E, Maya C, Peloso R, Rodríguez A. Lithium di- and trimethyl dimolybdenum(II) complexes with Mo–Mo quadruple bonds and bridging methyl groups. J Am Chem Soc 2015;137(38):12378–87.

[86] Curado N, Carrasco M, Campos J, Maya C, Rodríguez A, Ruiz E. An unsaturated four-coordinate dimethyl dimolybdenum complex with a molybdenum-molybdenum quadruple bond. Chem − A Eur J 2017;23(1):194–205.

[87] Pérez-Jiménez M, Campos J, López-Serrano J, Carmona E. Reactivity of a trans-[H−Mo≡Mo−H] unit towards alkenes and alkynes: bimetallic migratory insertion, H-elimination and other reactions. Chem Commun 2018;54(66):9186–9.

[88] Pérez-Jiménez M, Curado N, Maya C, Campos J, Ruiz E, Álvarez S. Experimental and computational studies on quadruply bonded dimolybdenum complexes with terminal and bridging hydride ligands. Chem A Eur J 2021;27(21):6569−78.

[89] Wilkinson G, Rosenblum M, Whiting MC, Woodward RB. The structure of iron biscyclopentadienyl. J Am Chem Soc 1952;74(8):2125−6.

[90] Fischer EO, Pfab W. Cyclopentadien-metallkomplexe, ein neuer typ metallorganischer verbindungen. Z Naturforsch B 1952;7(7):377−9.

[91] Wilkinson G. The iron sandwich. A recollection of the first four months. J Organomet Chem 1975;100(1):273−8.

[92] Helling JF, Braitsch DM. New fluxional organometallic compounds. Pseudoferrocene systems. J Am Chem Soc 1970;92(24):7209−10.

[93] Cotton FA. Fluxional organometallic molecules. Acc Chem Res 1968;1(9):257−65.

[94] Pike SD, Thompson AL, Algarra AG, Apperley DC, Macgregor SA, Weller AS. Synthesis and characterization of a Rhodium(I) σ-alkane complex in the solid state. Science 2012;337(6102):1648−51.

[95] Ricard L, Weiss R, Newton WE, Chen GJJ, McDonald JW. Binding and activation of enzymatic substrates by metal complexes. 4. Structural evidence for acetylene as a four-electron donor in carbonylacetylenebis(diethyldithiocarbamate)tungsten. J Am Chem Soc 1978;100(4):1318−20.

[96] Ndambuki S, Ziegler T. Analysis of the putative Cr-Cr quintuple bond in Ar'CrCrAr' (Ar' = C_6H_3-2,6(C_6H_3-2,6-Pr^i_2)$_2$) based on the combined natural orbitals for chemical valence and extended transition state method. Inorg Chem 2012;51(14):7794−800.

[97] Evans WJ, Davis BL. Chemistry of tris(pentamethylcyclopentadienyl) f-element complexes, $(C_5Me_5)_3$M. Chem Rev 2002;102(6):2119−36.

[98] Ortiz JV, Hoffmann R. Hydride bridges between $LnCp_2$ centers. Inorg Chem 1985;24 (13):2095−104.

[99] Berthet J-C, Thuéry P, Ephritikhine M. Bending of "Uranocene" ((η^8-C_8H_8)$_2$U): Synthesis and crystal structure of the cyanido complex [(η^8-C_8H_8)$_2$U(CN)][NEt_4]. Organometallics 2008;27(8):1664−6.

[100] Cantat T, Jaroschik F, Ricard L, Le Floch P, Nief F, Mézailles N. Thulium alkylidene complexes: synthesis, X-ray structures, and reactivity. Organometallics 2006;25 (5):1329−32.

[101] Brookhart M, Green MLH, Parkin G. Agostic interactions in transition metal compounds. Proc Natl Acad Sci USA 2007;104(17):6908−14.

[102] Ferretti F, Rimoldi M, Ragaini F, Macchi P. Reaction of arylhydroxylamines with [Pd (Neoc)(NO_3)$_2$] (Neoc = neocuproine). Non-innocent behavior of the nitrate anion. Inorganica Chim Acta 2018;470:284−9.

[103] Gardner BM, Seechurn CCCJ, Colacot TJ. Industrial milestones in organometallic chemistry. In: Colacot TJ, Seechurn CCCJ, editors. Organometallic chemistry in industry. Wiley-VCH Verlag GmbH & Co. KGaA; 2020. p. 1−22.

[104] Eisch JJ. Henry Gilman: American pioneer in the rise of organometallic chemistry in modern science and technology. Organometallics 2002;21(25):5439−63.

[105] Thomas AA, Wang H, Zahrt AF, Denmark SE. Structural, kinetic, and computational characterization of the elusive arylpalladium(II)boronate complexes in the Suzuki−Miyaura reaction. J Am Chem Soc 2017;139(10):3805−21.

[106] Lennox AJJ, Lloyd-Jones GC. Transmetalation in the Suzuki-Miyaura coupling: the fork in the trail. Angew Chem Int Ed 2013;52(29):7362−70.

[107] Carrow BP, Hartwig JF. Distinguishing between pathways for transmetalation in Suzuki − Miyaura reactions. J Am Chem Soc 2011;133(7):2116−19.
[108] Chinchilla R, Nájera C. Recent advances in Sonogashira reactions. Chem Soc Rev 2011;40(10):5084−121.
[109] Pu X, Li H, Colacot TJ. Heck alkynylation (copper-free Sonogashira coupling) of aryl and heteroaryl chlorides, using Pd complexes of t-Bu$_2$(p-NMe$_2$C$_6$H$_4$)P: Understanding the structure−activity relationships and copper effects. J Org Chem 2013;78(2):568−81.
[110] Handa S, Smith JD, Zhang Y, Takale BS, Gallou F, Lipshutz BH. Sustainable HandaPhos-*ppm* palladium technology for copper-free Sonogashira couplings in water under mild conditions. Org Lett 2018;20(3):542−5.
[111] Gazvoda M, Virant M, Pinter B, Košmrlj J. Mechanism of copper-free Sonogashira reaction operates through palladium-palladium transmetallation. Nat Commun 2018;9(1):4814−23.
[112] Scott JM, Molloy AM. The discovery of vitamin B$_{12}$. Ann Nutr Metab 2012;61(3):239−45.
[113] Wang Y, Astruc D, Abd-El-Aziz AS. Metallopolymers for advanced sustainable applications. Chem Soc Rev 2019;48(2):558−636.
[114] Kondratskyi A, Kondratska K, Vanden Abeele F, Gordienko D, Dubois C, Toillon R-A. Ferroquine, the next generation antimalarial drug, has antitumor activity. Sci Rep 2017;7(1):15896.
[115] Gasser G, Ott I, Metzler-Nolte N. Organometallic anticancer compounds. J Med Chem 2011;54(1):3−25.
[116] Tacke M. The interaction of titanocene Y with double-stranded DNA: a computational study. Lett Drug Des Discov 2008;5(5):332−5.
[117] Feng X, Zhang K, Hempenius MA, Vancso GJ. Organometallic polymers for electrode decoration in sensing applications. RSC Adv 2015;5(129):106355−76.
[118] Shi P, Qu Y, Liu C, Khan H, Sun P, Zhang W. Redox-responsive multicompartment vesicles of ferrocene-containing triblock terpolymer exhibiting on−off switchable pores. ACS Macro Lett 2016;5(1):88−93.
[119] Smieja JM, Kubiak CP. Re(bipy-tBu)(CO)$_3$Cl − improved catalytic activity for reduction of carbon dioxide: IR-spectroelectrochemical and mechanistic studies. Inorg Chem 2010;49(20):9283−9.
[120] Wang F, Neumann R, de Graaf C, Poblet JM. Photoreduction mechanism of CO$_2$ to CO catalyzed by a three-component hybrid construct with a bimetallic rhenium catalyst. ACS Catal 2021;11(3):1495−504.
[121] Takeda H, Cometto C, Ishitani O, Robert M. Electrons, photons, protons and earth-abundant metal complexes for molecular catalysis of CO$_2$ reduction. ACS Catal 2017;7(1):70−88.

Chapter 2

Organometallic compounds: bonding and spectral characteristics

Kirandeep[1], Kushal Arya[1], Richa[1], Vinod Kumar[2] and Ramesh Kataria[1]
[1]Department of Chemistry and Centre for Advanced Studies in Chemistry, Panjab University, Chandigarh, India, [2]Department of Chemistry, Central University of Haryana, Mahendergarh, Haryana, India

2.1 Introduction

Organometallics is defined as the chemistry of compounds that comprises at least one metal−carbon bond in which the carbon is amongst the organic groups, such as carbonyls, olefin complexes, cyclopentadienyl, π-complexes, cyanide, and fulminate compounds. Usually, the bond between metal and carbon is polarized as $M^{\delta+}-C^{\delta-}$ in organometallic compounds but their properties can amend with a change in bond type. For instance, ionic bonds containing organometallic compounds like organosodium and organopotassium burn spontaneously when exposed to air due to high reactivity whereas organometallic compounds having covalent bonds are less reactive. Therefore we can say that the type of bonds involved in organometallic compounds also play a crucial role in providing a unique platform for the researcher in the field of medicinal chemistry, catalysis, sensing and synthesis of organic compounds etc. The organometallic compounds are capable to provide a unique opportunity to researchers due to the availability of a source of a nucleophilic carbon atom which may react with an electrophilic carbon to make a new C−C bond, photophysical properties and tunable cytotoxicity due to ligand exchange at metal centers.

Therefore we may expect that any advancement in the field of organometallics will provide a new platform for researchers for social outcomes. As we mentioned, the properties of organometallics move around the nature of M−C bonds including covalent, ionic, donor−acceptor bonds and other types of bonds like M−M and hydrogen bonds. To keep in mind the above facts, we emphasize discussing the bonding and spectral characteristics of various organometallic compounds used for various applications [1−9].

2.2 Bonding in organometallic compounds

The organometallic compounds have M−C bonds of which some are polar ionic bonds and accordingly metal center possesses a positive charge whereas carbon has a negative charge. But as per literature, the majority of M−C bonds in organometallics are covalent by nature, which is further classified in σ, π, polynuclear and dative bonds. The detail of the different types of bonds involved in organometallic compounds are as follows:

2.2.1 Ionic bond

A considerable electronegativity difference is required to make polar $M^{\delta+}$−$C^{\delta-}$ bond as partially positive and partially negative charge respectively on metal and a carbon atom. The value of electronegativity of C-atom is 2.5, whereas for metals it stands between 0.8 and 2.5 for cesium and gold metal respectively [10]. Consequently, the polarity induced on the M−C bond and their magnitude of polarizability depends on the electronegativity of the metal. Based on the electronegativity index of alkali and alkaline earth metals form an ionic M−C bond in organometallic compounds [11]. But the metal belonging to the late transition series forms a less polar M−C bond due to their electronegativity behavior and incline in the ionic nature of the M−C bond with polarity. In addition, the polarity of the M−C bond is also dependent on the ligand's nature attached to the metal atom and the different substituents of the C-atom bonded to the metal of the complex. It is highly prominent if the carbanion in the complex is stabilized, which is influenced by the delocalization of the negative charge placed on hydrocarbon anion over numerous carbon atoms in aromatic or unsaturated rings. As shown in Fig. 2.1. the bond between alkali metals (Na, K) and cyclopropyl (Cp) ring is stable due to the presence of a negative charge on the Cp ring, which leads to the increase in the aromaticity of the Cp ring [12,13]. It is also governed by the hybridization of the C-atom attached to the metal center. Based on the value of electronegativity (2.75 for sp^2 C-atom and 3.29 for sp carbon) the acidity order of alkane, alkene, and alkyne groups can be shown as $C_2H_6 < C_2H_4 < C_2H_2$ [10]. A carbanion can easily abstract a proton from H_2O due to high basicity that also imitates the reactivity towards oxygen which results in the formation of a hydrocarbon and alkali metal hydroxide. Due to this, ionic organometallic compounds are found to be enormously moisture sensitive [11].

2.2.2 Covalent bond

All the elements used in ionic bond formation can form covalent bonds. The compound comprises the covalent bond by interatomic linkage due to the sharing of an electron pair between the metal and a carbon atom of the organic part. They can show high solubility in organic solvents and are

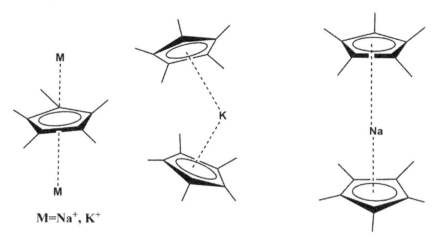

FIGURE 2.1 Ionic metal–Cp interaction [12].

insoluble in water. There are two types of covalent bonds present in organometallic compounds, that is, σ bonded organometallic compounds and π bonded organometallic compounds [14].

2.2.2.1 σ-bonded organometallic compounds

In these compounds, metal forms two centered two electron [2e−2c] covalent bonds with the carbon atom of the organic ligand. Generally, most of the elements comprising electronegativity value more than one can easily participate in this type of bonding [14]. Commonly bonding between transition metal and alkane represents a σ-covalent bond. For example, the Fe-CH_3 bond in $(\eta^5\text{-}C_5H_5)Fe(CO)_2CH_3$ complex is a σ-covalent bond [15].

2.2.2.2 Π-bonded organometallic compounds

A system holding electrons in π-orbitals is present in orbitals of metal atoms bringing about an arrangement where the metal atom is attached to several C-atoms instead of one [14]. For example, in $K[PtCl_3(\eta^2\text{-}C_2H_4)]$ (Zeise's salt), C_2H_4 bonded to the platinum metal through π* orbitals as shown in Fig. 2.2 [16,17].

Transition metal compounds rarely form stable homoleptic compounds of MRn type where M stands for transition metal and R for an alkyl group. The cause of their instability is based on kinetic studies and generally occurs because of β-elimination reactions or partial occupation of d-orbitals because it results in the decomposition of the complex into metal hydrides and olefin. For example, $Ti(CH_3)_4$ is not stable at room temperature [18]. According to Raman and NIR spectroscopic studies of the complex $[SWNT]_2Cr$ as shown in Fig. 2.3, the interaction between chromium (Cr) and the single-walled carbon nanotube (SWNTs) does not lead to considerable charge transfer, hence

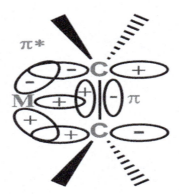

FIGURE 2.2 M–C Π bond [16].

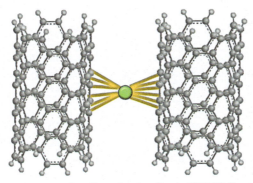

FIGURE 2.3 Pictorial representation of covalent bonding in [SWNT]$_2$Cr complex [19].

contributing to the covalent nature of the bond [19]. In some studies the spectroscopic data indicate that the interaction of carbon derivatives, such as nanotubes and graphene with metals, possess covalent bonding as shown in Fig. 2.4 [20,21].

2.2.3 Electron-deficient bond (polycentric localized bond)

Electron deficient or polycentric localized bonds are defined as the bonds in which metals like Li, Be, Al etc. exist with the coordination of loosely bonded electron species. For example, organometallic derivatives of beryllium (BeR$_2$) and aluminum (Al$_2$Me$_6$) [22].

As shown in Fig. 2.5, trimethylaluminium is present in the form of a dimer, where the organic group CH$_3$ plays the role of bridging ligand between aluminum atoms to accomplish dimerization. Al in each Al(CH$_3$)$_3$ group, contains three SP^3-orbitals forming three sigma Al-CH$_3$ bonds and has vacant d-orbitals. During dimerization, one sigma bond from both units

Organometallic compounds: bonding and spectral characteristics Chapter | 2   35

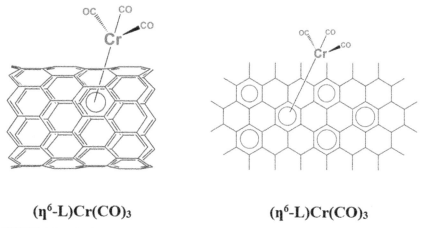

(η^6-L)Cr(CO)$_3$ (η^6-L)Cr(CO)$_3$

FIGURE 2.4 Metal–carbon covalent bonding in case of graphene and carbon nanotubes [20].

FIGURE 2.5 Structure of Al$_2$(Me)$_6$.

donates its electrons to vacant orbitals of the neighboring Al atom and makes two bridging bonds [22]. Due to less electron density between the atoms, these localized, three-center, bi-electronic [3c–2e] bonds are weaker than the usual two-center, two-electron [2c–2e] covalent bonds.

2.2.4 Delocalized bond in polynuclear systems

Polynuclear system with delocalized bonds is formed when it becomes difficult to differentiate between the electron pairs and the binding pairs of the atoms present in multiatom clusters. We must deliberate many e$^-$s as belongings to a set of atoms and carry the cluster conjointly. In the case of Li(R)$_4$ and Li(R)$_6$, the configuration of Li(R)$_4$ contains Li$_4$ tetrahedron [23]. The structure of the Penta-aurated indolium compound holds a coplanar indolyl trianion framework, wherein both the C-anions are connected by a gem-digold component [24]. As shown in Fig. 2.6, the Penta-aurated complex exhibited the binary hyper-conjugation of two adjacent gem-diaurated compounds that give rise to a better electron delocalization and more remarkable aromaticity [24].

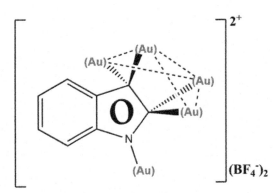

FIGURE 2.6 Structure of Penta-aurated Indolium compound [24].

2.2.5 Dative bonds

This is the most common M—C bond type of organometallic compound. As this type of bond is formed by electron donation in two opposite directions: from ligand to metal and from metal to ligand (back bonding) [11]. For example, bonds between transition-metal atom and unsaturated organic molecule and bond between CO and metal atom. But provided the metal atom should be in a lower oxidation state and must have a vacant d-orbital. The electron donation between metal and ligand also influences the strength of the bond between carbon and oxygen in the coordinated CO molecule as well as the bond between carbon atoms of organic ligand bonded to the metal [25]. The energy decomposition analysis (EDA) [26] combined with the natural orbital for chemical valance (NOCV) [27] gives rise to the EDA-NOCV scheme which is considered a powerful tool that gives numerical outcomes for the efficacy of the orbital interaction and Pauli repulsion or electrostatic interaction as well. The EDA-NOCV calculations unveiled that transition metal complexes with carbene [$(CO)_5W(CPh)_2$], [$(CO)_5Cr$-$(cyc$-$C_3Ph_2)$], carbyne [$(Br(CO)_4Cr\equiv(CPh)$], alkene [$(CO)_5W(C_2H_4)$], and alkyne [$(CO)_5W(C_2H_2)$] ligands, which are typically organized as donor–acceptor compounds, possess dative bonds as shown in Fig. 2.7 [28].

2.2.6 Hydrogen bonding in organometallics

Hydrogen bonding (H-bonding) is known to play a leading role in catalytical, chemical and supramolecular chemistry as well as in biochemical processes [29]. Based on important functions of organometallic compounds and coordination sites, hydrogen bonds can be classified into three categories:

1. Cationic organometallic hydrides acting as H^+ donors;
2. Transition elements behaving as H^+ acceptors; and
3. Hydride ligands acting as a typical H^+ acceptors.

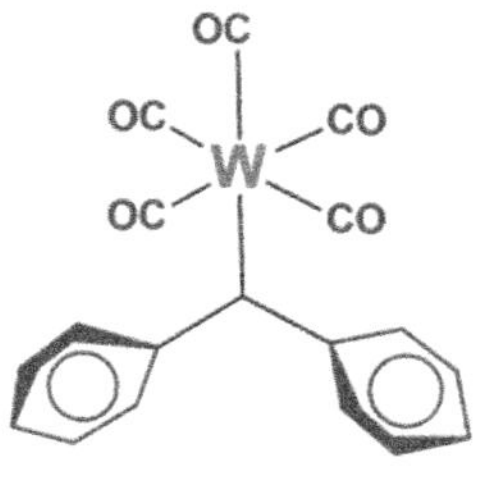

Organometallic compounds: bonding and spectral characteristics **Chapter | 2** **37**

(CO)₅W(CPH₂)

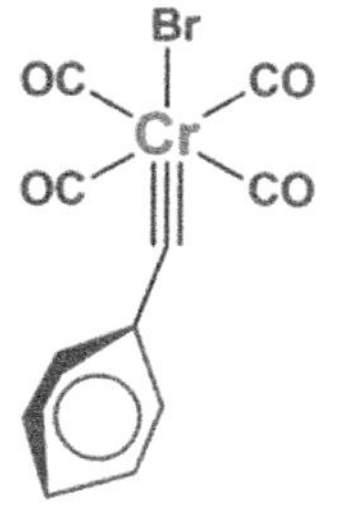

Br(CO)₄Cr≡(CPh)

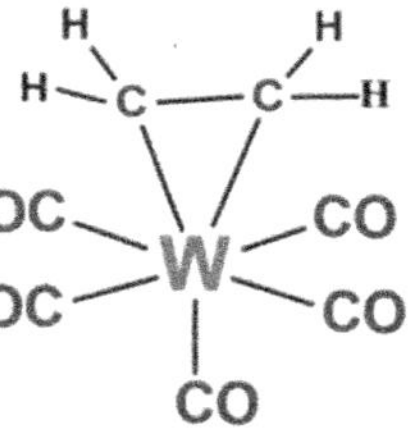

(CO)₅W(C₂H₄)

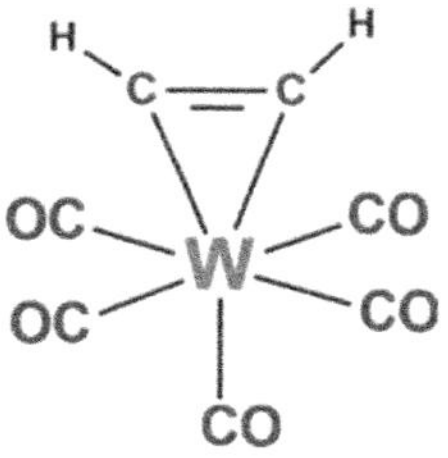

(CO)₅W(C₂H₂)

FIGURE 2.7 Structures of compounds containing dative bonds.

2.2.6.1 Cationic organometallic hydrides acting as H⁺ donors

A cationic hydride may act as H^+ donor constructing an ion-molecular ($MH^+\ldots B$) or ionic ($MH^+\ldots X^-$) hydrogen bond [30]. The ion-molecular interactions are formed for phosphine oxides that have exemplary H^+ acceptor ability to design hydrogen bonds along frail basicity. The ionic class of H-bonding was perceived where cationic hydrides like $(Cp_2{}^*OsH)^+$ and $(WH_5(dppe)_2)^+$ act as H^+ donors and CF_3COO^- as H^+ acceptors. In case of the strongest base $(Me_2N)_3PO$, there was not only the formation of a hydrogen bond but rescindable partial proton transfer also occurs. The structure of $[Os(\eta^2-H_2)(CH_3CN)(dppe)_2](BF_4)_2$ (shown in Fig. 2.8) showed that from two BF_4^- ions, fluorine forms a hydrogen bond with the dihydrogen ligand [31]. Thermodynamical aspects showed that $Ru(dppe)^2(\eta^2-H_2\ldots OSO_2CF_3)(CN)$ forms when $HOSO_2CF_3$ protonates $Ru(dppe)_2H(CN)$ in dichloromethane. Therefore, initial ionic hydrogen bonds involve classical or nonclassical cationic hydrides as proton donors.

2.2.6.2 Transition elements acting as H⁺ acceptors

The metal atoms containing lone pairs of d-type electrons are supposed to act as acceptors in hydrogen bonds, such as heteroatoms, that have sp-lone pairs. The intermolecular $M\cdots\cdots H-X$ bonds are not found initially, only

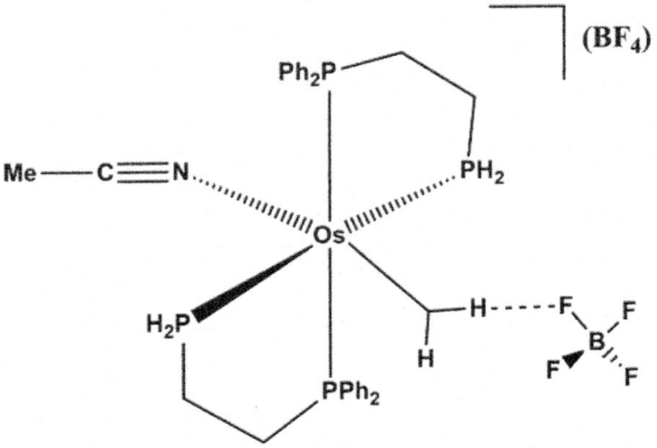

FIGURE 2.8 Structure of $[Os(\eta^2\text{-}H_2)(CH_3CN)(dppe)_2](BF_4)_2$ complex.

intramolecular hydrogen bonding with transition metals in organometallic compounds has been known, which can be due to steric factors. The first report of intermolecular hydrogen bonding appeared in 1990, it was found that electron-rich transition metal atoms like Rh of norbornadienyl-cyclopentadienyl derivatives and Ru and Os of metallocene form intermolecular hydrogen bonding with 4-fluorophenol [32].

2.2.6.3 Hydride ligands acting as atypical H^+ acceptors

The concept of H-bonding amid acid hydrogen atoms and hydride was recommended for $IrH(OH)(PMe_3)_4$. The crystal structure of the complex suggested the H·····H distance was 2.40 [1] which was "too long" for the normal hydrogen bond. Thus intramolecular hydrogen bonding to hydride was discovered. Later on, dihydrogen bond complexes with a hydrogen bond between the hydride ligand of the polyhydride ligand and the NH^+ group of crown ethers. The properties of the H—H bond are the same as that of a classical hydrogen bond [33—35].

2.3 Organometallic compounds containing metal—metal bond

Relatively all metallic elements are capable of forming metal—metal bonds but group d elements are notably preferred for doing so. Even lithium (lightest metal) and uranium (heaviest metal) go through metal—metal interaction but only a few compounds are known in the literature for early main group metals and lanthanoids. This is because for the formation of a stable organometallic compound with metal—metal bonds, several conditions have to be satisfied, that is, metal (preferably heavy transition metals) should be in a low oxidation state, relevant coordination sphere and good bonding partners. For instance, a dimeric organometallic compound like $[Cp_2V_2(CO)_4(\mu\text{-}ER)_2]$ having a metal—metal bond can be synthesized with the reaction of mononuclear complexes of [CpV(CO)$_4$] and [Cp'V(CO)$_4$] in the presence of disulfides and diselenides [36]. Similarly, chalcogen bridged organometallic compounds with metal–metal bond are also synthesized with reactions of [Cp$'_2$V$_2$E$_2$] and poly-sulfides resulting in a complex containing sulfur and selenium [37]. The reaction of white phosphorus in the presence of [CpCr(CO)$_3$]$_2$ yields a known complex [Cp$_2$(CO)$_4$(μ-P$_2$)] [38] and a new μ-alkylidene complex [Cr(PMe$_3$)(MeC$_6$H$_4$)(μ-MeC$_6$H$_4$)]$_2$ which separated in low yield. The crystallographic study explains that structural constraints in organometallic compounds are responsible for metal–metal bond formation. As [CpCr(μ-O)]$_4$ is obtained on the addition of trimethylamine-N-oxide to chromocene, bearing a distorted cubane core [39]. The organometallic compound [Cp$_2$Mo$_2$(MeC$_2$Me)(CO)(μ-PPh$_2$)$_2$] having a double metal–metal bond may be obtained with the reaction of [Cp$_2$Mo$_2$(CO)$_4$(μ-MeC$_2$Me)] and tetraphenyl-diphosphine [40]. Even for the formation of a versatile organometallic compound like [W(CO)$_2$(SnR$_3$)(S$_2$CNR$'_2$)$_2$], double desulphurization is also preferred [41,42].

2.4 Spectral characteristics of organometallics

Spectroscopy is a field which has a huge impact on the chemistry of organometallics and researchers in this field use blend of concepts of spectroscopic methods and bonding to understand the various properties of organometallic compounds. The spectroscopic techniques used for the structure elucidation of organometallics cover over ten orders of magnitude in photon energy. The different energy regions arrange for different information about the properties and structure of compounds. One can achieve electronic and geometric structures by choosing the right combination of methods for any system. The characterization of organometallics includes finding a comprehensive understanding of the compound from its identification, purity content and explication of the spectrochemical features. The understanding of the detailed structure of organometallics is critical to know the properties that are based on the standard structural properties.

There are various spectroscopic methods used in structure elucidation of organometallics, such as nuclear magnetic resonance (NMR), electron paramagnetic resonance (EPR), infrared (IR) spectroscopy, ultraviolet (UV) spectroscopy, electron nuclear double resonance (ENDOR), electron spin echo envelope modulation, Raman spectroscopy (RS), and X-ray absorption spectroscopy (XAS). The brief description of each technique is summarized below one by one.

2.5 Nuclear magnetic resonance spectroscopy

NMR spectroscopy is commonly used by chemists to measure ^{1}H, ^{13}C, or ^{31}P NMR spectra of diamagnetic organometallic compounds. NMR takes very less time to confirm whether the reaction has taken place or not. Mostly, NMR measurement is associated to,

1. Confirm whether the reaction has occurred or not, just by counting the number of signals and further deciding to continue the reaction.
2. Recognize the novel features of structures by remarking changes in chemical shift or J values.
3. A unique and sufficient probe is required for the resolution of the structure to follow the development or kinetics of the reaction.

NMR is the most frequently used technique to characterize organometallics as it is one of the simplest tools used as one can obtain spectra quickly and only a few peaks are observed [43]. In evaluating NMR spectra, coupling constants are exceptionally valuable as these authorize the connection between various nuclei in a molecule, so provide basic information about the proximity of a nucleus to another. Additionally, nuclear overhauser enhancement (NOE) provides information about connectivity, and direct information about the connectivity between nuclei through space is obtained. So, information obtained from NOE is complementary to that obtained from coupling constants. There are various types of NMR techniques which are generally being used for characterization as follows:

2.5.1 Decoupling difference nuclear magnetic resonance spectroscopy

Homonuclear decoupling is a well-recognized and appreciated tool used to determine the coupling of nuclei. This tool is best applied when all the signals are resolved and the molecule is simple. But in case of complex and complicated structures, signals may overlap. Then, it is difficult to find out which signal is changed on decoupling. To solve this problem, difference spectroscopy is used. In this technique, two spectra are recorded. The decoupling frequency of the first spectrum is placed in a region where no signal is found. Then decoupling frequency of the second spectrum is placed on a signal in the spectrum. Then, one peak is deducted to the another and all the unaffected resonances cancel out. And all the decoupled signals are left, these are the signals generated from the nuclei coupled to the decoupled nucleus and signals which show NOE.

2.5.2 Nuclear overhauser enhancement difference spectroscopy

One of the most important tools in defining connectivity between protons or protons and other nuclei is NOE. It allows for determining which of the nuclei are physically close to each other. This effect is based on the relaxation of protons, which is mainly due to dipole—dipole interactions. This effect is dependent on the distance between the nuclei, it falls off rapidly with distance. It helps in determining the *cis* or *trans* configuration of ligands on the metal.

The maximum NOE observed is given by $\gamma_A/2\gamma_B$, where γ_A and γ_B are the gyromagnetic ratios of the observed nuclei and irradiated nuclei respectively. Usually, when ^1H is both irradiated and observed nucleus, the maximum NOE is 0.5. In the case of ^{13}C, with ^1H decoupling, NOE is 1.99. For ^{15}N and ^{29}Si, γ is negative and gives maximum values of -4.93 and -2.52. In the case of ^{13}C, ^{15}N, and ^{29}Si, slow tumbling happens due to which maximum NOE approaches zero.

2.5.3 Heteronuclear experiments

In such cases, ^1H and X nucleus couple to give an improved signal based on the number of attached protons. Here, coupling between a sensitive nucleus, such as ^{19}F, ^{31}P, and the X nucleus, is considered. The two main types under this category which have gained the most attention are INEPT (insensitive nuclei enhanced by polarization transfer) and DEPT (distortionless enhancement by polarization transfer).

2.5.3.1 J Modulation test

In this case, a study with the Carbon and CH$_2$ groups 180° out of phase with CH and CH$_3$ groups is produced. This method works well if the coupling constant $^1J(^{13}C-^1H)$ lies between 125 and 165 Hz. The trial provides the explicit obligation of many H^+ with C atoms, by uniting the phase of the signal with the chemical shift and intensity. This experiment when combined with chemical shift C, CH$_2$, and CH$_3$ groups can be sorted.

2.5.3.2 Insensitive nuclei enhanced by polarization transfer

This method is beneficial to increase the response of the nucleus and enable ^1H relaxation time dominant over the relaxation time of the nucleus, which is generally observed for a lengthier period. The responsiveness of NMR study depends upon the population alteration amid the energy levels. If I = 1/2 as in the case of ^1H or ^{13}C, there are two energy levels, $m_I = 1/2$ and $m_I = -1/2$. As the energy breach amongst the two energy levels is small as equated to thermal energy, kT, the population dissimilarity is proportionate to the energy gap. The energy gap and NMR frequency for ^1H are four times as compared to ^{13}C leading to higher sensitivity of ^1H NMR. This advantage of population difference is taken in INEPT experiments to enhance the sensitivity of coupled nuclei. These experiments found a ^{13}C nucleus in the proximity of an excess population of H nuclei which results in a fourfold enhancement of the intensity of the signal of ^{13}C NMR. In the case of INEPT, there is no gain of intensity from the NOE. Most of the organometallic compounds show at least twofold NOE for carbon atoms that bear protons, this gain is immediately reduced to signal doubling. This technique is predominantly useful when NOE is negligible. This happens for compounds with high molecular weight, at 400 MHz ^{1}H observation, this technique is expedient for molecular weights greater than 1000 and at 80 MHz molecular weight greater than 5000 is beneficial. But if the compound contains paramagnetic contamination, NOE is quenched and INEPT or DEPT changes expedient. One of the disadvantages of this technique is that carbon atoms which are not coupled to ^{1}H are invisible to this technique. The factual significance of method is for another nucleus with $I = 1/2$, for example in the case of ^{103}Rh NOE is not detected, but the latent gain due to INEPT or DEPT is 31.8 times. This technique is useful for ^{57}Fe, ^{183}W, ^{29}Si, and ^{195}Pt.

2.5.3.3 Distortionless enhancement by polarization transfer

DEPT is a preferred technique for spectral editing, it is less sensitive to the magnitude of $^1J(^{13}C-^1H)$, but it needs a longer duration of time as compared to the basic INEPT spectra, so it is less fit for nuclei where the coupling constant is small.

2.5.3.4 Incredible natural abundance double quantum transfer experiment

It is used to detect $^{13}C-^{13}C$ connection in the absence of the mix-up of the ^{13}Carbon singlets of the species without adjacent ^{13}C atoms. It has considerable significance in creating the relationship amongst nuclei of low abundance and can be useful for ^{29}Si, ^{77}Se, and ^{183}W also [44–47].

NMR is useful for the study of rapid intermolecular exchange processes in organometallic systems. Here we are taking the example of photolysis of Re(^{i}PrCp)(CO)(PF$_3$)Xe which was examined using ^{19}F NMR spectroscopy (Fig. 2.9). The peak at -3.13 due to the presence of free PF$_3$ proposed that it may be photochemically lost from Re(^{i}PrCp)(CO)$_2$(PF$_3$), which leads to the synthesis of Re(^{i}PrCp)(CO)$_2$Xe.

In heteronuclear 2D ^{19}F − ^{129}Xe multiple quantum coherence shift correlated spectra of ^{129}Xe-labeled Re(^{i}PrCp)(CO)(PF$_3$) (^{129}Xe), it was found that the Xe ligand is not swapping by the free Xe. The quick exchange of free and bound Xe effactually randomizes the spin state of the ^{129}Xe that is sensed by the ^{19}F and ^{31}P nuclei hence, coupling was not witnessed. The occurrence of coupling in ^{19}F and ^{129}Xe indirectly detects the ^{129}Xe chemical shift with this experiment (Fig. 2.10) [48] (Fig. 2.11).

In Fig. 2.12, ^{31}P NMR spectra of the Pt(0) and Pt(II) complexes Pt (TPPTS)$_3$ and Pt(H)(TPPTS)$_3^+$ respectively (where TPPTS is the triphenylphosphine derivative P(m-NaSO$_3$C$_6$H$_4$)$_3$) as a function of pH are shown. At pH 13, the Pt(0) complex is found to be stable, whereas, at pH 4, the hydride

Organometallic compounds: bonding and spectral characteristics Chapter | 2 43

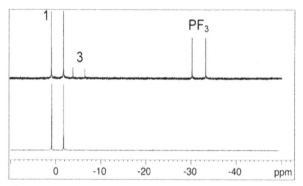

FIGURE 2.9 ^{19}F NMR spectra of the Re(iPrCp)(CO)$_2$(PF$_3$) (Lower) Before photolysis. (Upper) After photolysis [48].

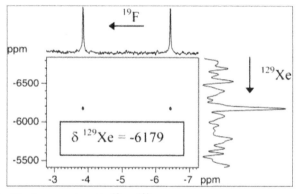

FIGURE 2.10 The heteronuclear 2D ^{19}F – ^{129}Xe multiple quantum coherence shift correlated spectrum of ^{129}Xe-labeled Re(^{i}PrCp)(CO)(PF$_3$) (^{129}Xe) [48].

cation is preferred. Using ^{31}P rather than ^{1}H or ^{13}C offers an easy and rapid overview of the changes in the chemistry [50].

In Fig. 2.13, the ^{1}H NMR spectrum of the deuterated rhodium pyrazolyl borate isonitrile complex, RhD(CH$_3$)(Tp')(CNCH$_2$Bu^t) is provided, in this graph it is shown how gradually the methyl region changes to the isomer where the deuterium atom has shifted to the methyl group to obtain RhH(CH$_2$D)(Tp')(CNCH$_2$Bu^t) [51].

In the case of the 19-electron complex (η^5-C$_5$Ph$_4$H)Mo(CO)$_2$(bis-diphenyl-phosphinomaleic anhydride), where the electrons exist primarily on the ligand, for rotation of the ring the barrier was found to be $\sim$12 kJ mol^{-1} at 25°C. Adequately presence of bulky substituents on the ring as well as metal can upsurge the barrier in diamagnetic complexes significantly due to which shapes of NMR lines are affected. For the two-site exchange, resonances will

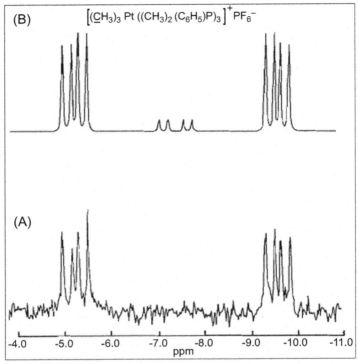

FIGURE 2.11 Observed (A) and calculated (B) ^{13}C NMR spectra for the non-^{195}Pt coupled platinum-methyl region of *fac* [(CH$_3$)$_3$Pt(CH$_3$)$_2$(C$_6$H$_5$)]$^+$PF$_6^-$ [49].

widen by increasing the rate; then the resonances will conjoin and lastly sharpen to a narrow resonance (see Fig. 2.14) [52].

2.6 Electron paramagnetic resonance

Organometallic chemistry is primarily ruled via the analysis of diamagnetic complexes, for which NMR spectroscopy is used as the significant method for their characterization. So, the techniques used to characterize paramagnetic complexes and the intermediates are essential, such as EPR spectroscopy. It is a useful spectroscopic tool to describe the SOMO (Singly occupied molecular orbital) of the paramagnetic complexes which, in turn, aids in understanding the reactivity of an open shell organometallic compound. For $S = 1/2$ system, which contains one unpaired electron only, the EPR spectrum is vastly indicative to decide the spin density distribution of a compound [53].

EPR studies are performed on frozen solutions which contain all orientations of the complex relative to the magnetic field. In the standard EPR

Organometallic compounds: bonding and spectral characteristics **Chapter | 2** 45

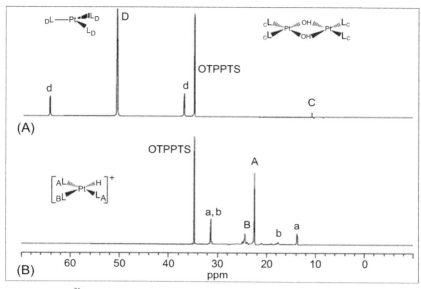

FIGURE 2.12 ^{31}P NMR spectra recorded between pH 4 and 13, (A) at pH 13 and (B) at pH 4 [50].

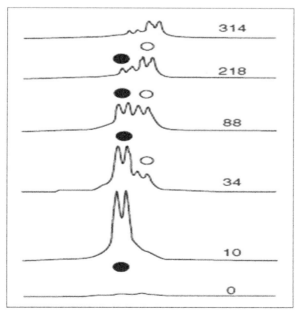

FIGURE 2.13 ^{1}H NMR spectra for rearrangement of RhD(CH$_3$)(Tp')(CNCH$_2$Bu^t) to RhH(CH$_2$D)(Tp')(CNCH$_2$Bu^t) [51].

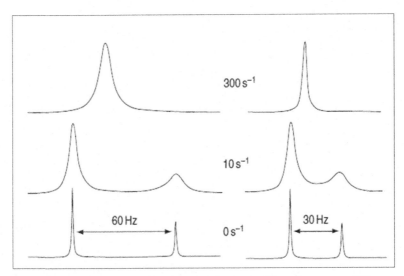

FIGURE 2.14 The effect of rate of exchange on the shape of lines of resonances [52].

experiment, the sample is in an oscillating field of microwaves which have a fixed frequency (X-band: $h\nu \sim 9$ GHz $= 0.30$ cm^{-1}), and a perpendicular magnetic field is varied. The splitting between the $M_s = +1/2$ and $-1/2$ sublevels of the ground state increases linearly with the magnetic field until the resonance condition, $h\nu = g\beta H$, is satisfied and microwave absorption occurs. This results in a peak in the microwave absorption spectrum at the specific magnetic field which defines the g value (g = $h\nu/\beta H$ = 0.71448 × microwave frequency in MHz/magnetic field in gauss). The g factors are often anisotropic which means that they have different values when the complex is oriented in different directions relative to the magnetic field, and consequently, the EPR experiment shows the absorption of microwaves at different magnetic field values (Fig. 2.15).

The interfaces which majorly impact upon EPR spectrum of any classes with $S = 1/2$ are the Zeeman effect that is due to the interaction between the unpaired electron with an applied magnetic field and the electron-nuclear hyperfine interaction which is the contact of the electrons with magnetic nuclei in the molecule. These interactions are dependent on the orientation of the molecule in an applied field. There are some other interactions which may affect the EPR spectrum, such as nuclear quadrupole coupling and, electron−electron interactions for the species that have more than one unpaired electron [54−61].

An encapsulation of assembly of tetra(4-pyridyl)metalloporphyrins (M[II](TPyP), where M = Co, Zn) was done to form cages and their reactivity

Organometallic compounds: bonding and spectral characteristics Chapter | 2   47

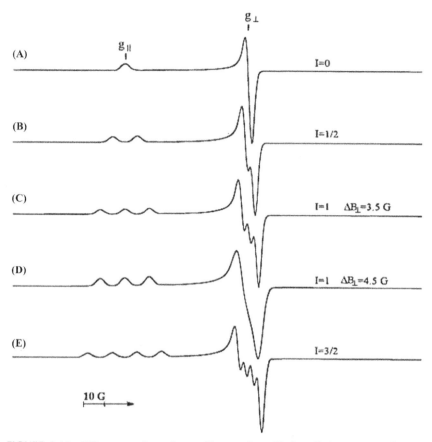

FIGURE 2.15 EPR spectra for polycrystalline species with $S = 1/2$ in presence of single nucleus with nuclear spin $I = 0$ (A), 1/2 (B), 1 (C, D) and 3/2 (E). Figure (D) displays the effect of broadening of the signal (C) on the resolution of the spectrum [54].

in radical-type transformations including diazo compounds. The encapsulated Co porphyrin showed considerably increased activity as compared to the free Co^{II}(TPyP) catalyst. The free Co^{II}(TPyP) showed a broad peak in the EPR investigation owing to intermolecular accretion which involves the management of pyridine to cobalt(II). After encapsulation, the building block behaves utterly different, showing sharper signals for the same species missing solid binding to axial ligands [53].

2.7 Electron nuclear double resonance

ENDOR spectroscopy combines the EPR experiment with an NMR experiment which reviews the nuclear spin splitting within an electron spin sublevel and consequently offering a high resolution method of obtaining nuclear hyperfine couplings to an electron spin. An EPR transition at a fixed magnetic field corresponding to a specific g value of the site will saturate at low temperature and high microwave power (i.e., it loses signal intensity of the spin sublevels involved in the transition owing to slow spin relaxation). Under such conditions, the system is scanned in energy (MHz) with a radio-frequency source until nuclear transitions come into resonance (i.e., a double resonance condition is satisfied). Then, saturation is reduced allocated to the presence of additional relaxation pathways, and microwave absorption is perceived. Thus in ENDOR spectroscopy, the EPR intensity is used to sensitively detect low energy transitions between nuclear spin sublevels (Fig. 2.16).

A nucleus with I greater than or equal to 1 will have an electric quadrupole moment which interacts with the electric field gradient in the proximity of the nucleus (i.e., the local chemical environment when the symmetry is less than O_h or T_d), two limits of the above equation need to be considered. For protons ($I = 1/2$) the nuclear Zeeman effect is large compared to the hyperfine coupling due to its large g_N value. Protons exhibit two peaks in the ENDOR spectrum centered on their nuclear Zeeman energy and split by the hyperfine coupling constant. For most other nuclei, the hyperfine coupling is larger than the nuclear Zeeman term.

To distinguish among possible assignments of peaks it is often beneficial to perform the ENDOR experiment at a second microwave frequency. This changes the magnetic field mandatory to satisfy the electron resonance condition for a given g value and therefore shifts the nuclear Zeeman frequency as this is dependent on the applied magnetic field. Instead, the hyperfine coupling is not dependent on the magnetic field and doesn't change with microwave frequency. Further, the anisotropy in the hyperfine coupling constant could be measured by obtaining ENDOR spectra at the different magnetic

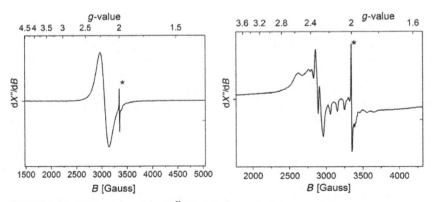

FIGURE 2.16 EPR spectra of the Co^{II}(TPyP) before and after encapsulation.

Organometallic compounds: bonding and spectral characteristics **Chapter | 2** **49**

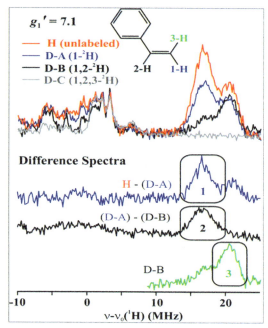

FIGURE 2.17 (Top) ^{1}H ENDOR spectra of L^{Me}Fe(styrene) with unlabeled and ^{2}H-labeled styrene. (Bottom) Difference ENDOR spectra [62].

fields associated with the different electron *g* values of the metal site. Lastly, it must be noted that ENDOR intensities usually decrease toward the low-frequency side of the spectrum. To study the low energy spectral region (associated with small hyperfine interactions with high sensitivity), one utilizes pulsed EPR methods.

The ENDOR characterization (Fig. 2.17) of the biomimetic Fe complex L^{Me}Fe(styrene) (L^{Me} = 2,4-bis(2,6-diisopropylphenylimido)-3-pentyl), offers an association to hyperfine and structural aspects of an alkene part on iron for the first time.

2.8 Electron spin echo envelope modulation

The pulsed EPR methods offer a highly sensitive mechanism for spotting nuclear spins that are feebly interacting with a paramagnetic metal center (i.e., very low super hyperfine coupling constants). An intense microwave pulse is applied to a compound in an external magnetic field of $\pi/2$ duration with the pulse perpendicular to the external field. The spins centered on the resonance field are rotated 90° into the *y*-axis. The short pulse would excite spin packets in a range of frequencies that could be greater or lesser than the central resonance frequency. After an assured time, a second intense perpendicular microwave pulse is applied that inverts the spins in such a way that their time behavior is reversed and they refocus on the central frequency. Another period after the second pulse is back in alignment which produces a spun echo. The amplitude of electron spin echo is then perceived as a function of the time between pulses.

The amplitude of the echo decays swiftly with time due to spin relaxation. Though, this decay is modulated by constructive and destructive interferences associated with nearby nuclear spins interacting with electron spin. This interference is caused by the microwave pulses which cover an energy range extending over the M_I components of a given spin sublevel. At low hyperfine couplings, these wave functions become slightly varied, such that normally forbidden microwave transitions ($\Delta M_S = \pm 1$, $\Delta M_I = \pm 1$) become weakly allowed and these can interfere with the allowed $\Delta M_S = \pm 1$, $\Delta M_I = 0$ transitions.

2.9 Infrared spectroscopy

IR spectroscopy also known as molecular spectroscopy is well known to provide information about the structure of molecules that exist in the sample. It is an easily applicable and flexible form of spectroscopic analysis. It is the examination of IR light interacting with a molecule. It could be examined in various ways by determining absorption, emission and reflection. It is used to find various functional groups in molecules. This technique measures the vibration of atoms and determines the functional groups.

There are some elementary requirements for IR spectroscopy such that any deformation or vibration should be linked to a permanent charge separation or electric dipole. The vibrations amongst atoms in the molecular group as in the case of the eight-atom S ring in pure sulfur would not be detected. In contrast, the motion of a C−H bond is typically easy to detect.

An IR spectrum is a plot of measured IR intensity versus wavelength (or frequency) of light obtained in this technique. When the sample is exposed to IR radiation, its molecules selectively absorb radiations of specific wavelengths resulting in the change of dipole moment of sample molecules. As a result, the vibrational energy levels of sample molecules transfer from the ground to the excited state. Most of the molecules are IR active excluding some homonuclear diatomic molecules like N, O, and Cl because of the zero dipole change in the rotation and vibration of these molecules. The most commonly used range in IR spectroscopy is $4000 \sim 400 \text{ cm}^{-1}$ as the absorption radiation of most inorganic ions and organic compounds lie within the specified region [63−66].

In Fig. 2.18 the IR spectra which are found upon dissolution of the solution of $W(CO)_5$ and N_2O in liquid Kr and photolyzed at 351 nm are shown. This spectrum is obtained as the change in spectrum before and after

Organometallic compounds: bonding and spectral characteristics Chapter | 2 51

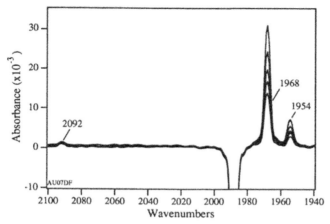

FIGURE 2.18 Infrared spectra of complex $W(CO)_5(N_2O)$ at 180 K [67].

photolysis at intervals of ∼1 min is observed. A new species formed is confirmed by the presence of crests at 1954, 1968, and 2092 cm^{-1} which gradually decays to form $W(CO)_6$ [67].

2.10 Ultraviolet–visible spectroscopy

Ultraviolet–visible (UV–Vis) spectroscopy is a spectroscopic technique that is much like FTIR and is beneficial for the identification of pure compounds. Several molecules comprise chromophores which absorb a particular wavelength of UV or visible region. The absorption obtained from the samples at a given wavelength is directly related to the concentration of the sample as per Beer–Lambert law (A = εbC). UV–Vis is an easy and cheap technique which allows the recovery of samples and good insight into pure compounds (Fig. 2.19).

This technique is useful for the evaluation of aromatic compounds. This technique helps to detect and separate compounds that are generally destroyed in other techniques. The absorption of UV–Vis radiations relates to the excitation of outer electrons. There are three types of electronic transitions, first are transitions involving π, σ, and n electrons, second is transitions involving charge-transfer electrons and third are transitions involving d and f electrons [68,69].

In UV-Vis spectra of Mb–CO and deoxy-Mb (where Mb is myoglobin), it was observed that if a complex releases CO into the solution, deoxy–Mb would convert into Mb–CO. Such changes can be monitored by observing the shift in the Q bands of the heme group in deoxy–Mb as well as Mb–CO [70].

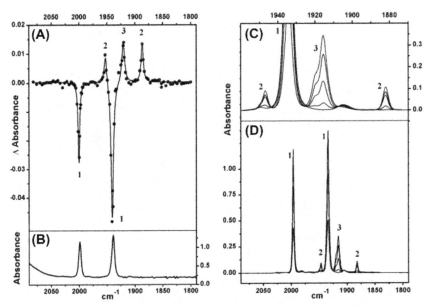

FIGURE 2.19 Infrared spectra obtained after photolysis of Re(iPrCp)(CO)$_2$(PF$_3$) (A) TRIR spectrum 20 μs after laser flash of a solution of Re(iPrCp)(CO)2(PF3)in scXe, doped with CO at room temperature. (B) FTIR spectrum of the solution. (C and D) FTIR spectra in IXe solution with no added CO at 166 K at various time intervals before and after 1, 6, 12 and 36 min of irradiation with UV light [48].

2.11 Raman spectroscopy

RS is another form of vibrational spectroscopy where UV or visible light is used to illuminate samples and then examine the light that is scattered from the samples. This illumination is generally provided by laser, with this illumination, molecules and their functional groups in the laser beam elastically scatter light and shift the wavelength of scattered light by the wavelengths that are equivalent to absorption bands of IR. These bands are stated as Raman spectra or Raman shifts. This form of vibrational spectroscopy is advantageous cause of the ability to examine symmetrical bonds and molecules which do not display permanent charge separation or dipole and is somewhat similar to IR spectroscopy. The modern instrument uses laser illumination, monochromator and Fourier transform the dispersive system, CCD detector, a wide range of sampling devices including microscopes, and computers have collectively made RS a multipurpose analytical tool [71–73].

Fig. 2.20 shows the Raman study of pure tetrahydrofuran and solution of Et$_3$Al, Et$_2$AlCl, and EtAlCl$_2$ in THF. The strongest spectral bands are found between 250 and 430 cm^{-1}, which correspond to aluminum−chlorine bonds, and those in the range 450−650 cm^{-1}, correspond to aluminum−alkyl bonds. The bands of the aluminum−alkyl bonds of the series display a trend

Organometallic compounds: bonding and spectral characteristics Chapter | 2   53

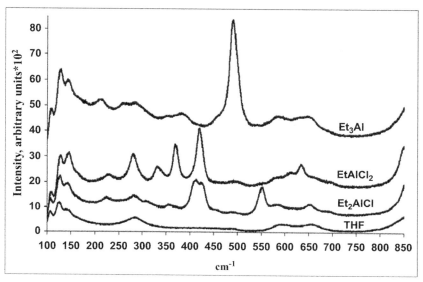

FIGURE 2.20 Raman spectra of alkyl-chloro aluminum compounds [71].

dependent on the Cl/R (chlorine/alkyl) ratio. So higher the Cl/R ratio, the higher the scattering wavenumber of the Al—R bonds. Similarly, a higher R/Cl (alkyl/chlorine) ratio leads to stronger bands for the Al—R bond.

Similarly, the adjusted spectra in Fig. 2.21 demonstrates the interesting trends. A peak corresponding to an aluminum—chlorine stretching vibration appeared at 367 cm^{-1} in all cases. The highest wavenumber is obtained for the EtAlCl$_3^-$ ion, at 611 cm^{-1}, and the position of the Al-R bands showed a linear correlation to the Cl/R ratio. Here, the aluminum—alkyl versus aluminum—chlorine peak height ratio corresponds to the ratio between the numbers of the Cl versus the organic ligands.

2.12 X-ray absorption spectroscopy

XAS is broadly used for the determination of geometric structure as well as element-specific electronic structure. Spectroscopic techniques like UV—Vis and IR are quite similar to XAS. Though in XAS, core electrons like 1s or 2p electrons are excited and the energy of these core electrons is element specific.

A standard form of the XAS spectrum of Pd foil is presented in the literature. The spectrum is divided into two parts: X-ray absorption near edge spectrum (XANES) and extended X-ray absorption fine structure spectrum (EXAFS). The region from absorption edge (Eo, i.e., onset of X-ray absorption), to around 30 eV is the XANES and the scattering portion of the

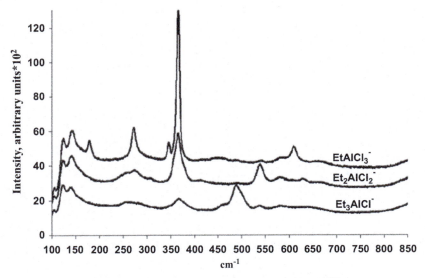

FIGURE 2.21 Raman spectra of Li organo-chloro-aluminum complexes [71].

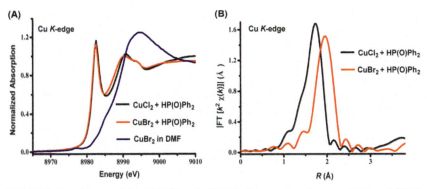

FIGURE 2.22 Spectra of the XANES and EXAFS after reactions between CuX$_2$ and HP(O)Ph$_2$ [79].

spectrum which continued to nearly 1000 eV above the edge is called EXAFS. The intensity and shape of peaks in the XANES region are based on the oxidation state, type and number of ligands and other species that are bonded to the excited atom. EXAFS region gives the type and number of adjacent atoms with their distance to the absorbing atom [74−78].

XAS was useful to examine the reaction of diphenylphosphine oxide (HP(O)Ph$_2$) with CuBr$_2$ under an inert condition (Fig. 2.22). The XANES information of the CuBr$_2$ solution gave a preedge at 8976.6 eV and after adding

Organometallic compounds: bonding and spectral characteristics Chapter | 2   55

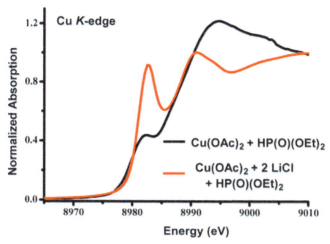

FIGURE 2.23 XANES spectra of various copper species [79].

HP(O)Ph$_2$ to this solution, a new copper species with an edge energy of 8981.7 eV was found. Similarly, when diphenylphosphine oxide and CuCl$_2$ were mixed at 80°C, a species with edge energy of 8981.7 eV was discovered, which is similar to the reaction between CuBr$_2$ and HP(O)Ph$_2$. EXAFS was also explored to describe the structure of the synthesized copper compounds. The EXAFS also exposed that the halide ion might be coordinated to Cu(I) species [79].

In the same reaction system, the XAS was also utilized to expose the halide influence to Cu(II) reduction development. It was found that copper acetate might be rapidly reduced by dialkyl phosphites and in presence of LiCl, the Cu(I) species were produced. Therefore the results discovered that the reduction of Cu(II) by P−H complexes was mediated by LiCl (Fig. 2.23).

2.13 Summary

This chapter focuses on the bonding and spectral characteristics of organometallic compounds. Here, various types of bonds found in organometallics are explained, such as a covalent bond, ionic bond, and hydrogen bond. Most of the properties of organometallic compounds are based on the type of M−C bond involved. Some are polar ionic bonds in which metal possesses a positive charge whereas carbon has a negative charge. The majority of M−C bonds in organometallics are covalent by nature, which is further classified into σ, π, polynuclear, and dative bonds. Together with M−C bonds, here bonding also comprises metal−hydrogen (M−H) and metal−metal (M−M) bonds. To study the structure of these compounds, various spectroscopic techniques are used. Out of these techniques, NMR is the oldest and most frequently used for structure elucidation. But now these days other advanced techniques like EPR, UV, and XAS are also used for the determination of the structure of compounds and intermediates in a short time.

References

[1] Ang WH, Casini A, Sava G, Dyson PJ. Organometallic ruthenium-based antitumor compounds with novel modes of action. J Organomet Chem 2011;696(5):989–98.
[2] Abbott JKC, Dougan BA. Synthesis of organometallic compounds. In: Xu R, Pang W, Huo Q, editors. Modern inorganic synthetic chemistry. Elsevier; 2011. p. 269–93.
[3] Noffke AL, Habtemariam A, Pizarro AM, Sadler PJ. Designing organometallic compounds for catalysis and therapy. Chem Commun 2012;48(43):5219–46.
[4] Pfennig BW. Principles of inorganic chemistry. Hoboken, NJ: John Wiley & Sons, Inc; 2015. p. 627–8.
[5] Alama A, Tasso B, Novelli F, Sparatore F. Organometallic compounds in oncology: implications of novel organotins as antitumor agents. Drug Discov Today 2009;14:500–8.
[6] Jana R, Pathak TP, Sigman MS. Advances in transition metal (Pd,Ni,Fe)-catalyzed cross-coupling reactions using alkyl-organometallics as reaction partners. Chem Rev 2011;111:1417–92.
[7] Pruchnik FP, Duraj SA. Introduction to organometallic chemistry. Organometallic chemistry of the transition elements. Modern inorganic chemistry. Springer; 1990.
[8] Greenwood FLT. An introduction to organometallic complexes in fluorescence cell imaging: current applications and future prospects. Organometallics 2012;31:5686–92.
[9] Crabtree RH. An organometallic future in green and energy chemistry. Organometallics 2011;30:17–19.
[10] Astruc D. Ionic and polar metal-carbon bonds: alkali and rare-earth complexes. In: Organometallic Chemistry and Catalysis, Springer; 2007. p. 12–3.
[11] Chris E.P. Basic principles of organic chemistry. 2015:28–37.
[12] Mishra S. Synthesis and reactivity of cyclopentadienyl based organometallic compounds and their electrochemical and biological properties. 2017:1–195.
[13] Hanusa TP. Non-cyclopentadienyl organometallic compounds of calcium, strontium and barium. Coord Chem Rev 1999;210(2000):329–67.
[14] Weiss E., Lucken E.A.C. Organometallic compounds. 1964:197.
[15] Mcfarlane KL, Ford PC. Room-temperature reactions of the intermediate(s) generated by flash photolysis of (η5-C5H5)Fe(CO)2CH3. Organometallics 1998;17:1166–8.
[16] Kubas GJ. Metal-dihydrogen and σ-bond coordination: the consummate extension of the Dewar-Chatt-Duncanson model for metal-olefin π bonding. J Organomet Chem 2001;635 (2001):37–68.
[17] Nobel, T.6.4:organometallic chemistry of d block metals (part 1). 1973:1–7.
[18] Klei E, Teuben JH. On the nature and reactivity of Cp2TiCH3. J Organomet Chem 1980;188:97–107.
[19] Bekyarova EB, et al. Effect of covalent chemistry on the electronic structure and properties of carbon nanotubes and graphene. Synth Met 2015;210(1):80–4.
[20] Tian X, et al. Effect of atomic interconnects on percolation in single-walled carbon nanotube thin film networks. Nano Lett 2014;14(7):3930–7.

[21] Brennan LJ, Gun'ko YK. Advances in the organometallic chemistry of carbon nanomaterials. Organometallics 2015;34(11):2086−97.

[22] Mason R. Structural inorganic chemistry and diffraction methods: metal-ligand bonds in polynuclear complexes and on metal surfaces. Coord Chem Rev 1973;1:513−26.

[23] Zerger R, Rhine W, Stucky G. Stereochemistry of polynuclear compounds of the main group elements. The bonding and the effect of metal-hydrogen-carbon interactions in the molecular structure of cyclohexyllithium, a hexameric organolithium compound. J Am Chem Soc 1974;96:6048−55.

[24] Tang J, Zhao L. Polynuclear organometallic clusters: synthesis, structure, and reactivity studies. Chem Commun 2020;56(13):1915−25.

[25] Brammer L, Espallargas GM. Structure and bonding in organometallic compounds: Diffraction methods. Compr Organomet Chem 2007;1(1):573−603.

[26] Su P, Li H. Energy decomposition analysis of covalent bonds and intermolecular interactions. J Chem Phys 2009;131(1):1−15.

[27] Mitoraj M, Michalak A. Natural orbitals for chemical valence as descriptors of chemical bonding in transition metal complexes. J Mol Model 2007;13(2):347−55.

[28] Jerabek P, Schwerdtfeger P, Frenking G. Dative and electron-sharing bonding in transition metal compounds. J Comput Chem 2019;40(1):247−64.

[29] Jeffrey GA, Jeffrey GA. An introduction to hydrogen bonding. New York: Oxford University Press; 1997.

[30] (a) Shubina ES, Krylov AN, Kreindlin AZ, Rybinskaya MI, Epstein LM. Hydrogen-bonded complexes involving the metal atom and protonation of metallocenes of the iron subgroup. J. Organomet Chem. 1994;465:259−62.
(b) Epstein LM, Shubina ES, et al. The role of hydrogen bonds involving the metal atom in protonation of polyhydrides of molybdenum and tungsten of $MH_4(dppe)_2$. J. Organomet Chem. 1995;493:275−7.

[31] Epstein LM, Shubina ES. New tes of hydrogen bonding in organometallic chemistry. Coord Chem Rev 2002;231:165−81.

[32] Shubina ES, Belkova NV, Epstein LM. Novel types of hydrogen bonding with transition metal π-complexes and hydrides. J Organomet Chem 1997;536−537:17−29.

[33] Park S, Ramachandran R, Loyp AJ, Morris RH. A new type of intramolecular H ··· H ··· H interaction involving N-H ··· H(Ir) ··· H-N atoms. Crystal and molecular structure of [irh(η^1-$SC_5H_4NH)_2(\eta^2$-$SC_5H_4N)(pcy_3)]BF_4 \cdot 0.72CH_2Cl_2$. J Chem Soc Chem Comm 1994;309:2201−2.

[34] (a) Lee JC, Peris E, Crabtree RH, Rheingold AL. An unusual type of H···H interaction: Ir-H···H-O and Ir-H···H-N hydrogen bonding and its involvement in σ-bond metathesis. J Am Chem Soc 1994;116:11014−19.
(b) Peris E., Lee J.C., Rambo J.R., et al., Factors affecting the strength of X—M hydrogen. 1995. 3485−3491.

[35] (a) Gusev DG, Lough AJ, Morris RH. New polyhydride anions and proton-hydride hydrogen bonding in their ion pairs. X-ray crystal structure determinations of Q[mer-Os$(H)_3(CO)(p^ipr_3)_2]$, Q = [K(18-crown-6)] and Q = [K(1-aza-18-crown-6)]. J Am Chem Soc 1998;120:13138−47.
(b) Abdur-Rashid K, Gusev DG, Lough AJ, Morris RH. Intermolecular proton-hydride bonding in ion pairs: synthesis and structural properties of $[K(Q)][MH_5(p^ipr_3)_2]$ (M = Os, Ru; Q = 18-crown-6, 1-aza-18-crown-6, 1,10-diaza-18-crown-6). Organometallics 2000;19:834−42.

[36] Hogarth G. Organometallic compounds containing metal-metal bonds. Organomet Chem 1991;20:145.

[37] Herberhold M, Kuhnlein M, Schrepfermann M, Ziegler ML, Nuber B. Zweikernige Pentamethylcyclopentadienyl-Vanadium-Komplexe mit unterschiedlichen Chalkogenen im Brückensystem. 51V NMR-Spekstroskopische Charakterisierung der Verbindungen und Röntgenstrukturanalyse von Cp$_2$*V$_2$Se$_2$S$_2$ und Cp$_2$*V$_2$Se$_3$O. J Organomet Chem 1990;398:259−74.

[38] Goh LY, Wong RCS, Chu CK, Hambley TW. Reaction of [{Cr (cp)(CO) 3} 2](cp = η 5-C 5H 5) with elemental phosphorus. Isolation of [Cr 2 (cp) 2 (P 5)] as a thermolysis product and its X-ray crystal structure. J Chem Soc Dalton Trans 1990;977−82.

[39] Bottomley F, Paez DE, Sutin L, White PS, Kohler FH, Instituí A, et al. Organometallic oxides: preparation and molecular and electronic structure of antiferromagnetic [(η-C$_5$H$_5$)Cr(μ$_3$-O)]$_4$ and [(η-C$_5$H$_5$)Cr]$_4$(μ$_3$-η^2-C$_5$H$_4$)(μ$_3$-O)$_3$. Organometallics 1990;9:2443−54.

[40] Conole G, Mcpartlin M, Mays MJ, Morris MJ. Chemistry of phosphido-bridged dimolybdenum complexes. Part 5. Synthesis and protonation of a phosphido-bridged dimolybdenum complex containing a terminal alkyne ligand: X-ray crystal structures of [Mo$_2$(μ-pph$_2$)$_2$(CO)(η-C$_2$Me$_2$)(η-C$_5$H$_5$) $_2$] and [Mo$_2$(μ-Co)(μ-pph$_2$){μ-Ph$_2$PC(Me) = chme}(η-C$_5$H$_5$)$_2$][BF$_4$]. J Chem Soc Dalton Trans 1990;2359−66.

[41] Hitchcock PB, Lappert MF, Mcgeary MJ. Stannadesulfurization of a bis(diethyldithiocarbamato)tungsten(II) complex: formation of an (aminocarbyne)tungsten complex. Organometallics 1990;2645−6.

[42] Hitchcock PB, Lappert MF, Mcgeary MJ. Stannadesulfurization of a bis(dimethyldithiocarbamato)tungsten(II) complex: formation of a coordinated Me$_2$NCCNMe$_2$ complex. J Am Chem Soc 1990;112:5658−60.

[43] Pregosin PS. NMR in organometallic chemistry. Wiley; 2012.

[44] Mann BE. Recent developments in NMR spectroscopy of organometallic compounds. Adv Organomet Chem 1988;397−457.

[45] Sanders JKM, Mersh JD. Nuclear magnetic double resonance; the use of difference spectroscopy. Prog Nucl Magn Spectrosc 1982;15:353−400.

[46] Mersh JD, Sanders JKM. The suppression of Bloch-Siegert shifts and subtraction artifacts in double-resonance difference spectroscopy. J Magn Reson Imaging 1982;9:289−98.

[47] Grode SH, Mowery MR. A method to observe NOE from regions of spectral overlap. Nuclear-overhauser-enhanced J-resolved difference spectroscopy. J Magn Reson Imaging 1997;126:142−5.

[48] Ball GE, Darwish TA, Geftakis S, George MW, Lawes DJ, Portius P, et al. Characterization of an organometallic xenon complex using NMR and IR spectroscopy. PNAS. 2005;102:1853−8.

[49] Clark HC, Manzer LE, Ward JEH. ^{13}C nuclear magnetic resonance studies of organometallic compounds. V. Di- and trimethylplatinum(IV) derivatives. Can J Chem 1974;52:1973−82.

[50] Helfer DS, Atwood JD. Interconversion between platinum(II) and platinum(0) with change of ph: aqueous reactions of Pt(H)(TPPTS)$^{3+}$ (TPPTS = P(m-C$_6$H$_4$SO$_3$Na)$_3$). Organometallics 2002;21:250−2.

[51] Wick DD, Reynolds KA, Jones WD. Evidence for methane σ-complexes in reductive elimination reactions from Tp'Rh(L)(CH$_3$)H. J Am Chem Soc 1999;121:3974−83.

[52] Faller JW. Yale University, New Haven, CT Dynamic NMR spectroscopy in organometallic chemistry. Elsevier; 2007. p. 407−24.

[53] Goswami M, Chiril A, Rebreyend C, deBruin B. EPR spectroscopy as a tool in homogeneous catalysis research. Top Catal 2015;58:719−50.

[54] Dyrek K. EPR as a tool to investigate the transition metal chemistry on oxide surfaces. Chem Rev 1997;97:305−31.

[55] Rieger AL, Rieger PH. Chemical insights from EPR spectra of organometallic radicals and radical ions. Organometallics 2004;23:154–62.

[56] Van Doorslaer S, Caretti I, Fallis I, Murphy DM. The power of electron paramagnetic resonance to study asymmetric homogeneous catalysts based on transition-metal complexes. Coord Chem Rev 2009;253:2116–30.

[57] Van Doorslaer S, Murphy D. EPR spectroscopy in catalysis. Top Curr Chem 2012;321:1–40.

[58] De Bruin B, Hetterscheid DGH, Koekkoek AJJ, Grutzmacher H. The organometallic chemistry of Rh-, Ir-, Pd-, and Pt based radicals: higher valent species. New York: Wiley; 2007. p. 247–354.

[59] Goodman BA, Raynor JB. Electron spin resonance of transition metal complexes. Adv Inorg Chem Radiochem 1970;13:135–362.

[60] Carter E, Murphy DM. The role of low valent transition metal complexes in homogeneous catalysis: an EPR investigation. Top Catal 2015;58:759–68.

[61] Adams CJ, Connelly NG, Rieger PH. EPR and NMR spectroscopic studies of $[mol_2(mec\equiv cme)Cp]^z$ (L = P-donor ligand, $z = 0$ and 1): fluxionality in a metal–alkyne redox pair. Chem Commun 2001;2458–9.

[62] Horitani M, Grubel K, Mcwilliams SF, Stubbert BD, Mercado BQ, Yu Y, et al. ENDOR characterization of an iron–alkene complex provides insight into a corresponding organometallic intermediate of nitrogenase. Chem Sci 2017;8:5941–8.

[63] Zecchina A, Aréan CO. Structure and reactivity of surface species obtained by interaction of organometallic compounds with oxidic surfaces: IR studies. Catal Rev Sci Eng 1993;35(2):261–317.

[64] Perkins WD. Fourier transform-infrared spectroscopy. Instrumentation. Topics in chemical instrumentation. J Chem Edu 1986;63:A5–10.

[65] Ball DW. Field guide to spectroscopy. Bellingham: SPIE Publication; 2006.

[66] Saptari V. Fourier-transform spectroscopy instrumentation engineering. Bellingham: SPIE Publication; 2003.

[67] Weiller BH, Time-Resolved IR. Spectroscopy of transient organometallic complexes in liquid rare-gas solvents, laser. In: chemistry of organometallics, ACS symposium series, Washington, DC: American Chemical Society; 1993. p. 164–76.

[68] Zhao YJ, He JC, Chen Q, He J, Hou HQ, Zheng Z. Evaluation of 206nm UV radiation for degrading organometallics in wastewater. Chem Eng 2011;167:22–7.

[69] Eyring MB, Martin P. Spectroscopy in forensic science, reference module in chemistry. In: Molecular sciences and chemical engineering, Elsevier; 2013.

[70] Atkin AJ, Lynam JM, Moulton BE, Sawle P, Motterlini R, Boyle NM, et al. Modification of the deoxy-myoglobin/carbonmonoxy-myoglobin UV-vis assay for reliable determination of CO-release rates from organometallic carbonyl complexes. Dalton Trans 2011;40:5755–61.

[71] Vestfried Y, Chusid O, Goffer Y, Aped Y, Aurbach D. Structural analysis of electrolyte solutions comprising magnesium-aluminate chloro-organic complexes by Raman spectroscopy. Organometallics. 2007;26:3130–7.

[72] Gardiner DJ. ISBN 978-0-387--50254-0 Practical Raman spectroscopy. Springer-Verlag; 1989.

[73] Hammes GG. ISBN 9780471733546. OCLC 850776164 Spectroscopy for the biological sciences. Wiley; 2005.

[74] Nelson RC, Miller JT. An introduction to X-ray absorptio n to X-ray absorption spectroscopy and its in situ application to organometallic compounds and homogeneous catalysts. Catal Sci Technol 2012;2:461−70.

[75] Lee PA, Citrin PH, Eisenberger P, Kincaid BM. Extended X-ray absorption fine structure and its strength and limitations as a structural tool. RevMod Phys 1981;53:769−806.

[76] Bertagnolli H, Ertel TS. X-ray absorption spectroscopy of amorphous solids, liquids, and catalytic and biochemical systems—capabilities and limitations. Angew Chem Int Ed 1994;33:45−66.

[77] Singh J, Lambertib C, Van, Bokhoven JA. Advanced X-ray absorption and emission spectroscopy: in situ catalytic studies. Chem Soc Rev 2010;39:4754−66.

[78] Getty K, Delgado-Jaime MU, Kennepohl P. Assignment of pre-edge features in the Ru K-edge X-ray absorption spectra of organometallic ruthenium complexes. Inorg Chim Acta 2008;361:1059−65.

[79] Yi H, Yang D, Luo Y, Pao CW, Lee JF, Lei A. Direct observation of reduction of Cu(II) to Cu(I) by P−H compounds using XAS and EPR spectroscopy. Organometallics 2016;35:1426−9.

Chapter 3

Frontier organometallic catalysis: traditional methodologies and future implications

Payal Rani[1], Mohit Saroha[2], Jayant Sindhu[1] and Ramesh Kataria[3]

[1]*Department of Chemistry, COBS&H, CCS Haryana Agricultural University, Hisar, India,*
[2]*Department of Chemistry, University of Delhi, New Delhi, India,* [3]*Department of Chemistry and Centre for Advanced Studies in Chemistry, Panjab University, Chandigarh, India*

3.1 Introduction

Compounds having at least one metal-carbon bond are generally defined as organometallic compounds. Zerovalent metal complexes, dinitrogen complexes and hydrides are some of the compounds devoid of any metal-carbon bond, but still they are categorized under this class due to their close resemblance with this class [1]. Organosilicon compounds are also considered organometallic compounds even though they lie below carbon in the periodic table. Organometallic-assisted chemistry deals with those transformations, which are accompanied by the assistance of metal where organic compounds and their metalloid are used as substrates. Renowned chemist due to their contribution in the field has shared nine Nobel prizes [2]. A cadet achieved the first organometallic synthesis in Paris in 1760 while studying arsenic containing cobalt salt-based inks [3]. He observed a stinking fuming liquid, which was later identified as a mixture of organo-arsenium compounds. Prussian blue is one of the cyano-iron complex known since antiquity for its presence in pigments [4]. In 1827 a π-complex of ethylene with Pt also known as Zeise's salt (Na[PtCl$_3$(η^2-C$_2$H$_4$)]) was synthesized and considered to be the first reported organometallic compound [5]. In the middle of 19th century a systematic development in organometallic chemistry was observed with the work of Bunsen and Frankland [6]. They had prepared a diverse range of organometallic compounds by incorporating Zn, Hg, Pb, Al, Si, and B. Moreover, the first metal carbonyl was also introduced in the same era.

Exploitation of organometallic compounds in catalysis was witnessed for the first time with the beginning of the 20th century [7]. The organometallic chemistry influences the development of other science disciplines as well and provides an interface to them. Catalysis using organometallic compounds contributes toward sustainable development by offering superior selectivity under mild reaction conditions [8].

Organometallic catalysts are broadly categorized into homogenous and heterogeneous catalysts [9]. The catalysts soluble in the liquid phase are known as homogenous catalysts, and offers advantage in terms of understanding reaction mechanism by kinetic and spectroscopic studies [10]. However, the heterogeneous catalysts are insoluble in reaction mixture and have several advantages like easy separation and high temperature tolerance. As far as selectivity is concerned, homogenous catalysts offers high chemo-, regio-, and stereoselectivity than heterogeneous catalysts. Heterogeneous catalysis offers the efficient synthesis of various chemicals such as H_2SO_4, HNO_3 and NH_3. The scope of organometallic catalysis is not only restricted to the synthesis of simple chemical compounds, but can also be utilized for carboxylation, C—H activation, hydrogenation, oxidation, hydrogen/deuterium (H/D) exchange, dehydrogenation, homogeneous syngas conversion, hydrosilylation, carbonylation and polymerization reactions. In this chapter, various traditional methodologies that utilizes homogenous and heterogeneous catalysts in various chemical process have been discussed along with their mechanistic approach. Subsequently, the frontier research going in this field will be discussed along with their future implications.

3.2 Fundamental principles

A catalyst is a compound used in stoichiometric amount and facilitates a reaction by lowering down its activation energy. It does not appear in the reaction balance and is usually written on the reaction arrow.

3.3 Common methods of preparation

Organometallic compounds are usually prepared by two common methods: (i) Reaction of metals with organic halides and (ii) Reaction of metals with existing chemical compounds.

3.3.1 Reaction of metals with organic halides

Organometallic compounds can be prepared by reacting elemental metal by following common methods (i) Synthesis of Grignard reagent; (ii) Reaction of metal with hydrocarbon for the synthesis of cyclopentadienyl sodium

(NaCp) and (iii) Synthesis of metal carbonyls by direct reaction of metals with carbonyls.

1. $M + RX \rightarrow RMX$
2. $2M + 2RH \rightarrow 2RM + H_2$
3. $M + CO \rightarrow M(CO)_n$

Organic halides react with sodium metal to form alkyl sodium compounds. *n*-amyl chloride reacts with sodium sand in petroleum ether to form *n*-amyl sodium (Eq. 3.1) [11].

$$2Na + RX \rightarrow RNa + NaX \qquad (3.1)$$

Another example is the reaction of organic halides with Mg in presence of diethyl ether to form Grignard reagent, *i.e.*, RMgX (Eq. 3.2). It is tolerable on limited no. of functional group such as ketones, esters, aldehydes, epoxides and nitriles [12].

$$RX + Mg \rightarrow R\text{-}Mg\text{-}X \qquad (3.2)$$

An ether solution of alkyl or aryl halides react with Be powder in a temperature range of 80–90°C to form organoberyllium compound. Buckton *et al.* reported reaction of alkyl iodide with Sb in sealed tube at temperature 140°C to yield R_3SbI_2 (Eq. 3.3) [13]. Alkyl iodide on treating with tellurium in sealed tube upon heating at 80°C produced R_2TeI_2 (Eq. 3.4) [14].

$$Sb + C_2H_5I \rightarrow Sb(C_4H_5)_3I_2 \qquad (3.3)$$

$$Te + 2\ MeI \rightarrow Te(Me)_2I_2 \qquad (3.4)$$

3.3.2 Reaction of metals with existing chemical compounds

Some of the representative methods used in the preparation of organometallic compounds are (i) Metathesis of metal halides with alkylating agents; (ii) Metal-hydrogen/ metal-halogen exchange reaction; (iii) by insertion reaction like hydrometallation, carbometallation, oxymetallation and carbonylation; (iv) Oxidative additions; (v) Decarbonylation; (vi) β-hydride elimination; and (vii) reductive carbonylation of metal oxides [15].

1. $MX + RM' \rightarrow RM + M'X$
2. $RM + R'H \rightarrow R'M + RH$
3. $RM + A(\text{Olefin and CO}) \rightarrow R\text{-}A\text{-}M$
4. $ML_n + AB(RX \text{ and ester}) \rightarrow M(A)(B)L_n$
5. $RCOM \rightarrow RM + CO$

6. $CH_3CH_2M \rightarrow HM(C_2H_4)$
7. $MO_n + mCO \rightarrow M(CO)_x + nCO_2$

Bergmann *et al.* reported cleavage of methyl ether by treating it with lithium to form organolithium compound (Eq. 3.5).

$$\text{Ph}_2\text{C(fluorenyl)OCH}_3 + 2\text{Li} \longrightarrow \text{Ph}_2\text{C(fluorenyl)Li} + \text{LiOCH}_3 \quad (3.5)$$

Thielmann *et al.* reported reaction of various ethers with potassium to yield organopotassium compound (Eq. 3.6).

$$Ph_3C-OR + 2K \longrightarrow Ph_3C-K + KOR \quad (3.6)$$

Organometallic compounds can be formed by the direct displacement of hydrogen from hydrocarbon by a highly reactive metal. Acetylene reacts with sodium in liquid ammonia to yield sodium acetylene along with the evolution of H_2 gas (Eq. 3.7).

$$2\,Ph-\!\!\!\equiv\!\!\!-H + 2Na \longrightarrow 2\,Ph-\!\!\!\equiv\!\!\!-Na + H_2 \quad (3.7)$$

The same reaction has been used to synthesize phenylethynyl derivatives of K, Rb and Cs [16]. Calcium reacts with acetylene in liquid ammonia to give high yield of diethynylcalcium (Eq. 3.8).

$$Ca + 2(C_2H_2) \rightarrow CaC_4H_2 \quad (3.8)$$

In some cases, metal reacts slowly with organic halides. In that case implementation of alloy enhances the rate of reaction. Duppa *et al.* reported formation of R_2Hg by the reaction of organic halide with sodium amalgam [17]. In recent time alkyl sulfates have been used in place of organic halides (Eq. 3.9).

$$R_2SO_4 + NaHg \rightarrow R_2Hg + Na_2SO_4 \quad (3.9)$$

Similarly, alkyl halides reacts with aluminum-magnesium alloy to yield R_2AlX (Eq. 3.10) [18]. Alkyl iodide reacts with Na-Sn alloy to form R_3SnI or R_4SnI [19].

$$RX + AlMg \rightarrow R_2AlX + MgX_2 \quad (3.10)$$

3.4 Mechanism Involved

3.4.1 Oxidative addition

Oxidative addition is a reaction pathway in organometallic compounds which involves an increase in oxidation and coordination number of metal center. A low valent, coordinatively unsaturated transition metal undergoes insertion reaction with X−Y to form X−M−Y (Eq. 3.11). It can be generalized as follows:

$$ML_4 + X-Y \longrightarrow L_4M\begin{smallmatrix}X\\Y\end{smallmatrix} \qquad (3.11)$$

For oxidative addition to occur there should be availability of vacant coordination site. Six-coordinated metal center are not a good candidate for oxidative addition as they are devoid of site for interaction. Losing one of the ligand is the only possibility for metal complexes to be reactive and undergo oxidative addition. In this process, oxidation number of metal increase from n to $n + 2$ and coordination number from m to $m + 2$.

Oxidative addition proceeds either through concerted or SN^2 mechanism depending on the nature of X−Y. Concerted mechanism is involved when X−Y is nonpolar as in case of H_2 (Eq. 3.12).

$$\underset{Cl'}{\overset{L}{}}\!\!\!Ir\!\!\!\underset{L}{\overset{CO}{}} + H_2 \longrightarrow \underset{Cl'\,L}{\overset{L\ \ CO\ H}{Ir}}\!\!\!\underset{H}{} \longrightarrow \underset{Cl'\,L}{\overset{L\ \ CO\ H}{Ir}}\!\!\!\underset{H}{} \quad (3.12)$$

Three centred transition state

Another example of concerted addition is of dioxygen to Vaska's complex (Eq. 3.13).

$$\underset{Cl'}{\overset{L}{}}\!\!\!Ir\!\!\!\underset{L}{\overset{CO}{}} + O_2 \longrightarrow \underset{Cl'\,L}{\overset{L\ \ CO\ O}{Ir}}\!\!\!\underset{O}{} \qquad (3.13)$$

However, S_N^2 pathway is involved when X−Y is polar, as in case of methyl iodide (Eq. 3.14).

$$L_nM\!:\ + CH_3I \longrightarrow ML_n\text{--}\underset{H_3}{C}\text{--}I \longrightarrow [L_nM\text{-}CH_3]I \longrightarrow L_nM(CH_3)I$$

(3.14)

3.4.1.1 Reductive elimination

Reductive elimination is reverse of oxidative addition as it involves removal of X−Y from the metal complex and coordination number of metal complex decreases from 6 to 4. It involves the removal of two ligands that are *cis* to each other. Reduction elimination involves transfer of an electron pair back on to metal from two anionic ligands. So, the metal should be in higher oxidation state. It can be represented as follow (Eq. 3.15):

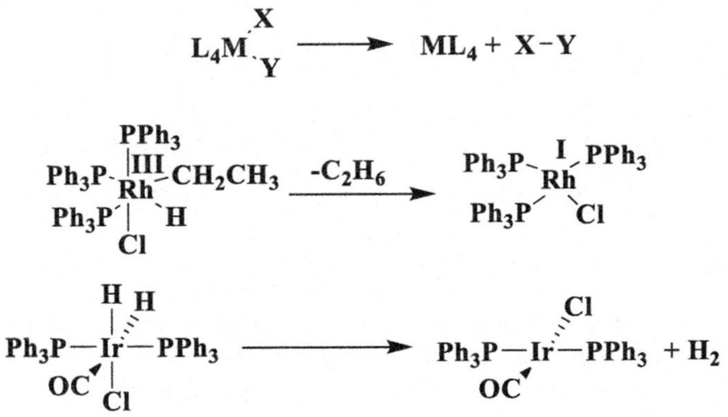

$$L_4M\overset{X}{\underset{Y}{\cdot}} \longrightarrow ML_4 + X-Y$$

$$\underset{Cl}{\overset{PPh_3}{\underset{Ph_3P}{Ph_3P\cdot\overset{III}{Rh}}\text{-}CH_2CH_3}} \xrightarrow{-C_2H_6} \underset{Ph_3P}{\overset{Ph_3P}{\underset{Cl}{\text{-}Rh\overset{I}{\text{-}}PPh_3}}} \quad (3.15)$$

$$\underset{\underset{Cl}{OC}}{\overset{H}{\underset{Ph_3P-Ir-PPh_3}{\text{-}}}} \longrightarrow \underset{OC}{\overset{Cl}{\underset{Ph_3P-Ir-PPh_3}{\text{-}}}} + H_2$$

3.4.1.2 Migratory Insertion

The introduction of ligand with unsaturation like CO, alkene between two other atoms that are bound together is known as insertion. CO ligand insert into M-CH₃ bond to form acyl complex is a well-known example of insertion (Eq. 3.16).

$$H_3C-Mn(CO)_5 + CO \rightleftharpoons H_3C-\overset{\overset{O}{\|}}{C}-Mn(CO)_5 \quad (3.16)$$

Following are the expected mechanisms for this reaction:

Mechanism I: In this mechanism, CO ligand directly insert into the M-CH₃ bond (Eq. 3.17).

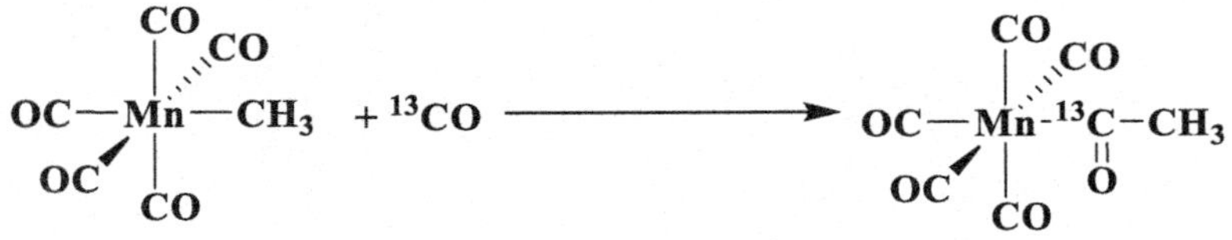

$$\underset{OC}{\overset{CO}{\underset{CO}{\text{OC-}Mn\text{-}CH_3}}} + {}^{13}CO \longrightarrow \underset{OC}{\overset{CO}{\underset{CO}{\text{OC-}Mn\text{-}{}^{13}C\text{-}CH_3}}}\overset{\|}{O}$$

(3.17)

Mechanism II: In this mechanism, the CO ligand laying *cis* to alkyl group migrates and insert into M−C bond leaving a vacant site for the incoming CO ligand (Eq. 3.18).

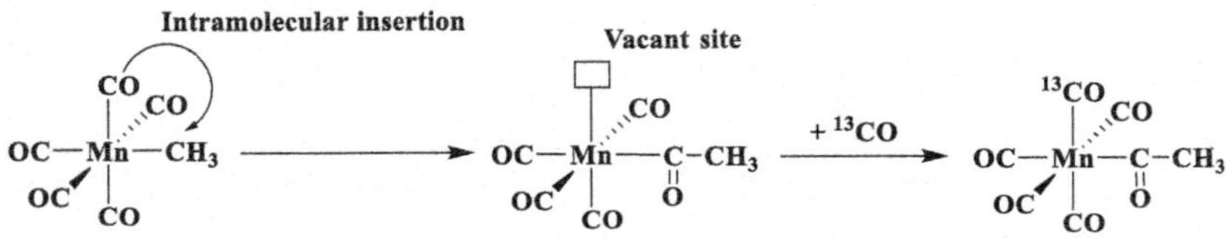

Intramolecular insertion   **Vacant site**

$$\underset{OC}{\overset{CO}{\underset{CO}{\text{OC-}Mn\text{-}CH_3}}} \longrightarrow \underset{OC}{\overset{\square}{\underset{CO}{\text{OC-}Mn\text{-}C\text{-}CH_3}}}\overset{\|}{O} \xrightarrow{+{}^{13}CO} \underset{OC}{\overset{{}^{13}CO}{\underset{CO}{\text{OC-}Mn\text{-}C\text{-}CH_3}}}\overset{\|}{O}$$

(3.18)

Mechanism III: The alkyl group migrates rather than CO group and attach to CO group thereby leaving vacant site for the incoming ligand (Eq. 3.19).

$$\begin{array}{c}\text{Methyl migration}\\ \text{OC-Mn-CH}_3 \xrightarrow{} \text{OC-Mn-}\square \xrightarrow{+^{13}\text{CO}} \text{OC-Mn-}^{13}\text{CO}\end{array}$$

Decarbonylation is the reverse of insertion reaction. (3.19)

3.4.1.3 β-Hydride elimination

It involves the transfer of β-H from the alkyl group to the metal center thereby converting alkyl group into alkene. In this reaction, β-H interacts with metal center forming cyclic intermediate and further hydrogen attached to metal. It can be represented as follows (Eq. 3.20):

$$L_nM{-}CH_2\text{ / }H{-}CH_2 \longrightarrow ML_n\text{ (alkene)}{-}H$$

(3.20)

Agostic interaction → β-H atom

3.5 Recent developments with examples

Garduno *et al.* reported the synthesis of Mn(I) based organometallic complex from commercially available $Mn_2(CO)_{10}$ (**2**). A reaction mixture containing 1,2-*bis*(diisopropylphosphaneyl)ethane (**1**) and $Mn_2(CO)_{10}$ (**2**) was refluxed for 60 min using *n*-pentanol (**3**) as hydride source and resulted into intermediate complex (**4**) in 67% yield. The intermediate (**4**) further treated with triflic acid for 60 min at room temperature to yield the desired complex (**5**) (Scheme 3.1). The same Mn(I) based organometallic complex (**5**) containing dippe ligand (dippe = $^iPr_2P(CH_2)_2P^iPr_2$) was utilized as catalyst in the hydrogenation of benzonitriles (**6**) into amines. The reaction was performed in the presence of 3 mol% of **5** and 10 mol% of KO^{t}Bu (**8**) under 100 *psi* of H_2 (**7**). It resulted into 91% of benzyl amine (**9**) along with 9% of byproducts (**10**), (**11**), and (**12**) when the content of the reaction was heated at 90°C for 15 min. It is noteworthy to mention here that a decrease in reaction yield was observed by decreasing the catalyst amount. The reaction pathway was

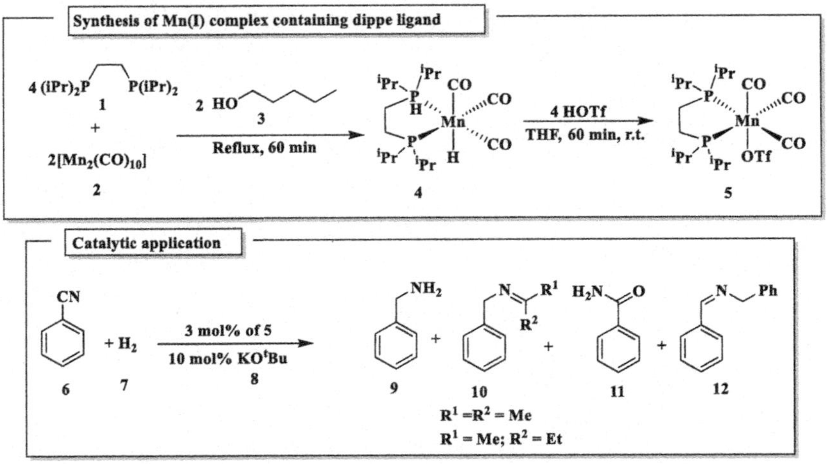

SCHEME 3.1 Synthesis of Mn complexes and its catalytic activity toward hydrogenation of benzonitriles.

investigated by performing various controlled experiments using TEMPO. No significant change in the product yield was observed when TEMPO was utilized as external additive. The developed protocol was also successfully applied to the hydrogenation of terephthalonitrile and adiponitrile into corresponding diamine [20].

The higher stability, better solubility, and brilliant catalysis associated with half-sandwiched moieties established them as a useful building block in the synthesis of organometallic complexes. It has been well evident from literature reports that half sandwich Ru(II) complexes with nitrogen holding ligands exhibit good catalytic activity in various organic transformations. In view of this, Yun et al. reported the synthesis of N,O coordinated half sandwich ruthenium complex (**19**) by the reaction of β-ketoamino ligand (**17**) with [(p-cymene)RuCl$_2$]$_2$ (**18**) in the presence of sodium acetate and methanol at 50°C for 8 h (Scheme 3.2). The ligand (**17**) utilized as substrate in the study was prepared by the reaction of ethylformate (**14**) and 1-dihydroindanone (**13**) using potassium tertiary butoxide (^{t}BuOK) (**8**) as base in diethylether followed by condensation with aromatic amine (**15**) using catalytic amount of formic acid (**16**). Reductive amination of aldehydes and amines was carried out in aq. solution using catalytic amount of **19** and a turn over frequency of 190 h^{-1} was observed. The developed methodology was investigated with differently substituted aromatic aldehyde and amine at 50°C for 5 h using 0.1 mol% of complex and resulted into corresponding amines (**21**) in 91–97% yield. Both electron donating and withdrawing groups are tolerated well on aldehyde and amine. The developed protocol offers various advantages in terms of broad substrate scope, low catalyst loading, and high catalytic activity (Scheme 3.2) [21].

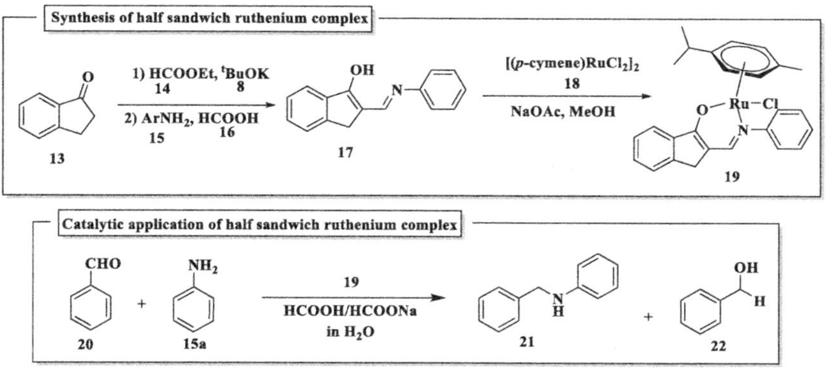

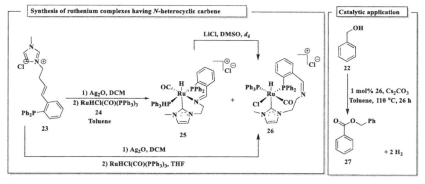

SCHEME 3.2 Synthesis of half sandwich Ru complexes and their catalytic activity toward reductive amination of aldehydes.

SCHEME 3.3 Synthesis of Ru complexes bearing N-heterocyclic carbene and their catalytic activity toward dehydrogenative coupling of alcohols.

He et al. reported two ruthenium complexes (**25**) and (**26**) having N-heterocyclic carbene nitrogen phosphine ligand. Both the developed catalysts were used in the dehydrogenating coupling of alcohol (**27**) to esters (**28**). Complex (**25**) was synthesized by stirring a solution of (E)-3-(4-(2-(diphenylphosphanyl)phenyl)but-3-en-1-yl)-1-methyl-1H-imidazol-3-ium chloride (**23**) with silver oxide in dichloromethane at room temperature for 2 h, followed by its reaction with RuHCl(CO)(PPh$_3$)$_3$ (**24**) in toluene at 60°C for 3 h. However, complex (**26**) was prepared by refluxing a mixture of **23**, silver oxide and RuHCl(CO)(PPh$_3$)$_3$ (**24**) in THF for 5 h (Scheme 3.3). It is noteworthy to mention here that complex (**25**) can be transformed into (**26**) by reacting it with LiCl in DMSO-d_6. Both complexes were utilized as catalyst in the dehydrogenative transformation of alcohols (**22**) into esters (**27**). All the reactions were carried out in presence of Cs$_2$CO$_3$ and Ru (1 mol%)

complex using toluene as solvent at 110°C for 26 h. A slight difference in the catalytic activity was observed as corresponding ester was formed in 97 and 94%, respectively, when **26** and **25** were used as catalyst. Excellent conversion was observed when *p*-substituted benzyl alcohol and long chain primary alcohols were utilized as substrates. The developed methodology exhibits significant steric effects as *o*-substituted substrates resulted into inferior yield, however no electronic effect was observed in the reaction (Scheme 3.3) [22].

Murugan *et al.* reported a series of novel Ru complexes by the reaction of RuHCl(CO)(APh$_3$)$_3$ (**29**) or RuH$_2$(CO)(APh$_3$)$_3$ (**30**) where A = P or As with (2-(pyren-1-ylmethylene) hydrazinyl)benzothiazole (**28**) and explored their catalytic activity for *N*-alkylation of aromatic amines (**21**). To investigate the catalytic activity of all the Ru complexes, reactions were carried out between aromatic amines (**33**) and alcohol. Excellent catalytic activity was observed in case of complex (**32a**) and resulted into desired product (**34**) in 93% yield. All other complexes yielded lesser amount of desired product under optimized reaction condition. Electron donating and withdrawing group on amine were well tolerated for the reaction (Scheme 3.4) [23].

Vijaypritha *et al.* reported the synthesis, structural characterization and catalytic activity of novel half-sandwich (η-6-*p*-cymene) Ru (II) complex (**37a**), (**37b**) and (**39**). The benzothiazole derived hydrazine (**35**) act as monobasic bidentate ligand and coordinated with Ru in N^N fashion. It has

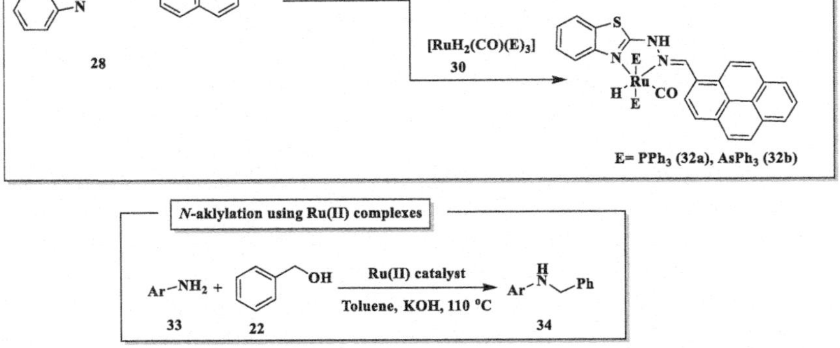

SCHEME 3.4 Synthesis of benzothiazole based Ru complexes and their catalytic activity toward *N*-alkylation of aromatic amines.

been revealed from X-ray single crystal studies that the developed complex exists in *pseudo*-octahedral geometry, which was compensated by one Cl, *p*-cymene ring and bidentate benzothiazole derived hydrazone. The developed catalyst was screened as catalyst in transamidation reaction of primary amides. The reaction was optimized with respect to solvent, time, substituents and catalyst loading. A mixture of amide and amine was refluxed for 8 h using 0.5 mol% of complex (**39**) in 1,4-dioxane to yield *N*-substituted amide. It is significant to mention here that the complex (**39**) exhibited best catalytic activity among all the synthesized complexes. Electron donating and withdrawing group on *o*-, *m*-, and *p*-position of aromatic amine were well tolerated in the reaction. The developed protocol offers various advantages in terms of readily available starting materials, benign reagents and devoid of waste products during reaction (Scheme 3.5) [24].

Wang *et al.* synthesized NHC pincer hydrido nickel complexes (**44**) having unsymmetrical pincer ligands amines and NHCs as donors. The hydrido nickel complex was synthesized by the hydrogenation of nickel halides (**43**). Single crystal X-ray diffraction was used for the characterization of developed NHC pincer hydrido nickel complex. The complex exhibits distorted square planar coordination geometry with central nickel atom surrounded by both six and five-membered chelate rings. The developed catalytic system was investigated in the dehydrohalogenation of alkyl halides (**45**) to alkanes (**46**). Complex (**44b**) showed best catalytic activity leads to >99% yield. Reaction was carried out in presence of (EtO)$_3$SiH, sodium tertiary butoxide as base, 3 mol% of (**44b**) in toluene at 80°C for 2 h (Scheme 3.6) [25].

Vielhaber *et al.* synthesized a series of halide-free organometallic complexes M(CO)$_4$(2-(diphenylphosphino)ethylamine) (**49**) based on group 6 metals, *i.e.*, Cr, Mo, and W. The organometallic complexes were synthesized by reacting commercially available (2-(diphenylphosphino)ethylamine) (**47**) and respective metal carbonyls (**48**). The synthesized complex was characterized using different characterization techniques like IR, NMR, HRMS and

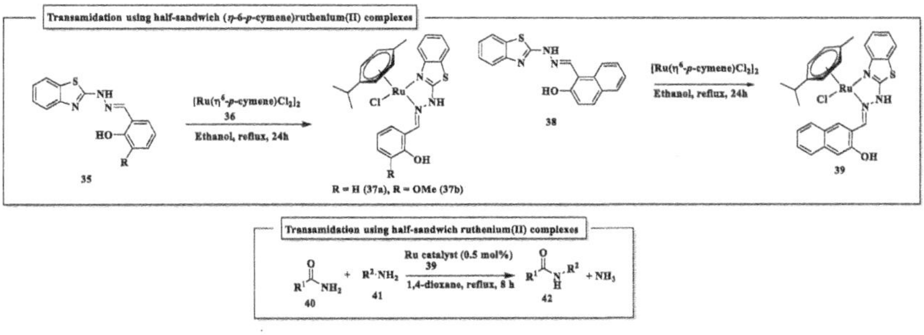

SCHEME 3.5 Synthesis of half sandwich Ru complexes and their catalytic activity toward transamidation of primary amides.

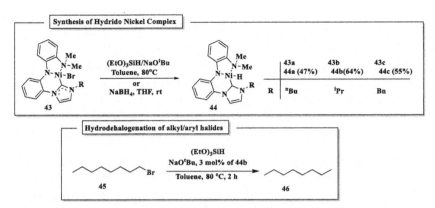

SCHEME 3.6 Synthesis of hydrido nickel complexes and their catalytic activity toward dehydrohalogenation of alkyl halides.

single crystal X-ray crystallography. The complexes were further investigated as catalyst in homogenous hydrogenation of acetophenones (**50**) to form corresponding alcohol derivatives (**51**). It is noteworthy to mention here that [Cr(CO)$_4$(PN)] (**49a**) demonstrated better catalytic activity in diglyme, provided an excess of *t*-BuOK as compared with catalyst used in the reaction. However, Mo and W complexes did not exhibit significant catalytic activity in this transformation. Thereafter, [Cr(CO)$_4$(PN)] (**49a**) was utilized as catalyst for the hydrogenation of aromatic aldehydes to corresponding primary alcohols. A significant decrease in catalytic activity was observed upon replacing acetophenones with aromatic aldehydes in the same catalytic system. The developed methodology is free from any auxiliary hydride reagent and hence avoids any outbreaks of related catalytic systems (Scheme 3.7) [26].

Pincolinamide is a well-known example for C-H functionalization in synthetic organic chemistry. Dolui *et al.* reported for the first time picolinamide directed and ferrocene (**53**) catalyzed oxidation of sp^3 C−H bond to the corresponding carbonyl compounds both for metal-sandwiched (**52**) and organic compounds (**55**). The developed method utilized tertiary butyl peroxybenzoate and Cu(OAc)$_2$ as oxidant. The developed methodology tolerated well both electron-withdrawing and donating groups on reactant. Picolinamides having *meta* and *para* substituted aryl groups as successful substrates, however *ortho* substituted substrates were unresponsive. A sp^3 C−H bond oxidation after sp^2 C−H functionalization was also demonstrated on the ferrocene backbone. It is worth mentioning here that the conversion of primary amines to acids by removing the directing group was reported for the first time by this group. The reaction was attempted on benzyl picolinamide at 80°C using ferrocene as catalyst, **TBPB** as oxidant in dichloroethane for 15 h. The corresponding carboxylic acids were obtained in good to excellent yield. The

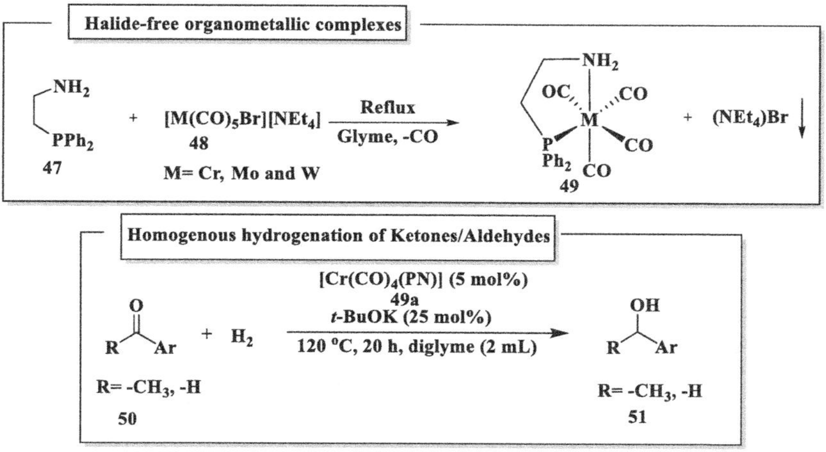

SCHEME 3.7 Synthesis of halide-free organometallic complexes and their catalytic activity toward hydrogenation of ketones.

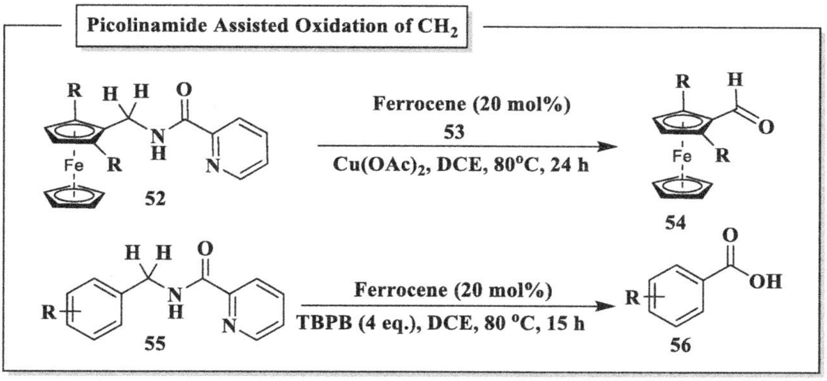

SCHEME 3.8 Pincolinamide assisted oxidation of CH_2 using ferrocene as catalyst.

developed methodology exhibits significant electronic effect as *p*-substituted substrates showed better yield. The presence of radical mechanism was revealed when the reaction was investigated in presence of TEMPO (Scheme 3.8) [27].

Bala *et al.* synthesized some Ru complex and explored their catalytic activity in the hydrogenation of ketones. The organometallic ruthenium (II) complexes was synthesized *via* C-H bond activation by the reaction of [Ru(PPh$_3$)$_2$Cl$_2$] (**58**) with ligand **57** and **60**. The structure of the developed complex was characterized by different spectroscopic techniques. Ruthenium hydrido carbonyl complexes (**59b**) and (**61**) were synthesized by reacting with [Ru(PPh$_3$)$_3$Cl$_2$] (**58**) with **57** and **60** ligands, respectively. Among all

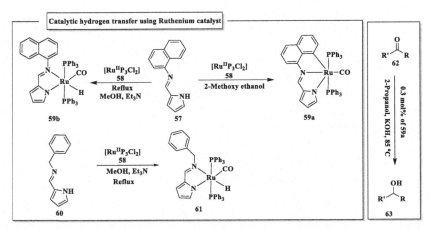

SCHEME 3.9 Synthesis of Ru complexes and their catalytic activity toward hydrogenation of ketones.

synthesized complexes, (**59a**) showed good catalytic activity and resulted into corresponding alcohol (**63**) in good to excellent yield. Reaction was carried out using 0.3 mol% of complex (**59a**), KOH in 2-propanol under reflux at 85°C for 6 h. It is noteworthy to mention here that the catalyst was inactive when 2-pyridyl acetone and 2-aminobenzophenone were utilized as substrates as they resulted into chelate formation with metal center (Scheme 3.9) [28].

The catalyst used for the hydrogenation of carbon dioxide to methanol (**67**) mainly based on precious metals. So, Lane *et al.* reported a homogeneous iron (II) pincer complex (**64**) of greater productivity (590 TON) which can synthesize methanol (**67**) from carbon dioxide and hydrogen. Initially, the reaction was performed using dimethylamine as shuttling agent where DMF formation was observed instead of methanol. This indicated that formylation step occurred but formamide hydrogenation did not go well. This indicated that the selected catalyst **64** was a competent catalyst in the hydrogenation of DMF. So, morpholine (**65**) such that *N*-formylmorpholine (**66**) was selected as shuttling agent. Still, significant amount of conversion was not observed as production of water in amine formylation step limited the catalysis. So, molecular sieves were added resulted into significant increase in the yield of methanol (Scheme 3.10) [29].

Jia *et al.* synthesized Ru complexes having phenolate-oxazoline ligand and investigated them as catalyst in hydrogenation of nitroarene (**71**) to aniline (**72**). Half sandwich Ru complexes were synthesized by the reaction of phenolate-oxazoline ligand (**68**) with [(*p*-cymene)Ru(μ-Cl)Cl]$_2$ (**69**). Initially, the catalytic activity of Ru complexes (**70**) was screened using 4-bromonitrobenzene as model substrate in presence of KOH and isopropanol as solvent. Ru complex (**70b**) showed best catalytic activity for the

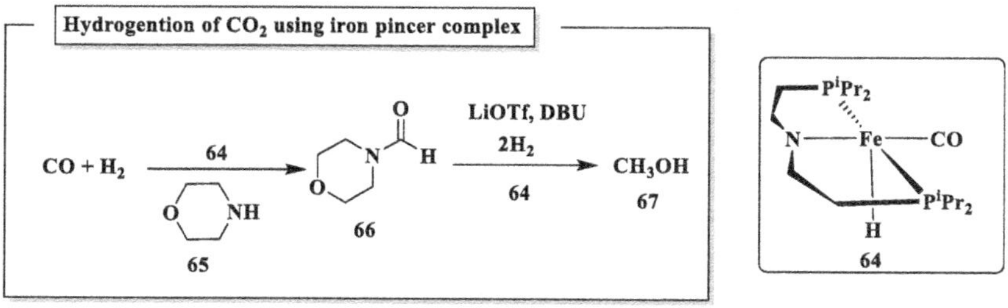

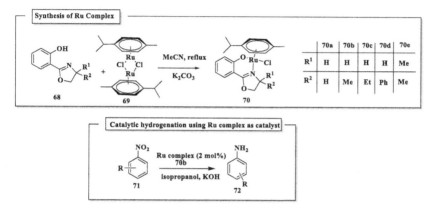

SCHEME 3.10 Hydrogenation of CO_2 using iron pincer complexes.

SCHEME 3.11 Synthesis of Ru complexes and their catalytic activity toward hydrogenation of nitroarenes.

hydrogenation of 4-bromonitrobenzene resulted into 95% yield of 4-bromoaniline. It showed good to excellent catalytic activity with isopropanol as hydride donor. With the optimized conditions, various nitroarenes bearing electron releasing and electron withdrawing group were explored as substrates and resulted into corresponding product in good to excellent yield. It is noteworthy to mention here that 1-chloro-3-nitrobenzene, 1-chloro-2-nitrobenzene, and 3-nitroaniline requires higher temperature for hydrogenation. In addition, nitro group in heterocyclic system afforded well as substrate (Scheme 3.11) [30].

Fan *et al.* reported the synthesis of half sandwich iridium complexes (**77**) as organometallic catalyst that showed catalytic activity for the amide synthesis. Initially, ligands were synthesized by the reaction of aryl amines (**74**) and 2-hydroxy-1-naphthaldehyde (**73**) in presence of ethanol using a few drops of acetic acid. Further iridium complexes were synthesized by the reaction of synthesized ligands (**75**) with $[Cp^*IrCl_2]_2$ (**76**) in the presence of sodium acetate as base and methanol as solvent at 50°C for 12 h. The

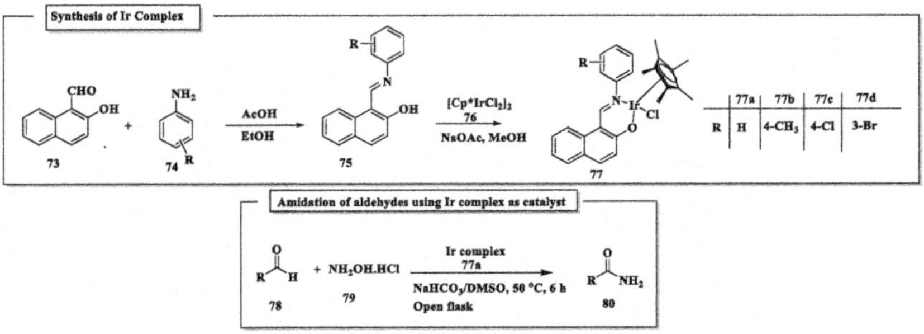

SCHEME 3.12 Synthesis of Ir complexes and their catalytic activity toward amidation of aldehydes.

catalytic activity of Ir complexes was screened further in the amidation of aldehydes (**78**). Benzaldehyde and NH$_2$OH.HCl was taken as model substrate and Ir complex (**77a**) as catalyst. A variety of bases and solvents were screened to investigate their influence on the feasibility of the reaction. It was found that using DMSO as solvent and NaHCO$_3$ as base resulted into desired product (**80**) in excellent yield. All the iridium complexes exhibited good catalytic activity. After establishing reaction condition, various aromatic aldehydes were explored under optimized conditions and resulted into corresponding amide products (**80**) in good to excellent yield. Electron donating and withdrawing group on aldehyde tolerated well in the established condition. It is noteworthy to mention here that poor results were obtained when ketones were utilized as substrates. Also, heterocyclic aldehydes bearing thiophene, pyridine, furan ring exhibited less reactivity (Scheme 3.12) [31].

Funk et al. reported the synthesis of iron complexes (**85**) and investigated their catalytic activity in the hydrogenation of carbonyl compounds and dehydrogenation of alcohols. Fe complexes (**85**) were synthesized by the reaction of ketone (**81**)/(**89**) with benzil (**82**) followed by its reaction with iron pentacarbonyl (**84**) at high temperature. Initially, the synthesized complexes were screened for catalytic activity for hydrogenation of acetophenone using 2-PrOH as hydrogen source. Complex **85d** found to be most active catalyst and resulted into >90% conversion. Further, catalytic activity of **85d** was explored with various carbonyl compounds (**91**) and resulted into corresponding products (**92**) in good to excellent yield. Electronic effect was observed as ketones bearing electron withdrawing undergo rapid reduction than ketones bearing electron releasing group. Moreover, the catalytic activity of complex **85d** was screened for the dehydrogenation of alcohol and leads to desired products in good to moderate yield (Scheme 3.13) [32].

Li et al. reported novel iridium complexes and evaluated them for hydrogen transfer in amination of alcohols. Iridium complexes (**97a–97c**) were

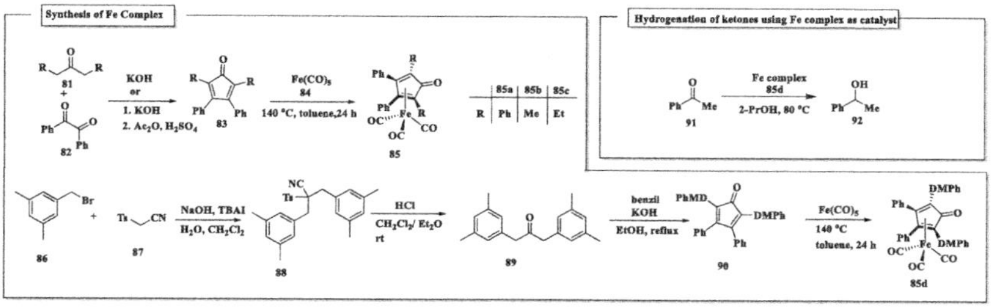

SCHEME 3.13 Synthesis of Fe omplexes and their catalytic activity toward hydrogenation of ketones.

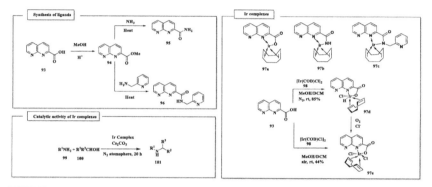

SCHEME 3.14 Synthesis of Ir complexes and their catalytic activity toward amination of alcohols.

synthesized by the reaction of ligands **93**, **95** and **96** with [(COD)Ir(OMe)]$_2$ (**98**) in anhydrous THF at room temperature. Moreover, reaction of **93** with [(COD)Ir(Cl)]$_2$ (**98**) under nitrogen atmosphere leads to the formation of complex **97d** in 85% yield and in presence of air resulted into the formation of complex **97e** in 44% yield. All the synthesized Ir complexes screened for their catalytic activity in the amination of alcohols using taking aniline and benzyl alcohol as model substrate in presence of Cs$_2$CO$_3$. Among all synthesized complex, **97b** and **97c** showed excellent catalytic activity for amination. After the establishment of reaction condition, various amines and alcohols (**100**) were explored using **97b** and **97c** as catalyst and leads to the formation of corresponding amines (**101**) in good to excellent yield. In the reaction, alcohol is dehydrogenated to carbonyl compound, that further reacts with amine to give imine intermediate. The metal hydride generated from the oxidation of alcohol reduce the C=N bond of imine to yield amine. Trace amount of imine left in the reaction along with major product (Scheme 3.14) [33].

3.6 Applications in emerging fields

3.6.1 Photoredox catalysis in medicinal chemistry

In past decade, the field of organic synthesis using organometallic-based photo redox catalysis has resurged to a new extent, resulting into diverse range of synthetic methodologies. Direct C—H functionalization can be achieved in a convenient way by using photo redox catalyzed disconnection strategies. Moreover, the applicability of these methodologies in the synthesis of structurally and functionally diverse molecules under mild reaction conditions is some of the added advantage. It is noteworthy to mention here that photo redox catalyzed protocols were not only restricted to simple organic frameworks, but can also be successfully applied in the synthesis of natural products and biologically relevant compounds. Utilizing visible light in organic synthesis offers an attractive alternative to existing conventional methodologies. Consequently, photo redox catalyzed organic synthesis is gaining much importance in the context of large scale production using flow chemistry and post-functionalization of medicinally relevant molecules.

In 1984, Deronzier and coworkers have developed a photochemical conversion of Pschorr reaction using photo-redox catalyst, which is one of the frequently cited example of photoredox catalysis [34]. Presently, much more sophisticated photo-redox catalyst has been developed with significant applications in organic synthesis. The past applications and future prospects of Ru^{II} polypyridine and cyclometalated Ir^{III} complexes in photo-redox catalysis-based organic synthesis are documented very well in the literature. Hexa-coordinated metal complexes offers long-lived MLCT excited states due to low-spin electron configurations. The strong light absorbing ability of these metal complexes along with long-lived excited state makes 2nd and 3rd row transition metals as suitable candidates for photo-redox catalysis. Ligand modification particularly in cyclometalated Ir^{III} complexes resulted into fine-tuning of redox and optical absorption properties. Continuous efforts have been devoted for modulating the excited state properties of photoredox catalyst.

Photoredox catalyzed transformations offers unexpected reactivities that are conventionally not allowed, which makes their use widespread in the field of natural product synthesis by offering distinct mode of photo-activation. Replacing UV light with visible light triggered photo-activation and offers selective excitation of photo-catalysts. Moreover, detrimental degradation of sensitive functionality in structurally complex molecules can also be avoided by using low energy light sources. Pioneering work from a number of groups demonstrated challenging bond disconnection for facilitating complex organic molecules. We herein, highlight some of the path-breaking application of photo-redox catalyst in medicinal chemistry.

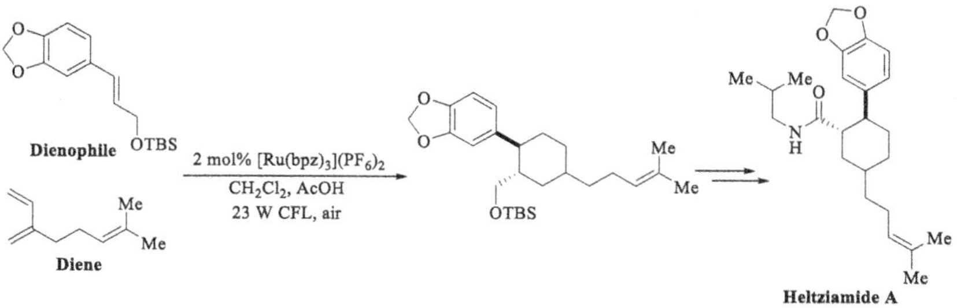

SCHEME 3.15 Photoredox catalyzed radical cation Diels alder synthesis of Heltziamide A.

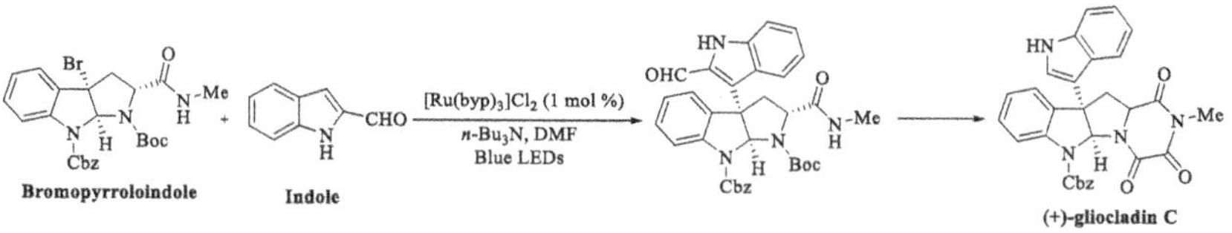

SCHEME 3.16 Photoredox catalyzed reductive coupling for the synthesis of (+)-gliocladin C.

Yoon *et al.* have reported the facilitation of thermally disfavored cycloaddition reactions through photo-redox catalysis. The developed strategy was applied to an electronically mismatched [4 + 2] cycloaddition reaction enabled by oxidation of dienophile component to a radical cation. The cycloadduct formed during the thermally disfavored cycloaddition leads to natural product in four steps, which was otherwise not accessible under thermal conditions. Reversal in activity in radical cation compared with parent alkene was highlighted in thermally disfavored cycloaddition reaction. The synthesis of a wide range of substituted γ-butyrolactone was achieved by utilizing the cycloaddition reactions with polar radical crossover. Moreover, methylenolactocin and protolichesterinic acid were successfully synthesized using same photoredox methodology [35] (Scheme 3.15).

In 2011 Stephenson *et al.* utilized photo-redox catalysis for the synthesis of (+)-gliocladin C, a key intermediate for diverse bisindole alkaloids, by facilitating indole-pyrroloindoline coupling step. The trapping of benzylic intermediate radical by indole was predicted in the developed strategy. It is worth mentioning here that unsubstituted indoles resulted into unexpected C_3–$C_{2'}$ isomer, however, desired C_3–$C_{3'}$ isomer was observed when substituted indole was coupled efficiently [36] (Scheme 3.16).

3.6.2 Organometallic bio-probes for cellular imaging

In addition to significant role in catalysis, organometallic compounds also control many biological processes. A large number of organometallic compounds have remarkable potential and extensively used as innovative tool in biological sciences. Organometallic compounds are considered to be toxic and unstable in physiological conditions. However, the outstanding physiochemical and photo-physical properties associated with some of them established themselves in lead role in the field of bio-organometallic chemistry. Bio-organometallic compounds endowed with specific photophysical properties and multidisciplinary research efforts led to the utilization of these compounds in the field of cellular imaging. The field of the development of organometallic compounds with luminescent properties and their use in cellular imaging has rapidly progressed over the past decades [37]. Metal-containing emitters and organometallic complexes have evolved as a family of molecular luminescent probes. To date, more than 100 organometallic complexes of Re(I), Ru(II), Os(II), Rh(II) and Ir(III) have been explored in different bioimaging applications [38]. The unique chemical stability, structural diversity, and spectroscopic properties are some of the promising features endowed due to the character of metal-carbon bond.

Organometallic compounds used in cellular imaging with promising luminescent properties can be divided into two main classes, (i) The *fac*-tricarbonyl derivatives of ruthenium with a polypyridine ligand and (ii) cyclometalated complexes of iridium and platinum containing 2-phenylpyridine (Scheme 3.17).

Long wavelength of emission along with long-lived excited state are the main photophysical parameters required for discriminating the emission of probe from autofluorescence of cellular system. In accordance with Hund's rule the emission from parallel spin, that is, triplet state is always of lower energy and characterized by large Stokes shift with negligible overlap between emission and absorption bands. Organic systems are devoid of triplet-singlet transitions due to spin forbidden selection rules. However, organometallic compounds have significant spin-orbit coupling which

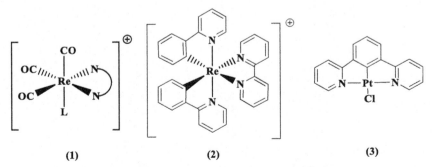

SCHEME 3.17 Structure of organometallic complexes with luminescent properties used for cell imaging.

overcomes the rigid distinction between triplet and singlet state, thus allowing fast and efficient intersystem crossing process from single excited state (S_0) to triplet excited state (T_1). Moreover, spin orbit coupling also promotes $T_1 \rightarrow S_0$ thus making phosphorescence feasible in case of metal complexes. Long-lived excited states and large Stoke shifts are some of the crucial features which differentiate luminescent organometallic compounds from organic dyes generally used in cell imagining. In addition to tunable and specific photophysical properties of organometallic complexes, the physiological features associated with it makes them attractive cellular imaging probes. Organometallic complexes with luminescent properties exhibits low photo-bleaching owing to the stability of their transition state toward degradation to non-emissive species [39]. Organometallic complexes have high kinetic stability and low toxicity owing to the presence of high degree of covalence in M—L bonds thus reducing their tendency to react with biomolecules. The chemical structure of a molecule not only decides its cytotoxicity but also governs its cellular uptake. The photophysical properties of metal complexes usually remain unaltered due to structural modifications on ancillary ligands, however these modifications alters lipophilicity and charge of a complex. Better cellular imaging probes can be developed by designing a wide range of structural analogs. It is worth mentioning here that the synthesis of these luminescent probes is not more demanding than the organic fluorophores due to high cost of iridium and platinum metals [40].

3.6.3 Bio-organometallic chemistry

Cis-platin is one of the oldest compounds known since antiquity for its anticancer properties. It has been extensively used for the treatment of testicular, ovarian cancers cells, and for the treatment of head and neck tumors [41]. Over the years, researchers have developed second and third generation *cis*-platins, that is, carboplatin and oxaliplatin for the treatment of colorectal cancers. With further development in the field, three derivatives of *cis*-platin namely loboplatin, nedaplatin and heptaplatin have been established as a drug. Moreover, dozen of other derivatives are under advanced clinical trials [42].

In mechanism, the N_7 site of two adjacent guanines bases attack on *cis*-platin which resulted into distortion of DNA in the cell nucleus and probably that of the mitochondria. Some drawbacks like high toxicity, susceptibility to resistance and renal problems are associated with Pt compounds. Along with Pt, other group metals like Ni and Pd show higher kinetic exchange rate reactivity as compared to Pt. In the 1980s Clarke suggested first Ru based anticancer compound, which will look like *cis*-platin [43]. The ferrocene derived penicillin (4) and chloroquine (5) shows promising behavior and is currently in use.

(4) **(5)**

3.6.4 Enzyme inhibition using organometallic compounds

Developing target based enzyme inhibitors is of prime interest for medicinal chemists as a high fraction of drugs currently present in market are related to this class. Atorvastatin sold under the name of Lipitor is an inhibitor of enzyme present in liver tissue. Similarly, Imatinib, an anticancer drug sold under the name of Gleevec are tyrosine kinase inhibitors. Metal-based drugs reaching market are supposed to have enzyme inhibition based mechanism of action. Some of the representative organometallic-based drugs based on gold complexes, antimony and arsenic are known to have antiarthritic, antileishmaniasis, and antisyphilis properties, respectively, however their mode of action is still not known. These findings enthused the workers to explore the future prospects of organometallic compound based enzyme inhibitors.

Enzymes having unique active sites and electrostatic surfaces offers incomparable catalytic activity and specificity for planned substrates. Designing enzyme inhibitors with great selectivity is still a big challenge for synthetic medicinal chemists. Developing molecules with similar interaction patterns in the binding pocket as shown by the natural inhibitor of enzyme is the prerequisite for blocking enzyme activity, which can be achieved by searching inhibitors in wide chemical spaces. In recent years organometallic complexes have been evolved as a structural class with great enzyme inhibition potential. The diverse range of geometries varying from square planar to octahedral along with unique bond polarities are some of the fascinating features unavailable with organic framework, further enhance the potential of organometallic compounds as enzyme inhibitors. The potential of organometallic compounds as enzyme inhibitors has been recently reviewed [44]. A class of kinase inhibitors developed by Meggers *et al.* is the most extensively studied organometallic complexes. Protein kinase are enzymes that regulate the functioning of proteins by catalyzing their phosphorylation and involved in many cellular pathways upregulated in cancer tissues leading to uncontrolled growth. Therefore protein kinase inhibitors represents attractive target in cancer related drug discovery programs. Staurosporine is one of the natural protein kinase inhibitor derived from alkaloids with low nanomolar IC_{50} value [45,46]. It exhibits similar ATP binding pattern as exhibited by adenine in case of ATP. Based on these observations, Meggers and coworkers have designed organometallic-based kinase inhibitors using staurosporine as structural template. The core of staurosporine remain unaltered and structural modifications were done considering the metal complexation prospective. Additional moieties were added to the open coordinating sites of developed organometallic complex by replacing glycosyl group in the parent framework, which shown to effect the specificity and affinity of metal containing certain kinase inhibitors. Some of the ruthenium complexes were synthesized and explored for their efficacy as Kinase inhibitors.

GSK3a inhibitor
0.9 nM

DAPK1 inhibitor
2.0 nM

Pim-1 inhibitor
0.075 nM

FLT4 inhibitor
123 nM

3.6.5 Organometallic compounds in solar cells

With a rapid increase in world population, the demand of energy has also increased largely since most of the human needs are energy dependent. Dependency on nonrenewable sources specifically oil, coal and gas [2] has raised alarming situation, as the nonrenewable sources are at the verge of extinction. Renewable resources, like solar, wind, geothermal, tidal and wave sources, contribute only 13% of the total energy demand of the globe. Global warming and environmental pollution are some of the serious issues faced by humanity, which raised due to excessive exploitation of fossil fuels. This situation has raised the concern of globe toward the development of renewable energy sources and sustainable development.

In recent years, researchers have invested their sincere efforts toward the development of organometallic-based dyes, which can be used as dye-sensitized solar cell (DSSC) [47]. In 1991 Michael Grätzel and coworkers reported the first of its kind ruthenium complexes based solar cell, which has photoanode and platinum in the counter electrode [48]. The organometallic complexes of ruthenium metal and its analogs act as dyes in combination with platinum metals and play a pivotal role in sustainable energy technology. In recent years organometallic compounds are inexpensive and promising alternative to the proven solid-state photovoltaic cells.

Ruthenium polypyridyl complexes and its scaffolds have received considerable attention in past few years due to high stability of oxidation states and feasible photoelectrochemical properties [49]. These complexes are categorized further based on different ligands attached to the main framework.

Carboxylate dyes, phosphonate ruthenium dyes, and polynuclear bipyridyl dyes are some of the examples from this established class [50]. Polypyridinic complexes of d^6 metal ions shows intense metal-to-ligand charge transfer bands in the visible region. The working efficiency of complexes can be fine-tuned by the varying the molecular structure of dye. Exploration of structural features reveals the fact that the photophysical and redox properties of metal complexes can be improved, as the energy of π^* orbital lowers substantially by substituting these metal complexes with carboxylic groups. The ruthenium (II)-polypyridyl DSSC complexes exhibits high efficiency as they have wide applicability in both visible and near infrared regions. The working efficiency of Ru and Cu metal complexes under standard condition is 11% under DSSCs principal. Thus ruthenium(II)-polypyridyl complexes, opens new horizons toward the development of DSSCs using organometallic complexes having broad applicability in visible range and high stabilities with redox mediators (Fig. 3.1 and Table 3.1).

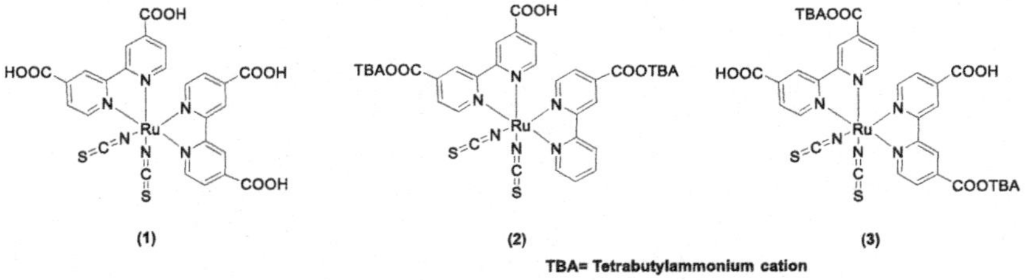

FIGURE 3.1 Structure of some ruthenium-based complexes used as photosensitizer.

TABLE 3.1 Properties of some ruthenium based complexes used as photosensitizer.

Dye	IPCE %	Absorption (nm) ε (10^3 m²/mol)	Short-circuit Current, J_{sc}, mA/cm²	Fill factor (FF)	Open-circuit voltage, V_{oc}, mV	Efficiency, η %
1	83	534 (1.42)	18.20	0.730	720	10.00 [15,16]
2	85	532 (1.40)	17.73	0.750	846	11.18 [10,17]
3	80		20.90	0.722	736	11.10 [18,19]

Metal complex dyes exhibits better results than organic dyes because of their excellent stability toward Photo-degradation. Among the complexes discussed above, organometallic complex **1** has two bipyridine and two thiocyanato (NCS) ligands, which are loosely attached to ruthenium account for its ability to absorb radiation up to 800 nm. However, complex **2** shows highest efficiency. DSSCs will remain at the center of ongoing research efforts to utilize clean and renewable solar energy, because the technology is inexpensive and offers promising alternative to the proven solid-state photovoltaic cells. Ru and Zn metal-based complexes, used as dyes for DSSCs, can give conversion efficiencies over 11%. Other metal complexes such Re, Pt, Cu, Os, and others have been proved to be useful as sensitizers for DSSCs. Metal complexes in DSSCs offers a future prospective toward the development of green energy alternatives.

3.6.5.1 Organometallic compounds in carbon dioxide fixation

Elevation of CO_2 level in air with the exploitation of fossil fuels leads to global warming, which aroused scientific community toward the development of environment benign methodologies for the direct conversion of CO_2 into the value added products for stabilizing CO_2 concentration in air. Only a small fraction of the total carbon dioxide (circa 750 Gt/y) is involved in natural cycle, and the atmospheric concentration of CO_2 increases with an increase in overall population of this globe. The fixation of carbon dioxide is not confined to the synthesis of simple molecules, however complex molecular structures can also be built using simple and inert CO_2 as substrates. Scientist has tried to convert CO_2 into more valuable products by utilizing artificial systems for its transformation. In past few years conversion of CO_2 into valuable products by chemical methods has developed into a new field. The climatic changes cannot be alleviated by utilization of CO_2 alone, however main emphasis is to replace a nonrenewable fossil-fuel based starting material with renewable carbon feedstock, that is, CO_2. Therefore developing efficient process for CO_2 conversion is viewed as the integral part of sustainable development [51].

Several technological applications are involved in CO_2 capturing and storage (CCS), however it is capturing and utilization can be done using several chemical processes (CCU). The chemical or industrial processes utilize nearly 170 Mt/y of CO_2 while technological processes associated with its storage fixes around 28 Mt/y of CO_2. In 1861 first CO_2 fixation process, that is, solvay process was developed, and used for the conversion of atmospheric CO_2 into $NaHCO_3$ and Na_2CO_3. The second CO_2 fixation method utilizes it as a main substrate along with phenol in the synthesis of salicylic acid. In 1870 CO_2 fixation methods were evolved with the largest application of CO_2 in the synthesis of urea [52].

Ishitani and coworkers have developed a supramolecular photocatalyst based on ruthenium (II) photosensitizer and a manganese (I) catalyst. The conversion of CO_2 into formic acid was reported to be higher for dinuclear complexes as compared to mononuclear ones [53]. In a similar attempt, an attractive hybrid consisting of rhodium (II) catalyst linked covalently with protein scaffold was studied by Shaw and coworkers [54]. They had studied the effect of amino acid mutations on the catalytic activity of the developed hybrid in CO_2 reduction activity. The study revealed that presence of positive charge near catalytic sites is beneficial for catalytic activity. Xi and coworkers had developed a novel protocol for titanium (II) catalyzed carboxylation of aryl chlorides by utilizing CO_2 as reactant under mild conditions [55]. A new dinuclear uranium complex was developed by Mazzanti and coworkers for promoting reductive disproportionation of CO_2 to carbonates [56]. In a similar attempt Meyer and coworkers elaborate the conversion of CO_2 into carbonates by a close synthetic cycle using *tri*-aryloxideligated uranium (IV) complex. Insertion of metal complex into polycarbonate is a conceptually fascinating approach which was utilized by Darensbourg *et al.* for the one-pot synthesis of polycarbonate sandwiched ruthenium complexes [57]. Many researchers around the globe are continuously investing their efforts for the development of novel and efficient methodologies for the organometallic catalyst catalyzed conversion of CO_2 into valuable products.

3.7 Future prospective and scope

The development of environmentally benign artificial photosynthesizer and dye-sensitized solar cells for conversion of carbon dioxide into some chemicals that can be used is a scientific challenge in front of organometallic catalysis. Hence, the carbon dioxide fixation (e.g., to prevent global warming) is the main challenge in front of organometallic chemistry. The main challenges in front of organometallic chemistry are

- The activation of small molecules using environment benign condition is a scientific challenge for example, functionalization of alkanes, olefins, carbon monoxide and carbon dioxide, and dihydrogen.
- Synthesis of small value added organic compounds with biological or pharmacological significance *via* catalytic processes.
- Synthesize the carbon materials which are biologically active against enzymes, present in organometallic active centers is the bio-organometallic chemistry with a diversity of potential applications.

3.8 Summary

Organometallic catalysis has been explored widely for chemical transformation that is not feasible by classical method. Here, the recent development in the synthesis of organometallic compounds such as Ru, Ir, Fe, Cr and Ni complexes with their catalytic applications has been summarized. In addition, their applications in emerging field have been explored. Recently organometallic complexes have been widely used for cell imaging, as dye sensitized solar cells, photo sensitizers and photo redox catalysts. In addition, organometallic complexes have been explored for CO_2 fixation by the conversion of CO_2 into value added products that leads to stabilize CO_2 concentration in air.

3.9 Problems with solutions

1. Is it possible to catalyze the dehydrogenation of alkanes, the reverse reaction of olefin hydrogenation?
2. Write the products of following gold mediated reactions.

Solution:

3. Why the maximum time the ruthenium-based complexes used as photosensitizers?
4. Explain the future of photosynthesis reaction for the fixation of CO_2 and synthesis of low and high energy molecules.
5. Explain how the electron receptor and donor combination help in generating the common ideology for the synthesis of polymer solar cells (PSCs).
6. Explain the application of methodology of green chemistry in the fixation of CO_2 and its conversion to small synthetic utility compounds.

3.10 Objective type questions

1. Increase in concentration of carbon monoxide in cobalt-catalyzed hydroformylation of 1-pentene .... the rate of reaction.
   a. Increase

b. Decrease
c. **Decrease by thousand**
d. Increase by thousand

Explanation: In cobalt catalyzed hydroformylation reaction, initial step is dissociation of carbon monooxide, so increase in concentration of carbon monoxide decrease the rate of dissociation.

2. In Wilkinson's catalyst, replacement of PPh$_3$ with P(n-Bu)$_3$ decrease the rate of reaction because ....
   a. P(n-Bu)$_3$ makes more stronger bonds then PPh$_3$
   b. Steric hindrance is less in case of P(n-Bu)$_3$ which leads to slow dissociation
   c. **Both a and b**
   d. None of the above

   Explanation: The replacement of PPh$_3$ with P(n-Bu)$_3$ leads to slow reaction because P(n-Bu)$_3$ makes stronger bonds then PPh$_3$ and due to less Steric hindrance dissociation of P(n-Bu)$_3$ from metal complex is very slow and dissociation is rate determining step.

3. What is (are) the product(s) of 1-pentene hydroformylation?
   a. 1-Hexanal
   b. 2-Hexanal
   c. Both 1-hexanal and 2-hexanal in equal amount
   d. **1-hexanal is major while 2-hexanal is minor**

4. [Ru(C$_2$H$_5$)Cl(PPh$_3$)$_3$] is stable only under a pressure of ethene because:
   a. It is 16-electron complex
   b. It forms an 18-electron adduct with ethene
   c. **One of the decomposition products is ethene**
   d. It prevents alpha- elimination of ethene

References

[1] Pruchnik FP, Duraj SA. Introduction to organometallic chemistry BT. In: Pruchnik FP, Duraj SA, editors. Organometallic chemistry of the transition elements. Boston, MA: Springer US; 1990. p. 1–21. Available from: https://doi.org/10.1007/978-1-4899-2076-8_1.

[2] Chavain N, Biot C. Organometallic complexes: new tools for chemotherapy. Curr Med Chem 2010;17:2729–45. Available from: http://doi.org/10.2174/092986710791859306.

[3] Meinema HA. A review of: "Organometallic compounds. Volume III. Compounds of arsenic, antimony and bismuth. Second edition, first supplement. Edited by M. Dub. Springer Verlag, New York, Heidelberg, Berlin, 1972. xxi + 613 pp. $24.80". Synth React Inorg Met Chem 1974;4:179–80. Available from: https://doi.org/10.1080/00945717408069648.

[4] Bhatt V. Essentials of coordination chemistry: a simplified approach with 3D visuals. Elsevier Science; 2015. Available from: https://books.google.co.in/books?id = 492ECgAAQBAJ.

[5] Thayer JS. Historical origins of organometallic chemistry. Part I, Zeise's salt. J Chem Educ 1969;46. Available from: https://doi.org/10.1021/ed046p442.

[6] Thayer JS. Historical origins of organometallic chemistry. Part II, Edward Frankland and diethylzinc. J Chem Educ 1969;46. Available from: https://doi.org/10.1021/ed046p764.

[7] Parshall GW, Putscher RE. Organometallic chemistry and catalysis in industry. J Chem Educ 1986;63. Available from: https://doi.org/10.1021/ed063p189.

[8] Dixneuf PH, Soulé JF. Organometallics for green catalysis. Springer International Publishing; 2019. Available from: https://books.google.co.in/books?id = IB2KDwAAQBAJ.

[9] Copéret C, Chabanas M, Petroff Saint-Arroman R, Basset J-M. Homogeneous and heterogeneous catalysis: bridging the gap through surface organometallic chemistry. Angew Chem Int Ed 2003;42:156−81. Available from: https://doi.org/10.1002/anie.200390072.

[10] Schrod M, Luft G, Grobe J. Investigation of the synthesis of acetic anhydride by homogeneous catalysis: II. Spectroscopic studies of the mechanism. J Mol Catal 1983;22:169−78. Available from: https://doi.org/10.1016/0304-5102(83)83023-5.

[11] Frame GF. NOTES 1940;445.

[12] Rieke RD. Preparation of organometallic compounds from highly reactive metal powders. Sci (80-) 1989;246:1260−4. Available from: https://doi.org/10.1126/science.246.4935.1260.

[13] Buckton GB. XII.—On the stibethyls and stibmethyls. Q J Chem Soc Lond 1861;13:115−21. Available from: https://doi.org/10.1039/QJ8611300115.

[14] Drew HDK. LXXXII.—Non-existence of isomerism among the dialkylelluronium dihalides. J Chem Soc 1929;560−9. Available from: https://doi.org/10.1039/JR9290000560.

[15] Komiya S. Chemistry synthesis of organometallic compounds. 1997:1−432.

[16] Gilman H, Young RV. Relative reactivities of organometallic compounds. XV. Organoalkali compounds. J Org Chem 1936;01:315−31. Available from: https://doi.org/10.1021/jo01233a001.

[17] Frankland E, Duppa BF. Ueber ein neues Verfahren zur Darstellung der Quecksilberverbindungen der Alkoholradicale. Justus Liebigs Ann Chem 1864;130:104−17. Available from: https://doi.org/10.1002/jlac.18641300110.

[18] Grosse AV, Mavity JM. Organoaluminum compounds: I. Methods of preparation. J Org Chem 1940;05:106−21. Available from: https://doi.org/10.1021/jo01208a004.

[19] Cahours A. Untersuchungen über die metallhaltigen organischen Radicale. Justus Liebigs Ann Chem 1862;122:192−221. Available from: https://doi.org/10.1002/jlac.18621220205.

[20] Garduño JA, García JJ. Non-Pincer Mn(I) organometallics for the selective catalytic hydrogenation of nitriles to primary amines. ACS Catal 2019;9:392−401. Available from: https://doi.org/10.1021/acscatal.8b03899.

[21] Yun XJ, Ling C, Deng W, Liu ZJ, Yao ZJ. Half-Sandwich Ru(II) complexes with N,O-chelate ligands: diverse catalytic activity for amine synthesis in water. Organometallics 2020;39:3830−8. Available from: https://doi.org/10.1021/acs.organomet.0c00554.

[22] He X, Li Y, Fu H, Zheng X, Chen H, Li R, et al. Synthesis of unsymmetrical N-heterocyclic carbene-nitrogen-phosphine chelated ruthenium(II) complexes and their reactivity in acceptorless dehydrogenative coupling of alcohols to esters. Organometallics 2019;38:1750−60. Available from: https://doi.org/10.1021/acs.organomet.9b00071.

[23] Murugan K, Vijayapritha S, Viswanathamurthi P, Saravanan K, Vijayan P, Ojwach SO. Ru(II) complexes containing (2-(pyren-1-ylmethylene)hydrazinyl)benzothiazole: Synthesis, solid-state structure, computational study and catalysis in N-alkylation reactions. Inorganica Chim Acta 2020;512:119864. Available from: https://doi.org/10.1016/j.ica.2020.119864.

[24] Vijayapritha S, Viswanathamurthi P. New half-sandwich (η6-p-cymene)ruthenium(II) complexes with benzothiazole hydrazone Schiff base ligand: synthesis, structural

characterization and catalysis in transamidation of carboxamide with primary amines. J Organomet Chem 2020;929:121555. Available from: https://doi.org/10.1016/j.jorganchem.2020.121555.
[25] Wang Z., Li X., Sun H., Fuhr O., Fenske D. Synthesis of NHC pincer hydrido nickel complexes and their catalytic applications in hydrodehalogenation. 2017. Available from: https://doi.org/10.1021/acs.organomet.7b00848.
[26] Vielhaber T, Faust K, Topf C. Group 6 metal carbonyl complexes supported by a bidentate PN ligand: syntheses, characterization, and catalytic hydrogenation activity. 2020. Available from: https://doi.org/10.1021/acs.organomet.0c00612.
[27] Dolui P, Hazra S, Deb M, Elias AJ. Picolinamide assisted oxidation of CH_2 groups bound to organic and organometallic compounds using ferrocene as a catalyst. Organometallics 2019;38:2015–21. Available from: https://doi.org/10.1021/acs.organomet.9b00085.
[28] Bala M, Ratnam A, Kumar R, Ghosh K. Naphthyl C8-H hydrogen activation and synthesis of organometallic ruthenium complex: crystal structure of hydride intermediates and catalytic transfer hydrogenation. J Organomet Chem 2019;880:91–7. Available from: https://doi.org/10.1016/j.jorganchem.2018.10.034.
[29] Lane EM, Zhang Y, Hazari N, Bernskoetter WH. Sequential hydrogenation of CO_2 to methanol using a pincer iron catalyst. Organometallics 2019;38:3084–91. Available from: https://doi.org/10.1021/acs.organomet.9b00413.
[30] Jia W, Ling S, Zhang H, Sheng E, Lee R. Half-sandwich ruthenium phenolate − oxazoline complexes: experimental and theoretical studies in catalytic transfer hydrogenation of nitroarene. 2017:1–8. Available from: https://doi.org/10.1021/acs.organomet.7b00721.
[31] Fan X, Deng W, Liu Z, Yao Z. Half-sandwich iridium complexes for the one-pot synthesis of amides: preparation, structure, and diverse catalytic activity. 2020. Available from: https://doi.org/10.1021/acs.inorgchem.0c02497.
[32] Funk TW, Mahoney AR, Sponenburg RA, Zimmerman KP, Kim DK, Harrison EE. Synthesis and catalytic activity of (3,4-diphenylcyclopentadienone)iron tricarbonyl compounds in transfer hydrogenations and dehydrogenations. 2018. Available from: https://doi.org/10.1021/acs.organomet.8b00037.
[33] Li M, Hsu YP, Liu YH, Peng SM, Liu ST. Iridium complexes with ligands of 1,8-naphthyridine-2-carboxylic acid derivatives-preparation and catalysis. J Organomet Chem 2020;927:1–7. Available from: https://doi.org/10.1016/j.jorganchem.2020.121537.
[34] Cano-Yelo H, Deronzier A. Photo-oxidation of some carbinols by the Ru(II) polypyridyl complex-aryl diazonium salt system. Tetrahedron Lett 1984;25:5517–20. Available from: https://doi.org/10.1016/S0040-4039(01)81614-2.
[35] Ischay MA, Anzovino ME, Du J, Yoon TP. Efficient visible light photocatalysis of [2 + 2] enone cycloadditions. J Am Chem Soc 2008;130:12886–7. Available from: https://doi.org/10.1021/ja805387f.
[36] Narayanam JMR, Tucker JW, Stephenson CRJ. Electron-transfer photoredox catalysis: development of a tin-free reductive dehalogenation reaction. J Am Chem Soc 2009;131:8756–7. Available from: https://doi.org/10.1021/ja9033582.
[37] Licandro E, Panigati M, Salmain M, Vessières A. Organometallic bioprobes for cellular imaging. Bioorganomet Chem 2014;339–92. Available from: https://doi.org/10.1002/9783527673438.ch11.
[38] Raszeja L, Maghnouj A, Hahn S, Metzler-Nolte N. A novel organometallic ReI complex with favourable properties for bioimaging and applicability in solid-phase peptide synthesis. ChemBioChem 2011;12:371–6. Available from: https://doi.org/10.1002/cbic.201000576.

[39] Thorp-Greenwood FL, Balasingham RG, Coogan MP. Organometallic complexes of transition metals in luminescent cell imaging applications. J Organomet Chem 2012;714:12−21. Available from: https://doi.org/10.1016/j.jorganchem.2012.01.020.
[40] Metzler-Nolte N. Bioorganometallic chemistry. 2007. Available from: https://doi.org/10.1016/s0022-328x(99)00303-4.
[41] Loehrer PJ, Einhorn LH. Cisplatin. Ann Intern Med 1984;100:704−13. Available from: https://doi.org/10.7326/0003-4819-100-5-704.
[42] Sazonova EV, Kopeina GS, Imyanitov EN, Zhivotovsky B. Platinum drugs and taxanes: can we overcome resistance? Cell Death Discov 2021;7:155. Available from: https://doi.org/10.1038/s41420-021-00554-5.
[43] Clarke MJ. The potential of ruthenium in anticancer pharmaceuticals. 2009. Available from: https://doi.org/10.1021/bk-1980-0140.ch010.
[44] Schatzschneider U, Metzler-Nolte N. New principles in medicinal organometallic chemistry. Angew Chem Int Ed 2006;45:1504−7. Available from: https://doi.org/10.1002/anie.200504604.
[45] Manns J, Daubrawa M, Driessen S, Paasch F, Hoffmann N, Löffler A, et al. Triggering of a novel intrinsic apoptosis pathway by the kinase inhibitor staurosporine: activation of caspase-9 in the absence of Apaf-1. FASEB J 2011;25:3250−61. Available from: https://doi.org/10.1096/fj.10-177527.
[46] Falcieri E, Martelli AM, Bareggi R, Cataldi A, Cocco L. The protein kinase inhibitor staurosporine induces morphological changes typical of apoptosis in MOLT-4 cells without concomitant DNA fragmentation. Biochem Biophys Res Commun 1993;193:19−25. Available from: https://doi.org/10.1006/bbrc.1993.1584.
[47] McConnell RD. Assessment of the dye-sensitized solar cell. Renew Sustain Energy Rev 2002;6:271−93. Available from: https://doi.org/10.1016/S1364-0321(01)00012-0.
[48] Chen C-Y, Wang M, Li J-Y, Pootrakulchote N, Alibabaei L, Ngoc-le C, et al. Highly efficient light-harvesting ruthenium sensitizer for thin-film dye-sensitized solar cells. ACS Nano 2009;3:3103−9. Available from: https://doi.org/10.1021/nn900756s.
[49] Nasr C, Hotchandani S, Kamat PV. Role of iodide in photoelectrochemical solar cells. Electron transfer between iodide ions and ruthenium polypyridyl complex anchored on nanocrystalline SiO_2 and SnO_2 films. J Phys Chem B 1998;102:4944−51. Available from: https://doi.org/10.1021/jp9811427.
[50] Hyde JT, Hanson K, Vannucci AK, Lapides AM, Alibabaei L, Norris MR, et al. Electrochemical instability of phosphonate-derivatized, ruthenium(III) polypyridyl complexes on metal oxide surfaces. ACS Appl Mater Interfaces 2015;7:9554−62. Available from: https://doi.org/10.1021/acsami.5b01000.
[51] Huang K, Sun CL, Shi ZJ. Transition-metal-catalyzed C−C bond formation through the fixation of carbon dioxide. Chem Soc Rev 2011;40:2435−52. Available from: https://doi.org/10.1039/c0cs00129e.
[52] Hazari N, Iwasawa N, Hopmann KH. Organometallic chemistry for enabling carbon dioxide utilization. Organometallics 2020;39:1457−60. Available from: https://doi.org/10.1021/acs.organomet.0c00229.
[53] Fabry DC, Koizumi H, Ghosh D, Yamazaki Y, Takeda H, Tamaki Y, et al. A Ru(II)−Mn (I) supramolecular photocatalyst for CO2 reduction. Organometallics 2020;39:1511−18. Available from: https://doi.org/10.1021/acs.organomet.9b00755.
[54] Laureanti JA, Ginovska B, Buchko GW, Schenter GK, Hebert M, Zadvornyy OA, et al. A positive charge in the outer coordination sphere of an artificial enzyme increases CO_2

hydrogenation. Organometallics 2020;39:1532−44. Available from: https://doi.org/10.1021/acs.organomet.9b00843.

[55] Hang W, Yi Y, Xi C. Cp2TiCl2-catalyzed carboxylation of aryl chlorides with carbon dioxide in the presence of n-BuMgCl. Organometallics 2019;39:1476−9. Available from: https://doi.org/10.1021/acs.organomet.9b00712.

[56] Jori N, Falcone M, Scopelliti R, Mazzanti M. Carbon dioxide reduction by multimetallic uranium(IV) complexes supported by redox-active schiff base ligands. Organometallics 2020;39:1590−601. Available from: https://doi.org/10.1021/acs.organomet.9b00792.

[57] Bhat GA, Rashad AZ, Folsom TM, Darensbourg DJ. Placing single-metal complexes into the backbone of CO2-based polycarbonate chains, construction of nanostructures for prospective micellar catalysis. Organometallics 2020;39:1612−18. Available from: https://doi.org/10.1021/acs.organomet.9b00704.

Chapter 4

Coordination cages and clusters as functional materials

Mohd Zeeshan, Farhat Vakil and M. Shahid
Functional Inorganic Materials Lab (FIML), Department of Chemistry, Aligarh Muslim University, Aligarh, Uttar Pradesh, India

4.1 Introduction

Three-dimensional organized structures known as coordination cages serve as hosts in host-guest chemistry. They frequently only rely on noncovalent interactions as opposed to covalent bonds and are self-assembled from organometallic precursors. Combinations of a coordination cage and a guest molecule known as inclusion compounds. Cage complexes can be formed with a wide variety of anionic ligands, such as [(H$_2$O)Cu(μ-CH$_3$CO$_2$)$_4$Cu(OH$_2$)] [1]. In this complex, two copper metals are coordinated together by bridging ligand acetate ion. For an instance, a coordination cage of Al$_{50}$(Cyclopentadienyl)$_{10}$ is shown in Fig. 4.1A and boron clusters B$_{80}$ based on icosahedron units in (B). The simplest cage type molecule is white phosphorous (P$_4$) which is more stable at room temperature. The bond length is about 222.3 pm and the molecule is tetrahedron. It requires a bond angle of 60 degrees. P$_4$ is the first industrial product that is crucial for the preparation of other phosphorous containing compounds. The high electrophilic reactivity of the P4 tetrahedron, the restricted selectivity for P−P bond rupture after the first P−P bond cleavage, and the P4 atoms' ineffective conversion of phosphorous into other elements are the causes of this [2,3]. P$_4$O$_6$, P$_4$O$_7$, P$_4$O$_8$, P$_4$O$_9$, and P$_4$O$_{10}$ are some of important organophosphorus compounds that are prepared from P$_4$ molecule. Other organophosphorus compounds also reported with S$_8$ such as P$_4$S$_9$ and P$_4$S$_{10}$ that are isostructural and isoelectronic with phosphorous oxides [3]. In general, coordination molecular cages (CMCs) regarded as metal organic polyhedrals (MOPs) made from metal ions and organic ligands. Their structures and composition affect their properties. For the development of novel CMCs with cluster organic polyhedral structures, it is crucial to replace the metal ions of MOPs with multinuclear metal clusters [4].

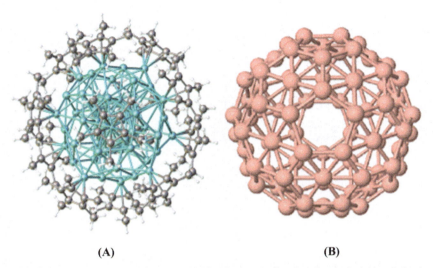

FIGURE 4.1 (A) A coordination cage of Al$_{50}$(Cyclopentadienyl)$_{10}$ showing a core of aluminum sheathed in ten pentamethyl cyclopentadienyl ligands [5]. (B) Boron clusters B$_{80}$ based on icosahedron units [6]. *Adapted with permission from Vollet J, Hartig JR, Schnöckel H. Al50C120H180: a pseudofullerene shell of 60 carbon atoms and 60 methyl groups protecting a cluster core of 50 aluminum atoms. Angew Chem Int Ed 2004;43(24): 3186−3189 and Jena P, Sun Q. Super atomic clusters: design rules and potential for building blocks of materials. Chem Rev 2018;118(11):5755−5870.*

The chemistry of metal boranes is diverse and various boron cage compounds also been reported, such as B$_2$H$_6$, B$_4$H$_{10}$, and B$_{12}$H$_{12}$. Diborane is the simplest of boron hydrides [1,3]. They are electron deficient compounds. Since they are strong reductants, they are unable to take electrons when they are provided by reducing agents. Each of the two terminal and two bridging hydrogen atoms in diborane surrounds a boron atom. The bond length of bridged B−H bond is about 137 pm while the bond length of terminal B−H bond is 119 pm. The bond angle of bridged H−B−H bond is about 97 degrees while the bond angle of terminal H−B−H is about 122 degrees. In a general exothermic process, BH$_3$ dimerizes to B$_2$H$_6$. The bonding in boron hydrogen compounds are described by the "two electron-three center" model created by W.N. Lipscomb. A bonding set of orbitals, two antibonding orbitals, or a bonding, nonbonding, and antibonding set of orbitals are frequently formed when three atomic orbitals are coupled in a "two electron, three center" bond.

For boron to resemble neon, the "two electron-three center" connection must form to finish the electron shell. This kind of link is common for metals with few electrons, such as AlH$_3$, in which aluminum coordinates with six hydrogen atoms following dimerization, comparable to AlF$_3$. Due to their various B−B bond types, molecular boranes (B$_n$H$_n$) resemble carbohydrides.

The typical boranes are typically molecular, diamagnetic, and optically transparent substances. Typical boranes fall under the well-known classes of closo, nido, arachno, hypho, and klado. Closo boranes are the most stable class of anions represented by $B_nH_n^{2-}$. Various types of borane clusters shown in Fig. 4.2 and their binding modes in Fig. 4.3.

Only closed n-boron polyhedral clusters show terminal hydrogen atoms. The polyhedral structures of nido (Latin: nidus; English: nest) and arachno boranes (Latin: arachne; English: spider) are, in contrast, missing one or two boron atoms at the corners of the polyhedron, respectively. Closo borane exhibits polyhedral structures with all corners occupied by boron atoms.

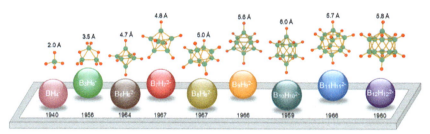

FIGURE 4.2 Structures of arachno-, nido-, and closo-borane [7] anion clusters and the year they were first synthesized. The longest distance between terminal H-atoms is listed above the structure. *Adapted with permission from Hansen BRS, Paskevicius M, Li HW, Akiba E, Jensen TR. Metal boranes: progress and applications. Coord Chem Rev 2016;323:60–70.*

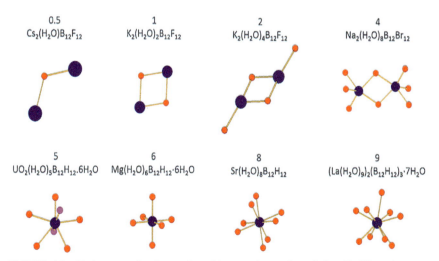

FIGURE 4.3 Various coordination modes of borane cluster (from 0.5 to 9). Water (oxygen: red) metal cations (blue) [8]. *Adapted with permission from Sama F, Ansari IA, Raizada M, Ahmad M, Nagaraja CM, Shahid M, et al. Design, structures and study of non-covalent interactions of mono-, di-, and tetranuclear complexes of a bifurcated quadridentate tripod ligand, N-(aminopropyl)-diethanolamine. N J Chem 2017;41(5):1959–1972.*

Nido boranes are shown by B_nH_{n+4} formula like B_2H_6 and $B_{10}H_{14}$ are well known nido boranes. The borohydride anion, BH^-_4, which can alternatively be thought of as the Lewis acid-base pair BH_3 and H^-, initiates the formation of nido structures, and then result in a sequence of anions with the general formula $B_nH^-_{n+3}$. B_nH_{n+6}, which can be illustrated as B_4H_{10} and the anions, $B_nH^-_{n+5}$, or $B_2H_7^-$ and $B_3H_8^-$, is the symbol for the arachno-borane series. Bridging hydrogen atoms, or B—H—B, are lost during the creation of anions as a result of their propensity to be more acidic than terminal hydrogen atoms. In general, the order of closo- nido-arachno-boranes exhibits an increase in reactivity and a decrease in stability [1]. Classification of different metal boranes clusters is given in Table 4.1 and Fig. 4.4.

All compounds with larger cyclic structures and triangular metal—metal bonds are clusters. Cluster formations often resemble the closely packed metal structures, where ligand bridges keep several metal atoms together. For the early d-block metals, organometallic compounds are uncommon, but for the f-block metals, they are virtually unknown. The elements of group six to ten have many metal carbonyl clusters (chromium to nickel family). The 18-electron rule and metal—metal and metal—ligand electron pair bonding can be used to explain the bonding in small clusters. $Mn_2(CO)_{10}$,

TABLE 4.1 Boron hydrides are categorized and counted in terms of electrons.

Type	Formula pairs	Skeletal electron	Examples
Closo	$B_nH_n^{2-}$	$n+1$	$B_5H_5^{2-}$, $B_{12}H_{12}^{2-}$
Nido	B_nH_{n+4}	$n+2$	B_2H_6, B_5H_9, B_6H_{10}
Arachno	B_nH_{n+6}	$n+3$	B_4H_{10}, B_5H_{11}
Hypho	B_nH_{n+8}	$n+4$	$B_5H_{12}^-$

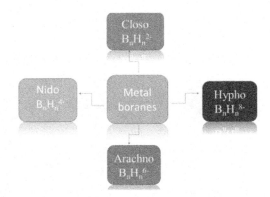

FIGURE 4.4 Classification of metal boranes and their binding units.

Os$_3$(CO)$_{12}$, Co$_4$(CO)$_{12}$, Fe$_2$(CO)$_9$, Co$_3$(CH)(CO)$_9$, and [Te$_6$](O$_3$SCF$_3$) (shown in Figs. 4.5 and 4.6) are some of the examples of carbonyl clusters with transition metals. Except for one Mn—Mn link, it is assumed that each Mn atom in Mn$_2$(CO)$_{10}$ has seventeen electrons (seven from manganese and ten from the five CO ligands). Since the two metal atoms in the Mn—Mn bond share two electrons, this boosts the electron counts of each by one, resulting in two metal atoms with 18 electrons, which results in a total electron count of 34 rather than 36. Before M—M bonding, in which each metal shares an additional two electrons with neighboring metals, is taken into account, each [Os(CO)$_4$] fragment in Os$_3$(CO)$_{12}$ has 16 electrons. This leads to an increase in the number of electrons surrounding each metal to 18, but only 48 total electrons rather than 54. Cluster valence electrons (CVEs) are the name given to these bonding electrons, which are required for x metal atoms and y M—M bonds (so, require $18x-2y$ electrons) [1].

4.1.1 Fundamental principles

In cages and clusters, metal atoms typically have direct connections to one another. Typically, cages and clusters of metal atoms are physically connected to one another. As a result, metal cluster compounds can take the shape of a number of arrays, such as triangular, tetrahedral, and octahedral arrays. Transition metals and main group elements form robust clusters. Metal atoms are indeed directly bonded or merely held quite close together by a framework of bridging ligands. The bonding between the metal clusters and coordination cages is decided by MO or VB theory. The metal atoms are arranged in such a way as to minimize repulsions between the atoms [1].

4.2 Common methods of preparation

The boranes, borohydrides and metal clusters, such as Dy$_6$ clusters, are prepared by using various reagents. Their methods of preparation are given below.

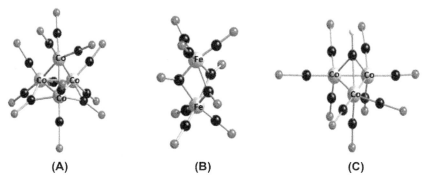

FIGURE 4.5 The structure of some metal clusters (A) Co$_4$(CO)$_{12}$, (B) Fe$_2$(CO)$_9$, and (C) Co$_3$(CH)(CO)$_9$.

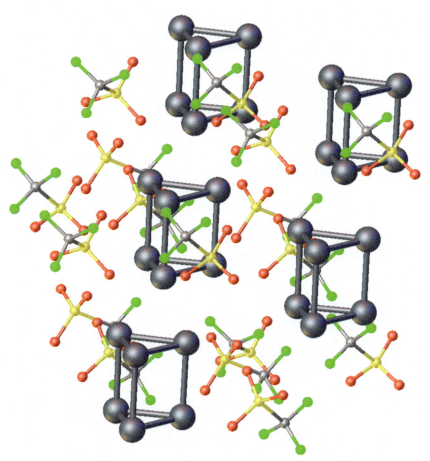

FIGURE 4.6 A portion of lattice of Te$_6$(O$_3$SCF$_3$)$_2$. The intra- and intertriangle Te—Te distances are measured as 2.70 and 3.06 angstrom respectively [9]. *Adapted with permission from Schulz, C, Daniels, J, Bredow, T, Beck, J. The electrochemical synthesis of polycationic clusters. Angew Chem Int Ed 2016;55(3):1173−1177.*

4.2.1 Synthesis of [Dy$_6$(μ_6-O)(μ_3-OH)$_8$(H$_2$O)$_{12}$(NO$_3$)$_6$](NO$_3$)$_2$(H$_2$O)$_2$

20 mL of MeOH/H2O (4:6 v/v) were used to dissolve a combination of Dy(NO$_3$)$_3$0.5H$_2$O (2 mmol; 0.88 mg) and 3-aminophenol (ampH) (8 mmol; 0.62 mL). After stirring for 40 minutes, Net$_3$ (2 mmol; 0.28 mL) was added, and the mixture was then refluxed for 1 hour at 90°C. It was allowed to slowly evaporate for 32 days in the dark after cooling to room temperature. After 32 days, white crystals were obtained. Its complete synthesis is shown in Scheme 4.1.

$$\text{Dy(NO}_3)_{3.5}\text{H}_2\text{O} + \text{3-aminophenol} \xrightarrow[\text{NEt}_3,\text{reflux}]{\text{MeOH}} [\text{Dy}_6(\mu_6\text{-}O)(\mu_3\text{-OH})_8(H_2O)_{12}(\text{NO}_3)_6](\text{NO}_3)_2(H_2O)_2]$$

Coordination cages and clusters as functional materials **Chapter | 4** 99

$$Dy(NO_3)_3 \cdot 5H_2O + 3\text{-aminophenol} \xrightarrow[NEt_3, \text{reflux}]{MeOH} [Dy_6(\mu_6\text{-}O)(\mu_3\text{-}OH)_8(H_2O)_{12}(NO_3)_6](NO_3)_2(H_2O)_2]$$

SCHEME 4.1 Synthesis of Dy$_6$ cluster.

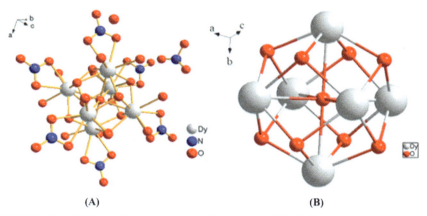

FIGURE 4.7 (A) Structural unit of Ln cluster [Dy$_6$(μ_6-O)(μ_3-OH)$_8$(H$_2$O)$_{12}$(NO$_3$)$_6$](NO$_3$)$_2$(H$_2$O)$_2$] and (B) core of the dysprosium molecule. *Adapted with permission from Raizada M, Shahid M, Hussain S, Ashafaq M, Siddiqi ZA. A new antiferromagnetic Dy6 oxido-material as a multifunctional aqueous phase sensor for picric acid as well as Fe^{3+} ions. Mater Adv 2020;1(9):3518–3531.*

Its structure was confirmed by single crystal X-ray diffraction [10]. The structure of the Dy$_6$ cluster is shown in Fig. 4.7.

Reactions of boranes and borohydrides [1]

$$BH_4^- + H^+ \rightarrow 1/2 B_2H_6 + H_2$$
$$B_{10}H_{14} + N(CH_3)_3 \rightarrow [HN(CH_3)_3]^+ [B_{10}H_{13}]$$
$$B_{10}H_{14} + Li[BH_4] \xrightarrow[R_2O]{ether} Li[B_{10}H_{13}] + R_2OBH_3 + H_2$$
$$Li[B_{10}H_{13}] + R_2OB \rightarrow Li[B_{11}H_{14}] + H_2 + R_2O$$
$$5K[B_9H_{14}] + 2B_5H_9 \xrightarrow[85°C]{polyrther} 5K[B_{11}H_{14}] + 9H_2$$
$$2[B_{11}H_{13}]^{2-} + Al_2(CH_3)_6 \xrightarrow{\Delta} 2[B_{11}H_{11}AlCH_3]^{2-} + 4CH_4$$

Reactions of transition metal clusters [1]: generally, boranes are quite reactive toward transition metal reagents where the attack can occur at different sites of the polyhedral cage. Thus a metallated analog of borane is formed when B$_5$H$_9$ is heated with Iron carbonyl, such as Fe(CO)$_5$.

$$B_5H_9 + Fe(CO)_5 \rightarrow [Fe(CO)_3B_4H_8]$$

This metal cluster can be represented as follows (see Fig. 4.8).

$$2Na_2[B_9C_2H_{11}] + FeCl_2 \xrightarrow{THF} 2NaCl + Na_2[Fe(B_9C_2H_{11})_2]$$
$$2Na[C_5H_5] + FeCl_2 \xrightarrow{THF} 2NaCl + Fe(C_5H_5)_2$$

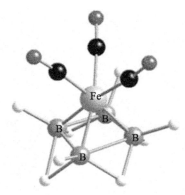

FIGURE 4.8 The structure of [Fe(CO)$_3$B$_4$H$_8$].

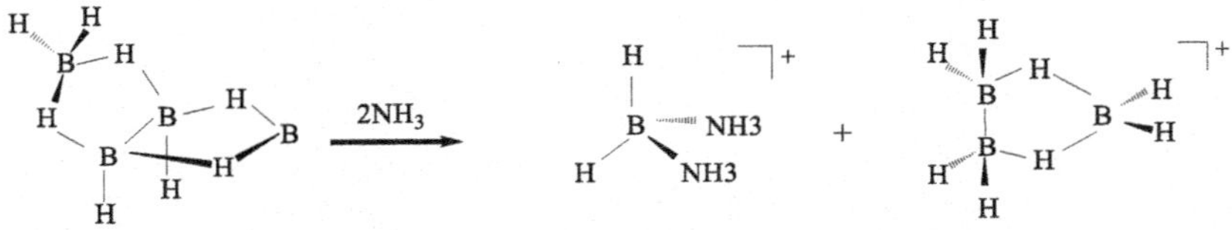

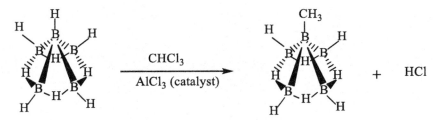

SCHEME 4.2 Partial fragmentation of B$_4$H$_{10}$ cluster.

SCHEME 4.3 Electrophilic substitution of B$_5$H$_9$ by Lewis acid (AlCl$_3$).

4.3 Mechanism involved

1. Lewis base cleavage [11] reaction, such as ammonia with higher borane clusters like B$_4$H$_{10}$ involves the cleavage of B—H—B bonds, cluster deprotonation, enlargement of the cluster and abstraction of one or more protons. It leads to a partial fragmentation of the cluster. Its mechanism is shown in Scheme 4.2.
2. Friedal-Crafts [1] is one of the most important reactions of borane clusters where electrophilic displacement of hydride ion catalyzed by a Lewis acid, such as AlCl$_3$. On the closed end of the boron clusters, the replacement typically occurs. A general pathway to alkylated and halogenated compounds is provided by this reaction. Its mechanism is shown in Scheme 4.3.

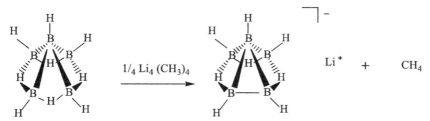

SCHEME 4.4 Deprotonation of decaborane.

3. Deprotonation of decaborane [1] requires weak base trimethylamine. Its mechanism given in Scheme 4.4.

4.4 Recent developments with examples

Significant advancements have been made in the design and synthesis of paramagnetic polymetallic cores with nucleus counts higher than five over the past 10 years. These cores are employed for a range of functional materials, such as fluorescent, optical, magnetic, and electrical materials. Significant advancements have been made in the design and synthesis of paramagnetic polymetallic cores with nucleus counts higher than five over the past ten years. These cores are employed for a range of functional materials, such as fluorescent, optical, magnetic, and electrical materials [7].

Out of the many homometallic clusters of the first-row transition series elements, only a few hydroxo or alkoxo bridged cobalt clusters found to exhibit single-molecule magnets (SMM) behavior. Some of them, such as Co_4 cubanes, Co_6 complexes, and Co_{12} wheels, may only have ferromagnetic characteristics. The majority of cubane-containing cobalt-based clusters that exhibit SMM behavior are the smallest and best defined. Generally, Co^{3+} is diamagnetic and Co^{2+} is paramagnetic ion in a mixed valence cobalt cluster with a single ion anisotropy and a significant spin orbit coupling [12].

The chemistry of cobalt salts with ampdH$_2$ (2-amino-2-methyl-1,3-propanediol) under different conditions of solvents is given in Scheme 4.5. Under the given scheme, different clusters are formed with a nucleus number of more than seven. With Kcat(MeOH) = 2026/cm, this Co_7 disk exhibits solvent-dependent catecholase activity in terms of its spectral, crystallographic, and magnetic properties [12].

Furthermore, N-methyldiethanolamine (MedeaH$_2$) or N-^{n}butyldiethanolamine (^{n}BudeaH$_2$) was allowed to react with Co(NO$_3$)$_2$0.6H$_2$O in the presence of auxiliary N-chelator, that is, 1,10-phenanthroline (Phen), or 2,2′-bipyridine (Bipy) resulting in stable crystalline products (see Scheme 4.6).

They have been characterized as mixed valence tetranuclear species [1–4] where 1, 2, 3, and 4 refers to the [Co$_4$(μ$_3$-OH)$_2$(Medea)$_2$(Phen)$_4$]

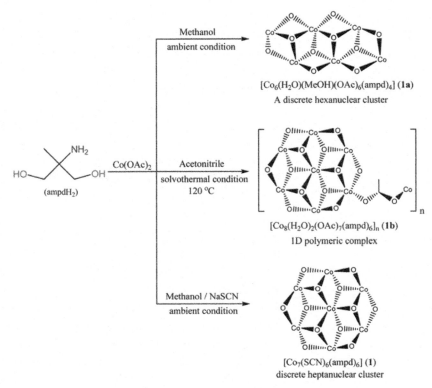

SCHEME 4.5 Synthetic route of discrete Co₆ cluster, Co₈ polymeric chain and discrete Co₇ disk using ampdH₂. *Adapted with permission from Ahamad MN, Sama F, Akhtar MN, Chen YC, Tong M-L, Ahmad M, et al. A disc-like Co7 cluster with a solvent dependent catecholase activity. N J Chem 2017;41(23):14057–14061.*

4NO$_3$0.6H$_2$O.2MeOH, [Co$_4$(μ_3-OH)$_2$(Medea)$_2$(Bipy)$_4$] 4NO$_3$0.8H$_2$O, [Co$_4$(μ_3-OH)$_2$(^{n}Budea)$_2$(Phen)$_4$]4NO$_3$0.9H$_2$O, and [Co$_4$(μ_3OH)$_2$(^{n}Budea)$_2$(Bipy)$_4$] 4NO$_3$ 0.6H$_2$O [8].

Amino alcohol ligands [8] are renowned for their special abilities to build metal organic frameworks with various topologies and architectural styles as well as mononuclear and polynuclear transition metal complexes. Transition metal complexes with tetra dentate ligands [8] are used to make high nuclearity coordination clusters, coordination polymers, supramolecular frameworks, and molecular magnetism in the fields of crystal engineering, molecular magnetism, molecular cooling, catalysis, and supramolecular chemistry. A dialcohol diamine ligand, H$_2$apdea (*N*-(3-aminopropyl)-diethanolamine) is known to form mono-, di-, and tetra nuclear transition metal clusters (see Scheme 4.7) [8].

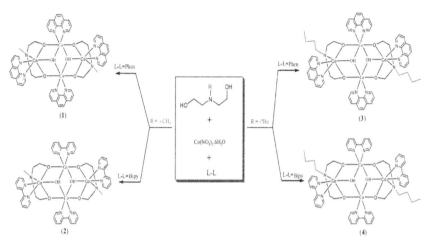

SCHEME 4.6 Synthetic procedure of different metal cluster [1–4] complexes. *Adapted with permission from Sama F, Ansari IA, Raizada M, Ahmad M, Nagaraja CM, Shahid M, et al. Design, structures and study of non-covalent interactions of mono-, di-, and tetranuclear complexes of a bifurcated quadridentate tripod ligand, N-(aminopropyl)-diethanolamine. N J Chem 2017;41(5):1959–1972.*

4.4.1 Applications in emerging fields

Vanadium complexes are crucial for self-assembling and guest-host chemistry. In aqueous solutions, the vanadium clusters maintain their distinctive geometrical characteristics but show nearly identical biological activities, such as dimer $\{[H_2V_2O_7]^{2-}, [HV_2O_7]^{3-}, [V_2O_7^{4-}]\}$, the tetramer $[V_4O_{12}]^{4-}$, the pentamer $[V_5O_{15}]^{5-}$, and the decamer $[H_2V_{10}O_{28}]^{4-}$, $[HV_{10}O_{28}]^{5-}$, $[V_{10}O_{28}]^{6-}$ that are essential for the functioning of enzymes in human beings and other higher animals. For instance, the structure–activity relationships of several antidiabetic vanadium complexes studied using vanadium complexes. Vanadium complexes are completely safe for use in clinical settings thanks to their use as antitumor medications and antiparasitic treatments as well as their augmentation of bioactive ligand activity through complexation with vanadium [13].

Luminescent property of Na{Cu$_{12}$Zn$_4$} as a cluster for the detection of Ca^{2+} ion: through ion exchange processes, different cations, such as the Ca^{2+} ion, may be enclosed in the Cu$_{12}$Zn$_4$ ring structure which has a cavity that captures the Na$^+$ ion (see Scheme 4.8).

The luminescent property resulting from the intra-ligand transition is indicated by the emission at 320 nm following stimulation at 260 nm [14]. These intra-ligand transitions are of the n→σ* type, which shows that there are several free hydroxyl groups with lone pairs present. During fluorescence titration, there is a visual color change of Na{Cu$_{12}$Zn$_4$} cluster upon addition

104 Recent Advances in Organometallic Chemistry

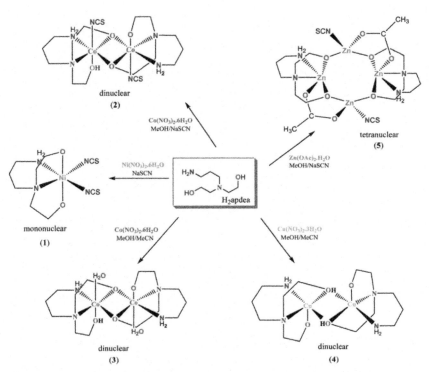

SCHEME 4.7 Transition metal clusters with H$_2$apdea (N-(3-aminopropyl)-diethanolamine) ligand [8]. *Adapted with permission from Sama F, Ansari IA, Raizada M, Ahmad M, Nagaraja CM, Shahid M, et al. Design, structures and study of non-covalent interactions of mono-, di-, and tetranuclear complexes of a bifurcated quadridentate tripod ligand, N-(aminopropyl)-diethanolamine. N J Chem 2017;41(5):1959–1972.*

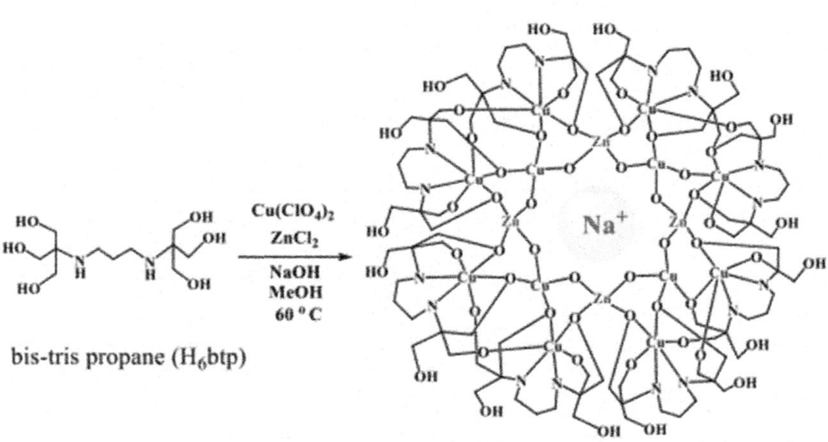

SCHEME 4.8 Synthetic route of Na{Cu$_{12}$Zn$_4$} cluster.

of cations. When other cations are added, no change is seen with the naked eye; nonetheless, this color change is discovered to be more pronounced and specific to the Ca^{2+} ion, that is, under visible light or UV light [14]. This clearly shows that Na{$Cu_{12}Zn_4$} possess chromogenic as well as florigenic sensor properties toward the Ca^{2+} ions [14]. Their binding modes is given in Scheme 4.9.

Na{$Cu_{12}Zn_4$} cluster as the selective adsorption [14] of organic dyes: the Na{$Cu_{12}Zn_4$} clusters have adsorption abilities for many organic pigments found in waste water. As an adsorbent, it works. Three different types of dyes; methyl orange (MO), methylene blue (MB), and rhodamine B(Rh-B) have been selected as adsorbates as a model system. The results of the UV-vis spectroscopy clearly show that the adsorbent, $NaCu_{12}Zn_4$, has an outstanding capacity to adsorb MB when it is added to the dye solution, further indicating a reduction in dye concentration in the solution (see Fig. 4.9). The two stages of the adsorption process may be seen easily from the plot of C/C_0 versus t. The first stage refers to the high initial dye concentration and opens active adsorption sites after a steady dye concentration where the adsorption equilibrium was reached. It is indicating that Rh-B and MB dye adsorption efficiencies can be attained in less time than 3 hours [14]. While the adsorption capacity of MO is discovered to be quite low, the color of the solution before and toward MB and Rh-B dyes. The percentage of dye removal from aqueous solution can be used to calculate the effectiveness of an adsorbent using the following equation:

$$\%\text{removal} = (C_0 - C_t) \times 100\%/C_0$$

where C_t denotes the dye concentration in mg/L at time t and C_0 denotes the dye concentration at the initial concentration in mg/L. For MB, MO, and Rh-B dyes, the absorption maxima are observed at 665, 460, and 550 nm of

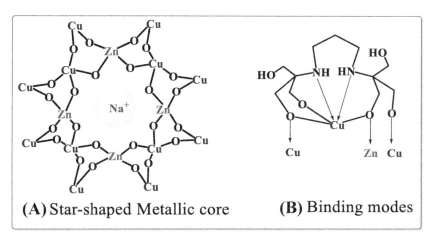

SCHEME 4.9 (A) Star-shaped metallic core of Na{$Cu_{12}Zn_4$} and (B) binding modes.

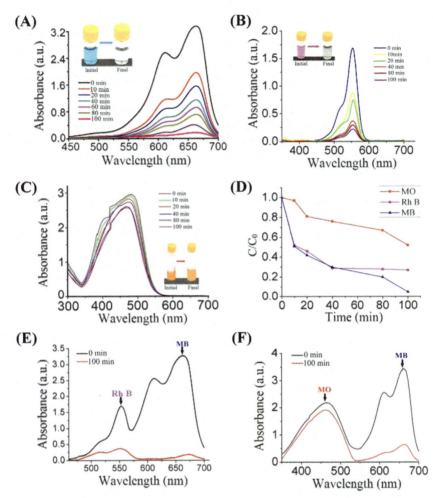

FIGURE 4.9 Modifications to the MB, Rh-B, and MO UV–vis absorption spectra following the addition of the adsorbent Na{Cu$_{12}$Zn$_4$} [(A), (B), and (C), respectively]; (D) effectiveness of MO, MB, and Rh-B removal variations in the UV–vis spectra of the MB and Rh-B combination (E) and the mixture of MB and MO (F) after the addition of Na{Cu$_{12}$Zn$_4$} [14].

wavelength. These peaks exhibit a considerable decline with increasing time, and during the first 15 minutes, 55%, 5%, and 50% adsorption, respectively, are attained for each dye. The performance of Na{Cu$_{12}$Zn$_4$} in terms of MB, MO, and Rh-B adsorption is determined to be 98%, 3%, and 95%, respectively, after 180 minutes, and the adsorption remains the same even after a day. Additionally, because it can absorb about 65% of the MB dye in just 30 minutes of testing, the Na{Cu$_{12}$Zn$_4$} cluster works well as an adsorbent for the removal of the dye (given in Fig. 4.10) [14].

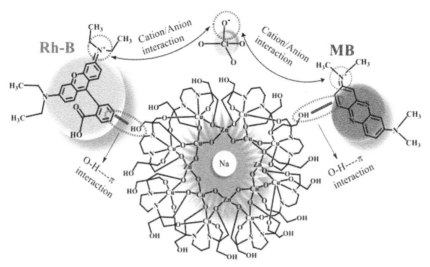

FIGURE 4.10 The mechanism to show dye adsorption via O—H···π and electrostatic interactions [14].

Application of [Dy$_6$(μ_6-O)(μ_3-OH)$_8$(H$_2$O)$_{12}$(NO$_3$)$_6$](NO$_3$)$_2$(H$_2$O)$_2$] cluster in nitroaromatic sensing: a prominent fluorescence band at 345 nm can be seen in the photoluminescence spectra of [Dy$_6$(μ_6-O)(μ_3-OH)$_8$(H$_2$O)$_{12}$(NO$_3$)$_6$](NO$_3$)$_2$(H$_2$O)$_2$] distributed in the water. Several noble-metal nanoclusters (NACs), including NB (nitrobenzene), MNP (3-nitrophenol), PNP (4-nitrophenol), 2,4-DNP (2,4-dinitrophenylhydrazine), and PA were added gradually to [Dy$_6$(μ_6-O)(μ_3-OH)$_8$(H$_2$O)$_{12}$(NO$_3$)$_6$](NO$_3$)$_2$(H$_2$O)$_2$] to perform fluorescence titration studies (2,4,6-trinitrophenol). With the addition of NB, MNP and PNP had little effect on the level of fluorescence. Interestingly, [Dy$_6$(μ_6-O)(μ_3-OH)$_8$(H$_2$O)$_{12}$(NO$_3$)$_6$](NO$_3$)$_2$(H$_2$O)$_2$] demonstrates exceptional sensing activity toward PA by reducing the initial fluorescence intensity by 58.6% after 0.91 ppb of PA is added to the aqueous phase [14]. This is shown in Fig. 4.11.

Metal sensing using the [Dy$_6$(μ_6-O)(μ_3-OH)$_8$(H$_2$O)$_{12}$(NO$_3$)$_6$](NO$_3$)$_2$(H$_2$O)$_2$] cluster: it was also looked on how well the current OM could detect various transition metal ions in water. It has been discovered that [Dy$_6$(μ_6-O)(μ_3-OH)$_8$(H$_2$O)$_{12}$(NO$_3$)$_6$](NO$_3$)$_2$(H$_2$O)$_2$] may preferentially detect Fe^{3+}. In a typical experiment, the gradual addition of Fe^{3+} in water solution was followed by the recording of the fluorescence spectra of [Dy$_6$(μ_6-O)(μ_3-OH)$_8$(H$_2$O)$_{12}$(NO$_3$)$_6$](NO$_3$)$_2$(H$_2$O)$_2$] (1 mM). As anticipated, the addition of 5.58 ppb of Fe^{3+} solution caused immediate and substantial fluorescence31 quenching (80%) with significant spectroscopic variation when added to [Dy$_6$(μ_6-O)(μ_3-OH)$_8$(H$_2$O)$_{12}$(NO$_3$)$_6$](NO$_3$)$_2$(H$_2$O)$_2$].

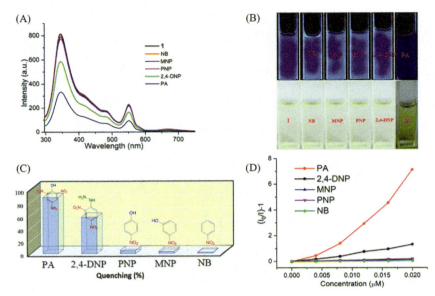

FIGURE 4.11 (A) The change in fluorescence intensity of [Dy$_6$(μ_6-O)(μ_3-OH)$_8$(H$_2$O)$_{12}$(NO$_3$)$_6$](NO$_3$)$_2$(H$_2$O)$_2$] (λ_{ex} = 296 nm) on addition of 0.91 ppb of different NACs. (B) Digital photographs of solutions in the presence of different nitroaromatic analytes under portable UV light (top) and under normal light (bottom). (C) Quenching efficiency of different analytes. (D) Stern–Volmer plot for various nitro analytes [10].

The hue of [Dy$_6$(μ_6-O)(μ_3-OH)$_8$(H$_2$O)$_{12}$(NO$_3$)$_6$](NO$_3$)$_2$(H$_2$O)$_2$] changes from colorless to bright orange when exposed to Fe^{3+}, which can be seen with the naked eye [10]. This is shown in Fig. 4.12. Due to the substantial neutron scatter cross section of B$_{10}$ and its safety, boron clusters and boranes are used in boron neutron capture therapy (BNCT) for the treatment of cancer. This makes them chemically more stable in biological systems. Disodium mercapto-undecahydro-closo-dodecaborate, Na$_2$B$_{12}$H$_{11}$SH and NaB$_{12}$H$_{11}$NH$_3$ are used for cancer treatment. Na$_2$B$_{12}$H$_{12}$ and Na$_2$B$_{10}$H$_{10}$ show properties like high stability, water solubility, low toxicity, and high boron content that makes them ideal candidates in BNCT. Furthermore, they are also involved in blood–brain barrier [7].

Higher boranes are also utilized as solid-state electrolytes, reducing agents, and boride precursors in organic chemistry. Li$_2$B$_{12}$H$_{12}$ is used as a luminescent dye in transparent head-up displays. Boron based polymers are also used in hydrogen storage [7].

4.5 Future prospective and scope

Atomically pure doped clusters synthesized at an advanced level. To create single crystals of doped materials, this atomic level purity is necessary.

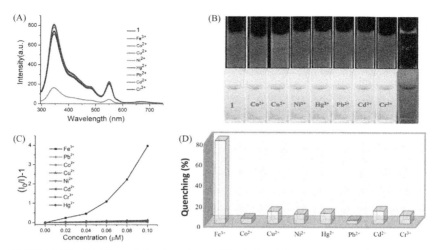

FIGURE 4.12 (A) The variation in cation addition-induced fluorescence intensity of 1 (ex = 296 nm). (B) Digital images of solutions of 1 with various metal ions in the presence of portable UV light (top) and regular light (bottom). (C) Stern–Volmer plots are used for different analytes. (D) The effectiveness of fluorescence quenching when several analytes are added (5.58 ppb) [10].

Mass spectrometry can verify the structural chemistry of clusters that are doped with tri- and tetrametallic elements. Clusters of noble metals lose a significant amount of their magnetic sensitivity without doping. A promising method for adding magnetic characteristics to such clusters is doping with ferromagnetic metals like Fe, Ni, and Co. It is commonly believed that doping clusters using such metals could result in the creation of novel synthetic pathways distinct from those now used for doping. Different foreign metal atoms may favor certain locations in the parent cluster during doping. This has the potential to modify the surface and catalytic reactivity of metal clusters. Indeed, there are still a lot of unexplored possibilities before the scientific community has a good hold on autonomously exact cluster doping with control over the type and quantity of foreign metal atoms [15].

4.6 Summary

We think the work presented here helps prepare new coordination cages, metal clusters, and other metal boranes as useful materials that can self-assemble from basic, inexpensive building pieces. These materials not only produce new cages and clusters, but they also have remarkable magnetic characteristics. The expansion of a family of heterometallic $Fe_x^{III} Ln_y^{III}$ clusters with various functional characteristics is the subject of ongoing research. Additionally, the chemistry of metal boranes is complex and divides into a wide range of related but highly distinct research disciplines.

Here, we outline and examine the intricate production processes as well as the complex coordination chemistry that is present in the majority of functional materials. The Dy_6 oxido material is the best dual-sensor because it can detect both picric acid (PA) and Fe^{3+} ions in aqueous media with detection limits of 0.03 and 0.09 ppb, respectively. In addition, the Co_7 disk exhibits spectroscopic, crystallographic, and magnetic characteristics that demonstrate solvent-dependent catalase activity, Kcat(MeOH) = 2026/cm. Additionally, boron clusters like $Na_2B_{12}H_{11}SH$ and $NaB_{12}H_{11}NH_3$ employed in the therapy of cancer. H_2apdea (*N*-(3-aminopropyl)-diethanolamine), a dialcohol diamine ligand, is also known to form mono-, di-, and tetra nuclear transition metal clusters. The specific binding of Ca^{2+} with $Na\{Cu_{12}Zn_4\}$ can also be determined by the change in emission spectra, which indicates the limit of detection (LOD) of 4.5 nM, one of the lowest values among the data so far reported, as well as by the visual color change. The HSAB idea and the same size of Ca^{2+} and Na^+ activate the calcium ion sensing. The exchange of cations with lower LOD is caused by these two elements. Additionally, it can be proven to be a powerful adsorbent for cationic dyes modified by electrostatic and O–H π interactions.

4.7 Problems with solution

The near impossibility of creating atomic clusters in the proportions required to synthesize new materials is one of the main problems with them. In most cases, when crystals are taken out of the high-vacuum environment, they become reactive and polluted. They will also change in terms of shape and magnetic properties in the gas phase. Thanks to studies on gas-phase clusters, structure-property interactions of nanoscale materials may now be examined, one atom and one electron at a time. The knowledge gained from hydrogen's interactions with metal atoms and their cations in the gas phase was used in the design and synthesis of the new class of hydrogen storage materials. Hydrogen can also join with an atom of a transition metal in a quasi-molecular bond with binding energies in between physisorption and chemisorption. For hydrogen to be released under ambient circumstances and then used as fuel in the transportation sector, these parameters must met by hydrogen storage materials. The ability of a metal cation to absorb hydrogen in quasimolecular form through a charge polarization mechanism later proven by Niu et al. In order to capture hydrogen in quasimolecular form, these approaches are currently being employed to dope porous materials with metal atoms. Most clusters investigated in the gas phase present a challenge since they are reactive and prone to changing when assembled or maintained on a surface. By manipulating the size and content of the material, it is feasible to produce magnetic particles from nonmagnetic materials. In order to effectively manage the "band gap" of nanoclusters for usage in optical devices, size and form can also be utilized [11].

Coordination cages and clusters as functional materials **Chapter | 4** **111**

4.8 Objective type questions

1. The oxidation state of Fe in the complex [Cp Fe(CO)$_2$]$_2$ is:
a. $+2$
b. $+1$
c. 0
d. -1

   Ans. b

2. The complex that obeys the 18 electron rule is:
a. [Mn(CO)5]
b. [Co(η5-C5H5)]
c. [Mo(CO)3(CH3CN)3]
d. [Ti(η5-C5H5)]

   Ans. c

3. Identify the complex which do not follow the 18 electron rule:
a. [Fe(H2O)6]
b. [Ru(η6-C6H6)(η6-C6H6)]
c. Na [Co(CO)3(PPh3)]
d. [Mn(CO)5] −

   Ans. b

4. The heptacity of nitrosyl in [Mo(η^1-allyl)$_3$(η^3-allyl)$_2$NO] is:
a. 1
b. 2
c. 3
d. 0

   Ans. a

5. The oxidation state of Mo in [(η^7-tropyllium)Mo(CO)$_3$]$^+$ is:
a. $+2$
b. $+1$
c. 0
d. -1

   Ans. c

6. The organometallic compound W(C$_5$H$_5$)$_2$(CO)$_2$ follows the 18 electron rule. The heptacities of the two cyclopentadienyl groups are:
a. 5 and 5
b. 3 and 5
c. 3 and 3
d. 1 and 5

   Ans. b

7. Which compound has four metal-metal bonds?
a. Fe2(CO)9
b. Co2(CO)8
c. [Re2Cl8]2−
d. [Ru3(CO)12]

    Ans. c

8. The number of M-M bonds in $Ir_4(CO)_{12}$ are:
a. 4
b. 6
c. 8
d. 0

    Ans. b

9. M-M quadrupole bonds are well known for the metal is:
a. Ni
b. Co
c. Fe
d. Re

    Ans. d

10. The cluster $Rh_6(CO)_{16}$, $Os_5C(CO)_{15}$ and $[Fe_4(CO)_{12}C]^{2-}$ has structure respectively:
a. Closo, nido, arachno
b. Nido, closo, arachno
c. Closo, arachno, nido
d. Arachno, closo, nido

    Ans. a

References

[1] Atkins PW, Overton TL, Rourke JP, Weller MT. Shriver and Atkins inorganic chemistry. 5th ed. Great Britain by Oxford University Press; 2010. p. 208−10.
[2] Du S, Yang J, Hu J, Chai Z, Luo G, Luo Y, et al. Direct functionalization of white phosphorus to cyclotetraphosphanes: selective formation of four P-C bonds. J Am Chem Soc 2019;141(17):6843−7.
[3] Huheey JE, Ellen AK, Richard LK. Inorganic chemistry: principles of structure and reactivity. New York: Harper Collins College Publishers; 1993. p. 583−8.
[4] Zhu ZK, Lin YY, Yu H, Li XX, Zheng ST. Inorganic−organic hybrid polyoxoniobates: polyoxoniobate metal complex cage and cage framework. Angew Chem Int Ed 2019;58 (47):16864−8.
[5] Vollet J, Hartig JR, Schnöckel H. Al50C120H180: a pseudofullerene shell of 60 carbon atoms and 60 methyl groups protecting a cluster core of 50 aluminum atoms. Angew Chem Int Ed 2004;43(24):3186−9.

[6] Jena P, Sun Q. Super atomic clusters: design rules and potential for building blocks of materials. Chem Rev 2018;118(11):5755–870.

[7] Hansen BRS, Paskevicius M, Li HW, Akiba E, Jensen TR. Metal boranes: progress and applications. Coord Chem Rev 2016;323:60–70.

[8] Sama F, Ansari IA, Raizada M, Ahmad M, Nagaraja CM, Shahid M, et al. Design, structures and study of non-covalent interactions of mono-, di-, and tetranuclear complexes of a bifurcated quadridentate tripod ligand, N-(aminopropyl)-diethanolamine. N J Chem 2017;41(5):1959–72.

[9] Schulz C, Daniels J, Bredow T, Beck J. The electrochemical synthesis of polycationic clusters. Angew Chem Int Ed 2016;55(3):1173–7.

[10] Raizada M, Shahid M, Hussain S, Ashafaq M, Siddiqi ZA. A new antiferromagnetic Dy6 oxido-material as a multifunctional aqueous phase sensor for picric acid as well as Fe 3 + ions. Mater Adv 2020;1(9):3518–31.

[11] Otto T, Zones SI, Iglesia E. Challenges and strategies in the encapsulation and stabilization of monodisperse Au clusters within zeolites. J Catal 2016;339:195–208.

[12] Ahamad MN, Sama F, Akhtar MN, Chen YC, Tong M-L, Ahmad M, et al. A disc-like Co7 cluster with a solvent dependent catecholase activity. N J Chem 2017;41(23):14057–61.

[13] Shahid M, Sharma PK, Anjuli, Chibber S, Siddiqi ZA. Isolation of a decavanadate cluster [H2V10O28][4-picH]4 · 2H2O (4-pic = 4-picoline): crystal structure, electrochemical characterization, genotoxic and antimicrobial studies. J Clust Sci 2014;25(5):1435–47.

[14] Iman K, Shahid M, Ahmad M. A novel self-assembled Na{$Cu_{12}Zn_4$} multifunctional material: first report of a discrete coordination compound for detection of Ca^{2+} ions and selective adsorption of cationic dyes in water. Dalt Trans 2020;49(11):3423–33.

[15] Ghosh A, Mohammed OF, Bakr OM. Atomic-level doping of metal clusters. Acc Chem Res 2018;51(12):3094–103.

Chapter 5

C—H activation: A sustainable approach for the synthesis of functionalized heterocycles

Indu Sharma[1], Mukesh Kumari[2,3], Gargi Poonia[1], Sunil Kumar[4] and Ramesh Kataria[1]
[1]*Department of Chemistry and Centre for Advanced Studies in Chemistry, Panjab University, Chandigarh, India,* [2]*K. M. Government College, Narwana, Jind, India,* [3]*Department of Chemistry (SBAS), Maharaja Agrasen University, Baddi, Solan, India,* [4]*Department of Chemistry, J. C. Bose University of Science and Technology, YMCA, Faridabad, India*

5.1 Introduction

In organic synthesis, the C—C bond formation is an important coupling reaction for the construction of synthetic polycyclic aromatic heterocycles, linear molecules as well as natural products or their analogs. Therefore various organic name reactions were discovered from time to time viz. aldol reaction, Corey House synthesis, Negishi Coupling, alkane metathesis, Friedal Crafts reaction, Sonogashira coupling, Stille reaction, etc. [1—3]. Among these, C—H activation is one of the major interesting approaches which have attracted the interest of chemists in the last decade.

The concept of Green chemistry encompasses the design of chemical products and development of processes which minimize the use or generation of hazardous substances. Among the twelve principles given by P. T. Anastas and J. C. Warner, the four principles viz. catalysis, atom economy, avoid derivatives and use of safer solvents and auxiliaries can overcome by the concept of C—H bond functionalization or activation [4].

The usage of catalysis in expansion of organic molecules has transformed the area of organic synthesis over the past decade. Due to the incomplete d-orbitals in the transition metals, they act as excellent catalysts depending on the reaction because they can easily withdraw or give electrons to other reagents. Generally the coupling of two prefunctionalized substrates afforded the products in highly selective and anticipated manner in the presence of transition metal catalysts. Therefore transition metal catalysts were implemented as convenient tools in synthetic and medicinal chemistry due to their diverse reactivity pattern in enabling numerous molecular conversions [5–9]. Owing to the Lewis acidity and lesser electronegativity, main group metal and metalloids exhibit characteristic electronic properties and coordination behavior which differentiate them from organic ligands [10–13].

In C–H functionalization a C–H bond is interchanged by a new C–M bond which is helpful for C–C bond formation in coupling reactions. Otto Dimroth reported the first C–H activation reaction in 1902 [14]. The functionalization of nonactivated C – H bonds may achieved via conventional methods by using free radical processes, controlling the site selectivity either by intramolecular hydrogen abstraction or by the relative strengths of the C – H bonds [15]. Presently following the principles of Green chemistry, scientists are committed to shield the environment through the judicious choice of synthetic methods. Therefore C–C bond formation reactions are now achieved in single step via transition metal catalyst-mediated coupling (C–H activation) instead of multistep conventional coupling reaction.

The primary functionalization approach emphasis on experimental processes for comparatively simple hydrocarbons, but recently, C – H activation/functionalization has been improved to the point where it is projected as a viable method for the synthesis of targeted diverse molecular scaffolds. Hence, the focus of C–H activation has been progressed to the development of influential novel synthetic approaches and their wide application in the field of heterocyclic synthesis [16,17].

In the present chapter, transition metal-mediated syntheses viz. representative Ruthenium, Iron-catalyzed and specifically Rhodium salt-catalyzed reactions are explained along with their mechanism. A well-known Rhodium catalyst named Wilkinson's catalyst (Chlorotris(triphenylphosphine)rhodium (I)) is a homogeneous hydrogenation catalyst and is also used in directed C–H activation reactions [18]. In various metal-catalyzed C–H functionalization reactions, Rhodium complexes are used due to their high reactivity and selectivity [19–23]. The accessibility of large number of rhodium salts in different oxidation states facilitates the diversity in reaction patterns to be obtained. Therefore the metal-catalyzed synthesis of numerous polycyclic heterocycles along with their mechanisms is discussed in this chapter.

5.2 Principle of C–H activation

In organic synthesis, the C–C bond formation via C–H activation is an important coupling reaction for the synthesis of diverse heterocyles (Fig. 5.1). In C–H functionalization a carbon–hydrogen bond is cleaved and it will be exchanged by a new C–M bond which is helpful for the formation of C–C bond in coupling reactions. Therefore carbon-carbon bond formation reactions which are achieved in number of steps through conventional cross

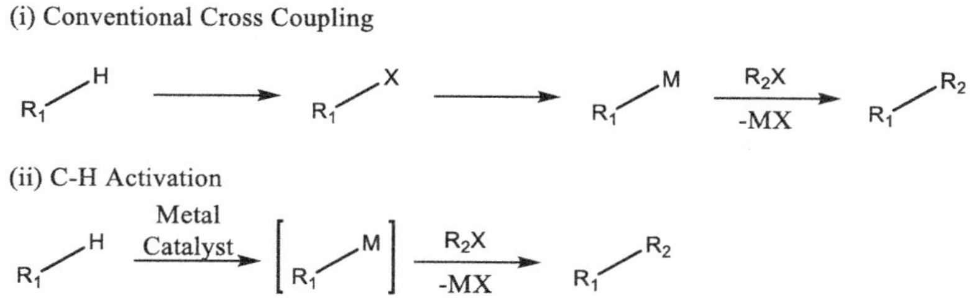

(i) Conventional Cross Coupling

$$R_1\text{—H} \longrightarrow R_1\text{—X} \longrightarrow R_1\text{—M} \xrightarrow[-MX]{R_2X} R_1\text{—}R_2$$

(ii) C–H Activation

$$R_1\text{—H} \xrightarrow{\text{Metal Catalyst}} [R_1\text{—M}] \xrightarrow[-MX]{R_2X} R_1\text{—}R_2$$

FIGURE 5.1 Conventional and C–H activation approach for C–C coupling reaction.

coupling are now achieved in single step via transition metal catalyst mediated coupling (C–H activation).

5.3 Method of preparation and mechanisms involved in C–H activation: recent applications in organic synthesis

5.3.1 Ruthenium-catalyzed C–H activation

Sandip Dhole and coworkers accomplished the regioselective synthesis of N-substituted-2-aminoisocoumarinoselenazoles by the direct insertion of internal alkyne (by C4-H and C6-H activation) on N-substituted-2-aminobenzoselenzaole-5-carboxylic acid by using Ru(II) catalyst [16]. 2-Amino benzoselenazoles **3** were synthesized by the cyclocondensation of isoselenocyanates with methyl 3-amino-4-fluorobenzoate (**2**) in the presence of pyridine as a base. Methyl 3-amino-4-fluorobenzoate (**2**) was obtained by the esterification and reduction reaction of 4-fluoro-3-nitrobenzoic acid (**1**) in the same solvent (methanol). Isoselenocyanates, were prepared by following a three-step procedure from amine and ethyl formate. N-Alkyl formamide, obtained by the reaction of alkyl amine and ethyl formate, was converted to isonitrile which yielded alkyl isoselenocyanate in refluxing dichloromethane using selenium powder at room temperature (Scheme 5.1).

In the next step, the alkylation of methyl 2-(alkylamino)benzoselenazole-5-carboxylate (**4**) with alkyl halide was carried out in basic medium. Now, as there are two sites which are available for alkylation in 2-amino-1,3-benzoselenazole, therefore alkyl group can either occupy the position-2 (**6**) or N-3 (**7**) of selenazole. In case of 2-aminobenzothiazoles, it was observed that the N-alkylation occurs on more basic endocyclic nitrogen (**6**). Ratio of regioselective products depend on nature of solvents and bases. Further, alkyne insertion was carried out using the Ruthenium catalyst [Ru(p-cymene)Cl$_2$]$_2$ (p-cymene = p-methyl isopropylbenzene) and the cyclized isocoumarinoselenazoles (**8–9**) were obtained within 4 hours. Both symmetric and unsymmetrical alkynes were used in the substrate scope. The addition of molecular oxygen as cocatalyst is crucial for the reaction and the reaction did not proceed without the presence of cocatalyst.

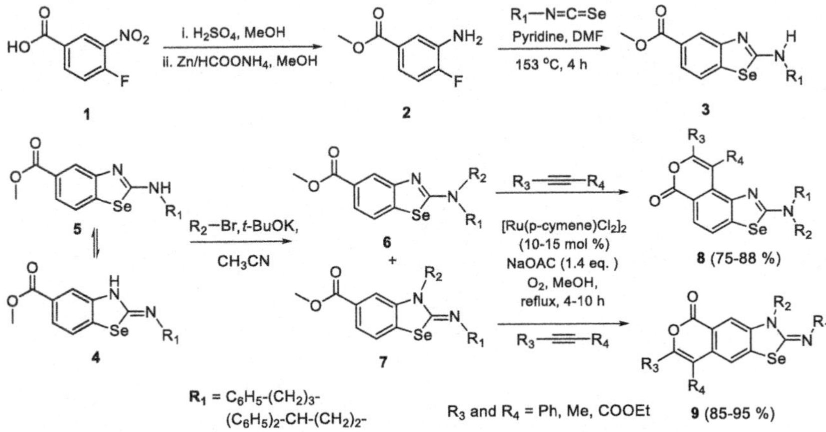

SCHEME 5.1 Regioselective synthesis of substituted 2-amino isocoumarinoselenazoles.

A proposed plausible mechanism for the reaction of *N*-substituted-2-aminobenzoselenazole-5-carboxylic acid (**10**) and alkyne is shown in Scheme 5.2. The complex formation of selenazole nitrogen to Ru followed by *ortho* C−H activation generated 5-membered ruthenacycle **11** and released two molecules of ethanoic acid. Further, alkyne insertion into the Ru-C of **11** produced the 7-membered ruthenacycle **12**. Protonation of **12** by acetic acid afforded **8** and regenerated Ru(II) species for the afterward reaction. Similarly, mechanism was proposed for the synthesis of linear isocoumarinselenazoles from another isomer **7** (Scheme 5.2).

5.3.2 Iron-catalyzed C−H activation

Nakamura et al. explored the Iron-catalyzed [2 + 2 + 2] annulation reaction of Grignard reagent **13** with alkynes through C−H activation under oxidative conditions at 0°C and provided an efficient route for polysubstituted naphthalenes [17]. The key coupling of phenylmagnesium bromide with diphenylacetylene was carried out by using Fe catalyst using 10 mol% 1,10-phenanthroline as ligand. The 1,2-dichloroisobutane (DCIB) was used as an oxidant in tetrahydrofuran solvent and was found to enhance the coupling of the **13** and diphenylacetylene to afford 1,2,3,4-tetraphenylnaphthalene (**14a**) in 73% yield at 0 °C (Scheme 5.3). A small amount of carbometalated product (**15**) was also obtained (Scheme 5.3). The electron-releasing and withdrawing substituents in substrates did not affect the reaction yields significantly, whereas diarylalkynes gave good yield as compared to dialkylalkynes.

In the proposed mechanism, a ferracycle (**17**) was generated via Fe(III)-catalyzed carbometallation of alkyne with arylmagnesium (**13**), followed by base-mediated C−H bond functionalization. Then, ferracycle **17** involved the insertion of another alkyne to afford another seven-membered ferracycle

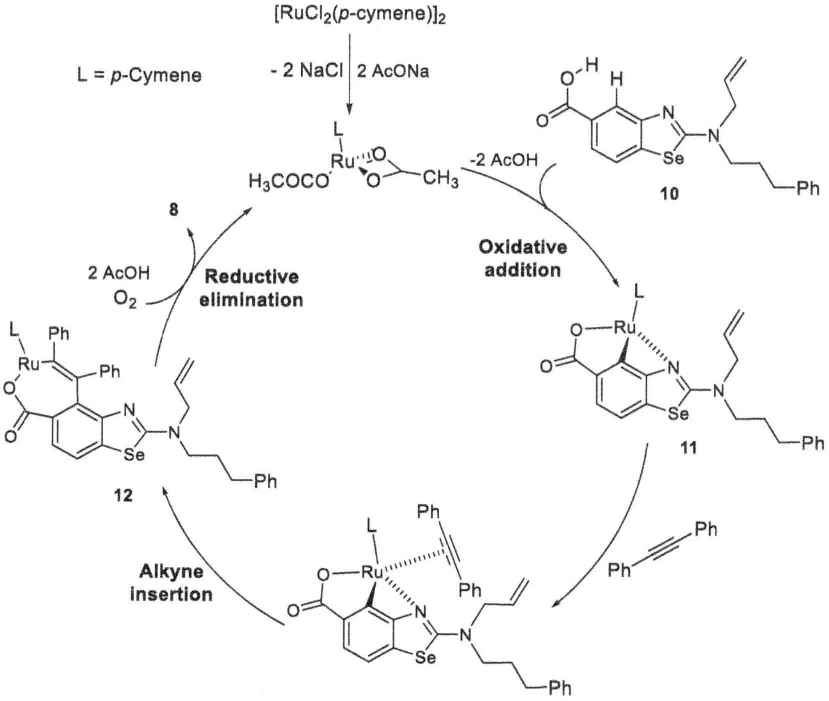

SCHEME 5.2 Proposed mechanism for Ru-catalyzed synthesis of isocoumarinoselenazoles.

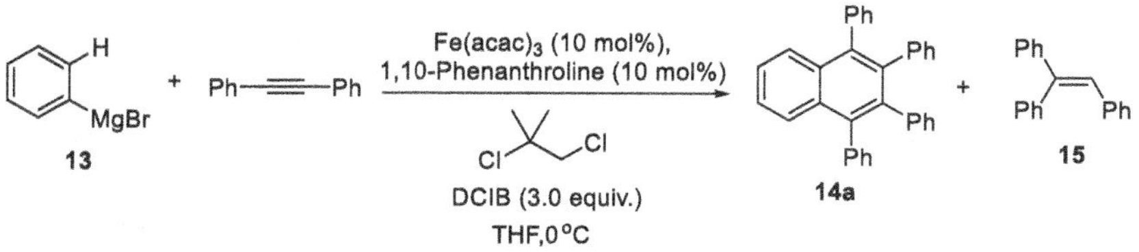

SCHEME 5.3 Synthesis of 1,2,3,4-tetraphenylnaphthalene via Fe-catalyzed [2 + 2 + 2] annulation.

intermediate **18**, which led to the formation of polysubstituted naphthalene (**14a**) after oxidatively induced reductive elimination and **15** was obtained by the addition reaction of **13** with diphenylacetylene (Scheme 5.4).

5.3.3 Rhodium-catalyzed C−H activation

Zhai et al. described the synthesis of various isoquinolines with the help of easily attainable α-chloro ketones [24]. The current approach was found to have broad functional group tolerance and using air as green oxidant rather than metals which was the limitation over other available methods.

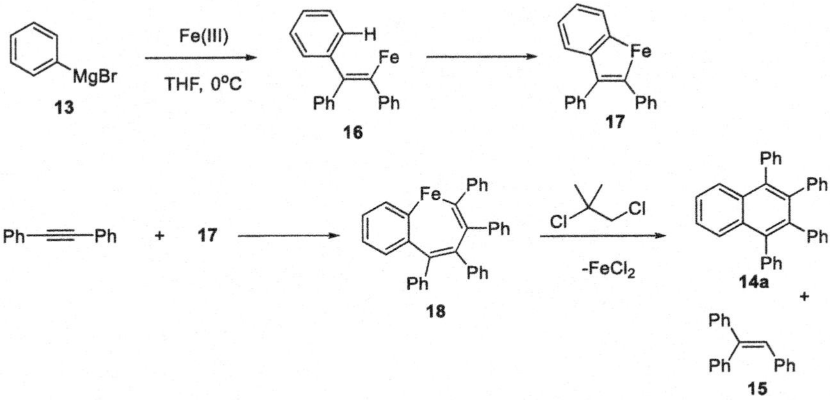

SCHEME 5.4 Iron-catalyzed [2 + 2 + 2] annulation reaction of alkyne and Grignard reagent (**13**).

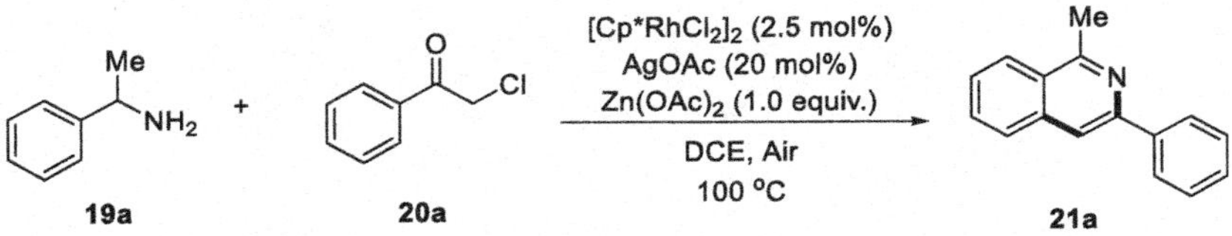

SCHEME 5.5 Rh-catalyzed synthesis of aryl-substituted isoquinolines.

The reaction was found to produce optimum results by the reaction of 1-phenylethanamine (**19a**) with α-chloroacetophenone (**20a**) using Rhodium catalyst [Cp*RhCl$_2$]$_2$, silver acetate and sodium acetate in dicholoroethane at 100°C in the presence of air. The additive Zn(OAc)$_2$ along with DCE solvent was found optimized being no changes with temperature. *Ortho-* and *para-* substituted benzylamine afforded good yields while in case of *meta*-OMe and F substitution two regioisomeric products have been furnished with good selectivity (Scheme 5.5).

Based on kinetic isotopic studies a plausible mechanism was presented. Initially, there is a formation of active species Cp*Rh(OAc)$_2$ using the precursor [Cp*RhCl$_2$]$_2$ and AgOAc by the mode of anion exchange (Scheme 5.6). *In situ* generated imine **22** underwent cyclometalation by oxidation to produce a five-membered rhodacyclic intermediate **23**. Then, oxygen atom from α-chloro ketone (**20a**) gets coordinated to the intermediate **23** to proceed toward an intermediate **24**. In **24**, methylene group from α-Cl ketone underwent nucleophilic attack by C$_{aryl}$-Rh to give α-aryl ketone species **25**. Eventually, the intramolecular attack of NH group on keto-group with the help of Cp*Rh(OAc)$_2$ resulted into intermediate **26** which yielded isoquinoline (**21a**) after dehydration process.

Chaudhary and coworkers explored the rhodium(III) complex assisted C−H activation during the reaction of aldimines (**27**) and benzimidates with

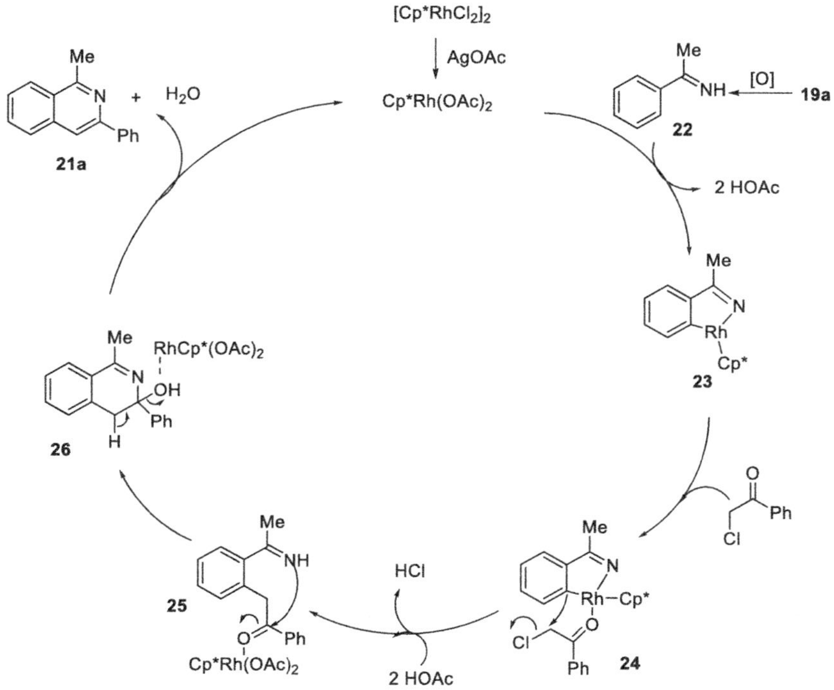

SCHEME 5.6 Mechanism for Rh-catalyzed synthesis of isoquinolines.

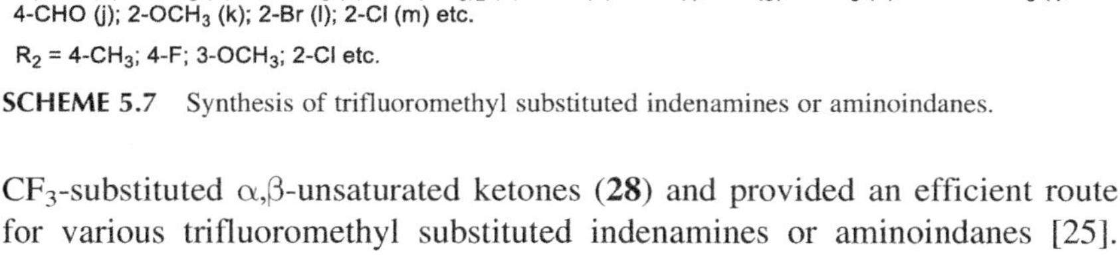

R$_1$ = H (a), 4-CH$_3$ (b); 4-OCH$_3$ (c); 4-N(CH$_3$)$_2$ (d); 4-Cl (e); 4-Br (f); 4-F (g); 4-CF$_3$ (h); 3-F-4-OCH$_3$ (i); 4-CHO (j); 2-OCH$_3$ (k); 2-Br (l); 2-Cl (m) etc.

R$_2$ = 4-CH$_3$; 4-F; 3-OCH$_3$; 2-Cl etc.

SCHEME 5.7 Synthesis of trifluoromethyl substituted indenamines or aminoindanes.

CF$_3$-substituted α,β-unsaturated ketones (**28**) and provided an efficient route for various trifluoromethyl substituted indenamines or aminoindanes [25]. The key coupling of the substrates benzimidate (**27**) and trifluoromethyl substituted α,β-unsaturated ketone **28** was optimized using rhodium catalyst and silver salts as additive. The combination of AgSbF$_6$ (10 mol%) and silver acetate as additive was found to promote the coupling of **27** and **28** by 80% in ethyl acetate (Scheme 5.7). Under the optimized C−H annulation condition, various electron-releasing and withdrawing groups at *meta*- and *para*- positions in the aromatic group of trifluoromethyl substituted enones

122 Recent Advances in Organometallic Chemistry

(**28a-i**) underwent coupling reaction resulted in the formation of desired **29a-i** in good yields. Because of steric bulk, moderate yields (**29j-m**) were obtained in case of *ortho*-substituted enones. In similar way, various electron-releasing and withdrawing groups were explored at the *meta* or *para*-positions of benzimidates and obtained the products in good yields.

In the proposed mechanism, active Rh(III) species were generated *in situ* in the presence of AgSbF$_6$ and coordinated to imine and afterward underwent C−H activation to provide five-membered rhodacycle **30**. In the next step, olefinic insertion of trifluoromethyl substituted enones **28a** generated rhodacycle **31**. Further intramolecular addition of imine takes place to afford rhodium complex **32**, which on protonation generates the intermediate **33** and regenerates the rhodium catalyst for further reaction. Subsequent 1,3-hydrogen shift in **33** provided the trifluoromethyl substituted indenamine **29** (Scheme 5.8).

Gi Hoon Ko et al. reported the synthesis of isochromenoindolones **37** using rhodium-catalyzed C−H activation involving the streamline formation of C−H and C−O bonds in one pot [26]. This technique was used to synthesize various isochromenoindolones along with the liberation of indolone-linked N-methoxybenzamide **36** and nitrogen gas (Scheme 5.9).

The optimization studies showed that alkylated product **37a** was found to have good yield by heating the combination of *N*-methoxybenzamide **34** (1 equiv.) and 3-diazooxinodole **35** (1.2 equiv) in the presence of rhodium

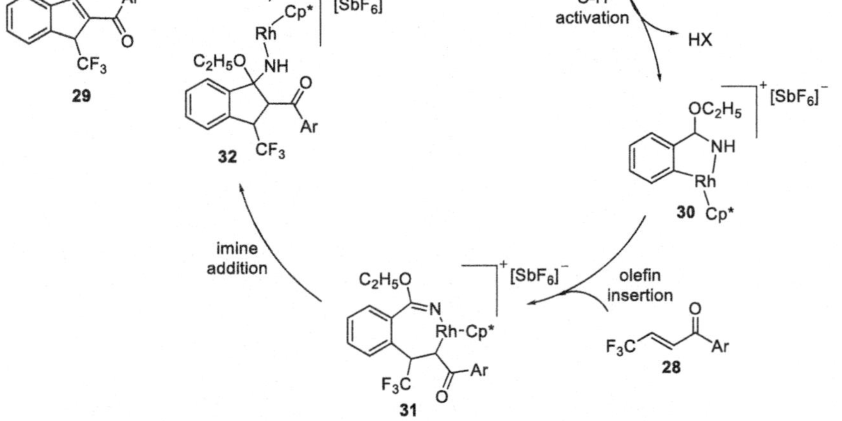

SCHEME 5.8 Proposed mechanism for the synthesis of indenamines or aminoindanes.

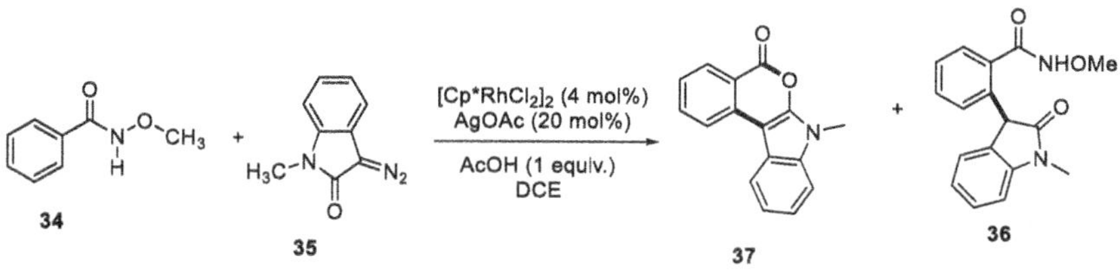

SCHEME 5.9 Synthesis of isochromenoindolones.

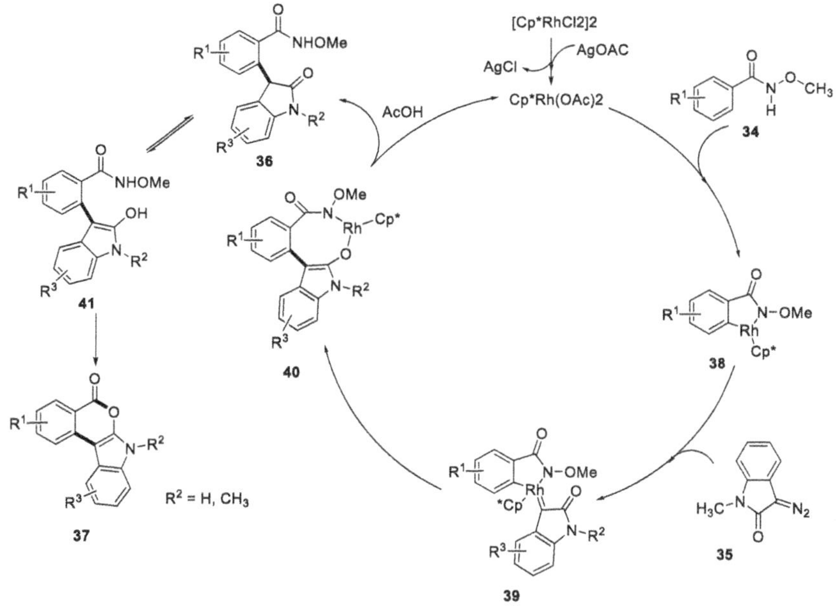

SCHEME 5.10 Proposed mechanism for the synthesis of isochromenoindolones (**41**).

catalyst [Cp*RhCl$_2$]$_2$, silver acetate and acetic acid using DCE as solvent at 40°C for 2 hours and afforded **37** in 65% yields. Further, the selective cyclization reaction for the formation of isochromenoindolone was optimized with the combination of **34a** with **35a** using benzoic acid (1 equiv.) by heating at 100 °C for 24 hours. N-Methoxybenzamide containing electron-releasing groups at positions 2 and 3 gave better yield rather than at position 4 where steric bulk dominates. Electron-withdrawing groups in **34** are also compatible for the reaction. Further, the alkyl, aryl, or aralkyl substituents on 3-diazo-oxindoles (**35**) were also consistent to deliver the desired products.

In mechanism, the rhodium species [Cp*Rh(OAc)$_2$] coordinated with 3-diazooxindole **34a** to provide rhodacycle **38** (Scheme 5.10). Then, molecular nitrogen gets released after the coordination of 3-diazooxindole to afford Rh-carbenoid intermediate **39**. In the next step, carbene undergoes migratory insertion of the Rh-Ar bond thus generating the rhodacycle **40**. Protonolysis of **40** leads to regeneration of the active rhodium catalyst and formation of **36** which underwent nucleophilic addition (intramolecularly) leading to the formation of desired isochromenoindolones **37**.

Chen et al. reported first time Rh(III)-catalyzed route to prepare 3,4-dihydropyrimido[1,6-a]indol-1(2*H*)-one derivatives by using silyl-enol ethers with *N*-carbamoylindoles (Scheme 5.11) [27].

The optimum conditions for the annulation reactions were obtained using indole derivative **42a** and ether **43a** in presence of Rh(III) catalyst (2.5 equiv.) and silver acetate as an additive in methanol and afforded **44a** in 85% yields. In the reaction, the electron-releasing as well as electron-donating groups at positions C-3, 4, 5, and 6 of indoles and substituted silyl enol ethers underwent smooth coupling to furnish dihydropyrimidoindolones.

In the mechanism, the rhodium species [Cp*Rh(OAc)$_2$], generated from [Cp*RhCl$_2$]$_2$ and AgOAc coordinated with **42a** to afford five-membered rhodacycle **45** via concerted metalation deprotonation (CMD) mechanism (Scheme 5.12). In the next step, the key intermediate **46** was obtained through the cooperation of trimethylsilyl enol ether **43a**. The olefinic insertion of alkene takes place with α-metalation via addition by AcOH to afford intermediate **47**, which underwent oxidation by silver acetate to afford Rh(V) complex **48**. After C(sp$_2$)-C(sp$_3$) reductive elimination of **48**, the complex **49** and the active rhodium species were obtained. Finally, the nucleophilic attack of the NH group to the carbonyl group in **49** resulted in the production of indolone **44a**.

Yin et al. developed an empirical method using Rh(III)-catalyzed annulation of phthalazinones (**52a**) and pyridazinones along with allenes, which induced the production of quaternary carbon-bearing indazole derivatives with good yields [28]. The synthesis of targeted products was followed certain steps sequentially, such as C−H functionalization and olefin insertion, ensuing β-hydride elimination and consequently intramolecular cyclization. The present synthetic procedure includes a wide substrate scope and highly selective toward Z-isomer (Scheme 5.13).

For the optimization, the model substrates, phthalazine-1,4-dione **50a** and (buta-2,3-dien-1-yl)benzene (**51a**) were reacted with [Cp*RhCl$_2$]$_2$ and silver acetate in acetonitrile at 100°C for 12 hours and afforded **52a** with 88% yield.

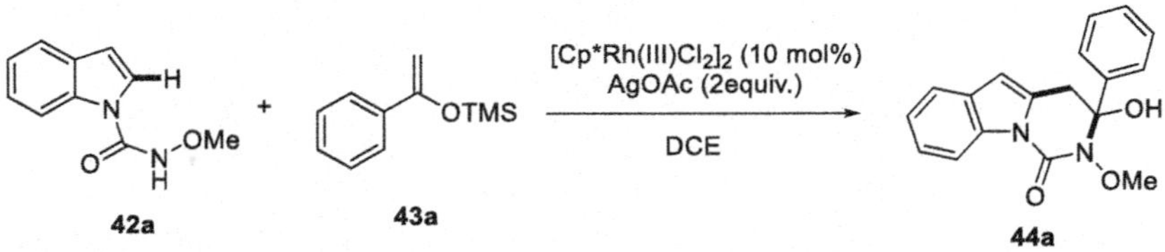

SCHEME 5.11 Rhodium assisted synthesis of pyrimidine fused indolones **44**.

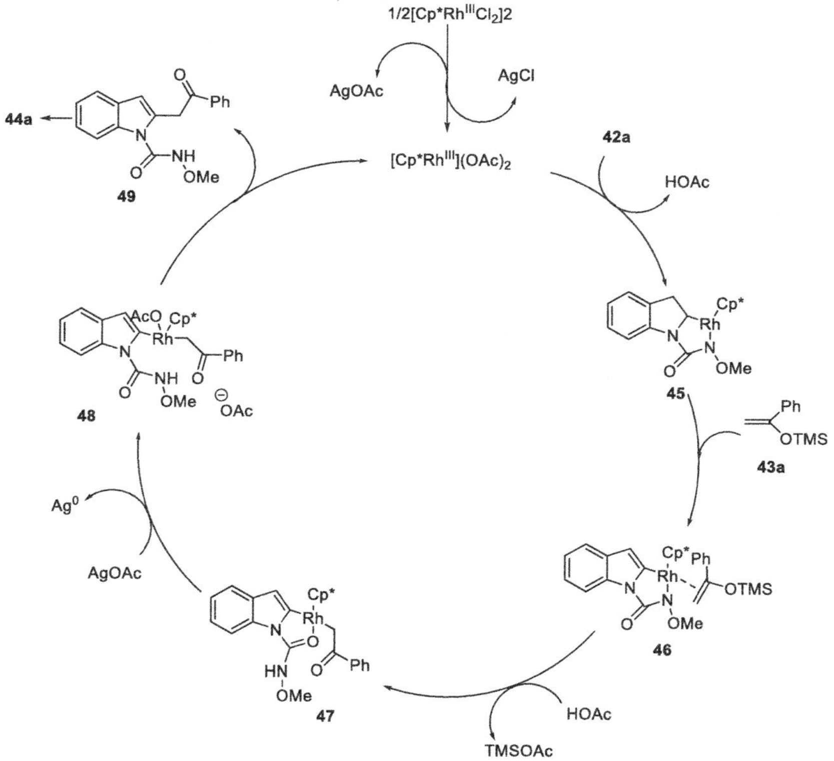

SCHEME 5.12 Proposed mechanism for the synthesis of 3,4-dihydropyrimido[1,6-a]indol-1(2H)-ones.

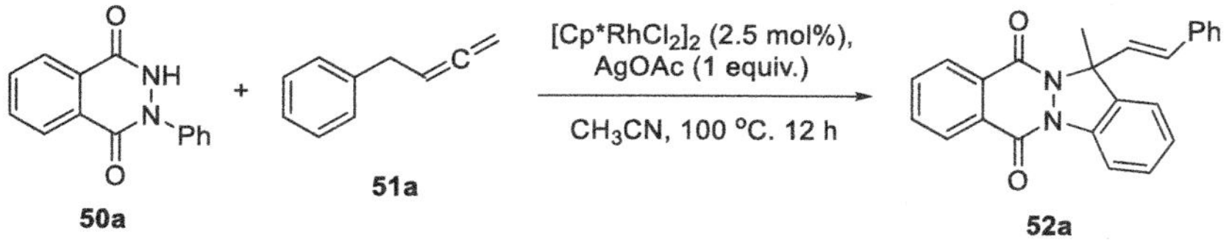

SCHEME 5.13 Rh-catalyzed preparation of indazole derivatives.

It was found that the yield was enhanced efficiently by increasing the temperature 120°C.

The study investigated that various electron-releasing or electron-withdrawing groups at various positions underwent [4 + 1] cyclization easily. Further, the fused cyclic products were separated in moderate to excellent yields. In addition, various electron-donating and withdrawing groups at position-2 of allenes gave desired product in excellent yield.

The mechanism for [4 + 1] annulation of N-aryl phthalazinones and pyridazinones with allenes under Rh(III) catalysis is shown in Scheme 5.14.

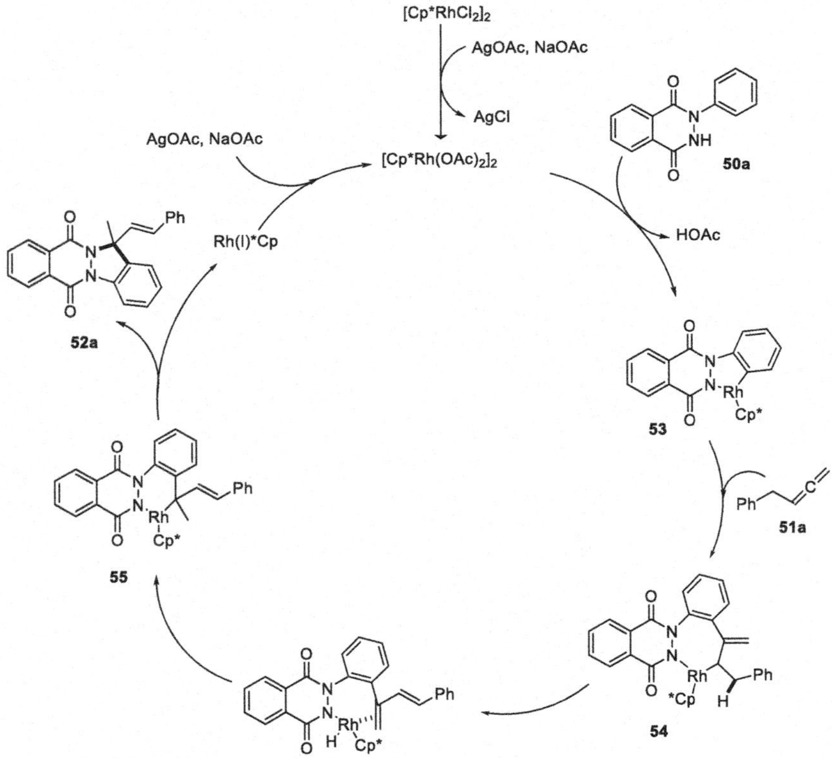

SCHEME 5.14 Rh-catalyzed synthesis of indazole derivatives.

In situ, via the ion exchange **50a** reacts with active rhodium of [Cp*RhCl₂]₂ and led to the formation of intermediate **53**. After that, allene **51a** get inserted and β-hydride elimination afforded intermediate **54**. In the next step, the Rh—H bond in intermediate **54** is inserted in the diene moiety to direct toward **55**, which then further gave **52a** after undergoing reductive elimination. The catalytic cycle get continued to give active rhodium catalyst by the use of silver acetate and sodium acetate.

Studies of Sheriker and coworkers established one-pot synthesis of trisubstituted furan derivative **59** by the use of (E)-4-phenylbut-3-en-2-one (**56**) and methyl acrylate (**57**) through C$_{(vinyl)}$—H activation [29]. Silver salt acts as Lewis acid catalyst as well as halide scavenger for getting a furan derivative by Paal—Knorr type cyclization (Scheme 5.15).

The Heck-type product was obtained by the reaction of (E)-4-phenylbut-3-en-2-one (**56a**) and methyl acrylate (**57a**) in the presence of [Cp*RhCl₂]₂ and AgSbF₆ as catalyst with oxidant and additives copper acetate and pivalolyl alcohol, respectively in dichloromethane. At the end, the intermediate **60** underwent Paal-Knorr type cyclization in the presence of the silver salt, and

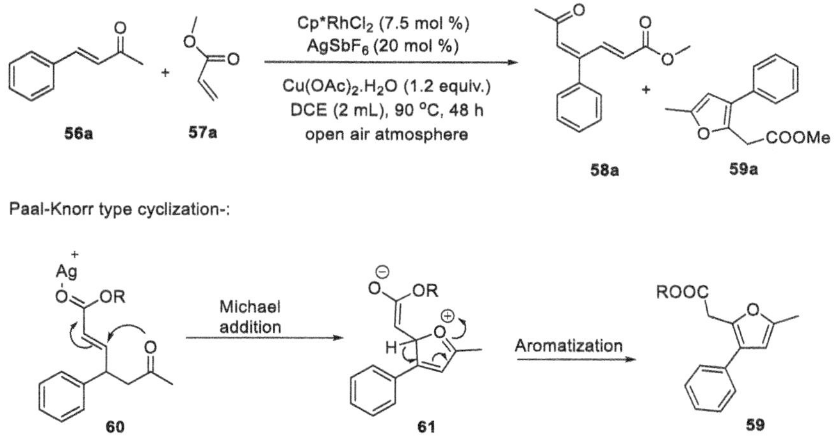

SCHEME 5.15 One-pot synthesis of tri-substituted furan derivatives.

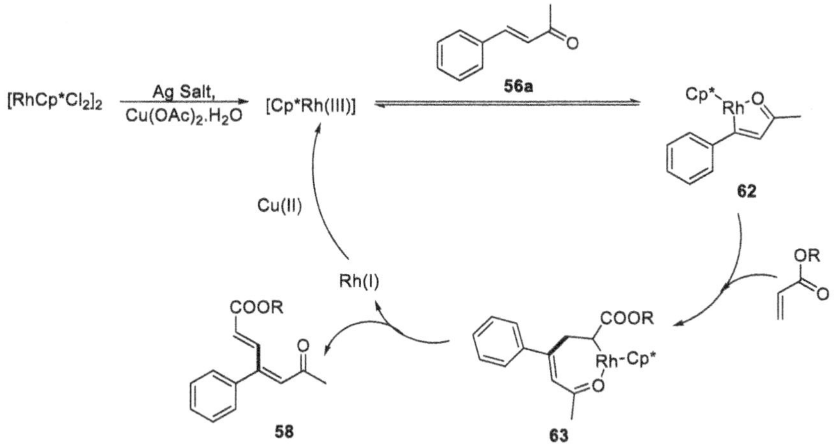

SCHEME 5.16 Rhodium-catalyzed C$_{(vinyl)}$–H activation.

resulted in the synthesis of furan derivative **59** through the intermediacy of unstable **61** (Scheme 5.15).

In the anticipated mechanism, the rhodium(II) catalyst generated active Rh(III) catalyst in the presence of AgSbF$_6$ and copper acetate. In next step, **56a** led to the formation of 5-membered rhodacycle **62**. Further, insertion of acrylate derivative produced the rhodium complex **63**. The oxidative Heck-type product **58** gets formed by β-hydride elimination of **63**. The copper acetate oxidizes Rh(I) to Rh(III) which completes the catalytic cycle (Scheme 5.16).

Wang and coworkers described the application of isoxazolones to synthesize biologically important compound 4-arylisoquinolines (**66**) and 4-arylisoquinolones

(**68**) by developing a Rh(III)-catalyzed regioselective C−H annulation of 3-arylisoxazolones (**64**) and propargyl alcohols (**65**) using double directing group activation approach [30].

The optimization studies revealed that catalyst [Cp*RhCl$_2$]$_2$ along with additive cesium acetate in the solvent DCE afforded 80% yields at room temperature. In the reaction, isoxazolone acted as first traceless directing group while −OH group of **65** upon interaction with metal catalyst acted as second directing group which assisted the regioselective synthesis of 3-hydroxyalkyl-4-arylisoquinoline derivatives (**66**). 4-Arylisoquinolones (**68**) were obtained via the same reaction using 3-aryl-1,4,2-dioxazol-5-one as starting precursor (Scheme 5.17).

In the substrate scope, the presence of electron-releasing and withdrawing groups in phenyl ring (*para*-position) of 3-arylisoxazolones (**64**) gave smooth reaction with good yields. Moreover, excellent regioselectivity with moderate yield was found in case of electron-releasing and withdrawing groups at *meta* and *ortho*-position in benzene ring. The existence of electron-withdrawing or -releasing group on phenyl ring in propargyl alcohols gave lesser yield as compared to the unsubstituted substrate.

In the proposed mechanism, N-atom of the **64a** coordinated to active catalyst CpRh(OAc)$_2$ and underwent C − H activation to provide the rhodacycle **69** (Scheme 5.18). In the next step, propargyl alcohol **65** underwent regioselective coordinative insertion into the Carbon − Rhodium bond of **69** to give seven-membered rhodacycle **70** which further get reductively eliminated to furnish intermediate **71** and active Rh(I) species. The intermediate **71** upon oxidation by the N-O bond produced rhodium(III) species **72**, which was protonated by acetic acid to afford 4-arylisoquinoline **66a** along with regeneration of the Rh(II) catalyst.

This methodology proved its synthetic utility by undergoing several transformations. For instance, reaction of disubstituted isoquinoline **66a** with

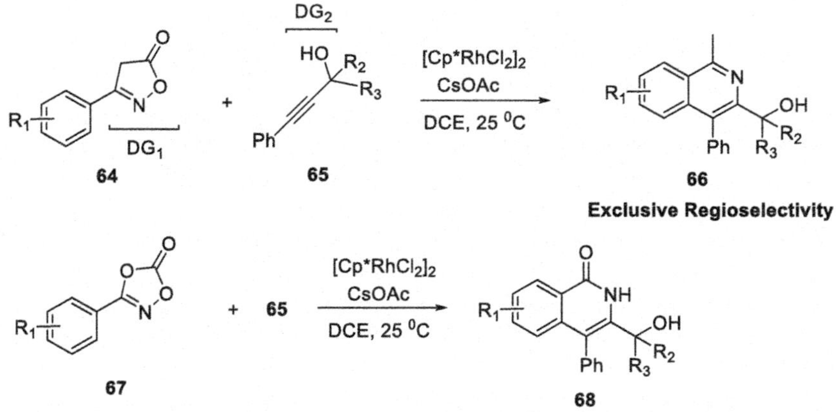

SCHEME 5.17 Regioselective synthesis of 4-arylisoquinolines and 4-arylisoquinolones.

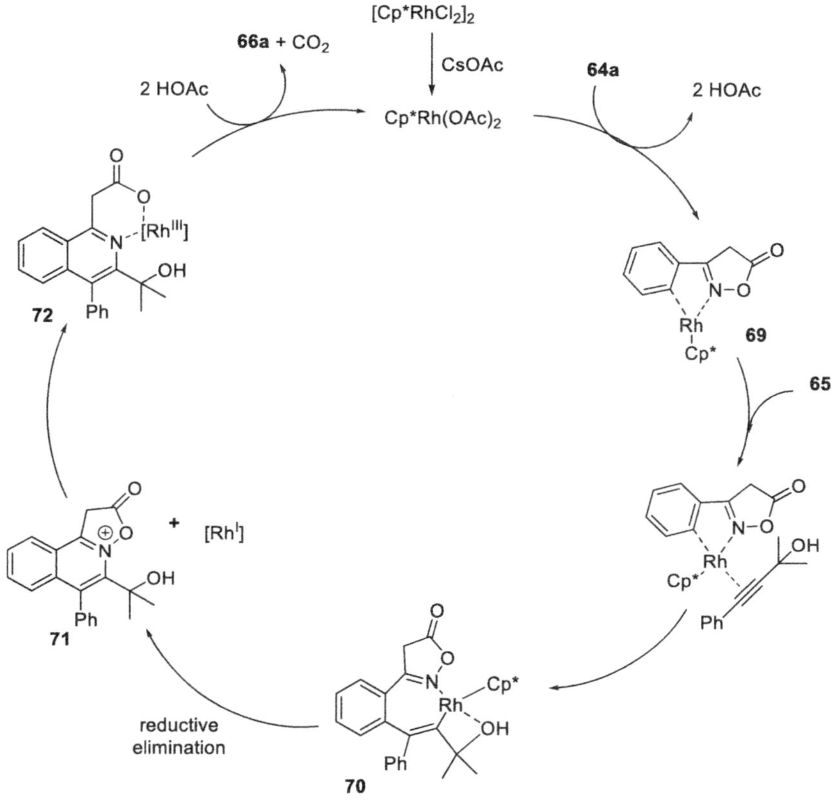

SCHEME 5.18 Proposed mechanism for the regioselective synthesis of 4-arylisoquinolines.

Lewis acid $BF_3 \cdot OEt_2$ afforded polycyclic isoquinoline derivative **73** (91% conversion). Further, N-oxide **74** was achieved by m-chloroperoxybenzoic acid mediated oxidation of **66a** in dichloromethane (96% yield). The cyclization of **68a** was obtained by using $BF_3 \cdot OEt_2$ in 1,2-dichloroethane to afford **75** (80% yield) (Scheme 5.19).

Zhong et al. reported Rh(III)-catalyzed C−H imidization/cyclization by the use of isoxazolones (**77**) as imidizating reagents for the preparation of dihydroquinazolin-4(1H)-ones (**78**) [31]. The process occurs with the assistance of a quaternary carbon and involves one-pot simultaneous construction of two new C−N bonds (Scheme 5.20).

The reaction was initiated by coupling reaction of N-methoxybenzamide (**76**) with isoxazol-5(4H)-one (**77**) and the optimization studies showed that the metal catalyst $[Cp^*RhCl_2]_2$ and sodium acetate along with polar protic solvent trifluoroethanol and base (KOPiv) as additive afforded the cyclized products in good yields. *Ortho*-substituted N-methoxybenzamides furnished lesser efficient yields while employment of *meta*-substitution gave

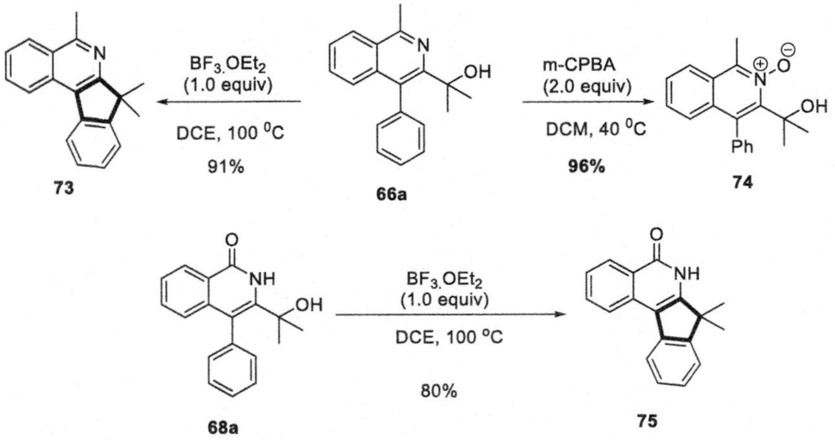

SCHEME 5.19 Synthetic application of 4-arylisoquinolines.

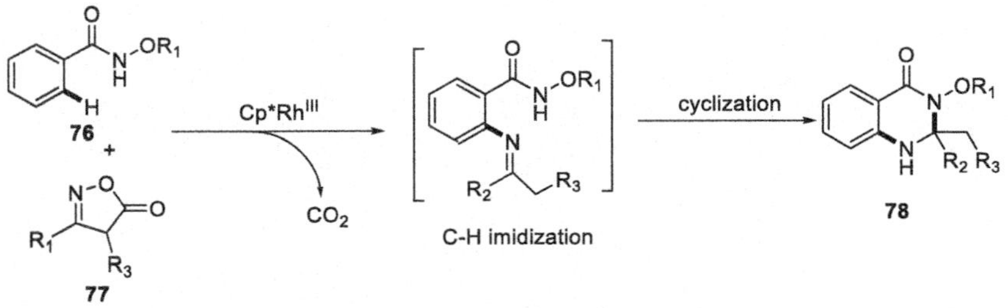

SCHEME 5.20 Rh(III)-catalyzed synthesis of dihydroquinazolin-4(1*H*)-one.

regioselective C—H metalation depending upon the nature of the substituents. On the other hand, C-3 position of isoxazol-5(4*H*)-ones having linear, branched alkyl and cycloalkyl substituents afforded good yields. The Gibbs free energy changes from the computational calculations put forward the sequential steps as well as the Rh(V) nitrenoid and imine as intermediates.

The proposed mechanism for the transformation gets started with the generation of Cp*Rh(OPiv)$_2$ by anion exchange mode, which acts as a mediator for reversible C—H/N—H cleavage to afford **79** from *N*-methoxybenzamide (**76a**). Isoxazolone (**77a**) gets coordinated with **79** following the oxidative addition and N—O bond cleavage and resulting in the zwitterionic Rh(V) intermediate **80**. Further, **80** get decarboxylated to proceed toward Rh(V) nitrenoid intermediate **81** and reductive elimination of **81** provided the intermediate **82**, a C—H amination product. The intermediate **82** get isomerized by acetic acid to generate intermediate **83**. Finally, the desired product (**78a**) was obtained by intramolecular nucleophilic addition (Scheme 5.21).

Deng et al. reported rhodium-catalyzed [4 + 2]-cycloaddition method for the selective construction of functionalized *cis*-dihydrobenzimidazo[2,1-a]

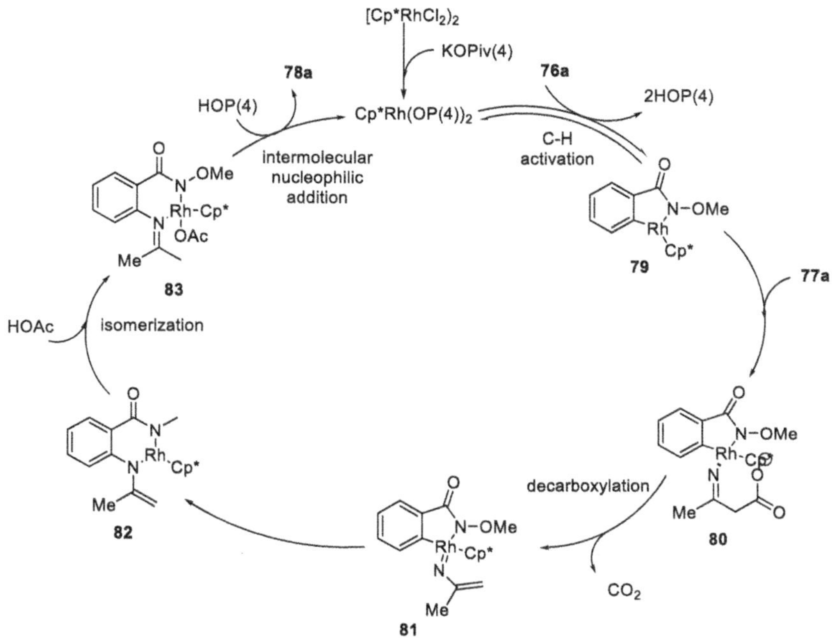

SCHEME 5.21 Proposed mechanism for the oxidant-free synthesis of dihydroquinazolin-4(1H)-one.

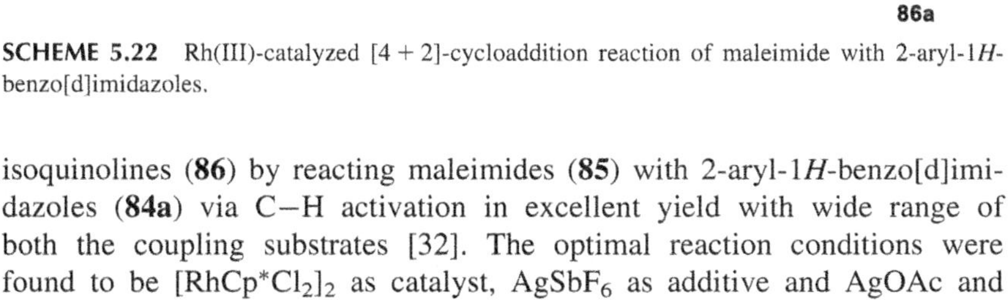

SCHEME 5.22 Rh(III)-catalyzed [4 + 2]-cycloaddition reaction of maleimide with 2-aryl-1H-benzo[d]imidazoles.

isoquinolines (**86**) by reacting maleimides (**85**) with 2-aryl-1H-benzo[d]imidazoles (**84a**) via C—H activation in excellent yield with wide range of both the coupling substrates [32]. The optimal reaction conditions were found to be [RhCp*Cl$_2$]$_2$ as catalyst, AgSbF$_6$ as additive and AgOAc and CsOAc as oxidant under nitrogen atmosphere in DCE under refluxing temperature for 12 hours (Scheme 5.22).

No electronic preference was shown by 2-aryl-1H-benzo[d]imidazoles (**84a**) containing different electron functional groups in the aromatic ring as it gave corresponding cyclized products in 52%–85% yields. Steric hindrance dominates upon the introduction of methyl group at *ortho*-position.

The reaction was gram-scale reaction with cyclic product (**86a**) in 71% yields. Further, maleimides with all protecting groups were found to undergo smooth cyclization with good yields. Upon shedding light on its mechanism, it was found that the active Rh(III) catalyst obtained by the reaction of rhodium catalyst [Cp*RhCl$_2$]$_2$ with additives AgSbF$_6$ and cesium acetate activates C−H bond of 2-aryl-1*H*-benzo[d]imidazole (**84a**) to afford five-membered rhodacycle intermediate **87**. Further, the intermediate **87** upon coordinating with maleimide (**85**) gave a coordination intermediate which underwent 1,2-migratory insertion and afforded **88**. Further reductive elimination of intermediate **88** gave desired product **86** along with Rh(I) species. The Rh(I) species get oxidized by AgOAc and regenerated Rh(III) species which completed the cycle (Scheme 5.23).

Lincong and coworkers proposed that under redox-neutral and acid/base-neutral conditions, chiral Rh (III) catalyst act as an efficient catalyst for the asymmetric [4 + 1] spiro annulation of O-pivaloyloximes (**89**) and α-diazo homophthalimides (**90**) with the help of Cramer-type Cp^{X}Rh(III) catalyst [33]. Further, resulting from N-O cleavage and C−H activation, it led to enantioselective synthesis of chiral spirocyclic imines. During optimization studies, an enantioselective formation of product (R)-**91** was achieved when (R)-Rh(II) was used in DCE (90%−92%ee) as compared to other solvents, such as chlorobenzene, TFE, ethyl acetate, and diethylether (Scheme 5.24).

Under the optimal conditions, an effective enantioselective synthesis of R-**91** was obtained using the electron-withdrawing, releasing, and halogen

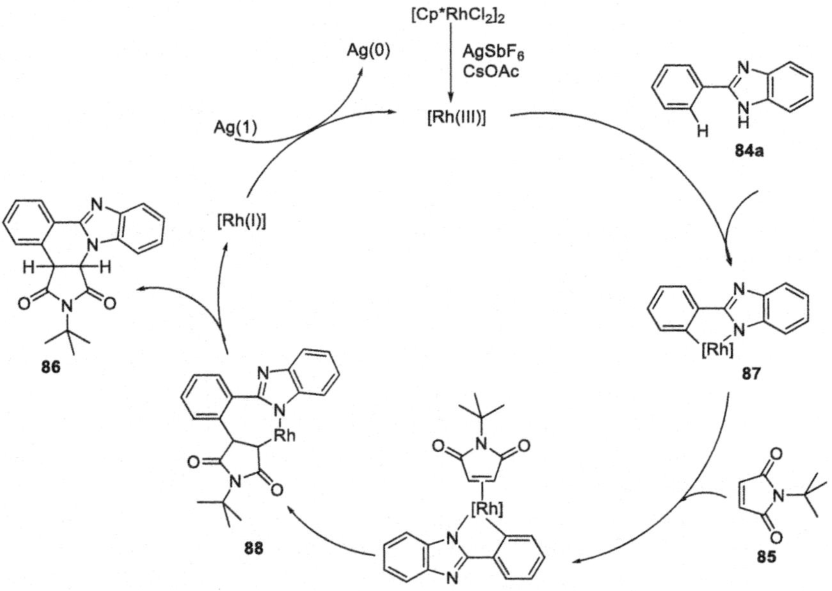

SCHEME 5.23 Proposed mechanism for Rh-catalyzed [4 + 2]-cycloaddition reaction of **84** and **85**.

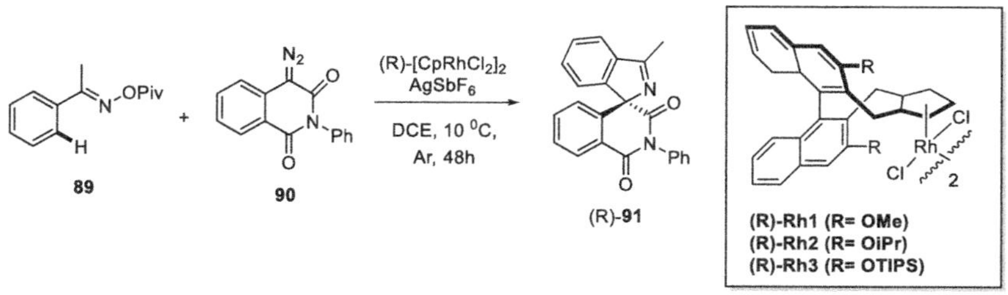

SCHEME 5.24 Asymmetric [4 + 1] spiro annulations of O-pivaloyl oximes with α-diazo compounds.

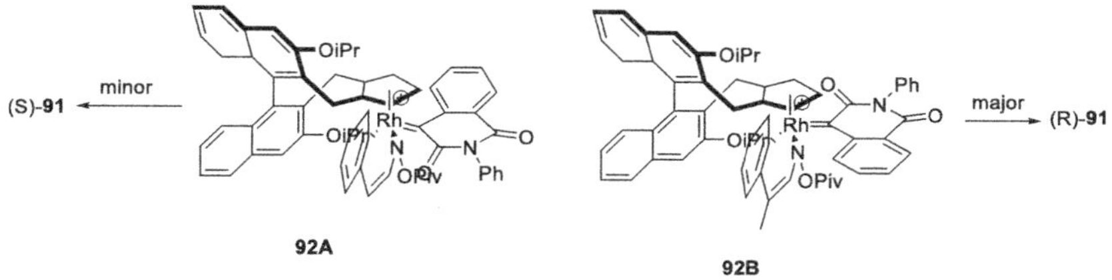

SCHEME 5.25 Proposed intermediates in asymmetric [4 + 1] spiro annulation reaction.

groups at *para*-position of phenyl ring from O-pivaloyloximes (**89**) and N-phenyl ring of **90** (87%–96%ee). In comparison to that, C–H annulation having slightly lesser enantioselectivity and reduced regioselectivity toward R-**91** was obtained using *ortho*-fluorine substituted O-pivaloyloxime while alkyl substituents causing no decrease in enantioselectivity. The enantioselectivity for the (R)-**91** formation may be explained by the stereochemical model. After C–H bond activation and carbene formation, both intermediates **92A** and **92B** are possible (Scheme 5.25). The conformation **92B** was more preferred for migratory insertion due to less steric hindrance between the O-Piv group and the carbene ligand.

This method was of greater synthetic utility as well as its product upon oxidation gave enantiopure *N*-oxide (**93**) in good yield while upon reduction with Pd/C afforded ring expansion product **94**. Further, the reduction of **91** with sodium borohydride in ethanol caused ring opening of imide and afforded product **95** in 55% yields (Scheme 5.26).

5.4 Future prospective and scope

From the literature reports, in terms of enhancement in catalytic properties, oxidatives, and solvents suggest that there are likely still gains to be found in the effectiveness of this system. The high cost of Ru and Rh complexes is one of the major drawbacks in such type of synthetic processes and needs cost-effective catalysts. The recognition of better systems which can be used in industry with cost-effective technique is highly awaited.

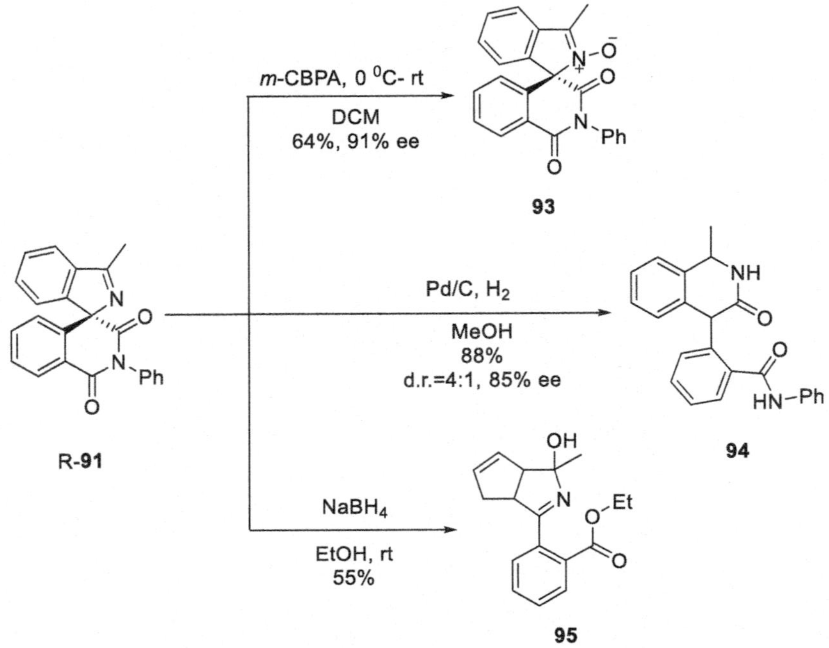

134 Recent Advances in Organometallic Chemistry

SCHEME 5.26 Oxidation and reduction reactions of R-**91**.

The chapter is of particular interest for academician as well as scientists from research and development sector working in the field of transition metal catalysis. The overall goal of report is to promote the application of C−H activation in the synthesis of polycyclic aromatic heterocycles via transition metal-catalyzed synthesis.

5.5 Summary

The carbon-carbon bond formation is an important coupling reaction for the construction of polycyclic aromatic heterocycles, linear molecules as well as natural products or their analogs. Therefore numerous organic name reactions for C−C bond formation viz. aldol reaction, Corey House synthesis, Negishi Coupling, alkane metathesis, Friedal Crafts reaction, Sonogashira coupling, Stille reaction, etc., were discovered from time to time in organic synthesis. Among these, C−H activation is one of the major interesting approaches which have attracted the interest of chemists in the last decade. The usage of catalysis in the expansion of organic molecules library has transformed the area of organic synthesis over the past decade. Generally the coupling of two prefunctionalized substrates afforded the products in highly selective and anticipated manner in the presence of transition metal catalysts.

Therefore transition metal catalysts were implemented as convenient tools in synthetic and medicinal chemistry due to their diverse reactivity pattern in facilitating various types of conversions. Nowadays, owing to lower electronegativity and Lewis acidity of main group, metals and metalloids exhibit coordination behavior and electronic properties and therefore act as supporting ligands in catalysis.

The functionalization of nonactivated C−H bonds may achieved via conventional methods by using free radical processes, controlling the site selectivity either by intramolecular hydrogen abstraction or by the relative strengths of the C−H bonds. Presently following the principles of Green chemistry, C−C bond formation reactions are achieved in single step via transition metal catalyst mediated coupling (via C−H functionalization) instead of multistep conventional coupling reaction. The primary approach emphasis on experimental processes for the functionalization of common

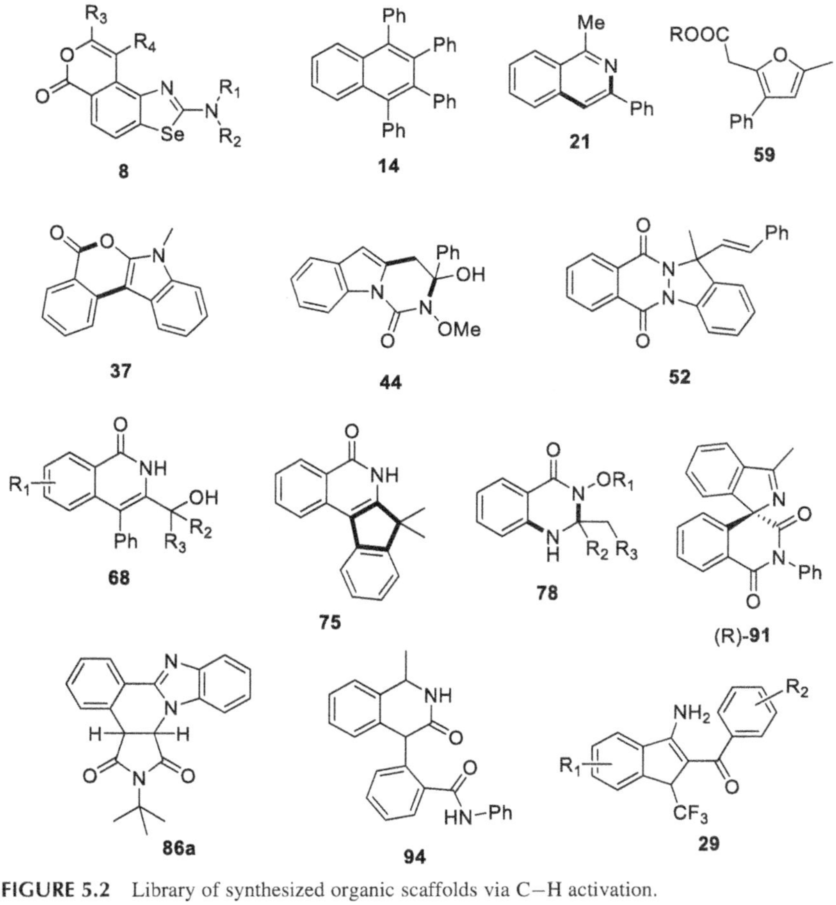

FIGURE 5.2 Library of synthesized organic scaffolds via C−H activation.

hydrocarbons, but recently, C−H activation/functionalization has been grown to the point where it can be projected as a viable approach for the synthesis of targeted diverse molecular scaffolds (Fig. 5.2). Hence, the focus of C−H functionalization has been moved to the development of influential novel synthetic approaches and their wide application in the field of heterocyclic synthesis.

From the above-discussed literature reports, C−H activation reactions are worth for the synthesis of heterocyclic scaffolds via C−C bond formation coupling reactions. The metal catalyst is used as catalyst which coordinated with substrate and after a sequential manner it was regenerated and further used to complete the reaction. In this chapter various transition metal-catalyzed reactions viz. Ruthenium, Iron, and specifically Rhodium salt-catalyzed reactions are explained along with their mechanism. These transition metal catalysts are used in synthesis of various heterocycles, such as *N*-substituted-2-aminoisocoumarinoselenazoles, polysubstituted naphthalenes, isoquinolines, indenamines/aminoindanes, isochromenoindolones, 3,4-dihydropyrimido[1,6-a]indol-1(2*H*)-ones, indazole derivatives, furan derivatives, and 4-arylisoquinolines/4-arylisoquinolones, and also used in various asymmetric [4 + 1] spiro annulation reactions (Fig. 5.2). Therefore, the metal-catalyzed C−H activation reactions are worth to synthesize various polycyclic heterocyclic scaffolds via C−C bond formation coupling reactions. This chapter will be certainly helpful for chemists working in the field of C−H functionalization for the synthesis of newer heterocyclic scaffolds.

References

[1] Varuna BVK, Kumar JD, Bettadapura KR, Siddaraju Y, Alagiri K, Prabhu KR. Recent advancements in dehydrogenative cross coupling reactions for C-C bond formation. Tetrahedron Lett 2017;58:803−24. Available from: https://doi.org/10.1016/j.tetlet.2017.01.035.

[2] Wang Z, Sun L, Tang J, Shi W, Slowing II. Ppm scale Pd catalyst applied in aqueous Sonogashira reaction. Tetrahedron Lett 2022;107. Available from: https://doi.org/10.1016/j.tetlet.2022.154107.

[3] Cattelle AD, Billen A, O'Rourke G, Brullot W, Verbiest T, Koeckelberghs G. Ligand-free, recyclable palladium-functionalized magnetite nanoparticles as a catalyst in the Suzuki-, Sonogashira, and Stille reaction. J Organomet Chem 2019;904. Available from: https://doi.org/10.1016/j.jorganchem.2019.121005.

[4] Anastas PT, Warner JC. Green chemistry: theory and practice. New York: Oxford University Press; 1998. p. 30.

[5] Gratz M, Backer A, Vondung L, Maser L, Reincke A, Langer R. Donor ligands based on tricoordinate boron formed by B−H-activation of bis(phosphine)boronium salts. Chem Commun 2017;53:7230−3. Available from: https://doi.org/10.1039/C7CC02335A.

[6] Morisako S, Watanabe S, Ikemoto S, Muratsugu S, Tada M, Yamashita M. Synthesis of a Pincer-IrV complex with a base-free alumanyl ligand and its application toward the dehydrogenation of alkanes. Angew Chem 2019;58:15031−5. Available from: https://doi.org/10.1002/anie.201909009.

[7] Backer A, Li Y, Fritz M, Gratz M, Ke Z, Langer R. Redox-active, boron-based ligands in iron complexes with inverted hydride reactivity in dehydrogenation catalysis. ACS Catal 2019;9:7300−9. Available from: https://doi.org/10.1021/acscatal.9b00882.

[8] Inagaki F, Nakazawa K, Maeda K, Koseki T, Mukai C. Substituent effects in the cyclization of Yne-diols catalyzed by gold complexes featuring L2/Z-type diphosphinoborane ligands. Organometallics 2017;36:3005−8. Available from: https://doi.org/10.1021/acs.organomet.7b00369.

[9] Marciniec B, Pietraszuk C, Pawluc P, Maciejewski H. Inorganometallics (transition metal − metalloid complexes) and catalysis. Chem Rev 2022;122:3996−4090. Available from: https://doi.org/10.1021/acs.chemrev.1c00417.

[10] Takaya J. Catalysis using transition metal complexes featuring main group metal and metalloid compounds as supporting ligands. Chem Sci 2021;12:1964−81. Available from: https://doi.org/10.1039/D0SC04238B.

[11] Yoshida H, Takemoto Y, Takaki K. Borylstannylation of alkynes with inverse regioselectivity: copper-catalyzed three component coupling using a masked diboron. Chem Commun 2015;51:6297−300. Available from: https://doi.org/10.1039/C5CC00439J.

[12] Grenet E, Das A, Caramenti P, Waser J. Rhodium-catalyzed C−H functionalization of heteroarenes using indole BX hypervalent iodine reagents. Beilstein J Org Chem 2018;14: 1208−14. Available from: https://doi.org/10.3762/bjoc.14.102.

[13] Song G, Wang F, Li X. C−C, C−O and C−N bond formation via rhodium(III)-catalyzed oxidative C−H activation. Chem Soc Rev 2012;41:3651−78. Available from: https://doi.org/10.1039/c2cs15281a.

[14] Goldman A.S., Goldberg K., editors. Organometallic C-H bond activation: an introduction ACS symposium series 885, activation and functionalization of C-H bonds; 2004. p. 1−43.

[15] Davies HML, Morton D. Recent advances in C − H functionalization. J Org Chem 2016; 81:343−50. Available from: https://doi.org/10.1021/acs.joc.5b02818.

[16] Dhole S, Liao JY, Kumar S, Salunke DB, Sun CM. Regioselective synthesis of angular isocoumarinselenazoles: a benzoselenazole-directed, site-specific, Ruthenium-catalyzed C (sp2)-H activation. Adv Synth Catal 2018;360:942−50. Available from: https://doi.org/10.1002/adsc.201701256.

[17] Ilies L, Matsumoto A, Kobayashi M, Yoshikai N, Nakamura E. Synthesis of polysubstituted naphthalenes by iron-catalyzed [2 + 2 + 2] annulation of grignard reagents with alkynes. Synlett 2012;23:2381−4. Available from: https://doi.org/10.1055/s-0032-1317077.

[18] Thalji RK, Ahrendt KA, Bergman RG, Ellman JA. Annulation of aromatic imines via directed C-H activation with Wilkinson's catalyst. J Am Chem Soc 2001;123:9692−3. Available from: https://doi.org/10.1021/ja016642j.

[19] Singam MKR, Babu US, Nagireddy A, Nanubolu JB, Reddy MS. Harnessing Rhodium-catalyzed C − H activation: regioselective cascade annulation for fused polyheterocycles. J Org Chem 2021;86:8069−77. Available from: https://doi.org/10.1021/acs.joc.1c00477.

[20] Rej S, Chatani N. Rhodium-catalyzed C(sp2)- or C(sp3)-H bond functionalization assisted by removable directing groups. Angew Chem 2019;58:8304−29. Available from: https://doi.org/10.1002/anie.201808159.

[21] Colby DA, Bergman RG, Ellman JA. Synthesis of dihydropyridines and pyridines from imines and alkynes via C H activation. J Am Chem Soc 2008;130:3645−51. Available from: https://doi.org/10.1021/ja7104784.

[22] Murai S, Kakiuchi F, Sekine S, Tanaka Y, Kamatani A, Sonoda M, et al. Efficient catalytic addition of aromatic carbon-hydrogen bonds to olefins. Nature 1993;366:529−31. Available from: https://doi.org/10.1038/366529a0.

[23] Thalji RK, Ahrendt KA, Bergman RG, Ellman JA. Annulation of aromatic imines via directed C-H bond activation. J Org Chem 2005;70:6775−81. Available from: https://doi.org/10.1021/jo050757e.

[24] Zhai R, Xu D, Bai L, Wang S, Kong D, Chen X. Synthesis of isoquinolines via Rh(III)-catalyzed C-H annulation of primary benzylamines with α-Cl ketones. Asian J Org Chem 2022;11. Available from: https://doi.org/10.1002/ajoc.202100662.

[25] Chaudhary BK, Auti P, Shinde SD, Yakkala PA, Giri D, Sharma S. Rh(III)-catalyzed [3 + 2] annulation via C−H activation: direct access to trifluoromethyl-substituted indenamines and aminoindanes. Org Lett 2019;21:2763−7. Available from: https://doi.org/10.1021/acs.orglett.9b00720.

[26] Ko GH, Maeng C, Jeong H, Han SH, Han GU, Lee K, et al. Rhodium(III)-catalyzed sequential C-H activation and cyclization from N-methoxyarylamides and 3-diazooxindoles for the synthesis of isochromenoindolones. Chem An Asian J 2021;16:3179−87. Available from: https://doi.org/10.1002/asia.202100797.

[27] Chen L, Wang Z, Pang B, Wang Y, Xu X, Wu G, et al. Rhodium(III)-catalyzed cascade C-H activation/ annulation of N-carbamoylindoles with Silyl enol ethers for the construction of dihydropyrimidoindoles skeletons. Asian J Org Chem 2021;10:2603−10. Available from: https://doi.org/10.1002/ajoc.202100415.

[28] Yin C, Zhong T, Zheng X, Li L, Zhou J, Yu C. Direct synthesis of indazole derivatives via Rh (III)-catalyzed C−H activation of phthalazinones and allenes. Org Biomol Chem 2021;19:7701−5. Available from: https://doi.org/10.1039/D1OB01458G.

[29] Sherikar MS, Bettadapur KR, Lankea V, Prabhu KR. Rhodium(III)-catalyzed synthesis of trisubstituted furans via vinylic C−H bond activation. Org Biomol Chem 2021;19:7470−4. Available from: https://doi.org/10.1039/D1OB01293B.

[30] Wang TT, Jin HS, Cao MM, Wang RB, Zhao LM. Rh(III)-catalyzed regioselective annulations of 3 arylisoxazolones and 3 aryl-1,4,2-dioxazol-5-ones with propargyl alcohols: access to 4 arylisoquinolines and 4-arylisoquinolones. Org Lett 2021;32:5952−7. Available from: https://doi.org/10.1021/acs.orglett.1c02049.

[31] Zhong X, Lin S, Xu H, Zhao X, Gao H, Yi W, et al. Rh(III)-catalysed cascade C−H imidization/ cyclization of N-methoxybenzamides with isoxazolones for the assembly of dihydroquinazolin-4(1H)-one derivatives. Org Chem Front 2022;9:1904−10. Available from: https://doi.org/10.1039/D1QO01935J.

[32] Deng C, Li CC, Yao J, Jin Q, Miao M, Zhou H. Rh(III)-catalyzed [4 + 2] cyclization of 2-aryl-1H-benzo[d]imidazoles with maleimides via C-H activation. Eur J Org Chem 2021;25:3552−8. Available from: https://doi.org/10.1002/ejoc.202100612.

[33] Sun L, Liu B, Zhao Y, hang J, Kong L, Wang F, et al. Rhodium (III)-catalyzed asymmetric [4 + 1] Spiro annulations of O-pivaloyl oximes with α-diazo compounds. Chem Commun 2021;57:8268−71. Available from: https://doi.org/10.1039/D1CC02888J.

Chapter 6

Organotransition metal chemistry in asymmetric catalysis mediated by different transition metal N-heterocyclic carbene complexes

Dakoju Ravi Kishore[1], Pankaj Kalita[2] and Naushad Ahmed[1]

[1]*Department of Chemistry, Indian Institute of Technology Hyderabad, Kandi, Sangareddy, Telangana, India,* [2]*Department of Chemistry, Nowgong Girls' College, Nagaon, Assam, India*

6.1 Introduction

The last two decades witnessed a lot of interest in the use of chiral N-heterocyclic carbenes (NHCs) in asymmetric organocatalytic transformations. After the first free and "bottelable" N-heterocyclic carbenes (NHCs) by Arduengo et al. in 1991, they have emerged as the most popular class of ligand systems [1]. Of late, NHCs are regarded as one of the most powerful tools in synthetic organometallic chemistry as they are being used as preferred ligands for the preparation of various transition and lanthanide metal complexes having active catalytic and magnetic behaviors [2–9]. In addition, NHCs were quickly explored as an alternative to phosphine ligands for their usage in metal-mediated homogeneous catalysis and as organocatalysts. Along with their steric characteristics, their strong sigma-donating activities play a critical role in defining how they interact with metal centers, which in turn affects the reactivity and selectivity of the transition-metal catalysts. Owing to their easy preparation, numerous NHC transition-metal complexes have been developed to utilize in various catalytic processes, in which the stereo electronic properties of the ligand scaffold can be altered by varying the backbone's nature (saturated or unsaturated), the substitution at the carbon and/or nitrogen atoms, and the ring size. Their stereochemical "topography" differs

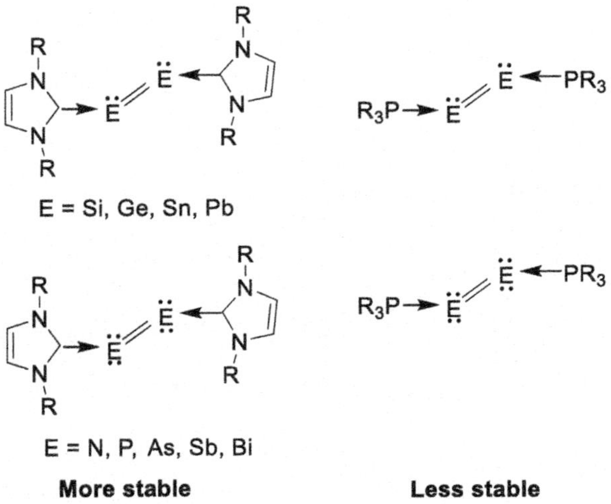

FIGURE 6.1 Comparative structures of common phosphines and NHC ligands.

prominently from that of the diarylphosphine units (Fig. 6.1) they seek to replace, whereas phosphines which have three substituents at the ligating atom are often more or less "cone-shaped" [10,11], the NHC-flat ligand's heterocyclic structure may be more correctly understood as a structural "wedge" that must be functionalized and therefore sculpted into a chiral ligand system [12].

These NHC ligands have unique donor qualities that enable them to stabilize a wide spectrum of metal complexes from soft late transition metals to strongly Lewis acidic metals like Group [1 and 2], early transition metals, F-block metals, and Group [13] metals [13–16]. The extension of normal achiral to chiral N-heterocyclic carbene ligand systems and their derived metal complexes proved to be very useful for enantioselectivity and to induce asymmetric catalysis. Enders and Herrmann pioneered the search for new chiral carbene ligands for asymmetric catalysis in 1996 [17,18]. However, the research group of Burgess published the first fully effective chiral catalyst using an NHC unit, in 2001 [19]. Currently, the field of stereoselective catalysis based on N-heterocyclic carbenes is rapidly expanding, but our understanding of the fundamental variables for successful ligand design remains limited. On the other hand, the chemical and pharmaceutical industries' demand for enantiopure chiral compounds drives the ongoing search for efficient chiral metal catalysts for a wide range of chemical reactions. Asymmetric transition-metal catalysis is among the most effective developments in modern synthetic chemistry, making it easier to access a diverse range of enantiopure chiral organic molecules.

6.2 Fundamental principles

Carbenes are highly reactive intermediate that consist of two unbonded or free electrons with two bonds connected to a carbon atom. This makes them a bit nucleophilic and easily coordinate with the metal centers. As soon as both or one of the carbon atoms (which are bonded to the carbene center) is replaced by the N atom, generates N-heterocyclic carbene (NHCs), which are more nucleophilic in nature. This is because each of the adjacent N atoms consists of one lone pair of electrons delocalizing to the carbene center (Scheme 6.1). The electrons present in the p-orbital of the sp^2 hybridized carbene carbon are conjugated to the lone pair of electrons in the p-orbital of the nitrogen atom at both or one of the ends thus improving the electron cloud density of the carbene carbon atom and providing significant nucleophilicity.

The presence of nitrogen atoms in NHCs not only increases the nucleophilicity of the carbene center but also readily available for various substituents for modification on demand. Since the NHCs provide more opportunity to modify or tune the electronic structures, the three common protocols followed to induce chiral environment (Fig. 6.2) about the carbenic carbon in NHCs framework are-

1. Inclusion of chirality on either two or one of the substituents on the nitrogen atoms of the NHC (**I** and **II**).

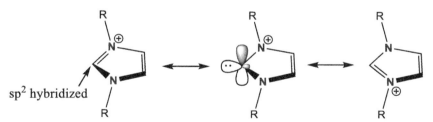

SCHEME 6.1 Resonance condition present in N-heterocyclic carbene ligand for their stability.

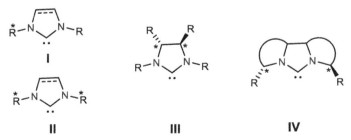

FIGURE 6.2 Example of a (**I**) common achiral N-heterocyclic carbene having R substituents (**II**) chiral inclusion in R (**III**) chiral inclusion in backbone of NHCs and (**IV**) chiral NHCs having fused rigid ring architecture.

2. Direct installation of chirality on the backbone of a saturated NHC framework (**III**).
3. Designing of chiral NHC ligands involves the development of structures where a central NHC ring is fused with one or two cyclic units to form rigid NHC architectures containing the necessary chiral information (**IV**).

The additional advantage of chiral N-heterocyclic carbenes over normal or achiral NHCs is due to their stereo-directing ability in asymmetric synthesis.

6.3 Common methods of preparation

In general, N-heterocyclic carbene metal complexes are synthesized from their precursor NHC ligands binding with a metal precursor upon deprotonation by an appropriate base in ethereal or hydrocarbon solvents, or via silver-routed transmetalation (Scheme 6.2) [11,20,21]. NHC complexes can also be synthesized using the metal precursor containing basic ligands, which leads to the formation of carbenes in situ or isolated states to bind with a suitable metal complex.

Numerous methods have been used to synthesize chiral precursor NHC ligands, one of which is the generation of N-substituents on the resultant NHC from chiral primary amine precursors or introducing a chiral environment to the backbone of the NHC ring. This method was used to synthesize chiral NHCs with C_2 symmetric or NHCs with one chiral N-substituent. The extensive range of N-substituents that can be used in this synthetic process (Scheme 6.3), including chiral amino acid substituents, steric bulk changes, and even extra donor groups, make it much more flexible.

The structural diversities around the NHC ring have been fine-tuned since the first synthesis of bicyclic triazolium chiral NHCs starting from γ-lactam or morpholinone [22]. Chiral bis(NHC)s can be synthesized by introducing a second NHC ligand as a linker using chiral diamines, such as dioxalone,

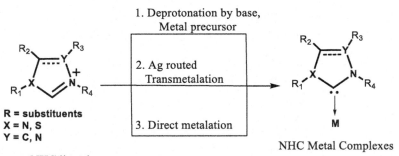

SCHEME 6.2 General approaches for synthesizing NHC metal complexes.

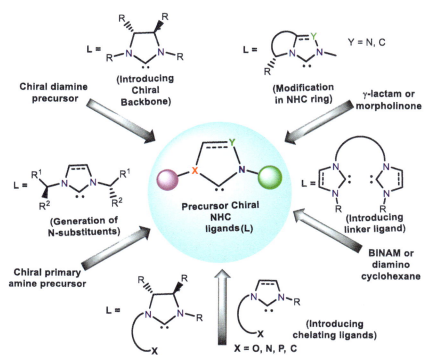

SCHEME 6.3 General selected approach for the synthesis of chiral-NHC ligand precursors.

BINAM, and diamino cyclohexane. The use of hybrid ligands including an NHC together with a donor atom, linked by an organic bridging unit, has been effectively developed and has led to the development of O/N/P/C-linked chiral NHCs [23–25]. Recently, cyclometalated chiral NHCs were synthesized via enantioselective metal-catalyzed *ortho*-C(sp^3)–H activation, which provides a rigid environment and locks carbon–metal bond rotation.

6.4 Mechanism involved

The mechanism of the formation of NHCs has been closely examined over the past century revealing that there are numerous unanswered questions and even contradictions in the field. We here describe the mechanism based on theoretical and experimental studies. As discussed in the previous section, the formation of NHCs from various azolium salts takes place in the presence of appropriate bases followed by the reaction of suitable metal precursors. The first step followed the conversion of azolium salts **I** into NHCs **II** by deprotonation. This step involved the association of the base with the more acidic hydrogen of azolium salts. In the second step the removal of the counter ion and the proton, which ultimately results in the formation of

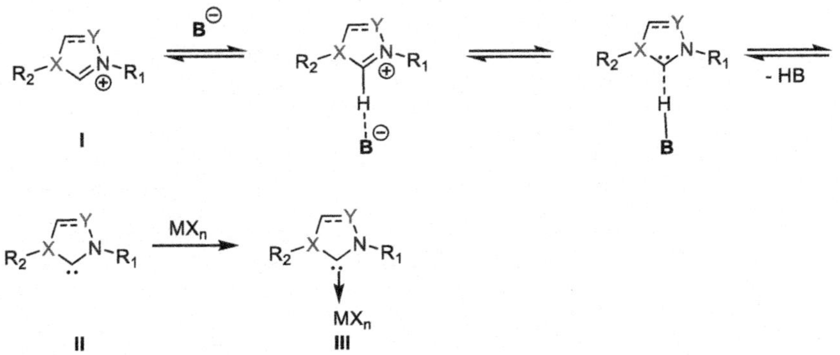

SCHEME 6.4 General mechanism for the chiral NHCs.

neutral carbene, a highly nucleophilic species. Then, these highly basic azolium salts (pK_a = 14−24) [26] react with electrophilic metal centers to afford NHC metal complexes, **III** (Scheme 6.4).

6.5 Recent developments with examples

Optically active pure organic compounds are under the limelight research area among synthetic chemists. This is due to their wide range of applications in pharmaceutical chemistry, agriculture, and fine useful chemicals for other industrial applications [27]. The last few years witnessed enormous development in asymmetric catalysis methodologies which include the fine-tuning/modification of catalysts and the synthetic strategy to get valuable products with increased chemical and enantiomeric yield [28]. Due to excellent sigma donor and strong metal−carbon (M−C) bond formation capability, the metal complexes having N-heterocyclic carbene ligand systems exhibited better air and thermal stability over phosphines [12]. N-heterocyclic carbenes (NHCs) are not only better donor ligands but also found to be flexible and unique to modulate steric and chiral environment around the metal ion which shapes its metal complexes specific for asymmetric catalytic activity. With all the above characteristics, chiral NHCs are found to rapidly emerging stereo-directing ligand systems. Herrmann et al. in 1996 reported the first homogenous asymmetric catalysis using chiral NHC-based stable Rh(I) catalyst **1** (Fig. 6.3) for the hydrosilylation of prochiral acetophenone [1].

Burgess et al. in 2001 developed chiral Ir(0) catalyst **2** for the effective asymmetric hydrogenation of *E*-aryl alkenes, such as *E*-1,2-diphenylpropene, with the increased enantiomeric yield 45%−95% by modifying the -R and -Ar substituents (Fig. 6.4) [19].

The progressive development in designing the new chiral NHCs ligands and their metal complexes with early (Ti, Zr, Hf, etc.) and late transition

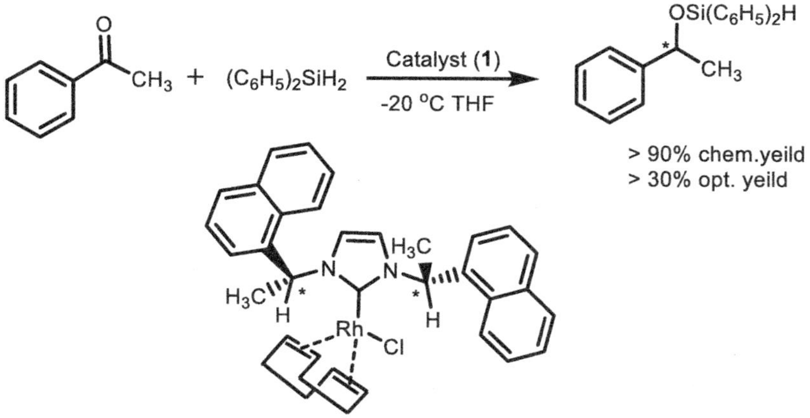

FIGURE 6.3 Asymmetric hydrosilylation reaction of acetophenone catalyzed by 1.

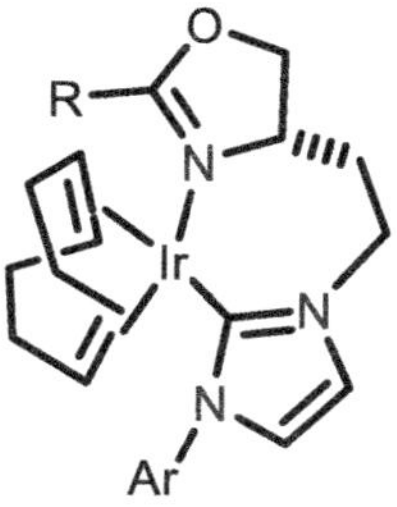

FIGURE 6.4 Molecular structure of catalyst 2 for asymmetric hydrogenation of E-aryl alkenes.

(Cu, Pd, Ir, Au, Ag, and Rh) metals were continued for better performance in asymmetric catalysis [11,29]. In the same line of interest the designed (Ag-NHCs) chiral catalyst **3−7** (Fig. 6.5) with C_2 symmetry reduces the number of diastereomers and transition states during the catalytic cycle which in turn helps in improving the chemical as well as enantiomeric yield [11].

A few catalysts have also shown their performance in more than one type of reaction, that is, catalyst **8** and found to be active in four types of asymmetric metathesis reactions (1) two-ring closure and (2) two-ring opening (Fig. 6.6). The better performance in terms of enantioselectivity (95%−98% ee) observed in the later cases of ring-opening reactions [20].

In some cases, the catalyst requires additives for better performance to improve the catalytic performance, and to get a better enantiomeric yield. For instance, catalyst **9** (Fig. 6.7) shows an enantiomeric yield of 23% while in the presence of NaI, the enantiomeric yield increases to 39% [20].

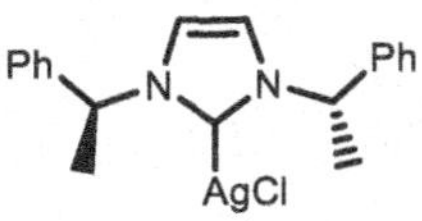

(3)

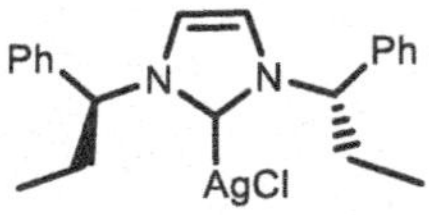

(4)

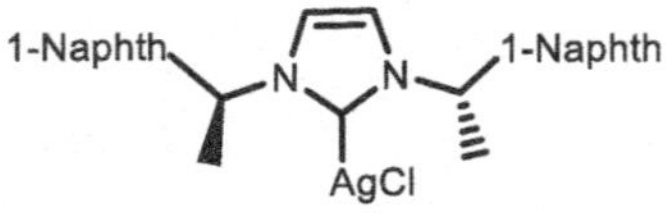

(5)

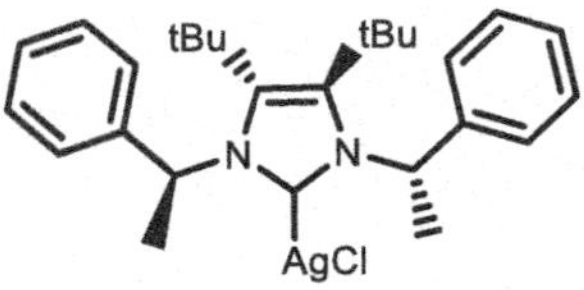

(6)

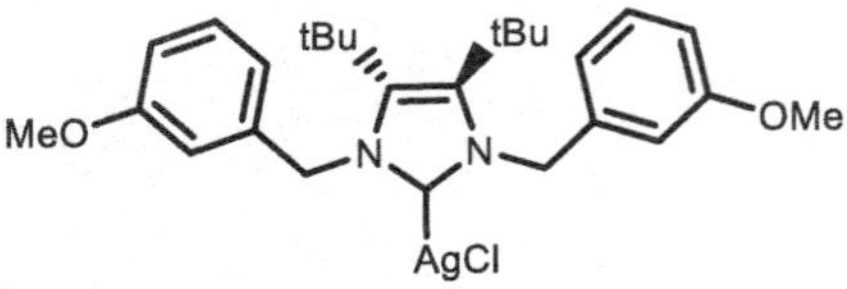

(7)

FIGURE 6.5 Few silver NHC catalysts with C_2 symmetry.

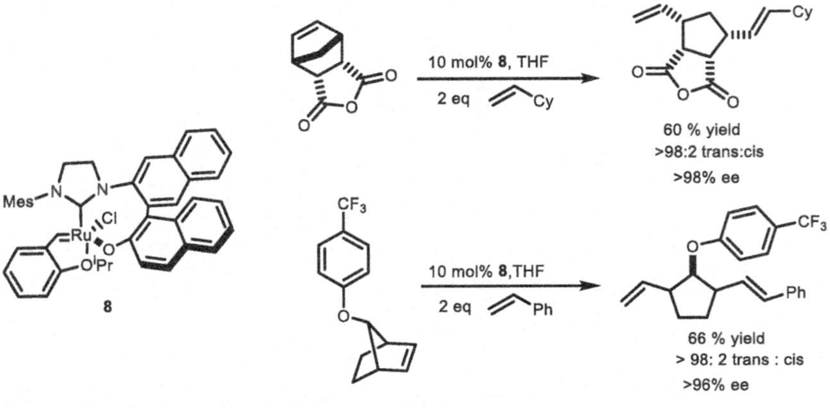

FIGURE 6.6 A few silver NHC catalysts with C_2 symmetry.

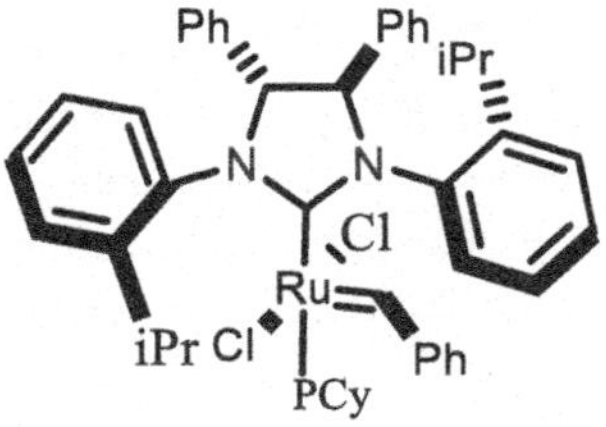

(9)

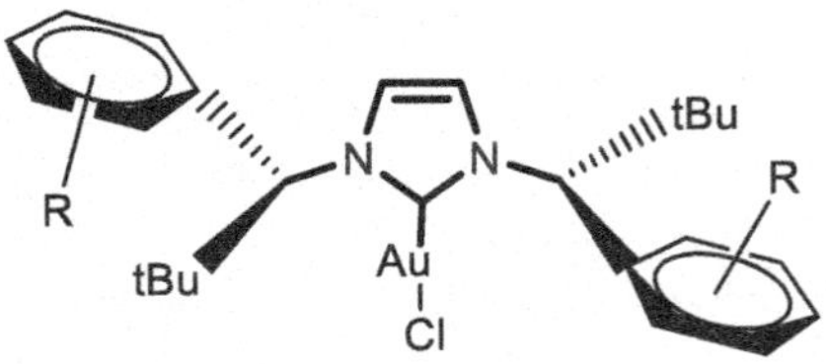

(10)

FIGURE 6.7 Structure of catalysts **9** and **10**. For catalyst **10**, R is the different substitutions (methoxy, aryl, naphthyl, etc.) on phenyl ring.

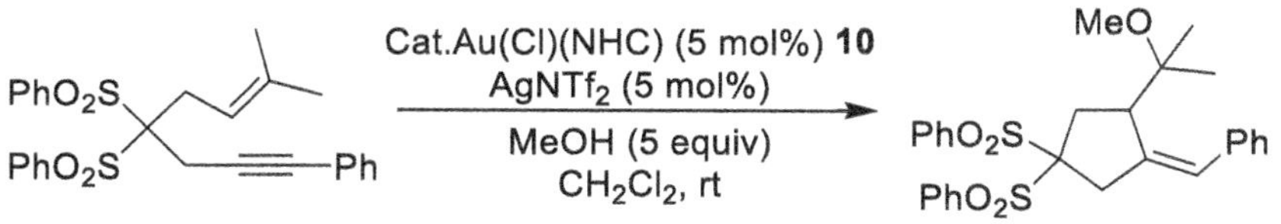

SCHEME 6.5 Methoxylation and intramolecular cyclization of 1,6-enyne by asymmetric catalysis.

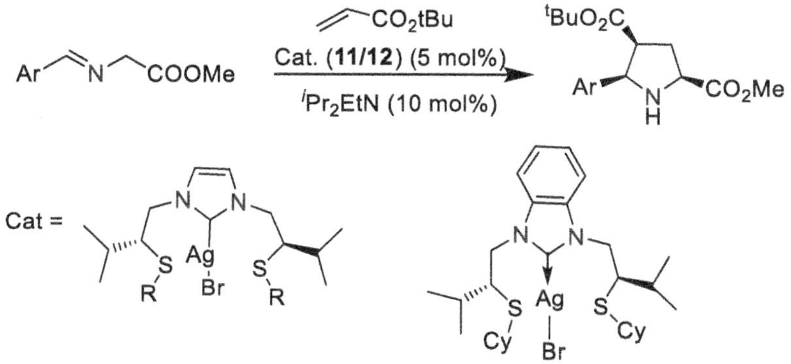

FIGURE 6.8 Molecular structure of catalysts **11** and **12**.

Catalyst **10** (Fig. 6.7) is found to be more active in the presence of AgNTf$_2$ {silver salt of bis-(trifluoromethanesulfonyl)imidate, NTf$_2^-$} via Au-triflimidate complex formation [30]. Catalyst **10** is involved in the methoxylation and intramolecular cyclization of 1,6-enyne in methanol with an enantiomeric yield of up to 72% (Scheme 6.5).

During the years 2003−2010, numerous *N*-heterocyclic-based chiral catalysts were developed for various asymmetric transformations, such as alkylation, hydrogenation, and hydrosylation reactions [31−34]. Lassaletta and coworkers in 2009, synthesized a new generation of chiral NHCs which consist of pincer type, C$_2$-symmetric S/C(NHC)/S neutral ligands and its silver complexes **11** and **12** proved to catalyze efficiently the 1,3-dipolar cycloaddition reactions of imino glycinates (Fig. 6.8) [35].

Very recently Clavier and coworkers isolated optically pure NHC−Cu complexes (Fig. 6.9, catalyst **13**) bearing C$_1$ and C$_2$ symmetric N-heterocyclic carbene (NHC) ligands from prochiral NHC precursors [36]. Interestingly transmetalation reaction strategies were followed to synthesize gold and palladium (NHC-Au/Pd) complexes (catalyst **14** and **15**) from NHC−Cu precursors with full stereoretentivity and 99%ee. Among them, Cu and Pd were found to show excellent performance in

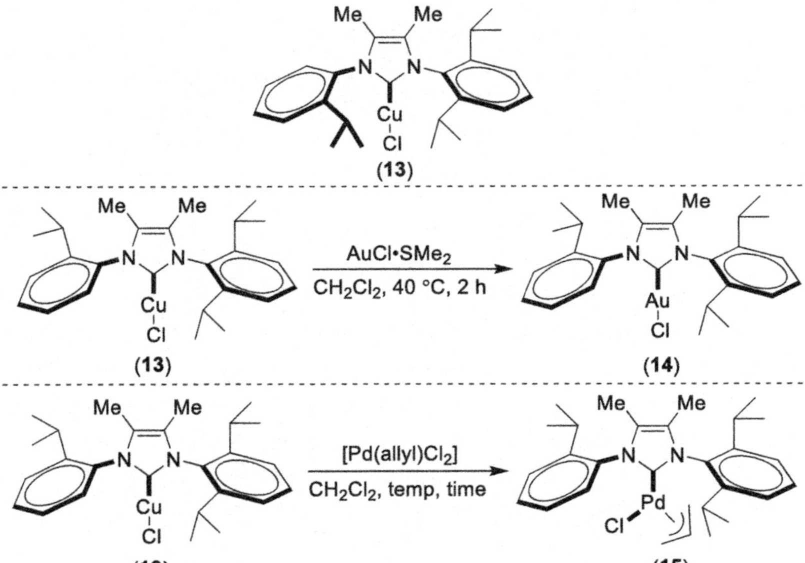

FIGURE 6.9 Structure of precursor catalyst 13 to derive catalysts 14 and 15 following transmetalation.

asymmetric allylic alkylation and intramolecular α-Arylation respectively with 84%−98%ee [36].

Although the metal complexes of chiral NHC-derived ligand systems widely used in asymmetric catalysis, Roland et al. report the involvement of the chiral N-heterocyclic ligand systems to promote the efficiency of metal-catalyzed Michael addition of diethyl zinc (II) reagent to prochiral cycloheptenone with 93%ee (Fig. 6.10) [29]. There are also a few reports on cooperative behavior of stereo-directing NHCs' carbene ligand systems to gain the enantioselectivity by promoting asymmetric catalysis [37−40].

6.6 Applications in emerging fields

Asymmetric catalysis results in the production of enantiopure pharmaceuticals and other useful industrial chemicals. The increasing demands for such optically active chemicals make the area of asymmetric catalysis an emerging and growing research field. The chiral N-heterocyclic ligand systems and their transition metal complexes are found to be very successful in the catalytic transformation of chemicals. The flexibility and suitability of the NHCs-TM (where TM = Transition metals) complexes for various asymmetric catalysis makes them explore/tune their activity by further modification. We have provided the graphical presentation in Fig. 6.11 for the various

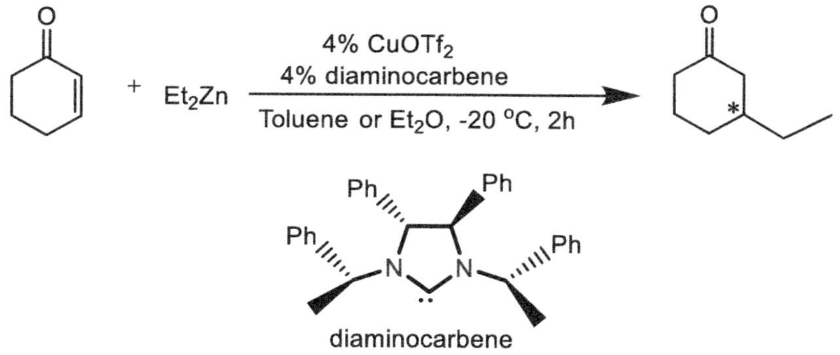

Organotransition metal chemistry **Chapter | 6** **149**

FIGURE 6.10 Chiral carbene used in metal-assisted asymmetric catalysis.

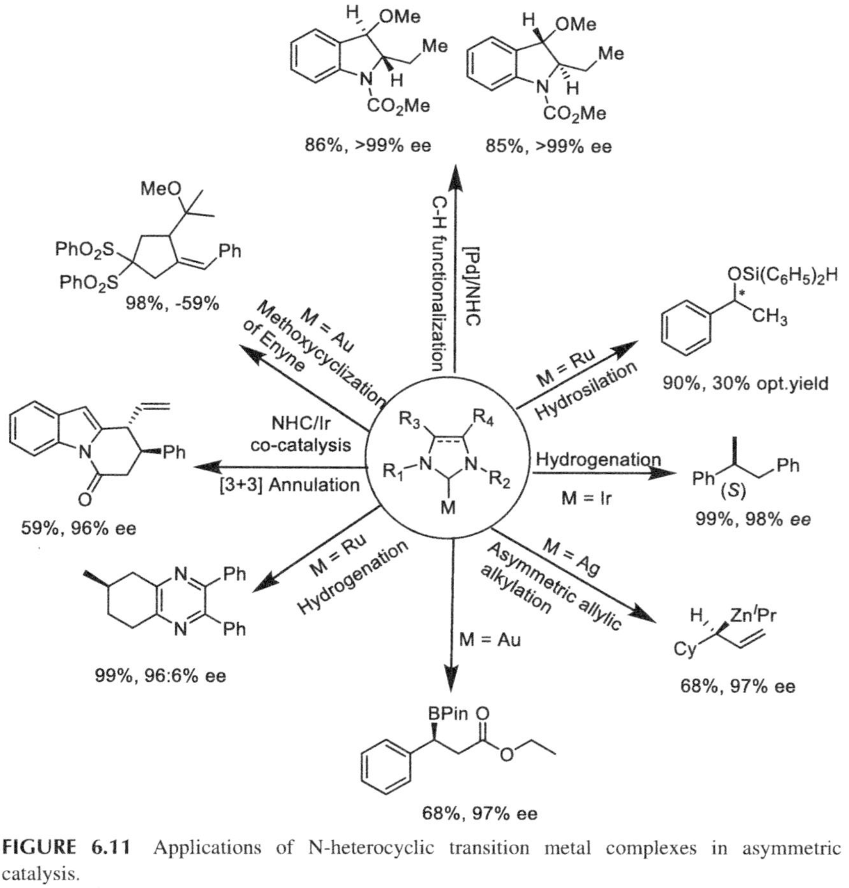

FIGURE 6.11 Applications of N-heterocyclic transition metal complexes in asymmetric catalysis.

emerging applications (selected) of asymmetric catalysis by N-heterocyclic transition metal complexes.

6.7 Future prospective and scope

The catalytic activity of NHCs-TM complexes was found to be not fully explored to date, while the demand for optically active chemicals is increasing. This opens the way to look into the future prospective and scope of further research on the chiral N-heterocyclic ligands and their transition metal complexes in asymmetric catalysis. The chiral/achiral N-heterocyclic ligand systems consisting of nitrogen atoms (adjacent to carbene center) are available for substitutions and further modifications and can further explore their coordination potential for a wide range of transition and lanthanide metals and finally their asymmetric catalytic activity.

6.8 Summary

In this survey, we have tried to explore the role and importance of chiral N-heterocyclic carbenes and their transition metal complexes (NHC-TM) in asymmetric catalysis reported to date. We covered a wide range of chiral NHC-TM complexes synthesized and their use in the production of valuable chemicals via asymmetric catalysis from the literature reports. We have also covered the common method of NHC-TM complexes preparation and the mechanism involved. Finally, we shed light on the prospective and further scope of NHC-TM complexes in asymmetric catalysis.

6.9 Problems with solutions

Q1. What are the common methods to induce chirality in the enantioselective synthesis from chiral NHC ligands?

Answer: In chiral NHC ligand-directed enantioselective synthesis, the most common methods to provide chiral induction ability are via the incorporation of chiral elements at the N-substituents or simply at the NHC core, or at the linkage.

Q2. How symmetry play a crucial role in chiral NHC?

Answer: The design of chiral catalysts is frequently based on C_2-symmetry to reduce the number of diastereomeric intermediates and transition states that play a role in the catalytic cycle.

Q3. Are NHC's strong sigma or pi donors?

Answer: NHCs are typically much stronger donors than phosphines. Since the NHC carbene protonates more readily than PR_3, the s-donor power of the NHC lone pair is undoubtedly much stronger.

Q4. What is the broad range of metals that are coordinated with NHC ligands?

Answer: The chemical stability and coordination versatility of NHC ligands have allowed almost infinite access to new organometallic topologies from alkaline-earth metals to rare-earth metals.

Q5. What is the order of electron-donating power of imidazole, benzimidazole, and imidazoline of the azole ring in NHC?

Answer: benzimidazole < imidazole < imidazoline (electronics can be altered by changing the nature of the azole ring).

1. NHCs are considered as an alternative to Phosphine ligands.
2. The first stable NHC is isolated in the year 1991 by Arduengo and coworkers.

6.10 Objective type questions

i. The first free and "bottelable" N-Heterocyclic carbenes (NHCs) was isolated by?
   (a) Burgess et al.
   (b) Raubenheimer et al.
   (c) S. Knowles et al.
   (d) Arduengo et al.
ii. Which year the stable NHC is isolated?
   (a) 1991
   (b) 1990
   (c) 1993
   (d) 1994
iii. Which research group reported the first highly enantioselective reaction in 2001?
   (a) Shi et al.
   (b) Arduengo et al.
   (c) Glorious et al.
   (d) Burgess et al.
iv. *N*-heterocyclic carbenes (NHCs) are
   (a) electron-poor neutral ligands
   (b) electron-rich charged ligands
   (c) electron-rich radicals
   (d) electron-rich neutral ligands
v. NHCs are usually better
   (a) σ-donors, but weaker π-acceptors.
   (b) σ-donors
   (c) weak π-acceptors.
   (d) π-donors, but weaker σ-acceptors.

References

[1] Arduengo III AJ, Harlow RL, Kline M. A stable crystalline carbene. J Am Chem Soc 1991;113(1):361–3.
[2] Wang N, Xu J, Lee JK. The importance of N-heterocyclic carbene basicity in organocatalysis. Org Biomol Chem 2018;16:8230–44.
[3] Zhong R, Lindhorst AC, Groche FJ, Kühn FE. Immobilization of N-heterocyclic carbene compounds: a synthetic perspective. Chem Rev 2017;117(3):1970–2058.
[4] Que Y, He H. Advances in N-heterocyclic carbene catalysis for natural product synthesis. Eur J Chem 2020;2020(37):5917–25.
[5] Meng Y-S, Mo Z, Wang B-U, Zhang Y-Q, Deng L, Gao S. Observation of the single-ion magnet behavior of d^8 ions on two-coordinate Co(I)–NHC complexes. Chem Sci 2015;6:7156–62.
[6] Long J, Lyubov DM, Gurina GA, Nelyubina YV, Salles F, Guari Y, et al. Using N-heterocyclic carbenes as weak equatorial ligands to design single-molecule magnets: zero-field slow relaxation in two octahedral dysprosium(III) complexes. Inorg Chem 2022;61(3):1264–9.
[7] Carroll XB, Errulat D, Murugesu M, Jenkins DM. Late lanthanide macrocyclic tetra-NHC complexes. Inorg Chem 2022;61(3):1611–19.
[8] Meihaus KR, Minasian SG, Lukens WW, Kozimor SA, Shuh DK, Tyliszczak T, et al. Influence of pyrazolate vs N-heterocyclic carbene ligands on the slow magnetic relaxation of homoleptic trischelate lanthanide(III) and uranium(III) complexes. J Am Chem Soc 2014;136(16):6056–68.
[9] Gupta T, Velmurugan G, Rajeshkumar T, Rajaraman G. Role of lanthanide-ligand bonding in the magnetization relaxation of mononuclear single-ion magnets: a case study on pyrazole and carbene ligated Ln(III)(Ln = Tb, Dy, Ho, Er) complexes. J Chem Sci 2016;128(10):1615–30.
[10] Crabtree RH. NHC ligands versus cyclopentadienyls and phosphines as spectator ligands in organometallic catalysis. J Organomet Chem 2005;690(24):5451–7.
[11] César V, Bellemin-Laponnaz S, Gade LH. Chiral N-heterocyclic carbenes as stereodirecting ligands in asymmetric catalysis. Chem Soc Rev 2004;33:619–36.
[12] Wilson DJD, Couchman SA, Dutton JL. Are N-heterocyclic carbenes "better" ligands than phosphines in main group chemistry? A theoretical case study of ligand-stabilized E2 molecules, L-E-E-L (L = NHC, phosphine; E = C, Si, Ge, Sn, Pb, N, P, As, Sb, Bi). Inorg Chem 2012;51(14):7657–68.
[13] Nesterov V, Reiter D, Bag P, Frisch P, Holzner R, Porzelt A, et al. NHCs in main group chemistry. Chem Rev 2018;118(19):9678–842.
[14] Arnold PL, Liddle ST. F-block N-heterocyclic carbene complexes. Chem Commun 2006;3959–71.
[15] Asay M, Jones C, Driess M. N-heterocyclic carbene analogues with low-valent group 13 and group 14 elements: syntheses, structures, and reactivities of a new generation of multitalented ligands. Chem Rev 2011;111(2):354–96.
[16] Zhang D, Zi G. N-heterocyclic carbene (NHC) complexes of group 4 transition metals. Chem Soc Rev 2015;44:1898–921.
[17] Enders D, Gielen H, Raabe G, Runsink J, Teles JH. Synthesis and stereochemistry of the first chiral (imidazolinylidene)- and (triazolinylidene)palladium(II) complexes. Chem Ber 1996;129:1483–8.

[18] Herrmann WA, Goossen LJ, Köcher C, Artus GRJ. Chiral heterocylic carbenes in asymmetric homogeneous catalysis. Angew Chem Int Ed 1996;35:2805–7.
[19] Powell MT, Hou DR, Perry MC, Cui X, Burgess K. Chiral imidazolylidine ligands for asymmetric hydrogenation of aryl alkenes. J Am Chem Soc 2001;123:8878–9.
[20] Perry MC, Burgess K. Chiral N-heterocyclic carbene-transition metal complexes in asymmetric catalysis. Tetrahedron Asymmetry 2003;14:951–61.
[21] Douthwaite RE. Metal-mediated asymmetric alkylation using chiral N-heterocyclic carbenes derived from chiral amines. Coord Chem Rev 2007;251:702–17.
[22] Chen J, Huang Y. Asymmetric catalysis with N-heterocyclic carbenes as non-covalent chiral templates. Nat Commun 2014;5:3437.
[23] Zhao D, Candish L, Paul D, Glorius F. N-Heterocyclic carbenes in asymmetric hydrogenation. ACS Catal 2016;6(9):5978–88.
[24] Gardiner MG, Ho CC. Recent advances in bidentate bis(N-heterocyclic carbene) transition metal complexes and their applications in metal-mediated reactions. Coord Chem Rev 2018;375:373–88.
[25] Fliedel C, Labande A, Manoury C, Poli R. Chiral N-heterocyclic carbene ligands with additional chelating group(s) applied to homogeneous metal-mediated asymmetric catalysis. Coord Chem Rev 2019;394:65–103.
[26] Hollóczki O. The Mechanism of N-heterocyclic carbene organocatalysis through a magnifying glass. Chem Eur J 2019;26(22):4885–94.
[27] Waldeck B. Biological significance of the enantiomeric purity of drugs. Chirality 1993;5(5):350–5.
[28] Xiang S-H, Tan B. Advances in asymmetric organocatalysis over the last 10 years. Nat Commun 2020;11:3786.
[29] Alexakis A, Winn CL, Guillen F, Pytkowicz J, Roland S, Mangeney P. Asymmetric synthesis with N-heterocyclic carbenes. Application copper-catalyzed conjugate. Addit Adv Synth Catal 2003;345(3):345–8.
[30] Banerjee D, Buzas AK, Besnard C, Kundig EP. Chiral N-heterocyclic carbene gold complexes: Synthesis, properties, and application in asymmetric catalysis. Organometallics 2012;31:8348–54.
[31] Veldhuizen JJV, Campbell JE, Giudici RE, Hoveyda AH. A readily available chiral Ag-based N-heterocyclic carbenecomplex for use in efficient and highly enantioselective Ru-catalyzed olefin metathesis and Cu-catalyzed allylic alkylation reaction. J Am Chem Soc 2005;127:6877–82.
[32] Gade LH, Bellemin-Laponnaz B. Mixed oxazoline-carbenes as stereodirecting ligands for asymmetric catalysis. Coord Chem Rev 2007;251:718–25.
[33] Liu LJ, Wang F, Shi M. Synthesis of chiral bis(N-heterocyclic carbene)palladium and rhodium complexes with 1,10-biphenyl scaffold and their application in asymmetric catalysis. Organometallics 2009;28:4416–20.
[34] Díez-Gonzalez S, Marion N, Nolan SP. N-heterocyclic carbenes in late transition metal catalysis. Chem Rev 2009;109:3612 367.
[35] Iglesias-Siguenza J, Ros A, Díez E, Magriz A, Vazquez A, Alvarez E, et al. C2-Symmetric S/C/S ligands based on N-heterocyclic carbenes: a new ligand architecture for asymmetric catalysis. Dalton Trans 2009;848:8485–8.
[36] Kong L, Morvan J, Pichon D, Jean M, Albalat M, Vives T, et al. From prochiral N-heterocyclic carbenes to optically pure metal complexes: new opportunities in asymmetric catalysis. J Am Chem Soc 2020;142:93–898.

[37] Zhang Z-J, Wen Y-H, Song J, Gong L-Z. Kinetic resolution of aziridines enabled by N-heterocyclic carbene/copper cooperative catalysis: carbene dose-controlled chemo-switchability. Angew Chem Int Ed 2020;60(6):3268–76.
[38] Ding Y-L, Zhao Y-L, Niu S-S, Wu P, Cheng Y. Asymmetric synthesis of multifunctionalized 2,3-benzodiazepines by a one-pot N-heterocyclic carbene/chiral palladium sequential catalysis. J Org Chem 2020;85(2):612–21.
[39] Zhang J, Gao Y-S, Gu B-M, Yang W-L, Tian B-X, Deng W-P. Cooperative n-heterocyclic carbene and iridium catalysis enables stereoselective and regiodivergent [3 + 2] and [3 + 3] annulation reactions. ACS Catal 2021;11(7):3810–21.
[40] Zhang Z-J, Zhang L, Geng R-L, Song J, Chen X-H, Gong LJ. N-heterocyclic carbene/copper cooperative catalysis for the asymmetric synthesis of spirooxindoles. Angew Chem Int Ed 2019;131:12318–22.

Chapter 7

Organophosphorus compounds: recent developments and future perspective

Vitthalrao Swamirao Kashid[1] and Azaj Ansari[2]
[1]Department of Humanities and Science (Chemistry), Malla Reddy Engineering College for Women, Hyderabad, Telangana, India, [2]Department of Chemistry, Central University of Haryana, Mahendergarh, Haryana, India

7.1 Introduction

Phosphorus-based compounds have played a significant role in the field of coordination/organometallic chemistry because of the versatile and diversity of their composition and of their coordination modalities [1,2]. These are perhaps the most widely used ligand systems in coordination chemistry specifically in organometallic chemistry and homogenous catalysis. The actual reason for their wide success lies in their ability to form a variety of compounds or complexes with late transition metals in their low oxidation states with intriguing geometries, which are very vital in metal-mediated homogenous catalysis to bioorganic chemistry [1–4]. The electronic and steric properties can be tuned by varying the substituents at the phosphorus atom and it also accommodates various geometries and shapes of compounds which can be very interesting for various studies and for utility. Till now several research groups have studied the coordination behavior of various phosphines (phosphorus compounds) and their applications in homogenous or heterogeneous catalytic applications. However, phosphines with different coordination sites, namely, mono-, di-, tri-, and tetradentate ligands which are significant contributions of versatile compounds but still having pincer moieties or chelating sites are less extensive and it can be a future scope for the researcher in the field [1–3]. This chapter gives a brief literature survey on multidentate P(III) ligands, their reactivity and applications in catalytic organic transformations.

7.2 Fundamental principles

7.2.1 Phosphorus-based ligands

Phosphorus compounds in trivalent oxidation state offer a unique opportunity to modify their steric and electronic properties. The phosphorus-based ligands on the basis of substituents attached to the phosphorus center, can be classified as phosphines, aminophosphines, phosphinites, phosphonites, or phosphites (Chart 7.1). The ligating ability of these ligands mostly depends on the steric effect, σ-donor ability, and π-acceptor capability. The steric demand comes from the bulkiness of the phosphorus substituents and determined by *cone angle* (θ). Tolman has instigated the idea of a *cone angle* (θ) to specify the approximate sum of space that a ligand occupies around the metal center [5,6]. The *cone angle* (θ) is described as the apex angle of a cylindrical cone centered 2.28 Å from the phosphorus that confines all atoms of its substituents. For bidentate ligands, an additional parameter *bite angle* (β_n) was introduced, and defined as the lowest energy chelation angle (P − M − P) of the metal complexes [7−9]. The σ-donor ability and π-acceptor abilities of phosphorus ligands mainly depend to the nature of phosphorus substitution. The electron-withdrawing substituents on phosphines make them good π-acceptor while the electron-releasing groups make them good σ-donors.

7.2.2 Classification of phosphorus-based ligands

Phosphorus(III)-based ligands are very important for homogeneous transition metal catalysis. Several methods of classifications are used to distinguish phosphorus ligands. The trivalent phosphorus ligands can be generally

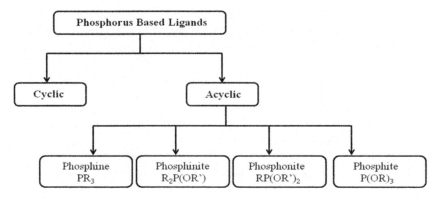

CHART 7.1 Classification of phosphorus-based ligands.

classified into two major classes; acyclic and cyclic phosphorus compounds. Phosphorus(III) compound can be subdivided into monophosphines, bisphosphines, trisphosphines, and polyphosphines depending on the number of phosphorus donor atoms present in the ligands [10]. They can be further subclassified into phosphines (PR$_3$), phosphinites (R$_2$P(OR′)). phosphonites (RP(OR′)$_2$) and phosphites (P(OR′)$_3$) based on the number of P—O bonds present on the phosphorus atom in consideration (Chart 7.1).

7.3 Common methods of preparation

7.3.1 Pincer ligands

From the first report on phosphine donor pincer also called as tridentate chelating complexes by Shaw and coworkers in 1976 [11], pincer ligands attracted huge attention in organometallic and more recently, in organic transformations. The word "pincer" was invented by Koten to describe monoanionic, tridentate meridional ligands. A classical pincer complex (Fig. 7.1). However, almost all tridentate meridional ligands are commonly termed as pincer ligand. The synthesis and applications of a wide variety of pincer complexes have been well reported in the literature [12]. The pincer ligands provide various modification sites for the fine tuning of both stereo-electronic properties around the central metal atom. High thermal stability, higher efficiency, functional group tolerance, and selectivity of pincer complexes make them excellent catalyst for many catalytic reactions. In recent years the utility of pincer complexes in homogeneous catalysis has evolved many folds [13,14].

7.3.2 Synthesis of pincer complexes

The pincer complex can be achieved by various reaction ways: via (1) C—H activation or electrophilic bond activation, (2) oxidative addition to an aryl halide bond, (3) *trans*-metallation, and (4) *trans*-cyclometallation [15]. Among all, C—H bond activation and oxidative addition over an aryl halide bond are the widely adopted methods.

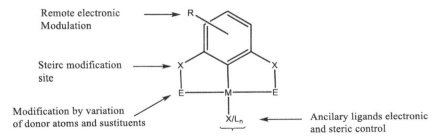

FIGURE 7.1 General example of pincer complex.

7.3.2.1 C−H bond activation

Metal-to-carbon bond formation through direct metallation by C−H activation is the most preferred method for the synthesis of pincer complexes. The process involves the formation of *trans* chelate complex followed by agostic interaction between M−X (metal-halide) and C−H bonds. It mostly requires drastic reaction conditions and also reaction time (Scheme 7.1) [16].

The mechanism of cyclopalladation reaction through C−H activation has been studied in detail [16,17]. Analogous to organic aromatic substitution reactions, an electron-donating group on aromatic ring facilitate the C−H activation. The cyclometallation proceeds with relatively milder reaction conditions with electron-donating substituents on central aromatic ring or on the phosphorus atom. The cyclometallation proceeds under relatively milder reaction conditions with electron-donating groups on central aromatic ring or on phosphorus atoms. In many cases, the reaction pathway is similar to aromatic electrophilic substitution reactions. The pincer formation through C−H bond activation is directed by preliminary heteroatom coordination. Eventually electrophilic aromatic substitution provides σ-bond (covalent bond) formation or arenium intermediate. The elimination of HX from the arenium intermediate results in the formation of a metal−carbon covalent bond (Scheme 7.1). The formation of the chelated complex is dependent on the steric and electronic properties of the two side arms of pincer ligands.

The orthometallation is not found to be very facile for all pincer ligand system. Sometimes ligands prefer to coordinate in bridging fashion rather than chelation, which is not a favorable condition mode for C−H activation. Baber and coworkers have reported the synthesis of insoluble polymer species by thermal reaction of PCP tridentate ligand with dichloropalladium(II) metal salts in a range of solvents [18]. Expected pincer complex was formed only after heating ligand and [PdCl$_2$(NCPh)$_2$] metal salt in 1,2-dichloroethane for few days. Unexpectedly, when the same reaction was performed under microwave condition, the rate of reaction was hugely enhanced with excellent yield of pincer complex within an hour (Scheme 7.2).

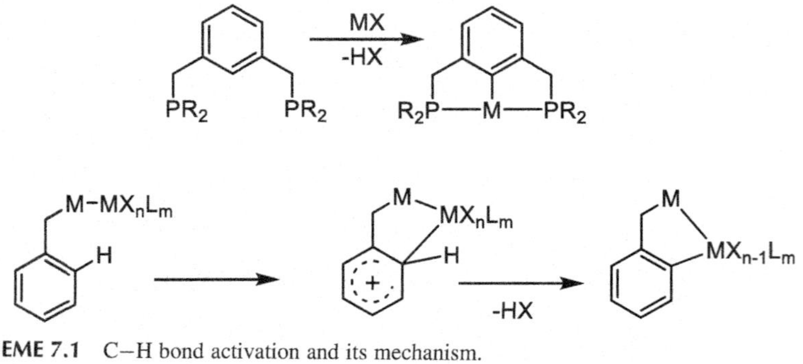

SCHEME 7.1 C−H bond activation and its mechanism.

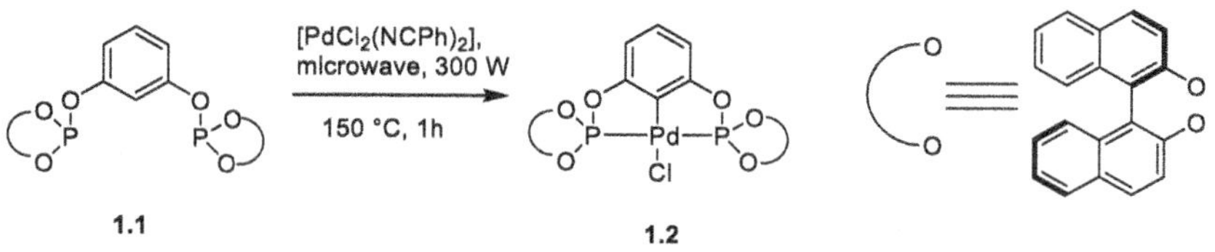

SCHEME 7.2 C−H bond activation in bisphosphite PCP pincer ligand.

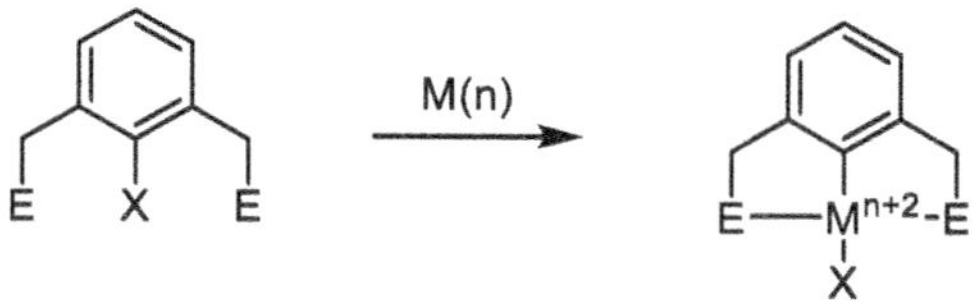

SCHEME 7.3 Pincer complex through oxidative addition.

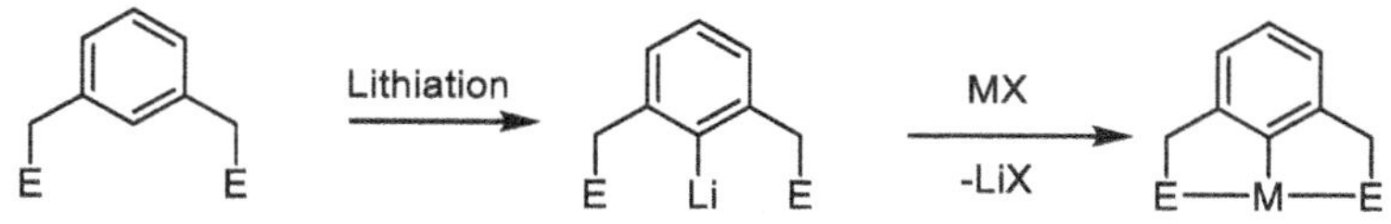

SCHEME 7.4 Pincer complex by *trans*-metallation.

7.3.3 Oxidative addition to aryl halide bond

An oxidative addition of C−X bond to the low-valent metal precursors is a preferred alternative way for C−H activation method. Prefunctionalization of ligand systems is required to achieve the pincer complexes with this method. However, some iron, rhodium and iridium pincer complexes were obtained by oxidative addition over C−H bond [19,20]. The oxidative addition procedure is mainly useful for the thermally unstable and acid sensitive functional group containing ligand systems and well applied to the zero-valent metal precursors, such as [Ni(COD)$_2$] or [Ni(PPh$_3$)$_4$] [21] and [Pd$_2$(dba)$_3$] (Scheme 7.3) [22].

7.3.4 *Trans*-metallation

Trans-metallation is very rarely used in the syntheses of pincer complexes. It involves the lithiation of ligands followed by the treatment with appropriate metal reagents (Scheme 7.4). This method does not work properly with pincer ligands, where the lithiation instantly occurs at benzylic position instead of aromatic ring [23,24]. Halide functionalized ligands may help to improvise the selectivity, but isomerization of aryl lithium species to more stable benzyl lithium is observed in a few cases.

7.3.5 *Trans*-cyclometallation

The *trans*-cyclometallation is an efficient method for the syntheses of pincer complexes. It involves the transfer of metals from a cyclometallated species to pincer ligand. In this whole process, replaced ligands work as a base and abstract the proton from the ligand to be cyclometallated. In these reactions, the driving force is the higher affinity of donor ligands toward the late transition metals. This method was initially popular for the synthesis of bidentate cyclometallated complexes and later extended to tetradentate pincer complexes [25]. The leading work in this field was carried out by Koten and coworkers, where the cyclometallated metal reagents with monoanionic terdentate ligand, *2,6*-$(CH_2NMe_2)_2H_6H_3$ were successfully used for the *trans*-cyclometallation of PCP pincer ligands as (Scheme 7.5) [26–28].

7.4 Classification of pincer complexes

The tridentate chelating pincer ligands are broadly classified into two classes: symmetric and asymmetric, on the basis of their coordinating side arms. Many of reported symmetric of the type; CCC [29], CNC [30], NNN [31], NCN [32], PCP [33,34], PNP [35–38], SCS [39], SPS [40], SNS [41], and OCO [42], and asymmetric, CNN [43], PCN [44,45], PNN [46] tridentate chelating pincer ligands (Chart 7.2). Among all tridentate chelating ligands reported in the previous literature, phosphorus-based compounds have drawn exceptional attention as they allow fine tuning of steric and electronic

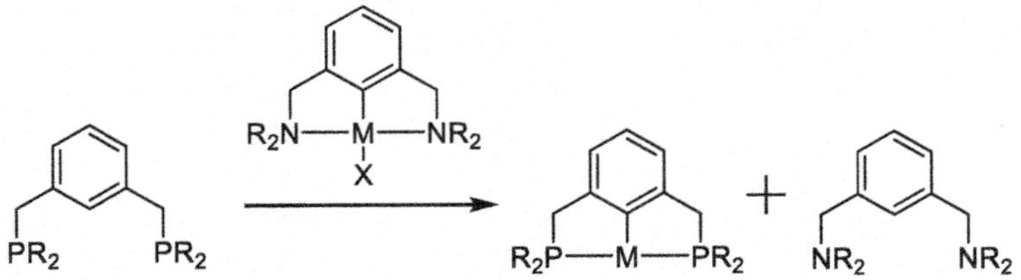

R2N—M—NR2
X
PR2
PR2
R2P—M—PR2
NR2
NR2

SCHEME 7.5 *Trans*-cyclometallation of pincer ligand.

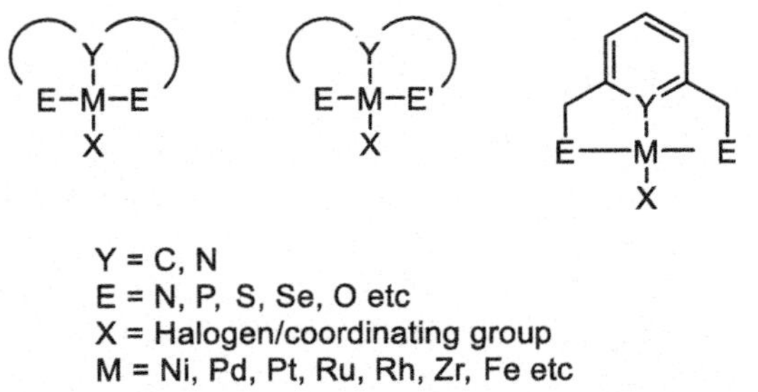

Y
E–M–E
X
Y
E–M–E'
X
Y
E—M— E
X
Y = C, N
E = N, P, S, Se, O etc
X = Halogen/coordinating group
M = Ni, Pd, Pt, Ru, Rh, Zr, Fe etc

CHART 7.2 Classification of tridentate chelating pincer complexes.

properties throughout the metal center thereby making them efficient and useful catalysts for various catalytic reactions.

7.4.1 Transition metal chemistry of pincer ligands

7.4.1.1 PCP pincer complexes

The symmetrical PCP pincer ligands **1.3** and **1.4** are largely used in the organometallic chemistry. Among these, the pincer ligand **1.4** has been widely used for the syntheses of pincer complexes. Balakrishna and coworkers have introduced the new bisphosphomide based pincer ligands and their transition metal chemistry which give a good amount of contribution in this field (Chart 7.3) [34,37].

The interesting five coordinated rhodium(II) and iridium complexes **1.5** and **1.6** were synthesized by the reaction of ligand **1.4** with MCl_3 (M = Rh, Ir) (Scheme 7.6) [11]. Dehydrochlorination of rhodium complex **1.5** obtained the 14 electron complex of the type **1.7** (Scheme 7.7). The presence of bulky

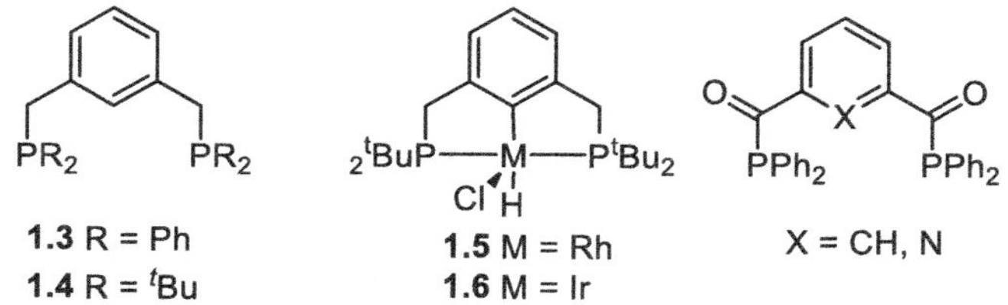

CHART 7.3 Tridentate-chelating pincer ligands and their complexes.

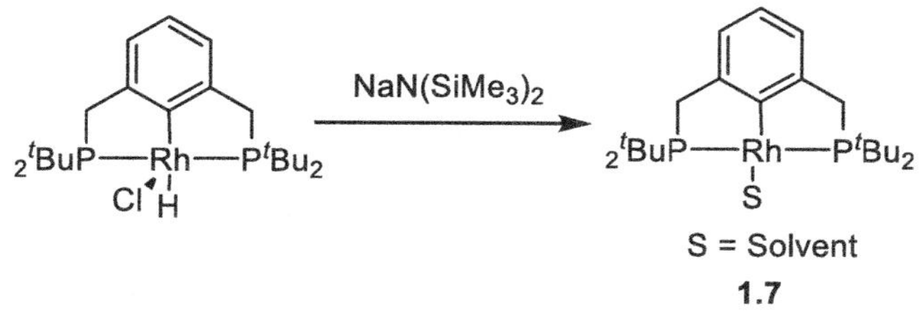

SCHEME 7.6 Dehydrochlorination of rhodium.

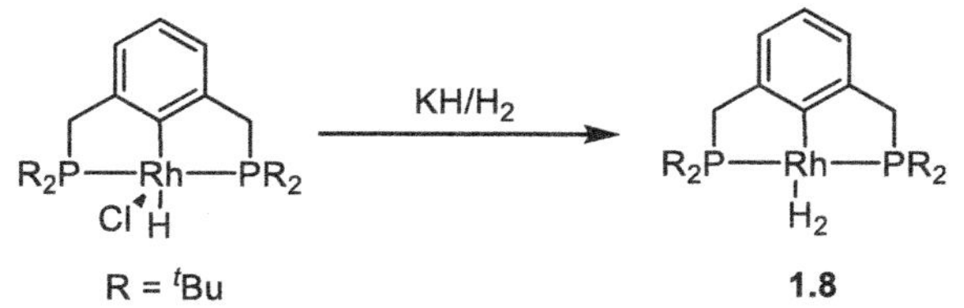

SCHEME 7.7 Dihydride complex.

tertiary butyl groups on phosphorus was considered to be crucial for the formation of such coordinatively unsaturated complexes [47].

The rhodium complex **1.5** reacts with H_2 with the replacement of HCl to form dihydrogen coordinated species **1.8**. Nemeh and coworkers have reported the reduction of CO_2 by rhodium complex **1.8** to produce carbon monooxide and the formation of a hydroxide derivative **1.10**. In the presence of carbon monooxide, it forms the complex **1.11** (Scheme 7.8) [47].

The interesting features of the bulky symmetrical PCP ligands is ability to stabilize the rhodium and iridium dihydride and tetrahydride pincer complexes. Such iridium complexes are successfully used for catalytic alkane dehydrogenation reactions. Jensen and coworkers have isolated the dihydride complexes **1.12** and **1.13** as shown in Scheme 7.9 [48–50].

Complexes **1.12** and **1.13** are highly efficient catalyst for dehydrogenation of alkene in the presence of *tert*-butylethylene (TBE) as dihydrogen acceptor [49]. It was observed that the catalytic efficiency of complexes **1.12** and **1.13** is getting suppressed by excess of alkene and thus to achieve higher turnover numbers a limited amount of tertiary butylethylene must be periodically added to the reaction mixture (Scheme 7.9).

Brookhart [51,52] and Goldman [53] have independently studied the effect of substituents on the ancillary phenyl ring of iridium complexes toward the transfer dehydrogenation of alkanes. The new catalysts of types **1.14** and **1.15** were more active than the established catalyst **1.13**. The reactivity of the complexes of types **1.14** and **1.15** were compared through computational studies. The striking difference observed was the affinity of PCP pincer complexes of the type **1.14** for TBE with the formation of the oxidative addition products. In contrast the affinity of complexes of the type **1.15** for TBE was rather low, where TBE simply binds to metals through π-coordination (Chart 7.4).

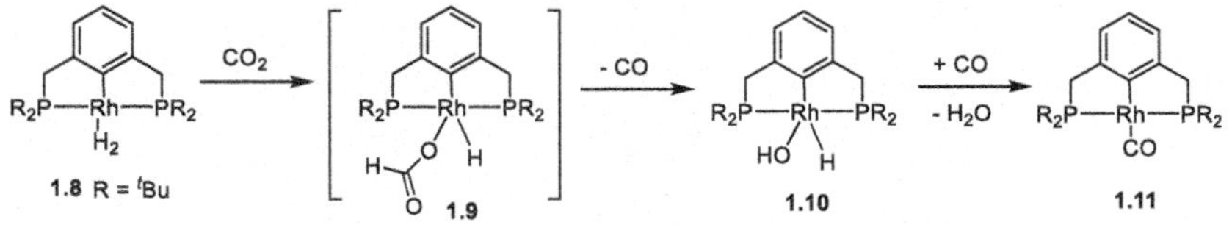

SCHEME 7.8 Activation of carbon monoxide.

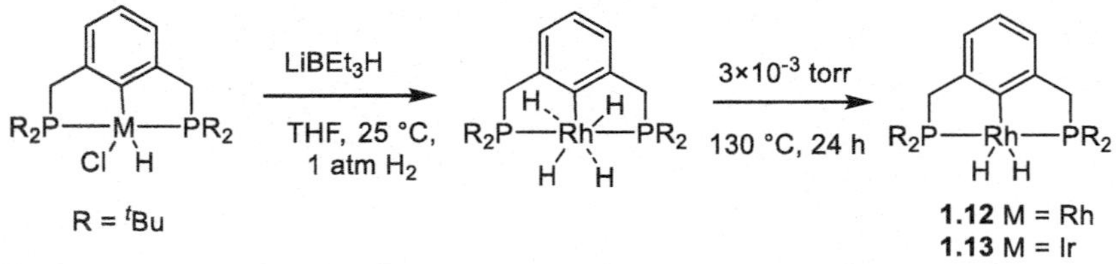

SCHEME 7.9 Dihydride and tetrahydride complexes.

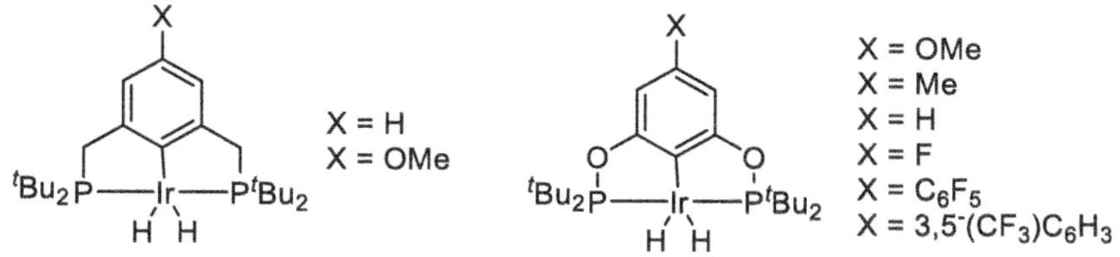

1.14 X = H, X = OMe

1.15 R = ᵗBu; X = OMe, Me, H, F, C₆F₅, 3,5-(CF₃)C₆H₃

CHART 7.4 Dihydride iridium(III) complexes.

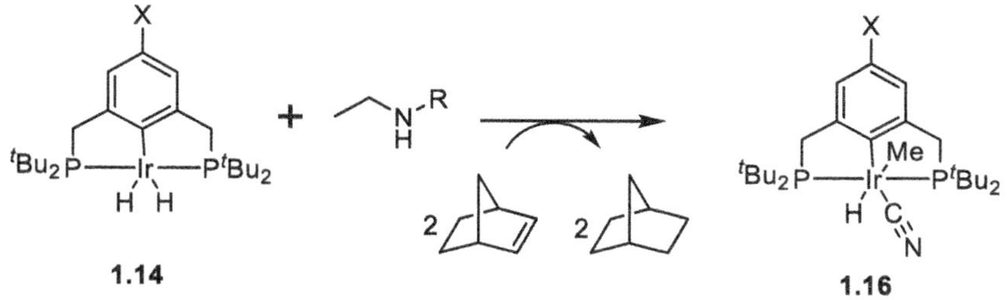

SCHEME 7.10 Hydrogenation of norbornene.

Iridium PCP pincer complexes are capable of activating not only the various C−H bonds but also the O−H bond of water and N−H bonds of aniline as well. In related studies, Zhang and coworkers have observed the cleavage of C−C bond on reacting N-ethylcyclohexylamine with **1.4** in the presence of norbornene (Scheme 7.10) [54].

Most notable work in the field of iridium catalyzed reactions is the alkane metathesis reported by Goldman and coworkers [55]. The complexes of types **1.14** and **1.15** in combination with complex **1.16** were used to carry out the organic transformations. The reaction initiates with the dehydrogenation of alkanes by iridium catalyst M to give terminal alkene and MH₂. The alkene obtained undergoes olefin metathesis promoted by complex **30** to yield ethylene and internal alkene. Finally, the internal alkene accepts hydrogen from MH₂ to form alkane with the subsequent regeneration of catalyst M (**1.17**; Scheme 7.11).

Recently, dehydroaromatization of *n*-alkanes has been obtained by iridium complexes **1.18−1.23** in the presence of TBE as hydrogen acceptor (Chart 7.5) [56]. The complexes **1.21** and **1.23** in combination with **1.19** were found to be the best catalytic systems. The selectivity for the formation of arenes with the same carbon number as the *n*-alkane substrates was very high for *n*-hexane and *n*-octane. Steps involved in the dehydroaromatization reaction are shown in Scheme 7.12.

The representative examples of group 10 metal pincer complexes are listed in Chart 7.6. Contrary to the prediction of Molten and Shaw regarding

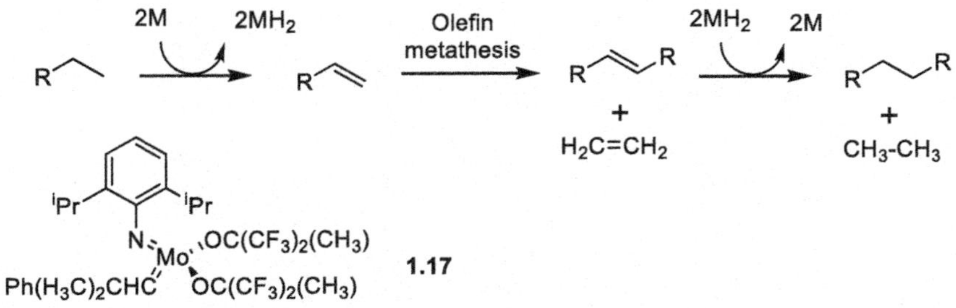

SCHEME 7.11 Alkane metathesis by iridium pincer.

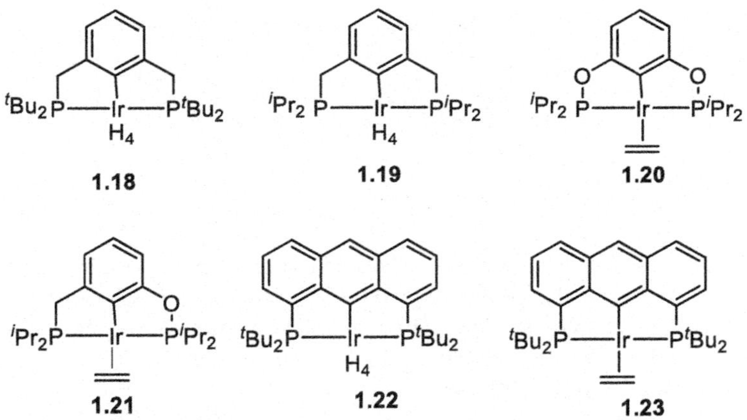

CHART 7.5 Iridium pincer complexes.

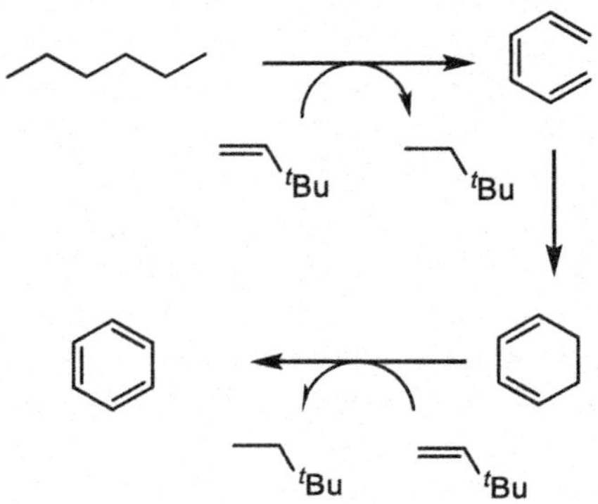

SCHEME 7.12 Dehydroaromatization reaction of hexane.

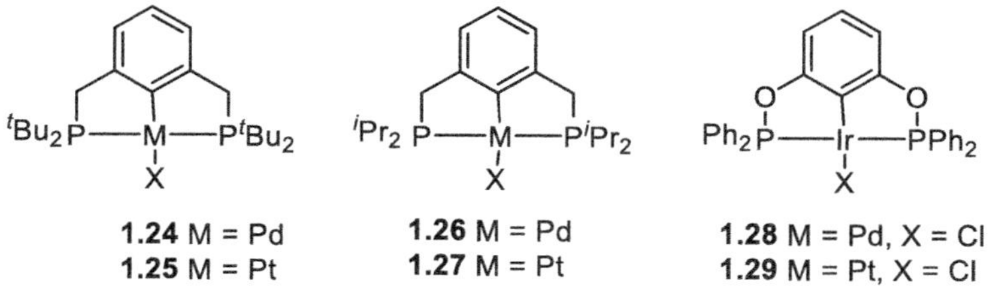

CHART 7.6 Palladium(II) and platinum(II) pincer complexes.

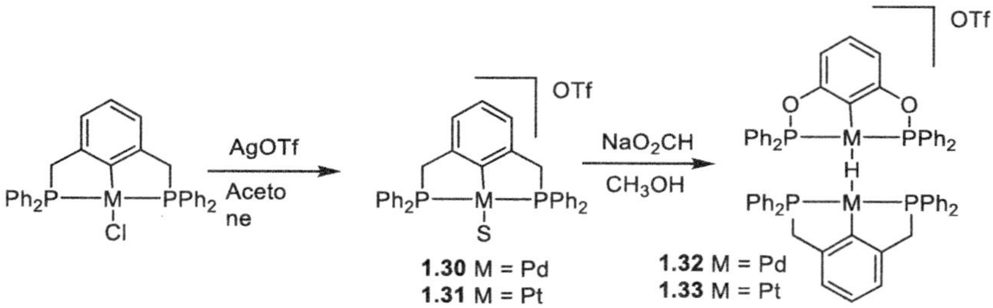

1.24 M = Pd
1.25 M = Pt

1.26 M = Pd
1.27 M = Pt

1.28 M = Pd, X = Cl
1.29 M = Pt, X = Cl

1.30 M = Pd
1.31 M = Pt

1.32 M = Pd
1.33 M = Pt

SCHEME 7.13 Synthesis of binuclear pincer complexes.

the necessity of bulky groups for the orthometallation, the pincer complexes **1.28** and **1.29** of ligand, 1,3-(CH$_2$PPh$_2$)$_2$C$_6$H$_4$ (**1.3**) were synthesized by Rimml and Venanzi in 1983 [57]. However, they have pointed out the role of steric bulky groups to drive the reaction toward the selective formation of *trans* species before cyclometallation, so that the C−H bond which is to be activated stay close to the metal center.

Palladium(II) and platinum(II) complexes **1.28** and **1.29** were further converted to the binuclear complexes **1.32** and **1.33** containing Pd-H-Pd bonds [58]. Bennett and coworkers have reported the syntheses of platinum pincer complex **1.29** by the reaction of **1.3** with [PtCl(CH$_3$)(COD)] or [PtCl$_2$(COD)] in good yield [59]. Treatment of complex **1.29** with AgBF$_4$ followed by KOH resulted in the formation of platinum complex, [Pt(OH){1,3-C$_6$H$_3$(CH$_2$PPh$_2$)$_2$}] (**1.34**) (Scheme 7.13).

Insertion of CO to the Pt−O bond is an interesting feature of the platinum hydroxo complexes. The reaction is not possible with platinum complexes of PPh$_3$ and PCy$_3$ ligands, where tertiary phosphines can be competitively displaced by carbon monooxide leading to the formation of platinum(0) carbonyl clusters. Platinum complex **1.34** reacts with carbon monooxide to give the corresponding carboxylate complex **1.35**, where cyclometallation of tertiary phosphine ligand provides the extra stability to the complex thus preventing the replacement of phosphine by carbon monooxide group (Scheme 7.14) [59].

Similar reaction of nickel complex **1.36** with 0.5 equiv of carbon monooxide obtained the binuclear pincer complex **1.38**, whereas the reaction of

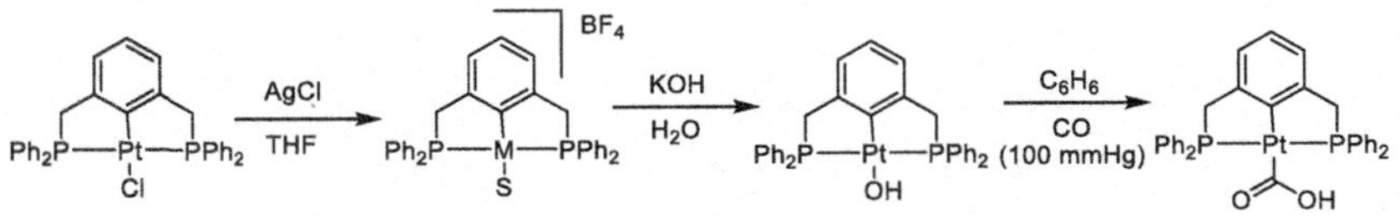

166 Recent Advances in Organometallic Chemistry

SCHEME 7.14 Reaction of CO insertion to platinum pincer complex.

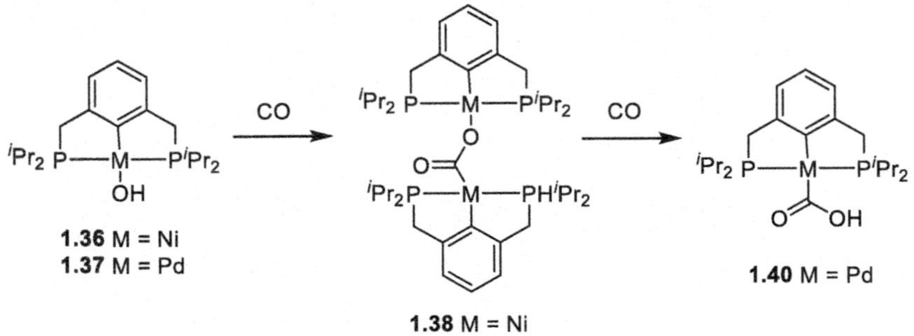

SCHEME 7.15 CO insertion to nickel pincer complex.

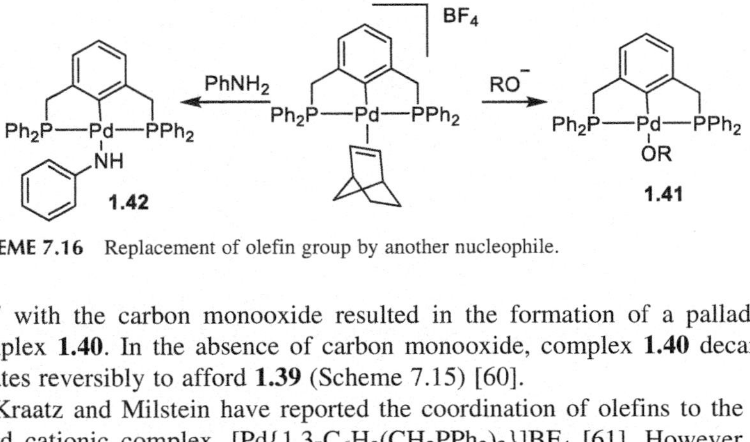

SCHEME 7.16 Replacement of olefin group by another nucleophile.

1.37 with the carbon monooxide resulted in the formation of a palladium complex **1.40**. In the absence of carbon monooxide, complex **1.40** decarbonylates reversibly to afford **1.39** (Scheme 7.15) [60].

Kraatz and Milstein have reported the coordination of olefins to the solvated cationic complex, [Pd{1,3-C$_6$H$_3$(CH$_2$PPh$_2$)$_2$}]BF$_4$ [61]. However, the olefins were only weakly bonded to the metals and insufficiently activated to allow the nucleophilic attack. Attempts to react them with nitrogen- and oxygen-based nucleophiles invariably lead to olefin replacement and formation of palladium complexes **1.41** and **1.42** (Scheme 7.16).

7.4.1.2 PNP pincer complexes

The chemistry of PNP pincer ligands was introduced by Nelson and coworkers [62]. The synthesis of PNP pincer complexes is straightforward, which involves the reaction of appropriate metal reagents with PNP ligands under normal reaction conditions (Charts 7.7 and 7.8). Zhang and coworkers have reported the synthesis of distorted square pyramidal iron complexes **1.42–1.46** [63]. The reaction of iron dichloride complex **69** with an excess of 0.5% sodium amalgam in the presence of carbon monooxide atmosphere yielded the iron dicarbonyl complex **1.48**, whereas the treatment of **1.44** with two equiv of NaBEt$_3$H produced the dihydride complex **1.50** [64]. Similar reaction of **1.47** with one equivalent of NaBEt$_3$H in the presence of carbon monooxide yielded the hydrido complex **1.50**. The complex **1.51** shows good catalytic activity for hydrogenation of ketones [38]. Pelczar and coworkers have reported the reduction of complex **1.42** with [NEt$_4$][BH$_4$] in the presence of carbon monooxide to give complex **1.49** [65].

The iron complexes of ligands **1.53–1.55** (Chart 7.7) have been applied as catalysts for coupling of some aromatic aldehydes with ethyldiazoacetate (Scheme 7.17) [66]. Interestingly, the complex **1.59** in the solid state reacts readily with carbon monooxide to give *cis*-[Fe{2,6- (CH$_2$P^{i}Pr$_2$)$_2$(C$_5$H$_3$N)}(CO)(Cl)$_2$]

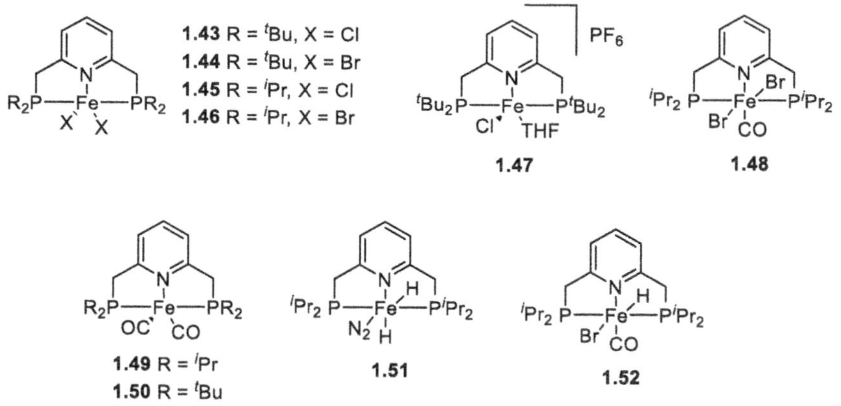

CHART 7.7 Some selected iron containing PNP pincer complexes.

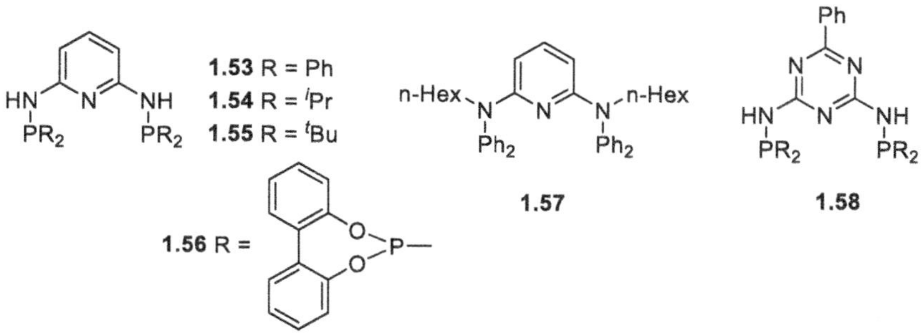

CHART 7.8 Some selected symmetrical PNP pincer ligands.

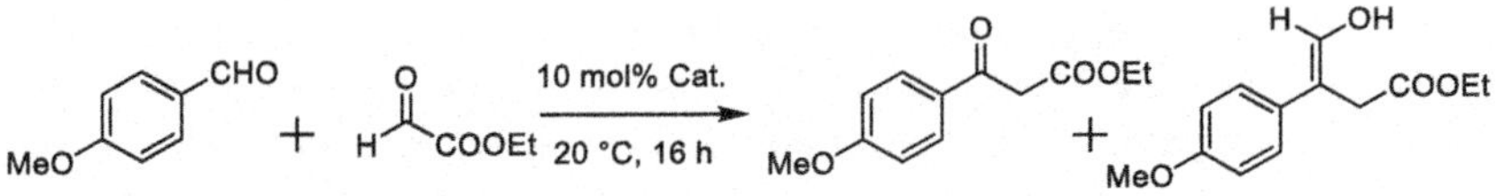

SCHEME 7.17 Example of C—C coupling reaction.

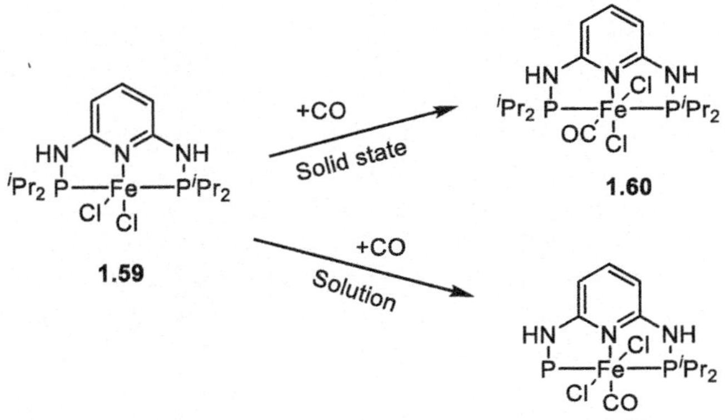

SCHEME 7.18 Reactions of CO with iron pincer complexes.

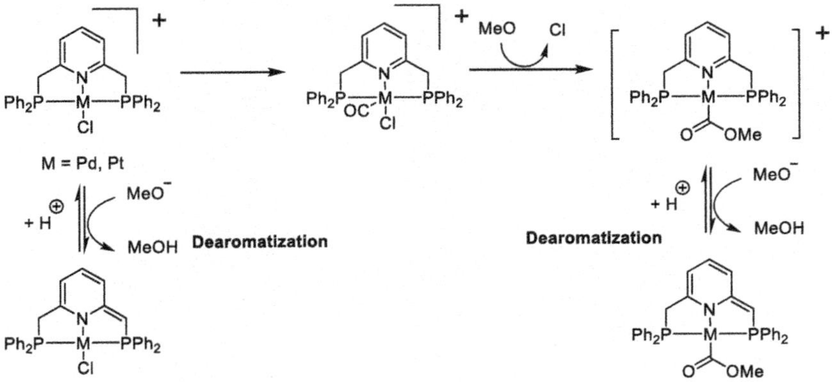

SCHEME 7.19 Dearomatization of pyridine rings.

(**1.60**), whereas the same reaction in solution provides exclusively the *trans*-[Fe{2,6-(CH$_2$P^{i}Pr$_2$)$_2$(C$_5$H$_3$N)}(CO)(Cl)$_2$] (**1.61**) (Scheme 7.18) [67,68].

Sacco and coworkers were first to observe the deprotonation and dearomatization sequences in PNP ligands while studying carbonylation of palladium and platinum PNP pincer complexes (Scheme 7.19) [69]. Palladium complexes **1.62** and **1.63** were synthesized by the reaction of in situ generated dicationic complex with ethylene and diethyl amine [70]. The coordinated ethylene was activated enough to react with secondary amines. Treatment of complex **1.62** with piperidine yielded corresponding β-aminoethyl complex **1.64** (Scheme 7.20). Hohn and coworkers have observed the similar addition reactions of water and alcohols to the coordinated olefins (Chart 7.9) [71].

van der Vlugt and coworkers reported the syntheses of cationic palladium pincer complexes **1.66**−**1.68** via transmetallation reaction utilizing the silver complex **1.65** (Scheme 7.21) [72]. The silver(I) complex **1.65** is quite reactive toward the various reagents and subsequent reaction with various reagents can give the different types of cationic silver complexes.

7.4.2 Application of tridentate chelating pincer complexes in catalysis

The tridentate-chelating pincer complexes have proven to be very useful in numerous catalytic reactions [13,73]. The donor atoms of the side arms and their substituents allow the fine-tuning of the steric and electronic properties around the central metal. The chelation effect provides extra thermal stability as well as stability toward air and moisture. The pincer complexes have been successfully employed in various catalytic organic transformations, such as Suzuki-Miyamura, Sonogashira, Stille, polymerization, dehydrogenation of alkenes, hydroformylation, hydrophosphination, and other C−C bond forming reactions [16,74,75]. Aldol and Michael types of reactions, electrophilic

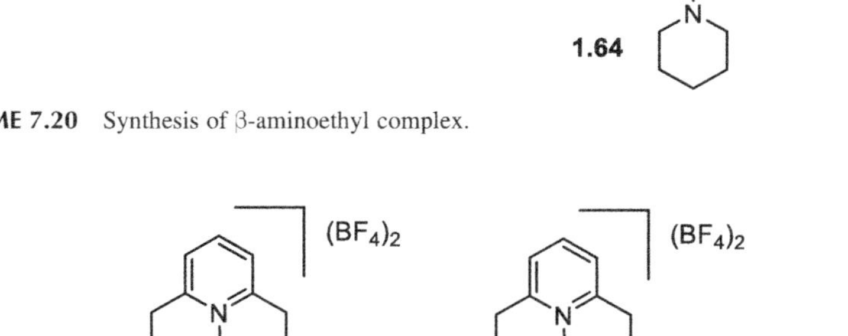

SCHEME 7.20 Synthesis of β-aminoethyl complex.

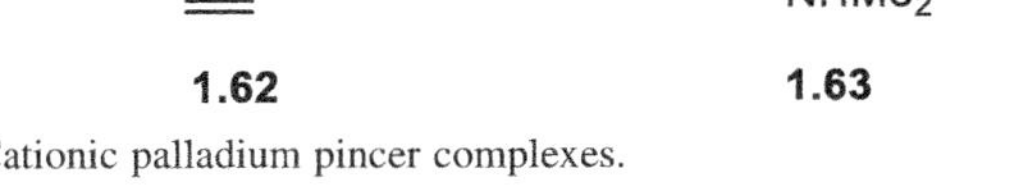

CHART 7.9 Cationic palladium pincer complexes.

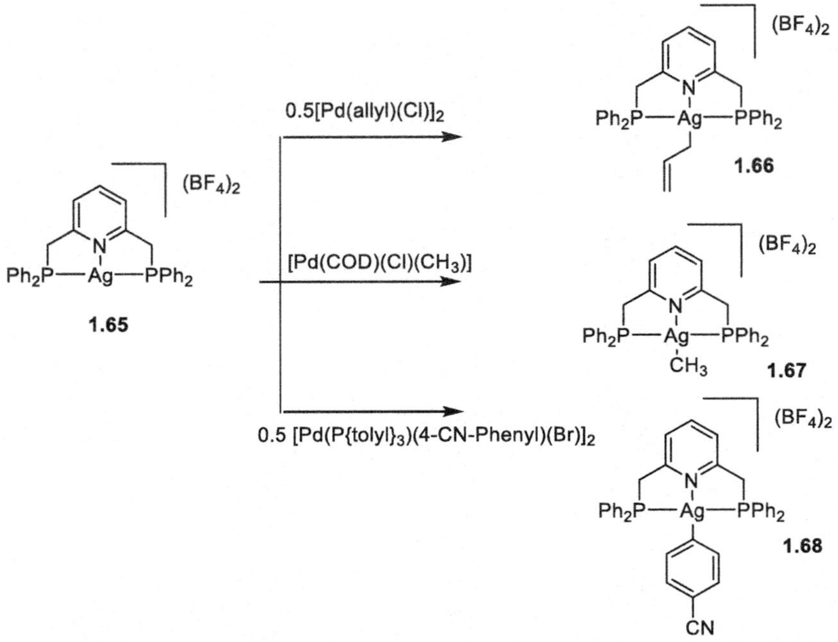

SCHEME 7.21 Cationic pincer complexes of silver(I).

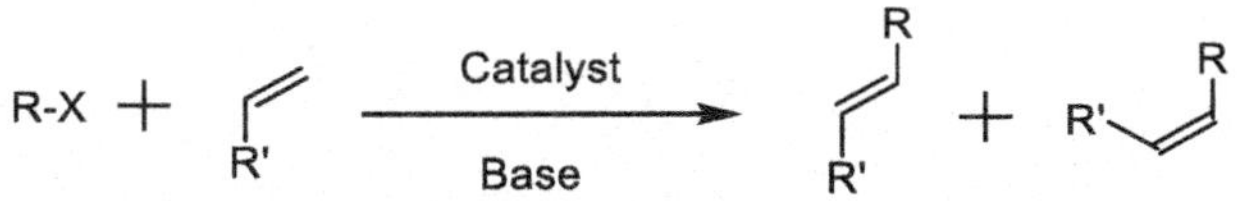

SCHEME 7.22 Heck-coupling reaction.

allylation of aldehydes, phosphonation are few other catalytic reactions explored using pincer complexes as catalysts [14,76,77]. Furthermore, chiral tridentate chelating pincer complexes have been extensively studied with high potential in asymmetric catalytic reactions.

7.4.2.1 Heck cross-coupling reactions

The Mizoroki-Heck reaction is one of the synthetically most important C−C bond forming methods. A Heck reaction is basically the coupling of activated alkenes with an aryl or vinyl halides catalyzed by an organopalladium (Pd(0) or Pd(II)) catalyst in the presence of a base (Scheme 7.22) [78].

Two mechanisms are proposed for palladium pincer catalyzed Heck coupling reactions and they are presently under debate [13,79,80]. The conventional palladium catalyzed Heck coupling reaction involves Pd^0/Pd^{II} catalytic cycle. In the case of pincer complexes it was believed that the Pd^{II} species is less likely to be reduced to Pd^0 but recent studies show that the palladium metal based pincer complexes decompose under catalytic reaction condition, to furnish colloidal Pd⁰ intermediate species at the end.

The pincer complexes act as precatalyst in Pd^0 to Pd^{II} catalytic cycle. It releases colloidal palladium in the form of bulky cluster or nanoparticles during course of catalytic reactions. In most of the cases, classical mercury drop test has proved the presence of naked metal species in reaction conditions. From Pd^{II} to Pd^{IV} mechanism involves alkene coordination to the pincer complex, followed by the loss of acid. The oxidative addition of the aryl halide forms an intermediate species as Pd^{IV}. The reductive elimination shows regeneration of the catalyst. Thus with this proof for the mechanism, the Heck coupling by pincer complexes may follow one of the path or both for the reaction. Typical reactions of Pd^0/Pd^{II} and Pd^{II}/Pd^{IV} catalytic cycles for Heck coupling reaction are shown in Chart 7.10.

7.4.2.2 Asymmetric allylation of sulfonimines

Szabo, Gebbnik and others have revealed the catalytic ability of palladium pincer complexes in electrophilic allylic substitution of imine substrates by allylstannanes or potassium trifluoro(allyl)borate as allylating reagents [81,82]. Electron deficient pincer complexes with weakly coordinating OTf group showed better activity compared to firmly attached halides. The catalytic reaction afforded allyl substituted amines in high yields with good enantio-selectivity. Mechanistic studies have shown that the pincer complexes undergo *trans*-metallation with allylating reagent to afford (η^1-allyal)-palladium complexes of the type **1.70** (Scheme 7.23). The reaction of intermediate **1.71** with sulfonimines resulted in the formation of allylated product. The catalytic results have shown that the utility of phosphate pincer

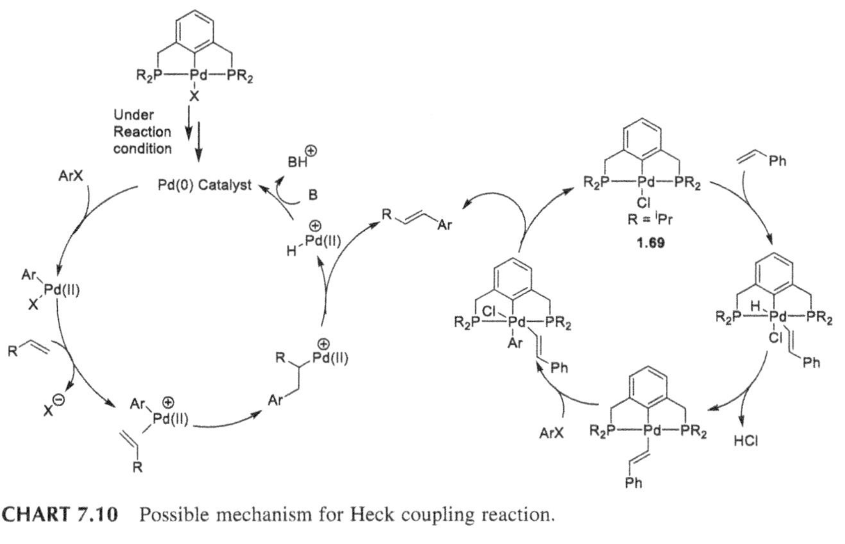

CHART 7.10 Possible mechanism for Heck coupling reaction.

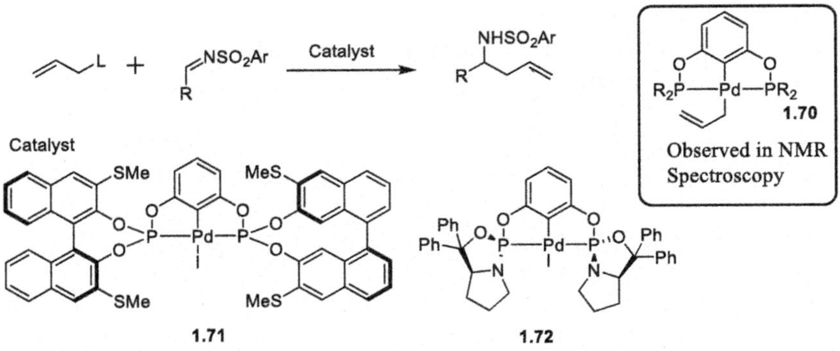

SCHEME 7.23 Catalytic allyalation through pincer complexes.

complexes bearing sterically bulky chiral substituents afforded higher catalytic activity in the allylation of imines with tributylallylstannane or potassium trifluoro(allyl)borate. The expected products were obtained in 40% − 85% yields with 50%−80% ee.

7.4.2.3 Hydrophosphination reaction

In recent years, the metal catalyzed P−C bond forming reactions drawn attention mainly because of their simple synthetic methods when compared to conventional methods [83] Duan and coworkers have shown the ability of palladium pincer complexes in catalyzing a new hydrophosphination reactions [84]. They have screened several chiral as well as achiral palladium pincer complexes for the hydrophosphination reaction and **1.73** showed good conversions with excellent enantioselectivity. The reaction shown the possibility of addition of diaryl phosphines across an activated C−C double bond (Scheme 7.24). The same group have also used the pincer complexes **1.73** for hydrophosphination reaction involving sulfonic ester [85]. Addition reaction of P−H bonds across nitroalkenes has been reported recently in the literature [86].

7.4.3 Chiral tridentate chelating pincer complexes

Numerous chiral pincer complexes have been synthesized and also utilized effectively as catalyst for enantioselective organic transformations. The stereochemical centers are introduced either through chiral substituents on phosphorus atom or by arising chirality at benzylic positions of central aromatic ring. They have reported that the methylation reaction can be done after cyclometallation as well (Chart 7.11) [87].

Longmire and coworkers studied the chemistry of chiral pincer ligands by synthesizing the **1.82** and **1.83** which, are found to be excellent catalyst for aldol condensation reaction between methylisocyanoacetate and aldehydes

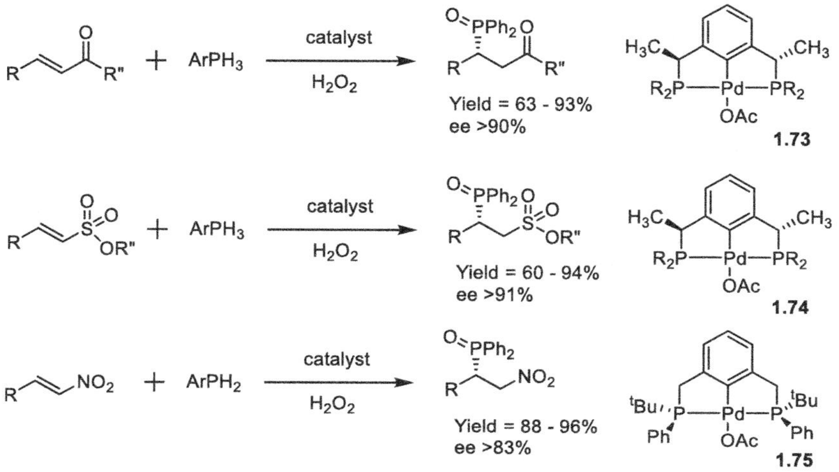

Organophosphorus compounds **Chapter | 7** 173

SCHEME 7.24 Pd-pincer complexes catalyzed hydrophosphination reactions.

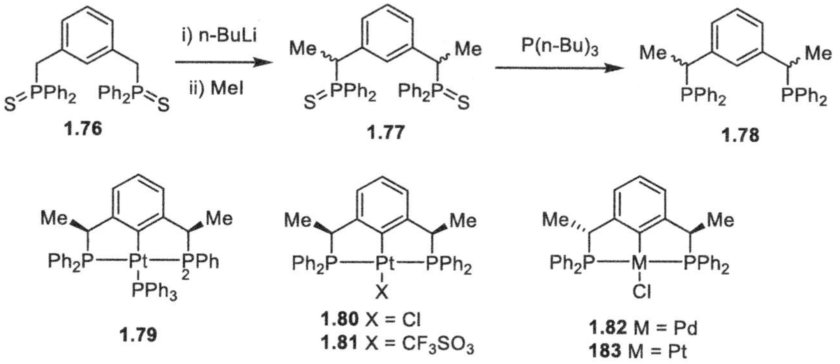

CHART 7.11 Examples of chiral PCP pincer complexes.

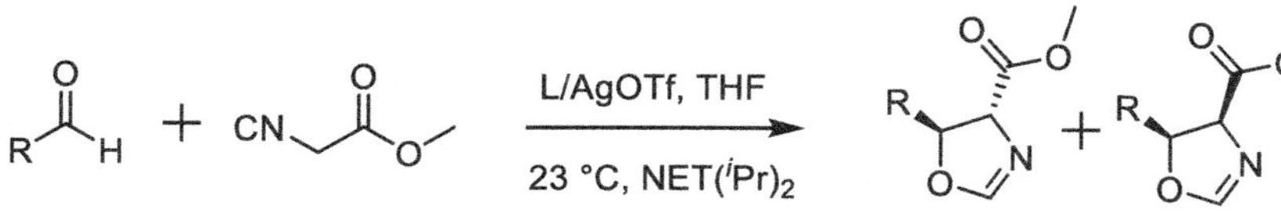

SCHEME 7.25 Chiral catalytic organic transformation by silver(I) complex.

(Scheme 7.25) [88]. The chiral ligand synthetic method was found to be better than that employed by Venanzi and coworkers.

The chiral palladium complexes containing BINOL (**1.86**) and TADOL (**1.87**) ligands were prepared by the oxidative addition of prefunctionalized ligands of the type **1.85** to Pd$_2$(dba)$_3$ [22] (Scheme 7.26).

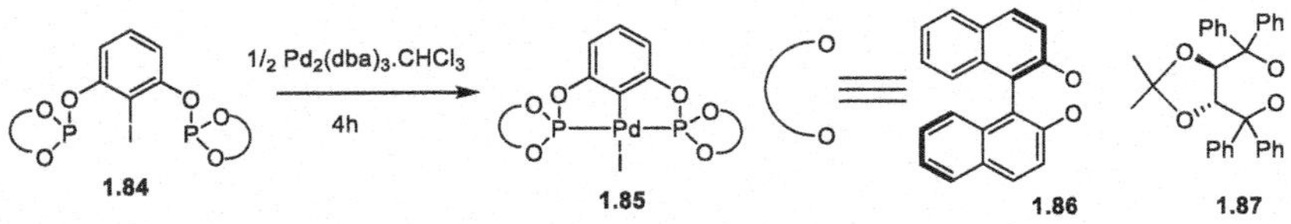

SCHEME 7.26 Bulky chiral catalytic organic transformation substituent on phosphorus.

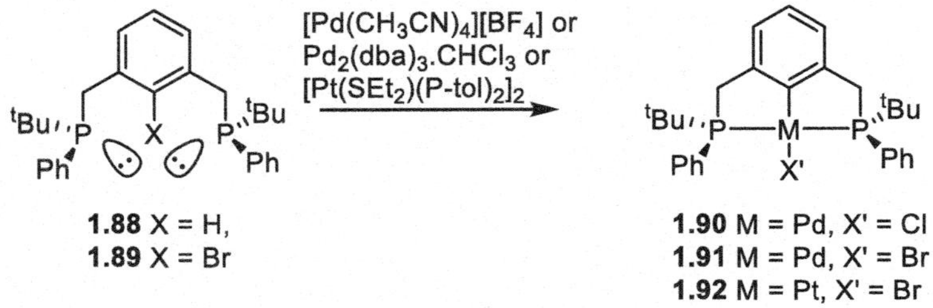

SCHEME 7.27 Pincer complexes through oxidative addition on aryl halide.

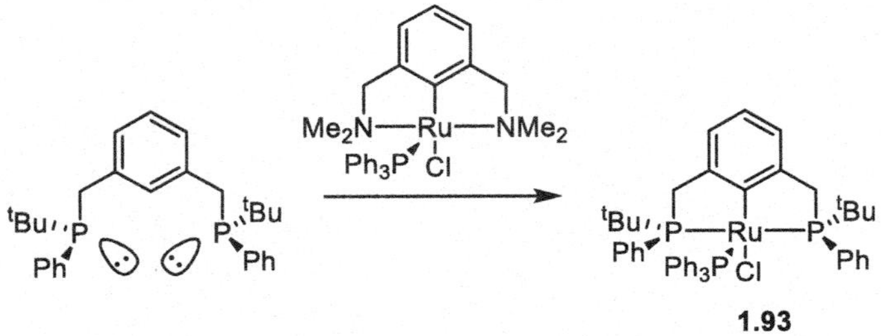

SCHEME 7.28 Transcyclometallation in chiral compound.

In the recent years the synthesis of enantiomeric pure transition metal pincer complexes of the type **1.90–1.92** and are reported (Scheme 7.27) [22]. The palladium pincer complexes were further transformed in to the aqua complexes and applied as catalyst for the aldol condensation of methyl 2-isocyanoacetate with the benzaldehyde. The synthesis of ruthenium(II) complex **1.93** involves transcyclometallation reaction (Scheme 7.28) [28].

7.4.4 Aminophosphines

The phosphorus and nitrogen elements form numerous compounds with large structural diversity [89,90]. The aminophosphines can be classified on the basis of degree of unsaturation in phosphorus and nitrogen bonds: phosphazanes (P−N), phosphazenes (P=N), and phosphonitriles (P≡N) [91]. The cyclophosphazanes are important class of saturated inorganic heterocyclic compounds in which the phosphorus and nitrogen atoms are arranged alternatively in a cyclic skeleton.

7.4.4.1 Cyclodiphosphazanes

Cyclodiphosphazanes of the type $\{XP(\mu\text{-}NR)\}_2$ are the major class of cyclic phosphorus-nitrogen compounds [92−96]. The importance of these four-membered cyclic compounds is mainly due to the application in diverse areas [97−103]. Cyclodiphosphazane $\{ClP(\mu\text{-}NPh)\}_2$ (**1.1**) was first reported by Michaelis and Schroeter in 1894 [104,105] and characterized in 1960s [106−108]. Bis(chloro)cyclodiphosphazane of the type cis-$\{ClP(\mu\text{-}NR)\}_2$ and mono(chloro)cyclodiphosphazane of the type cis-$\{ClP(\mu\text{-}NR)_2PN(H)R\}$ are important precursor for synthesizing a various types of various cyclodiphosphazane derivatives.

7.4.4.2 Cis/trans *isomerization*

The cyclodiphosphazanes can exist in two basic forms, namely, *cis* and *trans* isomers similar to cyclobutane derivatives (Chart 7.12) [109,110]. Majority of cyclodiphosphazanes exist in *cis* form both in solid state and in solution. But some of them exist as a mixture of *cis* and *trans* isomers in solution, whereas in the solid state, the *cis* isomer predominates. The $^{31}P\{^1H\}$ NMR data are very useful in distinguishing the isomers of cyclodiphosphazanes as they show distinct chemical shifts for both the isomers. The chemical shift value of *cis* isomers appears at high field compared to that of *trans* isomers [111]. A very less energy (*ca.* 4 to 40 kJ/mol) is required for the interconversion of *cis* and *trans* isomers [111,112].

7.4.4.3 Synthesis of cyclodiphosphazanes

The simplest moiety of cyclodiphosphazanes are the bis(chloro)cyclodiphosphazanes with general formula cis-$\{ClP(\mu\text{-}NR)\}_2$ (**1.94**). They can be conveniently synthesized by the reaction of phosphorustrichloride with primary amines (Scheme 7.29) [113,114]. The $^{31}P\{^1H\}$ NMR spectra of *cis*-bis(chloro) cyclodiphosphazanes show a single resonance around 200−210 ppm. The asymmetrically substituted cyclodiphosphazanes, cis-$\{ClP(\mu\text{-}NR)_2PN(H)R\}$ (**1.95**) are useful materials for the synthesis of acyclic oligomers, which can be prepared by reacting PCl_3 and primary amines in 1:4 molar ratio [115].

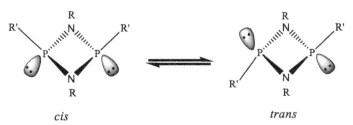

CHART 7.12 *Cis/trans* isomers of cyclodiphosphazanes.

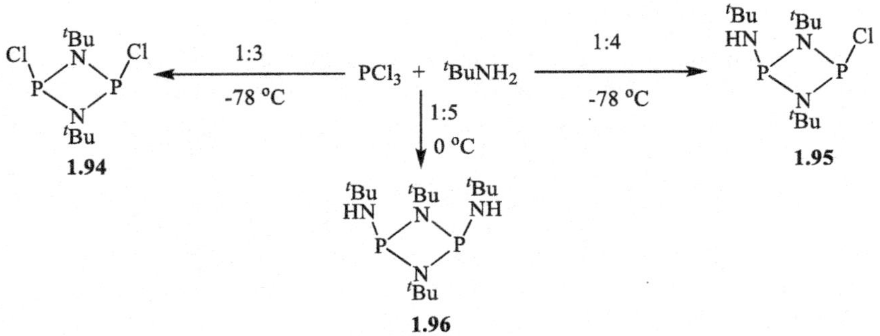

SCHEME 7.29 Halo- and aminocyclodiphosphazane.

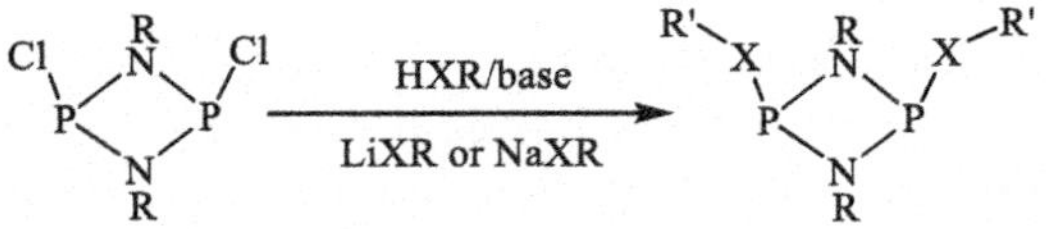

SCHEME 7.30 Functionalization of cyclodiphophazanes.

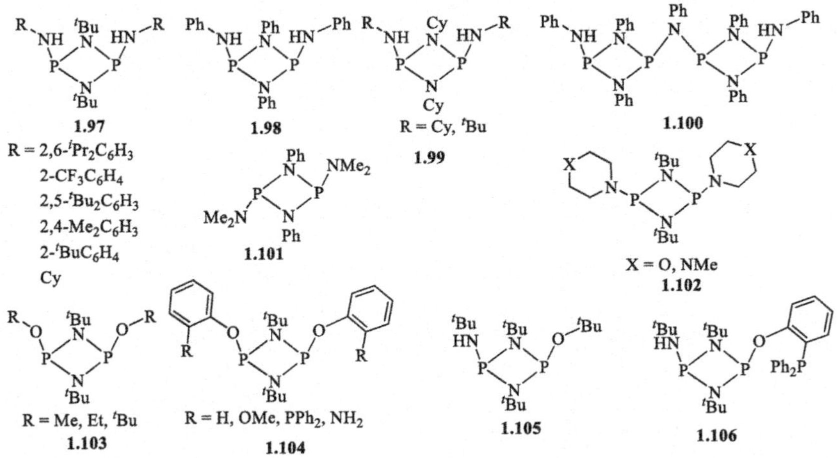

CHART 7.13 Examples of functionalized cyclodiphosphazanes.

The homo substituted bis(amino)cyclodiphosphazanes of the type **1.96** can be synthesized by the condensation reaction of appropriate amine with PCl_3 in 1:5 molar ratio, while the hetero-substituted derivatives were prepared by reacting the bis(chloro)cyclodiphosphazane **1.94** or mono(chloro) cyclodiphosphazane **1.95** with primary amines or alcohols in presence of a base or with lithium/sodium salts of amines or alcohols (Scheme 7.30). A few examples of cyclodiphosphazane derivatives (**1.96–1.105**) are shown in Chart 7.13 [113,116–126].

7.4.4.4 Reactivity of cyclodiphosphazanes

The reductive coupling of cis-{ClP(μ-N^{t}Bu)}$_2$ (**1.94**) using excess magnesium in THF under refluxing condition gave an interesting cage compound, {P(μ-N^{t}Bu)}$_4$ (**1.107**) [127]. In spite of the fact that the reaction appears very simple, several P−N bond breaking and forming steps must accompany the coupling process. The molecular structure of **1.107** has been determined by X-ray crystallography.

The reactions of cis-{ClP(μ-N^{t}Bu)}$_2$ (**1.94**) with H$_2$O in the presence of triethylamine in THF solvent at −78°C resulted in the formation of a P^{III}/P^V dimeric compound [(μ-O)-{P(μ-N^{t}Bu)}$_2$P(O)H}$_2$] (**1.108**), along with the P^V dimer {H(O)P(μ-N^{t}Bu)$_2$}$_2$ (**1.109**) as a minor product [128]. The bis-chloro derivative **1.94** can be converted into bis(fluoro)cyclodiphosphazane (**1.110**) by halogen exchange reaction with SbF$_3$ as shown in Scheme 7.31 [129]. Woods and coworkers reported the reduction of cis-{ClP(μ-N^{t}Bu)$_2$PN(H)^{t}Bu} (**1.95**) with LiBEt$_3$H to afford cis-{HP(μ-N^{t}Bu)$_2$PN(H)^{t}Bu} (**1.111**), whereas the same reaction using LiB^{s}Bu$_3$H as reducing agent gave trans-{HP(μ-N^{t}Bu)$_2$PN(H)^{t}Bu} (**1.112**) [130].

Chivers and coworkers found a planar P$_6$E$_6$ macrocycle incorporating cyclodiphosphazane groups by the oxidation of sodium salt **1.114** with molecular iodine (I$_2$) [131]. The 15-membered trimeric macrocycle **1.115** was obtained in very low yield. A cyclic tetra-selenide **1.116** incorporating P$_2$N$_2$ ring was isolated by the reaction of **1.14** with Se$_2$Cl$_2$. The P$_2$N$_2$ supported macrocycle **1.115** can be converted in to tetra-selenide (**1.116**) via two electron oxidation by I$_2$ as shown in Scheme 7.32. The treatment of cis-{Cl(Se)P(μ-N^{t}Bu)}$_2$ (**1.113**) with excess of sodium metal in toluene under refluxing conditions afforded a hexameric seleno macrocycle {(Se = P)(μ-N^{t}Bu)$_2$P(μ-Se)}$_6$ (**1.117**) formed by the formal head-to-tail cyclization [132].

Balakrishna and coworkers reported [1,3]-sigmatropic rearrangement of an amine functionalized cyclodiphosphazane into chalcogenophosphates with elemental chalcogens as shown in Scheme 7.33 [133]. The reactions of two equiv of sulfur or selenium with cis-{(μ-N^{t}BuP)(OCH$_2$CH$_2$NMe$_2$)}$_2$ (**1.118**)

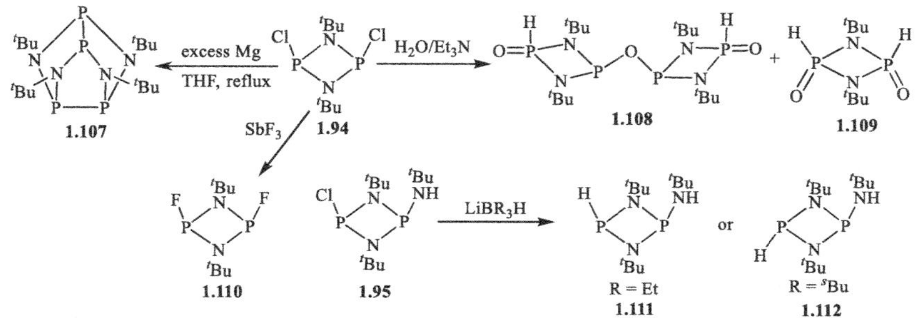

SCHEME 7.31 Reactivity of cyclodiphosphazanes with various reagents.

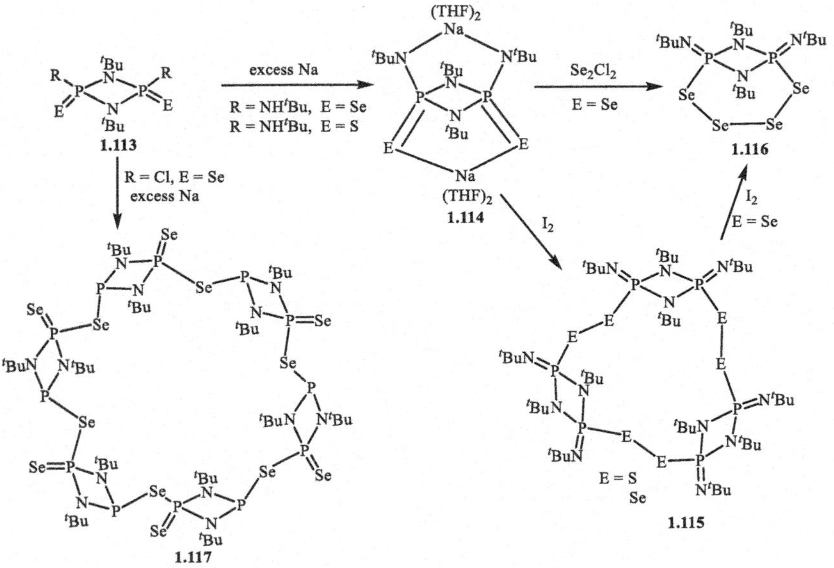

SCHEME 7.32 Reactivity of chalcogenide derivative of cyclodiphosphazenes.

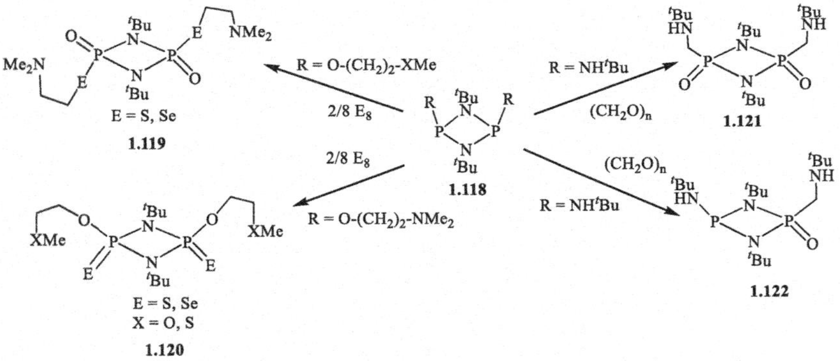

SCHEME 7.33 Reactivity of exocyclic P—N bonds.

afforded rearranged product *trans*-{(μ-N^{t}BuPO)(ECH$_2$CH$_2$NMe$_2$)}$_2$ (E = S or Se) (**1.119**). During the rearrangement, cyclodiphosphazane ring undergoes geometrical isomerization to form exclusively *trans* isomers. The interaction between N···E (E = S, Se) favours the rearrangement reaction by forming a six-membered transition state. Ether- and thioether-functionalized cyclodiphosphazanes lack corresponding strong O···E and S···E interactions and hence no rearrangement reaction was observed (Scheme 7.33). The reaction of bis(amino)cyclodiphosphazane with excess of paraformaldehyde produced exclusively the methylene inserted *cis*-cyclodiphosph(V)azane derivative **1.121** [134]. A similar reaction in 1:1 molar ratio produced monomethylene inserted *cis*-cyclodiphosph(III/V)azane derivative **1.122**. The insertion of methylene fragments was observed selectively into exo-cyclic P—N bonds of bis(amino)cyclodiphosphazane.

7.4.4.5 Macrocyclic cyclodiphosphazane derivatives

The bias toward the *cis* conformation of rigid four membered P_2N_2 rings provides a favorable preorganization for the formation of macrocyclic and cage compounds. Several bi-functional organic alcohols, amines or aminoalcohols (LL'H$_2$) have been utilized to gave a broad range of macrocycles $\{(\mu\text{-N}^t\text{Bu})_2(\text{LL}')\}_n$ ($n = 1-5$) [135]. The size of the macrocycle is basically depends on the size, position of reaction site and orientation of the organic linker. The reaction conditions also play an important role in producing selectivity. The monomers are generally formed with more flexible organic spacers, while rigid organic spacers favor higher oligomers in the compounds (Scheme 7.34 and Chart 7.14).

7.4.5 Phosphacyclophanes

A cyclic organic compounds containing aromatic groups and various linkers having supramolecular structures are commonly known as cyclophanes [136,137]. Cyclophanes containing donor atoms, such as O, N, and S, find numerous applications in synthesis [138–140], catalysis [141–144], biomimetic reactions [145,146], and ion sensing and molecular recognition [147,148].

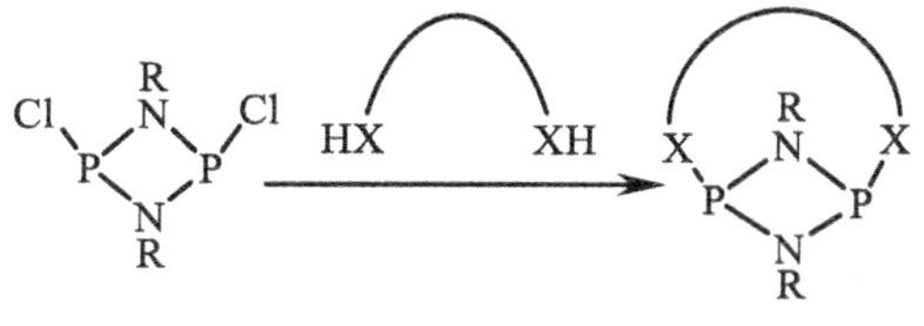

SCHEME 7.34 Bicyclic cyclodiphophazanes.

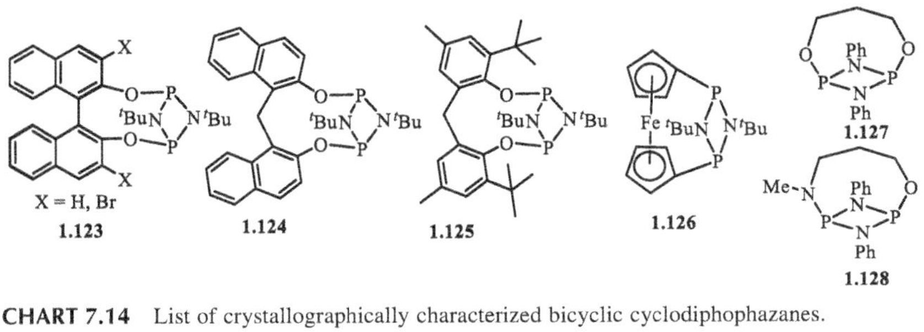

CHART 7.14 List of crystallographically characterized bicyclic cyclodiphophazanes.

Depending on the position of coupling of C−X (X = C or any hetero atom) of aromatic ring and linkers, the cyclophanes can be differentiated into three class ortho-, meta-, and paracyclophanes (Chart 7.15).

The added donor atoms can be the part of the cyclic systems as linkers or they can also be placed in the peripheral side as a part of exocyclic dangling side arms [149,150]. Crown ethers [151–154], cryptands [155–158], functionalized calixarenes [159,160], resorcinarenes [161], porphyrins, or cyclophanes containing thia- and aza-groups also belong to this class of cyclic systems [162]. Surprisingly, cyclophanes containing phosphorus(III) atoms ideal for soft-ion recognition and catalytic applications are scarce in the

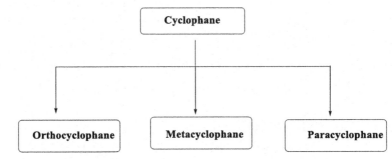

CHART 7.15 Classification of cyclophane.

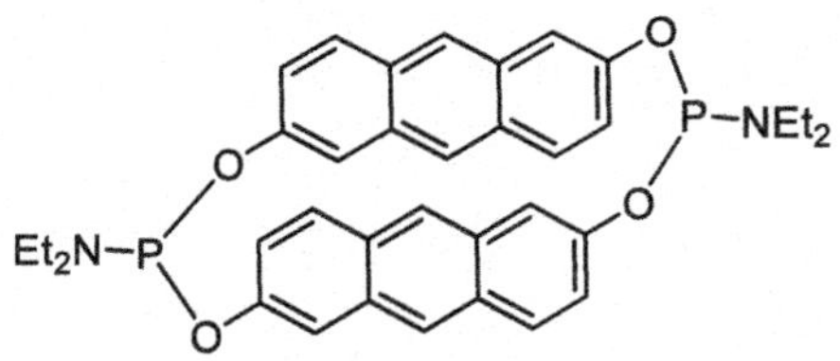

1.129

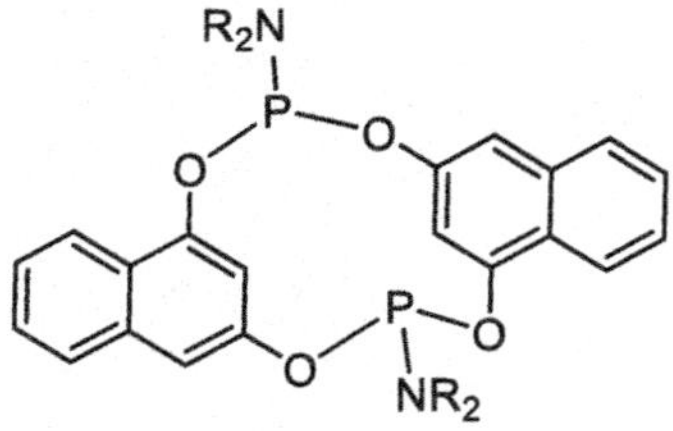

1.130

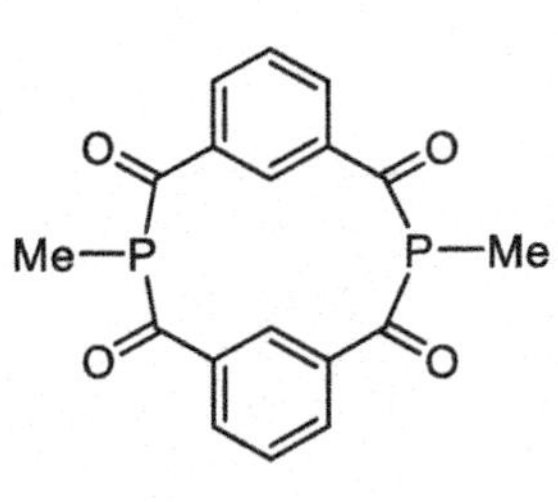

1.131

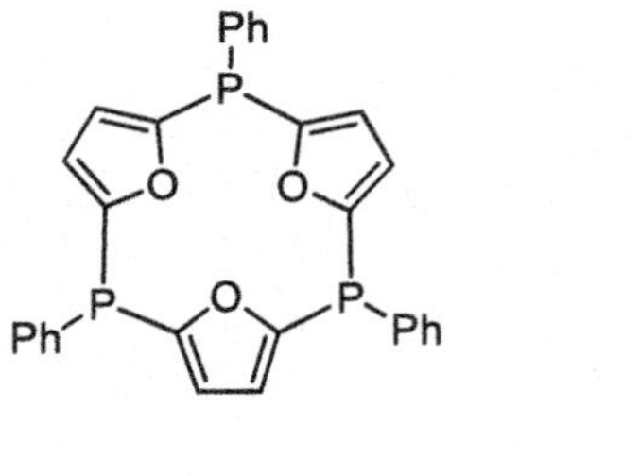

1.132

CHART 7.16 Examples of phosphacyclophanes.

literature which may be attributed to the sensitivity of phosphorus—carbon [163—166], phosphorus—oxygen [167—170], or phosphorus—nitrogen [171,172] bonds toward acids and bases or hydrolysis reactions. Compounds with phosphorus(III)—carbon bonds are sensitive to the oxidation as well (usually even more than to hydrolysis).

In spite of the fact that many phosphacyclophanes are known [150,164], phosphametacyclophanes are scarce and the couple of known examples are diphosphametacyclophane (**1.131**) [173] and phenylphosphinacalix[3]trifuran (**1.132**) [174] reported in literature (Chart 7.16).

7.5 Summary

This chapter includes the phosphorus compounds and its reactivity in detail with diverse examples. Phosphorus compounds are basically categorized in cyclic and acyclic phosphorus-based compounds in which it is further subcategorized depending on substitution over phosphorus. Number of phosphorus and other donor sites in one ligand it can be a good candidate for chelating ligand. Two donor phosphorus groups will be a good candidate for chelating ligand in coordination chemistry and also three donor groups will be a tridentate chelating ligand which can show interesting coordination compounds with great application in coordination or organometallic chemistry. There are numerous phosphorus compounds which show different types of structural geometries and shapes and its complexes also have huge importance due to catalytic applications of the compounds. The chelating ligands especially tridentate pincer compounds are very much interesting for the coordination chemistry and organometallic chemistry due to its interesting activity and utility in various organic transformations.

The cyclic phosphorus compounds are also gives interesting aspects for coordination chemistry. Cyclodiphosphazanes are one of the well explored cyclic phosphorus compound which show great attraction due to its wide scope. There are other cyclic phosphorus compounds have reported and its applications also explored. The phosphacyclophanes are one of the example of cyclic phosphorus compounds. These cyclic phosphacyclophanes have tremendous potential to explore it chemistry.

7.6 Problems with solutions

Q1. What are classification of phosphorus compounds?

Answer: The phosphorus compounds are basically classified in two main types that is cyclic phosphorus compounds and acyclic compounds. Further acyclic phosphorus compounds classified as phosphinite, phosphonite and phosphite depending on one, two and three alkoxy/aryloxy groups substituent on phosphorus respectively. The cyclic phosphorus compounds may consist of hetero atom in its ring compounds.

Q2. What are different types of pincer ligands?
Answer: Depending on the side arms of tridentate chelating pincer ligand there are two types of pincer ligands are symmetric and asymmetric pincer ligand.

Q3. What is Heck-cross coupling reaction?
Answer: A Heck reaction is basically the coupling of activated alkenes with an aryl or vinyl halides catalyzed by an organopalladium (Pd(0) or Pd (II)) catalyst in the presence of a base.

Q4. What are the different kinds of isomers in cyclodiphosphazane?
Answer: Cyclodiphosphazanes of the type $\{XP(\mu\text{-}NR)\}_2$ are the major class of cyclic phosphorus-nitrogen compounds which contains four membered alternate phosphorus and nitrogen ring.

Q5. What are the phosphacyclophane?
Answer: A cyclic organic compounds containing aromatic groups and various linkers having supramolecular structures are commonly known as cyclophanes. Cyclophanes containing donor atoms, such as O, N, and S, for connecting group.

7.7 Objective-type questions

i Who has first reported pincer compound?
  a Shaw and coworkers
  b Baber and coworkers
  c Koten and coworkers
  d None of them
     Answer: (a)

ii Phosphinite compound contains —————-number alkoxy/aryloxy groups
  a 1
  b 2
  c 3
  d 4
     Answer: (a)

iii Cyclodiphosphazane ring consist of —————groups/atom
  a Phosphorus and Carbon
  b Phosphorus and Nitrogen
  c Phosphorus and Halogen
  d Phosphorus and Oxygen
     Answer: (b)

iv Pincer ligand consist of —————number of donor sites
  a One
  b Two
  c Three
  d Four
     Answer: (c)

v In Heck-cross coupling reaction what are the possible oxidation states of palladium metal?
  a 0, +2
  b 0, +1
  c 0, +3
  d +1, +2
  Answer: (a)

vi What is the type of PNP pincer ligand?
  a Symmetrical
  b Unsymmetrical
  c Chiral
  d None of them
  Answer: (a)

vii Cis or trans cyclodiphosphazanes are depending on
  a Substitution on nitrogen
  b Substitution on phosphorus
  c Both nitrogen and Phosphorus
  d All of them
  Answer: (b)

viii —————-metal precursors are used in pincer complex formation?
  a Low valent
  b High valent
  c Both A & B
  d None of them
  Answer: (a)

ix Cyclophanes are not classified as
  a Orthocyclophane
  b Metacyclophane
  c Paracyclophane
  d Ipsocyclophane
  Answer: (d)

x Alternating phosphorus and nitrogen————— is cyclodiphosphazanes?
  a Four membered ring
  b Six membered ring
  c Two membered ring
  d None of them
  Answer: (a)

References

[1] Pignolt LH, editor. Homogeneous catalysis with metal phosphine complexes. New York: Plenum; 1983. p. 167.

[2] Kashid VS, Kote BS, Kunchur HS, Mague JT, Balakrishna MS. Transition metal complexes of N1, N1, N2, N2-tetrakis (diphenylphosphaneyl)-ethane-1, 2-diamine [(Ph2P) 2NCH2CH2N (PPh2) 2]. Polyhedron 2019;172:87–94.

[3] Hegedus LSS, Bjorn CG. Transition metals in the synthesis of complexes organic molecules. 3rd ed Sausalito: University Science Books; 2010. p. 459.

[4] Corbridge DCE. Phosphorus: an outline of its chemistry, biochemistry and technology. Amsterdam: Elsevier; 1995.

[5] Tolman CA. Steric effects of phosphorus ligands in organometallic chemistry and homogeneous catalysis. Chem Rev 1977;77(3):313−48.

[6] Niksch T, Görls H, Friedrich M, Oilunkaniemi R, Laitinen R, Weigand W. Synthesis and characterisation of 2,2-bis(hydroxymethyl) − 1,3-diselenolato metal(II) complexes bearing various phosphanes. Eur J Inorg Chem 2010;1:74−94.

[7] Freixa Z, van Leeuwen PWNM. Bite angle effects in diphosphine metal catalysts: steric or electronic? Dalton Trans 2003;10:1890−901.

[8] Kamer PC, van Leeuwen PW, Reek JN. Wide bite angle diphosphines: xantphos ligands in transition metal complexes and catalysis. Acc Chem Res 2001;34(11):895−904.

[9] van Zeist W-J, Bickelhaupt FM. Steric nature of the bite angle. A closer and a broader look. Dalton Trans 2011;40:3028−38.

[10] Dillon KBM, Francois/Nixon, John F. Phosphorus: the carbon copy: from organophosphorus to phospha-organic chemistry. Chichester: John Wiley; 1998. p. 366.

[11] Moulton CJ, Shaw BL. Transition metal-carbon bonds. Part XLII. Complexes of nickel, palladium, platinum, rhodium and iridium with the tridentate ligand 2,6-bis[(di-t-butylphosphino)methyl]phenyl. J Chem Soc Dalton Trans 1976;11:1020−4.

[12] Pijnenburg NJM, Dijkstra HP, van Koten G, Gebbink RJMK. SCS-pincer palladium-catalyzed auto-tandem catalysis using dendritic catalysts in semi-permeable compartments. Dalton Trans 2011;40(35):8896−905.

[13] Selander N J, Szabó K. Catalysis by palladium pincer complexes. Chem Rev 2011;111 (3):2048−76.

[14] van Koten G. Pincer ligands as powerful tools for catalysis in organic synthesis. J Organomet Chem 2013;730:156−64.

[15] Albrecht M, van Koten G. Platinum group organometallics based on "pincer" complexes: sensors, switches, and catalysts. Angew Chem Int Ed 2001;40(20):3750−81.

[16] van der Boom ME, Milstein D. Cyclometalated phosphine-based pincer complexes: mechanistic insight in catalysis, coordination, and bond activation. Chem Rev 2003;103 (5):1759−92.

[17] Albrecht M. Cyclometalation using d-block transition metals: fundamental aspects and recent trends. Chem Rev 2010;110(2):576−623.

[18] Baber RA, Bedford RB, Betham M, Blake ME, Coles SJ, Haddow MF, et al. Chiral palladium bis(phosphite) PCP-pincer complexes via ligand C-H activation. Chem Commun 2006;37:3880−2.

[19] Bhattacharya P, Guan H. Synthesis and catalytic applications of iron pincer complexes. Comments Inorg Chem 2011;32(2):88−112.

[20] Goldberg KI, Goldman AS. Organometallic C-H bond activation: An introduction. Activation and Functionalization of C-H Bonds, 885. ACS Symposium Series; 2004. p. 1−43.

[21] Grove DM, Van Koten G, Ubbels HJC, Zoet R, Spek AL. Organonickel(II) complexes of the tridentate monoanionic ligand o,o′-bis[(dimethylamino)methylphenyl (N-C-N)]. Syntheses and the x-ray crystal structure of the stable nickel(II) formate [Ni(N-C-N) O2CH]. Organometallics 1984;3(7):1003−9.

[22] Williams BS, Dani P, Lutz M, Spek AL, van Koten G. Development of the first P-stereogenic PCP pincer ligands, their metallation by palladium and platinum, and preliminary catalysis. Helv Chim Acta 2001;84(11):3519−30.

[23] Pape A, Lutz M, Müller G. Phosphane coordination to magnesium: synthesis and structure of bis[ortho, ortho′-bis{(dimethylphosphino)-methyl}phenyl]magnesium. Angew Chem Int Ed Engl 1994;33(22):2281–4.

[24] Pape A, Lutz M, Müller G. Phosphan-Koordination an Magnesium: Synthese und Struktur von Bis[ortho, ortho′-bis-{(dimethylphosphano)methyl}phenyl]magnesium. Angew Chem 1994;106:2375–7.

[25] Dupont J, Beydoun N, Pfeffer M. J Chem Soc, Dalton Trans 1989;1715–20.

[26] Dani P, Albrecht M, van Klink GPM, van Koten G. Transcyclometalation: a novel route to (chiral) bis-ortho-chelated bisphosphinoaryl ruthenium(II) complexes. Organometallics 2000;19(22):4468–76.

[27] Albrecht M, James SL, Veldman N, Spek AL, Koten GV. Can J Chem 2001;79:709–18.

[28] Medici S, Gagliardo M, Williams SB, Chase PA, Gladiali S, Lutz M, et al. Novel P-stereogenic PCP pincer-aryl ruthenium(II) complexes and their use in the asymmetric hydrogen transfer reaction of acetophenone. Helv Chim Acta 2005;88(3):694–705.

[29] Schultz KM, Goldberg KI, Gusev DG, Heinekey DM. Synthesis, structure, and reactivity of iridium NHC pincer complexes. Organometallics 2011;30(6):1429–37.

[30] Peris E, Loch JA, Mata J, Crabtree RH. A Pd complex of a tridentate pincer CNC bis-carbene ligand as a robust homogenous Heck catalyst. Chem Commun 2001;2:201–2.

[31] Peters JC, Harkins SB, Brown SD, Day MW. Pincer-like amido complexes of platinum, palladium, and nickel. Inorg Chem 2001;40(20):5083–91.

[32] Hartshorn CM, Steel PJ. Cyclometalated compounds. XI.1. Single and double cyclometalations of poly(pyrazolylmethyl)benzenes. Organometallics 1998;17(16):3487–96.

[33] Lee DW, Jensen CM, Morales-Morales D. Reactivity of iridium PCP pincer complexes toward CO and CO2. Crystal structures of IrH(κ2-O2COH){C6H3-2,6-(CH2PBut2)2} and IrH(C(O)OH){C6H3-2,6-(CH2PBut2)2} · H2O. Organometallics 2003;22(23):4744–9.

[34] Kumar P, Siddiqui MM, Reddi Y, Mague JT, Sunoj RB, Balakrishna MS. New bisphosphomide ligands, 1,3-phenylenebis((diphenylphosphino)methanone) and (2-bromo-1,3-phenylene)bis((diphenylphosphino)methanone): synthesis, coordination behavior, DFT calculations and catalytic studies. Dalton Trans 2013;42(32):11385–99.

[35] Jia G, Lee HM, Williams ID, Lau CP, Chen Y. Synthesis, characterization, and acidity properties of [MCl(H2)(L)(PMP)]BF4 (M = Ru, L = PPh3, CO; M = Os, L = PPh3; PMP = 2,6-(Ph2PCH2)2C5H3N. Organometallics 1997;16(18):3941–9.

[36] Cochran BM, Michael FE. Mechanistic studies of a palladium-catalyzed intramolecular hydroamination of unactivated alkenes: protonolysis of a stable palladium alkyl complex is the turnover-limiting step. J Am Chem Soc 2008;130(9):2786–92.

[37] Kumar P, Kashid VS, Reddi Y, Mague JT, Sunoj RB, Balakrishna MS. A phosphomide based PNP ligand, 2, 6-{Ph 2 PC (O)} 2 (C 5 H 3 N), showing PP, PNP and PNO coordination modes. Dalton Trans 2015;44(9):4167–79.

[38] Langer R, Diskin-Posner Y, Leitus G, Shimon LJW, Ben-David Y, Milstein D. Low-pressure hydrogenation of carbon dioxide catalyzed by an iron pincer complex exhibiting noble metal activity. Angew Chem Int Ed 2011;50(42):9948–52.

[39] South CR, Higley MN, Leung KCF, Lanari D, Nelson A, Grubbs RH, et al. Self-assembly with block copolymers through metal coordination of SCS–PdII pincer complexes and pseudorotaxane formation. Chem Eur J 2006;12(14):3789–97.

[40] Doux M, Bouet C, Mézailles N, Ricard L, Le Floch P. Synthesis and molecular structure of a palladium complex containing a λ5-phosphinine-based SPS pincer ligand. Organometallics 2002;21(13):2785–8.

[41] Klerman Y, Ben-Ari E, Diskin-Posner Y, Leitus G, Shimon LJW, Ben-David Y, et al. Pyridine-based SNS-iridium and -rhodium sulfide complexes, including d8-d8 metal-metal interactions in the solid state. Dalton Trans 2008;24:3226–34.

[42] Mehring M, Schürmann M, Jurkschat K. The first rigid O,C,O-pincer ligand and its application for the synthesis of penta- and hexacoordinate organotin(IV) compounds. Organometallics 1998;17(6):1227–36.

[43] del Pozo C, Corma A, Iglesias M, Sánchez F. Immobilization of (NHC)NN-pincer complexes on mesoporous MCM-41 support. Organometallics 2010;29(20):4491–8.

[44] Poverenov E, Gandelman M, Shimon LJW, Rozenberg H, Ben-David Y, Milstein D. Pincer "hemilabile" effect. PCN platinum(II) complexes with different amine "arm length.". Organometallics 2005;24(6):1082–90.

[45] Spasyuk DM, Gorelsky SI, van der Est A, Zargarian D. Characterization of divalent and trivalent species generated in the chemical and electrochemical oxidation of a dimeric pincer complex of nickel. Inorg Chem 2011;50(6):2661–74.

[46] Lindner R, van den Bosch B, Lutz M, Reek JNH, van der Vlugt JI. Tunable hemilabile ligands for adaptive transition metal complexes. Organometallics 2011;30(3):499–510.

[47] Nemesh S, Jensen C, Binamira-Soriaga E, Kaska WC. Organometallics 1983;2:1442–7.

[48] Jensen CM. Iridium PCP pincer complexes: highly active and robust catalysts for novel homogeneous aliphatic dehydrogenations. Chem Commun 1999;24:2443–9.

[49] Gupta M, Hagen C, Flesher RJ, Kaska WC, Jensen CM. A highly active alkane dehydrogenation catalyst: stabilization of dihydrido rhodium and iridium complexes by a P-C-P pincer ligand. Chem Commun 1996;17:2083–4.

[50] Gupta M, Hagen C, Kaska WC, Cramer RE, Jensen CM. Catalytic dehydrogenation of cycloalkanes to arenes by a dihydrido iridium P-C-P pincer complex. J Am Chem Soc 1997;119(4):840–1.

[51] Göttker-Schnetmann I, White P, Brookhart M. Iridium bis(phosphinite) p-XPCP pincer complexes: highly active catalysts for the transfer dehydrogenation of alkanes. J Am Chem Soc 2004;126(6):1804–11.

[52] Göttker-Schnetmann I, Brookhart M. Mechanistic studies of the transfer dehydrogenation of cyclooctane catalyzed by iridium bis(phosphinite) p-XPCP pincer complexes. J Am Chem Soc 2004;126(30):9330–8.

[53] Zhu K, Achord PD, Zhang X, Krogh-Jespersen K, Goldman AS. Highly effective pincer-ligated iridium catalysts for alkane dehydrogenation. DFT calculations of relevant thermodynamic, kinetic, and spectroscopic properties. J Am Chem Soc 2004;126(40):13044–53.

[54] Zhang X, Emge TJ, Ghosh R, Goldman AS. Selective cleavage of the C-C bonds of aminoethyl groups, via a multistep pathway, by a pincer iridium complex. J Am Chem Soc 2005;127(23):8250–1.

[55] Goldman AS, Roy AH, Huang Z, Ahuja R, Schinski W, Brookhart M. Catalytic alkane metathesis by tandem alkane dehydrogenation-olefin metathesis. Science 2006;312 (5771):257–61.

[56] Ahuja R, Punji B, Findlater M, Supplee C, Schinski W, Brookhart M, et al. Catalytic dehydroaromatization of n-alkanes by pincer-ligated iridium complexes. Nat Chem 2011;3 (2):167–71.

[57] Rimml H, Venanzi LM. The facile cyclometallation reaction of 1, 3-BIS [(diphenyl-phosphino) methyl] benzene. J Organomet Chem 1983;259(1):C6–7.

[58] Rimml H, Venanzi LM. A stable binuclear complex containing Pd-H-Pd bonds. J Organomet Chem 1984;260(2):C52–4.

[59] Bennett MA, Jin H, Willis AC. Preparation and X-ray structure of a platinum(II) hydroxycarbonyl, Pt(CO2H){C6H3(CH2PPh2) 2-2,6}, containing a trans-spanning, tridentate P,C, P-ligand. J Organomet Chem 1993;451(1):249–56.

[60] Campora J, Palma P, Del Rio D, Alvarez E. CO insertion reactions into the M-OH bonds of monomeric nickel and palladium hydroxides. Reversible decarbonylation of a hydroxycarbonyl palladium complex. Organometallics 2004;23(8):1652–5.

[61] Abugideiri F, Webster Keogh D, Kraatz H-B, Pearson W, Poli R. Instability of 15-electron Cp☆MoCl2L (L = 2-electron donor) derivatives. X-ray structure of Cp☆MoCl2 (PMe2Ph)2 and [Cp☆MoCl2(PMe2Ph)2]AlCl4. J Organomet Chem 1995;488(1):29–38.

[62] Dahlhoff WV, Nelson SM. Studies on the magnetic cross-over in five-co-ordinate complexes of iron(II), cobalt(II), and nickel(II). Part II. J Chem Soc A: Inorg Phys Theor 1971;0:2184–90.

[63] Zhang J, Gandelman M, Herrman D, Leitus G, Shimon LJW, Ben-David Y, et al. Iron(II) complexes based on electron-rich, bulky PNN- and PNP-type ligands. Inorg Chim Acta 2006;359(6):1955–60.

[64] Trovitch RJ, Lobkovsky E, Chirik PJ. Bis(diisopropylphosphino)pyridine iron dicarbonyl, dihydride, and silyl hydride complexes. Inorg Chem 2006;45(18):7252–60.

[65] Pelczar EM, Emge TJ, Krogh-Jespersen K, Goldman AS. Unusual structural and spectroscopic features of some PNP-pincer complexes of iron. Organometallics 2008;27(22): 5759–67.

[66] Benito-Garagorri D, Wiedermann J, Pollak M, Mereiter K, Kirchner K. Iron(II) complexes bearing tridentate PNP pincer-type ligands as catalysts for the selective formation of 3-hydroxyacrylates from aromatic aldehydes and ethyldiazoacetate. Organometallics 2007;26(1):217–22.

[67] Benito-Garagorri D, Puchberger M, Mereiter K, Kirchner K. Stereospecific and reversible CO binding at iron pincer complexes. Angew Chem Int Ed Engl 2008;47 (47):9142–5.

[68] Benito-Garagorri D, Alves LGa, Puchberger M, Mereiter K, Veiros LF, Calhorda MJ, et al. Striking differences between the solution and solid-state reactivity of iron PNP pincer complexes with carbon monoxide. Organometallics 2009;28(24):6902–14.

[69] Sacco A, Vasapollo G, Nobile CF, Piergiovanni A, Pellinghelli MA, Lanfranchi M. Syntheses and structures of 2-diphenylphosphinomethylenide-6-diphenylphosphinomethylenepyridine complexes of palladium(II) and platinum(II); crystal structures of [PtCl2-(CHPPH2)-6-(CH2PPh2)pyridine] and [Pd(COOMe)2-(CHPPh2)-6-(CH2PPh2)pyridine]. J Organomet Chem 1988;356(3):397–409.

[70] Hahn C, Vitagliano A, Giordano F, Taube R. Coordination of olefins and N-donor ligands at the fragment [2,6-bis((diphenylphosphino)methyl)pyridine]-palladium(II). Synthesis, structure, and amination of the new dicationic complexes [Pd(PNP)(CH2CHR)](BF4)2 (R = H, Ph). Organometallics 1998;17(10):2060–6.

[71] Hahn C, Morvillo P, Vitagliano A. Olefins coordinated at a highly electrophilic site-dicationic palladium(II) complexes and their equilibrium reactions with nucleophiles. Eur J Inorg Chem 2001;2001(2):419–29.

[72] van der Vlugt JI, Siegler MA, Janssen M, Vogt D, Spek AL. A cationic AgI(PNPtBu) species acting as PNP transfer agent: facile synthesis of Pd(PNPtBu)(alkyl) complexes and their reactivity compared to PCPtBu analogues. Organometallics 2009;28(24):7025–32.

[73] Kashid VS, Naik S, Balakrishna MS. Silver(I) complexes of bisphosphines PhN{P (OC6H4C3H5-o)2}2 (1) and 2,6-{Ph2PC(O)}2C5H3N (2). Proc Natl Acad Sci India Sect A 2016;86(4):601–4.

[74] Kashid VS, Balakrishna MS. Microwave-assisted copper (I) catalyzed A3-coupling reaction: Reactivity, substrate scope and the structural characterization of two coupling products. Cat Commun 2018;103:78−82.

[75] Choi J, MacArthur AHR, Brookhart M, Goldman AS. Dehydrogenation and related reactions catalyzed by iridium pincer complexes. Chem Rev 2011;111(3):1761−79.

[76] Bedford RB, Chang Y-N, Haddow MF, McMullin CL. Tuning ligand structure in chiral bis(phosphite) and mixed phosphite-phosphinite PCP-palladium pincer complexes. Dalton Trans 2011;40(35):9034−41.

[77] Ines B, SanMartin R, Churruca Ft, DomÃnguez E, Urtiaga MK, Arriortua MI. A nonsymmetric pincer-type palladium catalyst in Suzuki, Sonogashira, and Hiyama couplings in neat water. Organometallics 2008;27(12):2833−9.

[78] Dounay AB, Overman LE. The asymmetric intramolecular Heck reaction in natural product total synthesis. Chem Rev 2003;103(8):2945−64.

[79] Morales-Morales D, Redon R, Yung C, Jensen CM. High yield olefination of a wide scope of aryl chlorides catalyzed by the phosphinito palladium PCP pincer complex: [PdCl{CH(OPPr) − 2,6}]. Chem Commun 2000;17:1619−20.

[80] Eberhard MR. Insights into the Heck reaction with PCP pincer palladium(II) complexes. Org Lett 2004;6(13):2125−8.

[81] Aydin J, Kumar KS, Sayah MJ, Wallner OA, Szabó KJ. Synthesis and catalytic application of chiral 1,1'-bi-2-naphthol- and biphenanthrol-based pincer complexes: selective allylation of sulfonimines with allyl stannane and allyl trifluoroborate. J Org Chem 2007;72(13):4689−97.

[82] Niu J-L, Chen Q-T, Hao X-Q, Zhao Q-X, Gong J-F, Song M-P. Diphenylprolinol-derived symmetrical and unsymmetrical chiral pincer palladium(II) and nickel(II) complexes: synthesis via one-pot phosphorylation/metalation reaction and C-H activation. Organometallics 2010;29(9):2148−56.

[83] Schwan AL. Palladium catalyzed cross-coupling reactions for phosphorus-carbon bond formation. Chem Soc Rev 2004;33(4):218−24.

[84] Feng J-J, Chen X-F, Shi M, Duan W-L. Palladium-catalyzed asymmetric addition of diarylphosphines to enones toward the synthesis of chiral phosphines. J Am Chem Soc 2010;132(16):5562−3.

[85] Lu J, Ye J, Duan W-L. Palladium-catalyzed asymmetric addition of diarylphosphines to α,β-unsaturated sulfonic esters for the synthesis of chiral phosphine sulfonate compounds. Org Lett 2013;15(19):5016−19.

[86] Ding B, Zhang Z, Xu Y, Liu Y, Sugiya M, Imamoto T, et al. P-stereogenic PCP pincer-Pd complexes: synthesis and application in asymmetric addition of diarylphosphines to nitroalkenes. Org Lett 2013;15(21):5476−9.

[87] Gorla F, Venanzi LM, Albinati A. Synthesis of (1R,1'S)- and (1S,1'S),(1R,1'R)-bis[1-(diphenylphosphino)ethyl]benzene derivatives and their cyclometalation reactions with platinum(II) compounds. The x-ray crystal structures of [2,6-bis[(diphenylphosphino)methyl]phenyl]chloropalladium(II), [(1R,1'S) − 2,6-bis[1-(diphenylphosphino)ethyl]phenyl]chloroplatinum(II), and [(1R,1'R),(1S,1'S) − 2,6-bis[1-(diphenylphosphino)ethyl]phenyl]chloroplatinum(II). Organometallics 1994;13(1):43−54.

[88] Longmire JM, Zhang X, Shang M. Synthesis and X-ray crystal structures of palladium (II) and platinum(II) complexes of the PCP-type chiral tridentate ligand (1R,1'R) − 1,3-bis[1-(diphenylphosphino)ethyl]benzene. Use in the asymmetric aldol reaction of methyl isocyanoacetate and aldehydes. Organometallics 1998;17(20):4374−9.

[89] Agbossou F, Carpentier J-F, Hapiot F, Suisse I, Mortreux A. The aminophosphine-phosphinites and related ligands: synthesis, coordination chemistry and enantioselective catalysis1. Coord Chem Rev 1998;178–180(Part 2):1615–45.

[90] Agbossou-Niedercorn F, Suisse I. Chiral aminophosphine phosphinite ligands and related auxiliaries: recent advances in their design, coordination chemistry, and use in enantioselective catalysis. Coord Chem Rev 2003;242(1–2):145–58.

[91] Neilson RH, editor. Phosphorus–nitrogen compounds. Chichester, England: John Wiley & Sons; 1994.

[92] Stahl L. Bicyclic and tricyclic bis(amido)cyclodiphosph(III)azane compounds of main group elements. Coord Chem Rev 2000;210(1):203–50.

[93] Balakrishna MS, Reddy VS, Krishnamurthy SS, Nixon JF, Laurent JB. Coordination chemistry of diphosphinoamine and cyclodiphosphazane ligands. Coord Chem Re 1994;129(1–2):1–90.

[94] Balakrishna MS, Eisler DJ, Chivers T. Chemistry of pnictogen(III)-nitrogen ring systems. Chem Soc Rev 2007;36(4):650–64.

[95] Balakrishna MS, Chandrasekaran P, Venkateswaran R. Functionalized cyclodiphosphazanes cis-[tBuNP(OR)]2 (R = C6H4OMe-o, CH2CH2OMe, CH2CH2SMe, CH2CH2NMe2) as neutral 2e, 4e or 8e donor ligands. J Organomet Chem 2007;692(13):2642–8.

[96] Balakrishna MS. Cyclodiphosphazanes: options are endless. Dalton Trans 2016;45 (31):12252–82.

[97] Chandrasekaran P, Mague JT, Balakrishna MS. Tetranuclear rhodium(I) macrocycle containing cyclodiphosphazane [Rh2(μ-Cl)2(CO)2[(tBuNP(OC6H4OMe-o))2-κP]]2 and Its reversible conversion into trans-[Rh(CO)Cl{(tBuNP(OC6H4OMe-o))2-κP}2]. Organometallics 2005;24(15):3780–3.

[98] Chandrasekaran P, Mague JT, Balakrishna MS. Copper(I) coordination polymers [{Cu (μ-X)}2{RP(μ-NtBu)}2]n (R = OC6H4OMe-o; X = Cl, Br, and I) and their reversible conversion into mononuclear complexes [CuX{(RP(μ-NtBu))2}2]: synthesis and structural characterization. Inorg Chem 2006;45(17):6678–83.

[99] Chandrasekaran P, Mague JT, Balakrishna MS. One-dimensional silver(I) coordination polymers containing cyclodiphosphazane, cis-{(o-MeOC6H4O)P(μ-NtBu)}2. Dalton Trans 2007;27:2957–62.

[100] Balakrishna MS, Suresh D, Mague JT. Cyclodiphosphazane appended with thioether functionality: synthesis, transition metal chemistry and catalytic application in Suzuki-Miyaura cross-coupling reactions. Inorg Chim Acta 2011;372(1):259–65.

[101] Mohanty S, Balakrishna MS. Suzuki-Miyaura, Mizoroki-Heck carbon-carbon coupling and hydrogenation reactions catalyzed by PdII and RhI complexes containing cyclodiphosphazane cis-{tBuNP(OC6H4OMe-o)}2. J Chem Sci (Bangalore, India) 2010;122(2):137–42.

[102] Suresh D, Balakrishna MS, Rathinasamy K, Panda D, Mobin SM. Water-soluble cyclodiphosphazanes: synthesis, gold(I) metal complexes and their in vitro antitumor studies. Dalton Trans 2008;21:2812–14.

[103] Balakrishna MS, Suresh D, Rai A, Mague JT, Panda D. Dinuclear Copper(I) complexes containing cyclodiphosphazane derivatives and pyridyl ligands: synthesis, structural studies, and antiproliferative activity toward human cervical and breast cancer cells. Inorg Chem 2010;49(19):8790–801.

[104] Michaelis A. Untersuchungen Uber aromatische Borverbindungen. Ber Dtsch Chem Ges 1894;27(1):244–62.

[105] Michaelis M. Annalen 1903;326:129–32.

[106] Holmes RR, Forstner JA. Phosphorus nitrogen chemistry. VI. Preparation and properties of thiophosphorus tri-N-methylimide, P4S4N6(CH3)6. Inorg Chem 1963;2(2):377–80.

[107] Holmes RR. Phosphorus nitrogen chemistry. III. The preparation and properties of phosphorus tri-N-methylimide1,2. J Am Chem Soc 1961;83(6):1334–6.

[108] Muir KW. Stereochemistry of phosphorus compounds. Part II. Crystal and molecular structure of 1,3-di-t-butyl-2,4-dichlorodiazadiphosphetidine. J Chem Soc, Dalton Trans 1975;3:259–62.

[109] Davies AR, Dronsfield AT, Haszeldine RN, Taylor DR. Organophosphorus chemistry. Part XIII. The reaction of phosphorus trichloride with primary aromatic amines. J Chem Soc, Perkin Trans 1973;0:379–85.

[110] Kashid VS, Mague JT, Balakrishna MS. Macrocyclic cyclodiphosphazane [{P (μ-t BuN)} 2 (O − m − C 6 H 4 CHNCH 2) 2] 2: synthesis of chalcogen derivatives and gold (I) complex. J Chem Sci 2017;129(10):1531–7.

[111] Silaghi-Dumitrescu I, Haiduc I. On the geometry of 1,3-diazadiphosphetidines. The cistrans isomerism). Phosphorus Sulfur Silicon Relat Elem 1994;91:21–36.

[112] Silaghi-Dumitrescu I, Lara-Ochoa F, Haiduc I. "Edge" or "vertex" inversion at phosphorus in the cis − trans isomerization of diazadiphosphetidines? Model MNDO and ab initio molecular orbital calculations. Main Group Chem 1998;2:309–14.

[113] Chandrasekaran P, Mague JT, Balakrishna MS. Synthesis and derivatization of the bis (amido)-λ3-cyclodiphosphazanes cis-[R'(H)NP(μ-NR)]2, including a rare example, trans-[tBu(H)N(Se)P(μ-NCy)]2, showing intermolecular Se·H-O hydrogen bonding. Eur J Inorg Chem 2011;14:2264–72.

[114] Bashall A, Doyle EL, Tubb C, Kidd SJ, McPartlin M, Woods AD, et al. The tetrameric macrocycle [{P(μ-NtBu)}2NH]4. Chem Commun 2001;24:2542–3.

[115] Beswick M A, Hopkins A D, Kidd S J, Lawson Y G, Raithby P R, Wright DS, et al. A simple route to multifunctional phosphide and amide donor sets; syntheses and structures of [{ButPAs(NMe2)2}K.pmdeta]2 and [{CyNAs(NMe2)2K}2]∞. Chem Commun 1999;8:739–40.

[116] Balakrishna MS, Venkateswaran R, Mague JT. Transition metal chemistry of cyclodiphosphanes containing phosphine and amide-phosphine functionalities: formation of a stable dipalladium(II) complex containing a Pd-P σ-bond. Dalton Trans 2010;39(46):11149–62.

[117] Thompson ML, Haltiwanger RC, Norman AD. Synthesis and X-ray structure of a dinuclear cyclodiphosph (III) azane,[(PhNH) P 2 (NPh) 2] 2 NPh. J Chem Soc, Chem Commun 1979;15:647–8.

[118] Axenov KirillÂ V, Kotov VasilyÂ V, Klinga M, Leskelä, Markku, Repo T. New bulky bis(amino)cyclodiphosph(III)azanes and their titanium(IV) complexes: synthesis, structures and ethene polymerization studies. Eur J Inorg Chem 2004;2004(4):695–706.

[119] Axenov KirillÂ V, Klinga M, Leskelä M, Kotov V, Repo T. [Bis(amido)cyclodiphosph (III)azane]dichlorozirconium complexes for ethene polymerization. Eur J Inorg Chem 2004;2004(23):4702–9.

[120] Balakrishna MS, Suresh D, Mague JT. Mono-, bi-, tri- and tetranuclear palladium(II), copper(I), and gold(I) complexes of morpholine- and N-methylpiperazine-functionalized cyclodiphosph(III)azans, cis-[(tBuN-Î¼)2(PNC4H8X)2] (X = O, NMe). Eur J Inorg Chem 2010;2010(26):4201–10.

[121] Keat R, Rycroft DS, Thompson DG. Synthesis and properties of 2,4-dialkoxy-1,3-di-t-butylcyclodiphosph(III)-azanes. J Chem Soc, Dalton Trans 1979;7:1224–30.

[122] Balakrishna MS, Krishnamurthy SS. Organometallic chemistry of diphosphazanes: IV. Reactions of the cyclodiphosphazane, cis-[tBuNP(OPh)]2 with M(CO)6 (M = Cr, MO, W), rhodium(I), palladium(II) and platinum(II) derivatives. J Organomet Chem 1992;424 (2):243−51.

[123] Chandrasekaran P, Mague JT, Balakrishna MS. Cyclodiphosphazanes with hemilabile ponytails: synthesis, transition metal chemistry (Ru(II), Rh(I), Pd(II), Pt(II)), and crystal and molecular structures of mononuclear (Pd(II), Rh(I)) and bi- and tetranuclear rhodium (I) complexes. Inorg Chem 2005;44(22):7925−32.

[124] Chen HJ, Haltiwanger RC, Hill TG, Thompson ML, Coons DE, Norman AD. Synthesis and structural study of 2, 4-disubstituted 1, 3-diaryl-1, 3, 2, 4-diazadiphosphetidines. Inorg Chem 1985;24(26):4725−30.

[125] Schranz I, Lief GR, Carrow CJ, Haagenson DC, Grocholl L, Stahl L, et al. Reversal of polarization in amidophosphines: neutral-and anionic-κ P coordination vs. anionic-κ P, N coordination and the formation of nickelaazaphosphiranes. Dalton Trans 2005;20:3307−18.

[126] Bulloch G, Keat R, Thompson DG. Alkylaminocyclodiphosph(III)azanes. J Chem Soc, Dalton Trans 1977;1:99−104.

[127] DuBois D, Duesler EN, Paine RT. Synthesis and X-ray structural characterization of an eight-membered P 4 N 4 cage compuond analogue of S 4 N 4 and α-P 4 S 4. J Chem Soc Chem Commun 1984;8:488−9.

[128] Doyle EL, Garcia F, Humphrey SM, Kowenicki RA, Riera L, Woods AD, et al. Steric control in the oligomerisation of phosphazane dimers; towards new phosphorus-nitrogen macrocycles. Dalton Trans 2004;5:807−12.

[129] Burckett St. J, Laurent B, Hitchcock PB, Nixon JF. Syntheses and 31P NMR studies of some cycloocta-1,5-dienerhodium(I) complexes containing coordinated 1,3-DI-t-butyl-2,4-dihalogenocyclodiphosphazanes, [PXNtBu]2 (X = Cl, F) and related ligands. Crystal and molecular structure of bis(chloro)(cycloocta-1,5-diene)1,3-di-t-butyl-2,4-difluorocyclodiphosphazanedirhodium(I). J Organomet Chem 1983;249(1):243−54.

[130] Woods AD, McPartlin M. Synthesis and deprotonation of P−H-functionalised phosph (iii)azanes; borohydride dependency on the formation of cis- or trans-[HP(μ-NtBu) 2PNtBuH]. Dalton Trans 2004;1:90−3.

[131] Nordheider A, Chivers T, Thirumoorthi R, Vargas-Baca I, Woollins JD. Planar P6E6 (E = Se, S) macrocycles incorporating P2N2 scaffolds. Chem Commun 2012;48 (51):6346−8.

[132] Gonzalez-Calera S, Eisler DJ, Morey JV, McPartlin M, Singh S, Wright DS. The selenium-based hexameric macrocycle [(Se =)P(μ-NtBu)2P(μ-Se)]6. Angew Chem Int Ed Engl 2008;47(6):1111−14.

[133] Chandrasekaran P, Mague JT, Balakrishna MS. Intramolecular amine-induced [1,3]-sigmatropic rearrangement in the reactions of aminophosphinites or phosphites with elemental sulfur or selenium. Inorg Chem 2006;45(15):5893−7.

[134] Chandrasekaran P, Mague JT, Balakrishna MS. Methylene insertion into the exocyclic P-N bonds of bis(amido)cyclodiphosphazane, cis-[tBu(H)NP(μ-tBuN)]2. Tetrahedron Lett 2007;48(30):5227−9.

[135] Calera SG, Wright DS. Macrocyclic phosphazane ligands. Dalton Trans 2010;39(21): 5055−65.

[136] Jones CJ. Transition metals as structural components in the construction of molecular containers. Chem Soc Rev 1998;27(4):289−300.

[137] Kashid VS, Radhakrishna L, Balakrishna MS. First examples of tri- and tetraphosphametacyclophanes: synthesis and isolation of an unusual hexapalladium complex containing pincer units with Pd-P covalent bonds. Dalton Trans 2017;46(20):6510−13.

[138] Vögtle F, Lichtenthaler RG. Cyclophanes possessing cavities. Angew Chem Int Ed 1972;11(6):535−6.

[139] Jeletic MS, Ghiviriga I, Abboud KA, Veige AS. A new chiral di-N-heterocyclic carbene (NHC) cyclophane ligand and its application in palladium enantioselective catalysis. Dalton Trans 2010;39(28):6392−4.

[140] Ni Y, Lough AJ, Rheingold AL, Manners I. The first sulfur(VI)-nitrogen-phosphorus macrocycles. Angew Chem Int Ed Engl 1995;34(9):998−1001.

[141] Busch M, Cayir M, Nieger M, Thiel WR, Bräse S. Roadmap towards N-heterocyclic [2.2]paracyclophanes and their application in asymmetric catalysis. Eur J Org Chem 2013;2013(27):6108−23.

[142] Zhang Q, Cui X, Chen L, Liu H, Wu Y. Syntheses of chiral ferrocenophanes and their application to asymmetric catalysis. Eur J Org Chem 2014;2014(35):7823−9.

[143] Ananthnag GS, Mague JT, Balakrishna MS. Self-assembled cyclophane-type copper(I) complexes of 2,4,6-tris(diphenylphosphino) − 1,3,5-triazine and their catalytic application. Inorg Chem 2015;54(22):10985−92.

[144] Vorontsova NV, Bystrova GS, Antonov DY, Vologzhanina AV, Godovikov IA, Il'in MM. Novel ligands based on bromosubstituted hydroxycarbonyl [2.2]paracyclophane derivatives: synthesis and application in asymmetric catalysis. Tetrahedron Asymmetry 2010;21(6):731−8.

[145] Wack H, France S, Hafez AM, Drury WJ, Weatherwax A, Lectka T. Development of a new dimeric cyclophane ligand: application to enhanced diastereo- and enantioselectivity in the catalytic synthesis of β-lactams. J Org Chem 2004;69(13):4531−3.

[146] El-Sayed Ali T, Abdel-Aziz Abdel-Ghaffar S, El-Mahdy KM, Abdel-Karim SMSynthesis. characterization, and antimicrobial activity of some new phosphorus macrocyclic compounds containing pyrazole rings. Turk J Chem 2013;37(1):160−9.

[147] Baxter PNW. Cyclic donor − acceptor circuits: synthesis and fluorescence ion sensory properties of a mixed-heterocyclic dehydroannulene-type cyclophane. J Org Chem 2004;69(6):1813−21.

[148] Gago S, Gonzalez J, Blasco S, Parola AJ, Albelda MT, Garcia-Espana E, et al. Protonation, coordination chemistry, cyanometallate "supercomplex" formation and fluorescence chemosensing properties of a bis(2,2[prime or minute]-bipyridino)cyclophane receptor. Dalton Trans 2014;43(6):2437−47.

[149] Shimizu S, Shirakawa S, Sasaki Y, Hirai C. Novel Water-soluble calix[4]arene ligands with phosphane-containing groups for dual functional metal-complex catalysts: the biphasic hydroformylation of water-insoluble olefins. Angew Chem Int Ed 2000;39(7):1256−9.

[150] Zong J, Mague JT, Welch EC, Eckert IMK, Pascal Jr RA. Sterically congested macrobicycles with heteroatomic bridgehead functionality. Tetrahedron 2013;69(48):10316−21.

[151] Declercq J-P, Delangle P, Dutasta J-P, Van Oostenryck L, Simon P, Tinant B. Synthesis and molecular structure of new phosphorous-crown compounds containing the thiophosphoryl group. J Chem Soc Perkin Trans 2 1996;11:2471−8.

[152] Naidu KRM, Kumar MA, Dadapeer E, Babu KR, Raju CN, Ghosh SK. Synthesis and antioxidant activity of novel 15-alkyl/aryl-13,17-dihydro-15-5-dibenzo[e,k][1,3,7,10,2] dioxadiazaphosphacyclotridecin-15-selones/thiones/ones. J Heterocycl Chem 2011;48(2): 317−22.

[153] Tinant B, Delangle P, Mulatier J-C, Declercq J-P, Dutasta J-P. Synthesis and structure elucidation of large phosphorus macrocycles. J Incl Phenom Macrocycl Chem 2007;58(1-2):139−49.

[154] Caminade AM, Colombo-Khater D, Mitjaville J, Galliot C, Mas P, Majoral JP. Synthesis of main group elements containing macrocycles. Phosphorus Sulfur Silicon Relat Elem 1993;75(1-4):67−70.

[155] Caminade A-M, Majoral J-P. From phosphorus-containing macrocycles to phosphorus-containing dendrimers. Top Heterocycl Chem 2009;20:275−309.

[156] Bauer I, Habicher WD, Antipin IS, Sinyashin OG. Phosphorus macrocycles and cryptands. Russ Chem Bull 2004;53(7):1402−16.

[157] Mitjaville J, Caminade A-M, Majoral J-P. Synthesis of di- or tetrafunctionalized phosphorus macrocycles. Synthesis 1995;8:952−6.

[158] Magro G, Caminade A-M, Majoral J-P. Pseudo-halogen behavior of thiophosphoryl azides as a tool for the functionalization of phosphorus macrocycles. Tetrahedron Lett 2003;44(37):7007−10.

[159] Zhou J, Yang J, Hua B, Shao L, Zhang Z, Yu G. The synthesis, structure, and molecular recognition properties of a [2]calix[1]biphenyl-type hybrid[3]arene. Chem Commun 2016;52(8):1622−4.

[160] Avarvari N, Maigrot N, Ricard L, Mathey F, Le Floch P. Synthesis and X-ray crystal structures of silacalix[n]phosphinines: the first sp2-based phosphorus macrocycles. Chem - Eur J 1999;5(7):2109−18.

[161] Ananthnag GS, Kuntavalli S, Mague JT, Balakrishna MS. Resorcinol based acyclic dimeric and cyclic di- and tetrameric cyclodiphosphazanes: synthesis, structural studies, and transition metal complexes. Inorg Chem 2012;51(10):5919−30.

[162] Badri M, Majoral JP, Gonce F, Caminade AM, Salle M, Gorgues A. First phosphorus macrocycles incorporating tetrathiafulvalene (TTF) moieties. Tetrahedron Lett 1990;31(44):6343−6.

[163] Mathey F, Mercier F, Le, Floch P. New carbon-phosphorus macrocycles. Phosphorus, Sulfur Silicon Relat Elem 1999;144-146:251−6.

[164] Bolm C, Sharpless KB. Synthesis of a C3-symmetric phospha[2.2.2]cyclophane. Tetrahedron Lett 1988;29(40):5101−4.

[165] Tagne Kuate AC, Mohapatra SK, Daniliuc CG, Jones PG, Tamm M. Mono- and dilithiation of [(η7-C7H7)Ti(η5-C5Me5)] (pentamethyltroticene) for the synthesis of troticenyl monophosphanes and [2]troticenophanes with C−P and C−Si bridges. Organometallics 2012;31(24):8544−55.

[166] Velian A, Cummins CC. Facile synthesis of dibenzo-7λ3-phosphanorbornadiene derivatives using magnesium anthracene. J Am Chem Soc 2012;134(34):13978−81.

[167] Rasadkina EN, Slitikov PV, Pechkina MP, Vasyanina LK, Stash AI, Nifant'ev EE. 1,3-Dihydroxynaphthalene in the synthesis of phosphorus-containing macroheterocycles. Russ J Gen Chem 2005;75(12):1910−18.

[168] Nifant'ev EE, Rasadkina EN, Petrov AV. Phosphacyclophanes derived from anthracene-2,6-diol. Russ J Gen Chem 2005;75(4):660−1.

[169] Rasadkina EN, Slitikov PV, Mel'nik MS, Stash AI, Bel'skii VK, Nifant'ev EE. 2,6- and 1,6-dihydroxynaphthalenes in the synthesis of phosphacyclophanes. Russ J Gen Chem 2004;74(7):1080−6.

[170] Pudovik MA, Terentyeva SA, Kibardina LK, Pudovik EM, Burilov AR. Phosphacyclanes in reactions of bis(ortho-formylphenyl)phenylthioxophosphonate with diamines. Russ J Gen Chem 2014;84(7):1461−2.

[171] Park JK, Lee S. Sulfoxide and sulfone synthesis via electrochemical oxidation of sulfides. J Org Chem 2021;86(19):13790−9.
[172] Majoral JP, Badri M, Caminade AM, Delmas M, Gaset A. Facile synthesis of new classes of free and complexed polyaza phosphorus macrocycles. Inorg Chem 1988;27(21): 3873−5.
[173] Saunders AJ, Crossley IR, Coles MP, Roe SM. Facile self-assembly of the first diphosphametacyclophane. Chem Commun 2012;48(46):5766−8.
[174] Sun Y, Yan M-Q, Liu Y, Lian Z-Y, Meng T, Liu S-H, et al. Phenylphosphinacalix[3]trifuran: synthesis, coordination and application in the Suzuki-Miyaura cross-coupling reaction in water. RSC Adv 2015;5(87):71437−40.

Chapter 8

Synthesis of silicon and germanium organometallic compounds and their applications in catalysis

Manoj Kumar Pradhan[1], Ranjan Kumar Mohapatra[1], Mohammad Azam[2], Snehasish Mishra[3] and Azaj Ansari[4]
[1]Department of Chemistry, Government College of Engineering, Keonjhar, Odisha, India,
[2]Department of Chemistry, College of Science, King Saud University, Riyadh, Saudi Arabia,
[3]School of Biotechnology, KIIT Deemed-to-be University, Bhubaneswar, Odisha, India,
[4]Department of Chemistry, Central University of Haryana, Mahendergarh, Haryana, India

8.1 Introduction

Organometallic (OM) chemistry bridges organic and inorganic chemistry. Such compounds exhibit at least one metal−carbon (M−C) bond which are essentially polar ($M^{\delta+}-C^{\delta-}$) in nature. The bond may be ionic (as in potassium cyclopentadienyl), covalent (tetraethyl lead), coordinate covalent (silver (I)-N-heterocyclic carbine), or pi-dative (chromocene) in nature [1]. The aforementioned definition of OM compounds is not limited to M−C bond only; there are myriad of other examples where metal−hydrogen, metal-oxygen, metal-boron, metal-nitrogen, metal−sulfur, etc., bonds are also seen in this group of compounds. There are some complexes which do not contain any M−C bonds but are considered to be members of OM compounds. For example, Wilkinson catalyst, $Rh(PPh_3)_3Cl$ is used for hydrogenation of alkene and alkyne, has no M−C bond. Furthermore, the first OM compounds of main group elements are cacodyl oxide ($[(CH_3)_2As]_2O$-dimethylarsinous anhydride), whereas Zeise's salt, $K[Pt(C_2H_4)]Cl_3$ is the first OM compound of transition metal. OM compound is classified on the basis of hapticity which is defined as the number of carbon atom or (ligand atom) which are simultaneous attached to the metal. It is denoted by η^x ("eta").

The bonding pattern is uniquely different in such compounds as compared to coordination compounds, making them OM compounds instead of coordination compounds. Due to the enormous negative free energy of the metal oxide, carbon dioxide and water production such compounds are thermodynamically unstable to oxidation. This chapter discusses several synthetic routes for OM compounds of silicon (Si) and germanium (Ge). Both Si and Ge have four valence electrons, but germanium has more free electrons and higher conductivity at a given temperature. By far, silicon has been more widely applied in semiconductor industry. The first OM compound with transition metal, popularly known as Zeise salt, was synthesized by WC Zeise in 1825 [1].

$$K_2PtCl_4 + C_2H_4 \rightarrow K[PtCl_3(C_2H_4)] + KCl$$
Zeise's salt

The Group IV of the periodic table has two subgroups, the main group IVB, and the other IVA. Group IVA consists of titanium, zirconium and hafnium (generally considered as transition metals), and the main group IVB consists of carbon, silicon, germanium, tin, and lead. Out of these, silicon and germanium have gained greater attention in recent times. Organosilicon and organogermanium compounds are the OMs containing carbon-silicon (C-Si) bonds and carbon−germanium bonds. Silicon carbide is an inorganic compound. Silicon is the second most abundant element on Earth after oxygen. In nature, it is found mostly as an oxide (silica) or silicate (feldspar and kaolinite) in sand, rock, and clay. However, germanium occurs sparsely. Friedel and Crafts synthesized the organochlorosilane compound for the first time in 1863 [2]. Extensive research in the field of organosilicon compounds that followed was pioneered by Kipping [3]. Kipping was noted for synthesizing alkylsilanes and arylsilanes using Grignard reagents and preparing silicone oligomers and polymers for the first time [3]. Being colorless, flammable, hydrophobic, and stable in ambient conditions, the organosilicon compounds are mostly similar to ordinary organic compounds. Tetraethylgermane (obtained from germanium tetrachloride and diethylzinc) is the first synthesized organogermanium compound reported by Winkler in 1887 [4].

OM compounds or their intermediates derived from transition metal complexes catalyze a numerous reactions. For instance, Wilkinson's catalyst [RhCl(PPh$_3$)$_3$] catalyzes hydrogenation of alkenes. They are also popular in synthesizing several organic compounds. OMCs like trialkyl aluminum mixed with a transition metal halide (like titanium trichloride/tetrachloride) may be used as a heterogeneous catalyst to polymerize alkanes at low-temperature. To prevent infection in plantlets, seeds are pretreated with OMs like ethyl mercury chloride. Various organoarsenic compounds are also popular in the primary treatment of syphilis. Silicon-based rubbers are used as spare parts of a (human) body in modern surgery. Silicon and germanium share group 14 in the periodic table, along with tin and lead. The electronic configuration of silicon and germanium is shown in Table 8.1 and Fig. 8.1, and the physical properties of both the elements are presented in Table 8.2 [5].

Silicon and germanium are metalloids, that is, neither a metal nor a nonmetal. Metalloid category generally contains those elements that have properties of both metals and nonmetals and are hard at room temperature. Both are semiconductors with some key metal properties missing. Germanium

TABLE 8.1 Electronic configuration of silicon and germanium

Element	Symbol	Atomic number	Electronic configuration
Silicon	Si	14	[Ne] $3s^2\ 3p^2$
Germanium	Ge	32	[Ar] $3d^{10}\ 4s^2\ 4p^2$

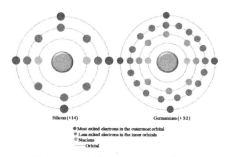

FIGURE 8.1 Electronic configuration of silicon and germanium.

TABLE 8.2 Physical properties of silicon and germanium.

Property	Silicon	Germanium
Atomic number	14	32
Molar mass (g/mol)	28.08	72.59
Density (g/cm^3)	2.49	5.35
Melting point (°C)	1410	937.5
Boiling point (°C)	2680	2830
Molar volume (cm^3/mol)	11.4	13.6
Atomic radius (A°)	1.11	1.22
Oxidation state (stable)	+4	+2, +4
Electronegativity	1.9	2.0
Ionization energy (kJ/mol)	786	761

looks metallic with a shiny silvery gray finish, and silicon has a transparent to translucent look. Although Ge is advocated as a nontoxic alternative, however due to its high costs reports of such applications of organogermanium compound are limited. Tetramethylgermanium and tetraethylgermanium are used in microelectronics industry as precursors to germanium dioxide chemical vapor deposition. A schematic two-dimensional representation for silicon and germanium showing covalent bonds at low temperature is furnished in Fig. 8.2.

The surface chemistry of Gr-IV semiconductors (like Si and Ge) is of growing interest these days due to the numerous technological applications (for instance, Si in integrated circuits and computing applications) [6–8]. These elements usually have wide significance because of their unusual properties. Germanium is a significant semiconductor used in electronic devices due to small bandgap and high intrinsic mobility of holes and electrons [9]. Ge films gained attention for their use in phase-change random access memory (PRAM) devices. Since its use to build the first transistor in 1947, germanium is currently being revived as a microelectronics material due to the unusual potential in high-speed and optoelectronic Si/Ge heterostructures [10–12]. Silicongermanium layers have the potential to provide next-gen performance in existing silicon-based complementary metal oxide semiconductor field effect transistors (CMOS FETs) [13]. Due to the appreciably high cost than silicon, germanium has a limited use. On a square-inch basis, prime grade Ge(100) is approximately 500 times costlier than prime

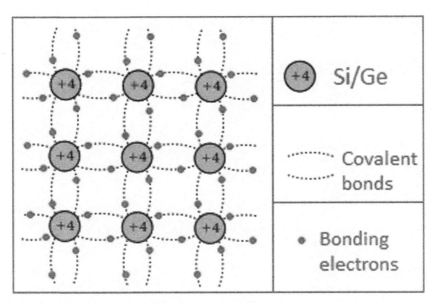

FIGURE 8.2 2-D structural representations of silicon and germanium.

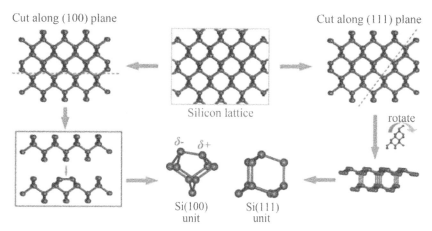

FIGURE 8.3 Structure of ordered Si surfaces; (100) plane (left) or along the (111) plane (right) [17]. *Adapted from Rodriguez-Reyes JCF, Teplyakov AV. Chemistry of organometallic compounds on silicon: the first step in film growth. Chem Eur J 2007;13:9164-9176.*

grade Si(100) [14]. Doped and intrinsic Ge(100), (110), and (111) wafers are also commercially available [15,16]. The structure of ordered silicon surfaces are presented in Fig. 8.3 [17]. Nanocrystalline porous Si reportedly could emit visible light through photoluminescence at room temperature [18]. The number of scientific reports on porous germanium (Fig. 8.4) is very less as compared to published papers on porous silicon [18]. Whether the properties of porous Ge are distinctly different from those of porous Si is still unclear.

8.2 Synthesizing of silicon and germanium organometallic compounds

Most OMCs could be synthesized by having one of the four M—C bond formation (reactions of a metal with an organic halide, metal displacement, metathesis, or hydrometallation). Carbon bonds with an electropositive atom in the OM compound. Due to the presence of the M—C bond, there is covalency. The M—C bonds vary widely by nature, ranging from bonds that are essentially ionic to primarily covalent bonds. The special feature of such compounds is that some of them are liquid even at ambient temperature. Transmetallation is favorable when the displacing metal is higher in the electrochemical series than the displaced metal. The more the electropositive character of metal is, that much more will be the ionic character of the M—C bond. Carbon has a negative charge and metal has a positive charge. As a result, the organic part is always nucleophilic and basic. As a result, OM compounds could act both as a nucleophile as well as a base. The section below discusses the various methods of OM compound synthesis [19].

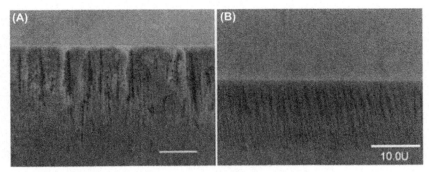

FIGURE 8.4 Cross-sectional SEM image of (A) porous silicon and (B) porous germanium [14]. *Adapted from Buriak JM. Organometallic chemistry on silicon and germanium surfaces, Chem Rev 2002;102(5):1271–1308.*

8.2.1 Direct synthesis

Organosilicon compounds of silicon are synthesized from other OM compounds with silicon halide and ester of orthosilicic acid. Tetraethylsilane, the first organosilicon compound, was prepared by Friedel Crafts by reacting tetrachlorosilane with diethylzinc. The bulk of organosilicon compounds is derived by direct process by reacting methyl chloride with silicon-copper alloy. Although the main product is dimethyldichlorosilane, however, a variety of other products including trimethylsilyl chloride and methyltrichlorosilane are also obtained.

$$2CH_3Cl + Si \rightarrow (CH_3)_2SiCl_2$$
(dimethyldichlorosilane)

A popular method, about a million tonne of organosilicon compounds are prepared annually worldwide through this.

8.2.2 Hydrosilylation

Another popular method for Si–C bonding is hydrosilation. In this, the compounds with Si–H bonds (hydrosilanes) are added to unsaturated substrates like alkenes (Fig. 8.5). Although other unsaturated substrates like alkynes, imines, ketones, and aldehydes could also be employed, but the reaction products of these are of little economic value.

8.2.3 Silenes

Organosilicon compounds with silene Si = C bonds (also known as alkylidenesilanes) are as follows (Fig. 8.6).

Silicon and germanium organometallic compounds Chapter | 8 201

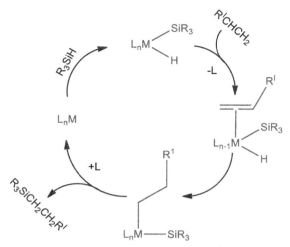

FIGURE 8.5 Metal-catalyzed hydrosilylation of alkene (R = , R¹ = alkyl groups).

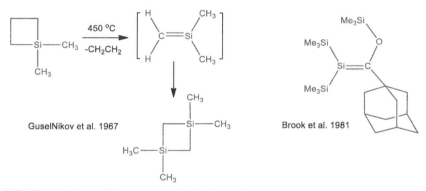

FIGURE 8.6 Organosilicon compounds with silene Si = C bonds.

8.2.4 Silicones, silyl ethers, silanols, siloxanes, and siloxides

Compounds with Si = O double bond are called silanones and are extremely unstable. These are long-chain polymers with repeating siloxane linkages. Silicones have silicon-oxygen backbone chain and two organic groups are attached to each silicon center (Fig. 8.7).

Silyl ethers with Si—O—C bond are prepared by reacting alcohols with silyl chlorides:

$$(CH_3)_3SiCl + ROH \rightarrow HCl + (CH_3)_3Si\text{-}O\text{-}R.$$

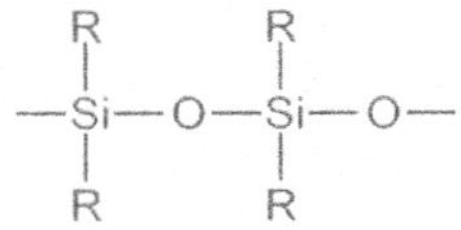

$$-\underset{R}{\overset{R}{\underset{|}{Si}}}-O-\underset{R}{\overset{R}{\underset{|}{Si}}}-O-$$

FIGURE 8.7 Silicones with silicon-oxygen backbone chain.

Silanols are analogs of alcohols, generally prepared by hydrolyzing silyl chlorides:

$$R_3SiCl + H_2O \rightarrow HCl + R_3SiOH.$$

Silanols may also be prepared by oxidizing silyl hydrides in presence of a metal catalyst: 2 $2R_3SiH + O_2 \rightarrow 2R_3SiOH.$

Silanols tend to dehydrate to siloxanes:

$$2R_3SiOH \rightarrow H_2O + R_3Si\text{-}O\text{-}SiR_3.$$

Siloxides are deprotonated derivatives of silanols:

$$R_3SiOH + NaOH \rightarrow H_2O + R_3SiONa.$$

8.2.5 Bonding through carbon

Yao et al. successfully isolated silylone (bis-NHC)Si and germylone (bis-NHC)Ge from monatomic zerovalent Si (silylone) and Ge (germylone) and bis-N-heterocyclic carbene (bis-NHC) ligand (Fig. 8.8). They established the electronic structures of these compounds through theoretical calculations and spectroscopic data [20].

Ochiai et al. prepared ditopic germanium complex from amino(imino)germylene by reacting it with $BF_3 \cdot OEt_2$. This monocation converts to germylene-germyliumylidene when treated with $Na[BAr^F_4]$. Tetrafluoroborate salt reacts with Me_3SiOTf to give the key compound (Fig. 8.9). Crystallographic study revealed that 4 + has a significant bis(germyliumylidene) dication character [21].

NHC-GeCl$_2$ and Mg[Ge(SiCH$_3$)$_3$]$_2$0.2THF or Mg[Sn(SiCH$_3$)$_3$]$_2$0.2THF were added to obtain NHC-Ge compounds (Fig. 8.10) [22].

Filippou et al. could synthesize [Cp(CO)$_2$Mo · GeC(SiCH$_3$)$_3$] complex from GeCl$_2$, Li[CpMo(CO)$_3$] and tetramethyl imidazolium salt at room temperature (Fig. 8.11) [23].

The final Ge-OMC synthesized by adding n-butyl lithium dropwise with constant stirring at $-78°C$ in THF followed by the addition of GeCl$_2$ · dioxane (Fig. 8.12). THF was added to the metal mixture at $0°C$ and stirred for 15 hours. It was further used to synthesize the final compound by adding THF at room temperature with 15-hour constant stirring [24].

Silicon and germanium organometallic compounds Chapter | 8 **203**

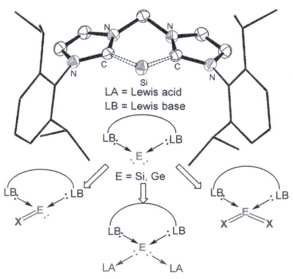

FIGURE 8.8 Synthesis of Si and Ge compounds involving bonding through carbon. *Adapted from Yao S, Xiong Y, Driess M. A new area in main-group chemistry: zerovalent monoatomic silicon compounds and their analogues. Acc Chem Res 2017; 50(8):2026−2037.*

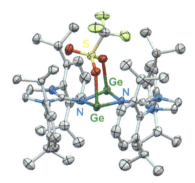

FIGURE 8.9 Structure of Germylene-triflatogermyliumylidene. *Adapted from Ochiai T, Szilvási T, Franz D, Irran E, Inoue S. Isolation and structure of germylenegermyliumylidenes stabilized by N-heterocyclic imines. Angew Chem 2016;128(38):11791−11796.*

Sodium cyclopentadienyl (NaCp) was added dropwise in THF at 0°C to obtain the final Ge-OMC (Fig. 8.13). The obtained solution was warmed at room temperature with 12-hour constant stirring. The yellow crystals of Ge-OMC (Fig. 8.13) were extracted with an average 74% yield [25].

[RhCl(cod)$_2$Cl]$_2$ was added with constant stirring at room temperature in THF to obtain C2 (Fig. 8.14). The mixture was filtered after 16-hour stirring, and dark purple crystals of 194°C melting point were obtained [26].

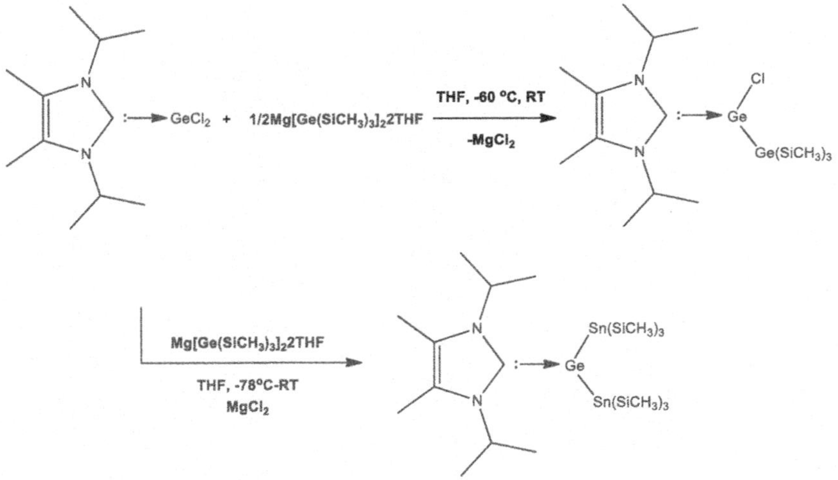

FIGURE 8.10 Synthesis of NHC-stabilized hypermetallyl germylene.

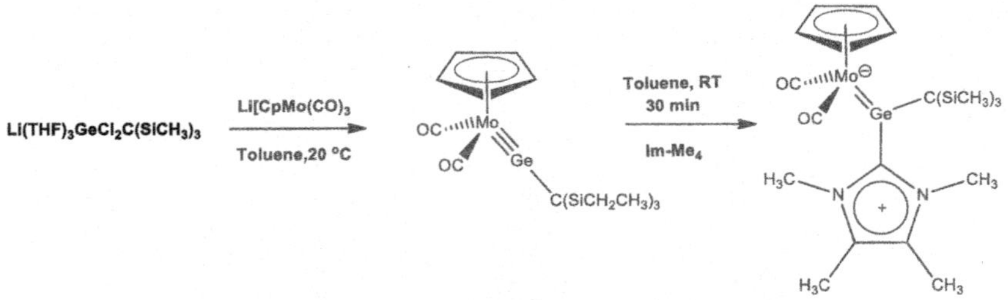

FIGURE 8.11 Synthesis of [Cp(CO)$_2$Mo · GeC(SiCH$_3$)$_3$] complex.

Further, C4 was obtained by adding [RhCl(cod)]$_2$ to C3 dropwise to toluene with constant stirring at room temperature. After 16-hour stirring, the filtered reaction mixture was an orange-colored transparent crystal with 85.7% yield [26].

The Ge-OMC was synthesized by adding n-butyl lithium to bis-sulfone in toluene at 40°C, stirred for 20 minutes at the same temperature (Fig. 8.15). The reaction mixture was added over a GeCl$_2$-dioxane suspension and stirred for 18 hours at room temperature. Colorless crystals of Ge-OMC were thus obtained through slow diffusion. Then, the final compound was obtained by adding Fe$_2$(CO)$_9$ to germylene solution in THF at -20°C, with constant stirring of the mixture for 24 hours [27].

The iron germylene complex having Fe—H and Ge—H bonds was synthesized in hexane with constant stirring at ambient temperature (Fig. 8.16). Pure red crystals were obtained with 42% yield [28].

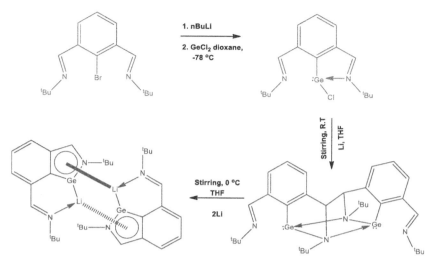

FIGURE 8.12 Preparation of germylidenide anion. *Adapted from Seow C, Xi H-W, Li Y, So C-W. Synthesis of a germylidenide anion from the C—C bond activation of a bis (germylene). Organometallics 2016;35(8):1060–1063.*

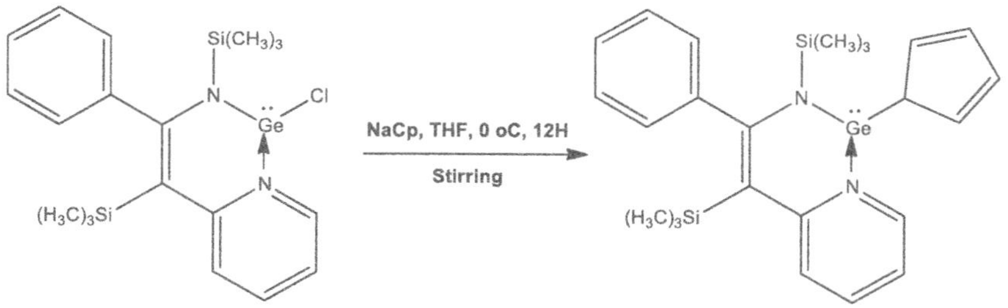

FIGURE 8.13 Metallogermylenes (Cp-substituted germylene). *Adapted from Leung W-P, Chiu W-K, Mak TC. Synthesis and structural characterization of metallogermylenes, Cp-substituted germylene, and a germanium (II)-borane adduct from pyridyl-1-azaallyl germanium (II) chloride. Organometallics 2012;31(19):6966–6971.*

A Ge—Ge compound was obtained in THF at room temperature with constant stirring (Fig. 8.17) [29].

A Ge—Mn compound was formed by adding a stirred suspension of [$Mn_2(CO)_{10}$] in toluene (Fig. 8.18). The reaction mixture was refluxed overnight and the filtrate was concentrated to obtain dark red crystals, with 46.3% yield [26].

8.3 Catalysis

OM compounds are gaining great interest as the world focuses on sustainable and eco-friendly catalytic synthesis systems, as these offer excellent

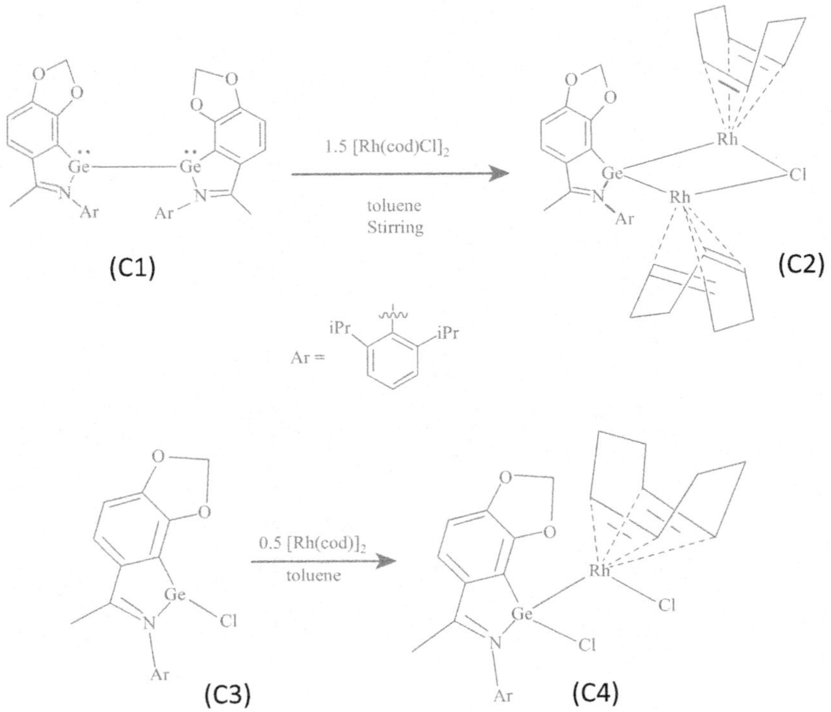

FIGURE 8.14 Synthesis of germanium(I) dimer. *Adapted from Ismail MLB, Liu F-Q, Yim W-L, Ganguly R, Li Y, So C-W. Reactivity of a base-stabilized germanium(I) dimer toward group 9 metal(I) chloride and dimanganese decacarbonyl. Inorg Chem 2017;56(9):5402–5410.*

opportunities for green organic synthesis of especially fine chemicals. In this context, OM compounds of p-block elements are also attractive candidates. Lewis-acidic compounds based on boron [30–33] and phosphorus [34–37] cations are in focus of catalysis. Si(II) and Ge(II) compounds are new-age catalysts for various organosilicon chemical transformations. Such compounds effectively catalyze the hydrosilylation of C—C multiple bonds and numerous siloxane-coupling reactions. Very few examples of catalysts based on electrophilic Si or Ge compounds are known till date. Numerous electrophilic Si(IV) compounds are known to catalyze hydrosilylation of aldehydes [38], hydrodefluorination reactions [39], and the cross-coupling of arylsilanes and fluorohydrocarbons [40]. Cationic Ge(II) compounds stabilized by iminato- and carbene-ligands [41,42] catalyze hydroboration reactions of aldehydes and ketones with pinacolborane. The hydrogen-bridged cationic bis(silylium) and bis(germylium) compounds could catalyze hydrosilylation of carbon-carbon double bonds [43,44]. The cyclopentadienyl-coordinated cationic Si(II) and Ge(II) compounds (Fig. 8.19) are good candidates for catalysis due to their open coordination sphere [45,46]. In 2011, Leszczynska et al. [47]

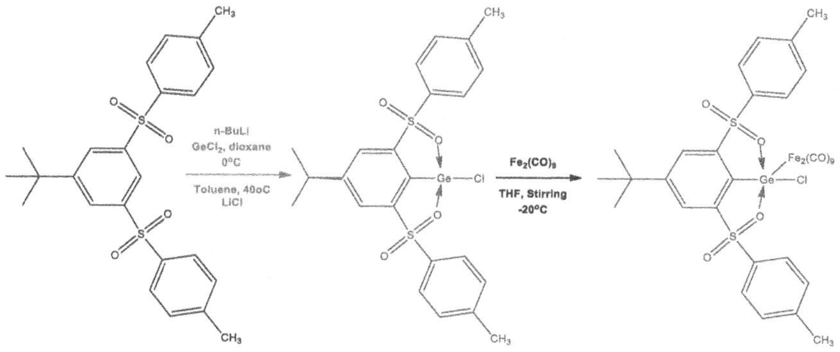

FIGURE 8.15 Synthesis of bis-sulphonyl O,C,O-chelated germylenes. *Adapted from Deak N, Petrar PM, Mallet-Ladeira S, Silaghi-Dumitrescu L, Nemeş G, Madec D, et al. O-chelated metallylenes (Ge, Sn) as adjustable ligands for iron and tungsten complexes. Chem Eur J 2016;22(4):1349−1354.*

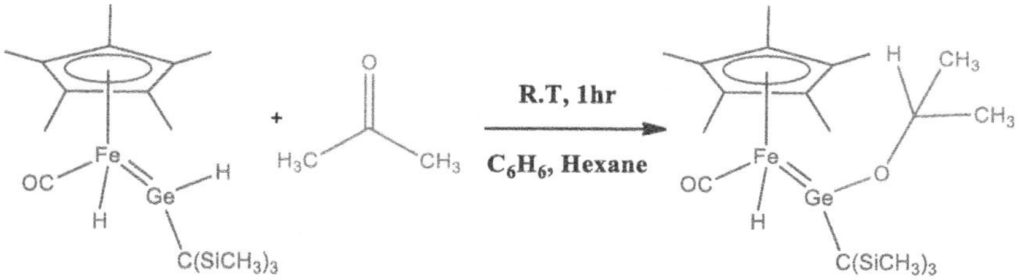

FIGURE 8.16 Synthesis of iron germylene complex having Fe−H and Ge−H bonds. *Adapted from Dhungana TP, Hashimoto H, Tobita H. An iron germylene complex having Fe-H and Ge-H bonds: synthesis, structure and reactivity. Dalton Trans 2017;46:8167−8179.*

reported the cyclopentadienyl-coordinated cationic Si(II) as a catalyst to degrade oligo(ethyleneglycol) ethers (Fig. 8.19) [48].

OM compounds find extensive applications as Reagents, such as *n*-BuLi, *s*-BuLi, *t*-BuLi, Grignard reagent (RMgX), and Gilman's reagent (R$_2$CuLi), are used as reagents, in catalysis, especially in hydrogenation, hydroformation, and olefin metathesis. In addition, OM compounds are also used as drugs, such as antimicrobial agents, antitumor agents, and antidiabetic agents. Moreover, OMC applications as *antiknocking agents* and also in the manufacture of some semiconductors are well known [49]. The radioprotective activity of organosilicon and organogermanium compounds is now well known in the literature [50] and is concerned the synthesis and evaluation in chemical radioprotection of new organosilylated and organogermylated structures, metallathiazolidines, and metalladithioacetals derived from 1-[*N*-(2-mercaptoethyl) − 2-aminoethyl] − 2-(1-naphthylmethyl) − 2-imidazoline and 1-[*N*-(2-mercaptopropyl) − 2-aminoethyl] − 2-(1-naphthylmethyl) − 2-imidazoline. In this work, the toxicity and radioprotective activity of metallathiazolidines and metalladithioacetals

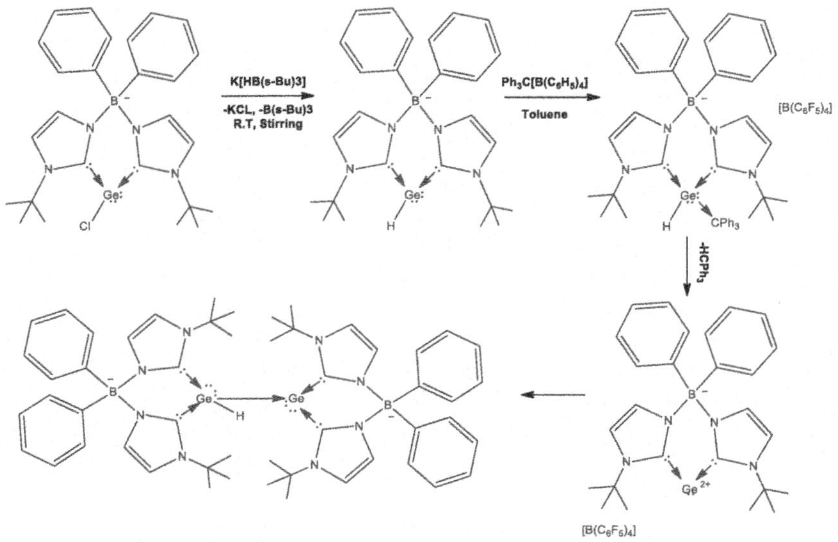

FIGURE 8.17 Synthesis of germyliumylidene hydride. *Adapted from Xiong Y, Szilvási T, Yao S, Tan G, Driess M. Synthesis and unexpected reactivity of germyliumylidene hydride [:GeH]+ stabilized by a bis (N-heterocyclic carbene) borate ligand. J Am Chem Soc 2014;136 (32):11300−11303.*

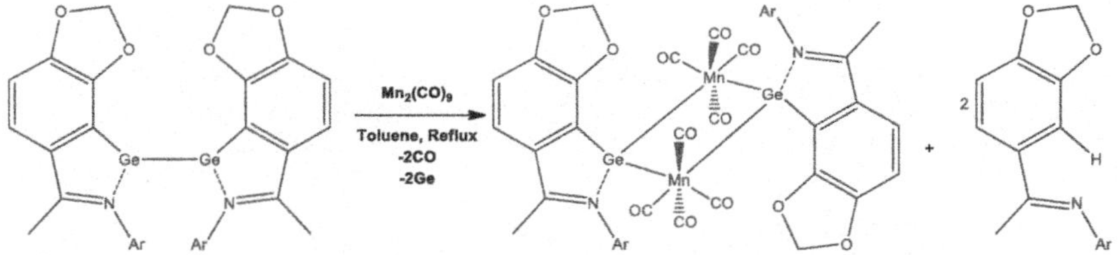

FIGURE 8.18 Synthesis of Ge(I) dimer stabilized by dimanganese decacarbonyl. *Adapted from Ismail MLB, Liu F-Q, Yim W-L, Ganguly R, Li Y, So C-W. Reactivity of a base-stabilized germanium(I) dimer toward group 9 metal(I) chloride and dimanganese decacarbonyl. Inorg Chem 2017;56(9):5402−5410.*

was presented (Fig. 8.20). These compounds exhibited significant pharmacological properties.

Organogermanium chemistry describes about the compounds containing at least one Ge−C bond. Over the years, such compounds are used in the manufacture of microelectronics (semiconductor), sensors, optics, glasses, optical fibers, infrared optics, polymers, ceramics, and catalysts. The biological activities especially in the field of bactericides, fungicides, psychotropic agents, antitumor agents as well as in radioprotective activity are notable. The organic derivatives of Ge display various structures and may be obtained from the reaction of halogenated organic compound with Ge (Fig. 8.21).

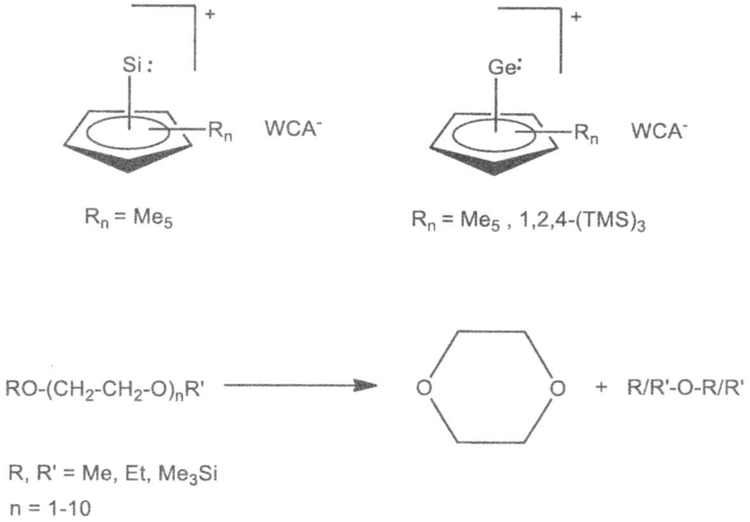

Silicon and germanium organometallic compounds Chapter | 8 **209**

$R_n = Me_5$

$R_n = Me_5$, 1,2,4-(TMS)$_3$

RO-(CH$_2$-CH$_2$-O)$_n$R' ⟶ [dioxane structure] + R/R'-O-R/R'

R, R' = Me, Et, Me$_3$Si
n = 1-10

FIGURE 8.19 Cyclopentadienyl (Cp)-coordinated cationic Si(II) and Ge(II) compounds (TMS = trimethylsilyl, Me = methyl, WCA = weakly coordinating anion); degradation of oligo(ethyleneglycol) ethers. *Adapted from Leszczynska K, Mix A, Berger RJF, Rummel B, Neumann B, Stammler H-G et al. The pentamethylcyclopentadienylsilicon(II) cation as a catalyst for the specific degradation of oligo(ethyleneglycol) diethers. Angew Chem Int Ed 2011;50:6843–6846; Fritz-Langhals E, Main group catalysis: cationic Si(II) and Ge(II) compounds as catalysts in organosilicon chemistry. Reactions 2021;2:442–456. https://doi.org/10.3390/reactions2040028.*

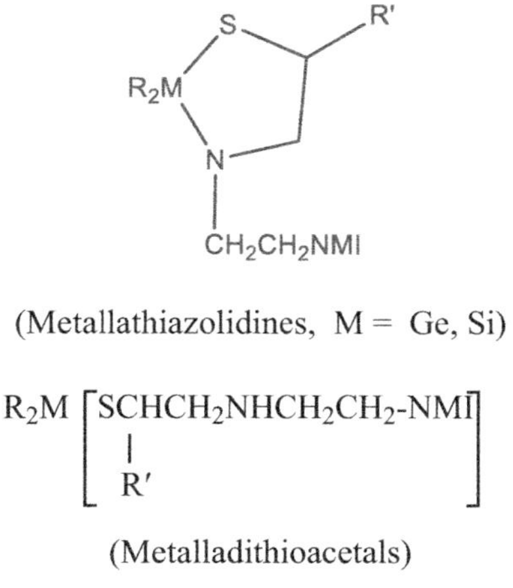

(Metallathiazolidines, M = Ge, Si)

$$R_2M \begin{bmatrix} SCHCH_2NHCH_2CH_2-NMI \\ | \\ R' \end{bmatrix}$$

(Metalladithioacetals)

FIGURE 8.20 Structure of metallathiazolidines and metalladithioacetals.

Germylenes react with *p*-benzoquinones leading to polymers in excellent yields as shown below and exhibit interesting semiconductor properties (Fig. 8.22).

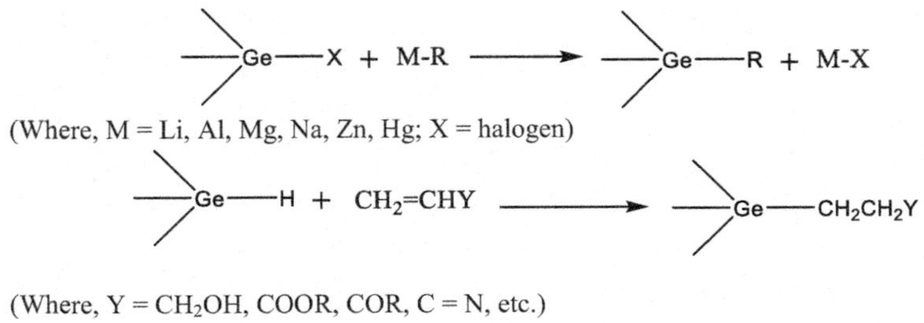

$$\text{\textbackslash Ge—X} + \text{M-R} \longrightarrow \text{\textbackslash Ge—R} + \text{M-X}$$

(Where, M = Li, Al, Mg, Na, Zn, Hg; X = halogen)

$$\text{\textbackslash Ge—H} + \text{CH}_2=\text{CHY} \longrightarrow \text{\textbackslash Ge—CH}_2\text{CH}_2\text{Y}$$

(Where, Y = CH$_2$OH, COOR, COR, C = N, etc.)

FIGURE 8.21 The Ge-OMC having germanium−carbon bond.

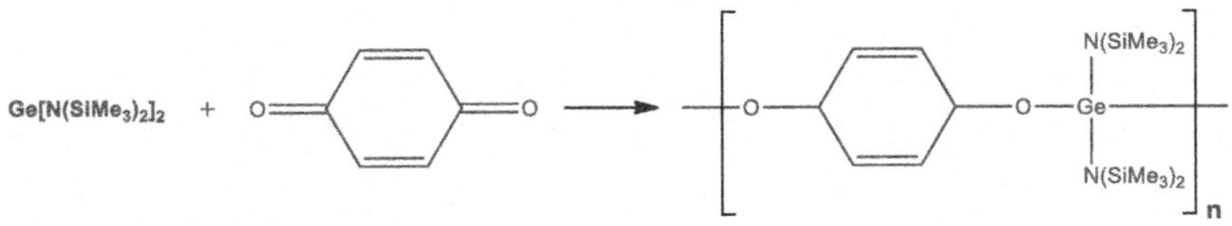

FIGURE 8.22 Ge-polymer by the reaction of germylenes and p-benzoquinones.

Both germanium and silicon react with halogen and hydrogen halides to give analogous products. Halides of both metals react with water to give respective metal oxide and hydroxide. Ge has higher optical densities than Si compounds, leading to the second most important application of Ge compounds. The controlled combustion of GeCl$_4$ with SiCl$_4$ in a hydrogen−oxygen flame in a chemical vapor axial deposition formation of ingots from which step index fiber optics are drawn.

8.4 Future prospective and scope

Group-IV (second group of transition metals in Periodic Table) nanocrystals have emerged as promising materials with extended areas of application of bulk diamond, silicon, germanium, and related materials beyond the traditional boundary. The potential to apply these in areas like optoelectronics and memory devices is being progressively deciphered through research over the recent couple of decades, while unparallel challenges have also arisen. An urgent need to consider as to what is achieved and the limitations in the growth, characterization, and modeling of silicon- and germanium-related materials is felt. While silicon and germanium present interesting diverse transport properties, their (SiGe) composites have breakthrough transport properties. Such alloys are applied in photovoltaic cells, microelectronic devices, integrated circuits, and thermoelectricity. Si0.8Ge0.2 alloy has significantly attracted attention in thermoelectricity as an energy harvesting material, to power space applications and other industrial applications in recent times. Si-based integrated device for monolithic and hybrid optoelectronic integration has been of interest as the on-chip optical interconnects and optical computing has progressed in the recent past. Epitaxial Ge-on-Si is a significant material for optoelectronic device applications owing to its narrow pseudo-direct gap behavior and compatibility. Epitaxial Ge-on-Si materials could be bigger players in silicon photonics.

Interest in the radioprotection of organosilicon and organogermanium compounds is growing as reflected through the growing number of scientific reports in recent times. Organo-silylated and -germylated derivatives have shown good radioprotection as compared to the native organic compounds. This is attributed to the increased hydrosolubility due to the OM groups, increased liposolubility due to the presence of organic ligands, and their biological activity that favors their passage through cell membrane.

Recent developments in smart technology that has revolutionized devices and gadgets primarily revolve around miniaturization, automation and microfluidics, and a shift from electrical to electronics. It aims at providing high-end hitherto remote solutions to the various issues of human day-to-day life. Silicon and germanium as elements have proved to be quite promising in helping achieve this. These two elements particularly have touched a wide array of human life, from the mundane household chores, office automations, quality industrial deliverables, to the life-saving medical diagnostics and therapeutics. Therefore the scope of applying the OM compound derivatives of silicon and germanium is huge. Various other combinations of more promising OM compounds involving silicon and germanium may be engineered and synthesized.

8.5 Summary

Many generalizations about the OM surface chemistry of silicon and germanium could be made. Investigations still remain to be extended to especially the packing and bonding nature of organic groups in dissolved chemical reactions, the Si−Si and Ge−Ge dimers bonding and the reactivity on Si (100) − 2 × 1 reconstructed surface, the long-term stability of the surface and their biocompatibility. Due to the proximity of basic science and applied technology, the field is extremely exciting. It is a matter of time and the need before an array of tailored organic interfaces are available and they reach their commercial potential. As the new-age devices and gadgets reach greater heights of miniaturization and automation, silicon and germanium OMs shall be in more demand. Thus the scope of applying the OM compound derivatives of silicon and germanium is huge and shall have a greater defining role in advanced research and application in the future.

Although the promises of the silicon and germanium OM compounds in advanced technology are quite high, the near impossibility of creating atomic clusters in the proportions required to synthesize novel materials is a major bottleneck. When crystals are recovered out of the high-vacuum environment, they often become reactive and are polluted thereby losing their desired properties in quality as well as quantity. Their shape and magnetic properties also tend to change in the gaseous phase. The structure-property interactions of such nanomaterials could now be examined (one atom and one electron at a time), thanks to the gas-phase cluster studies. The knowledge gained on the gaseous-phase interactions of hydrogen with metal atoms and their cations can be used in designing and synthesizing novel class of hydrogen energy storage materials. The hydrogen storage material parameters must be met for the hydrogen to release under ambient conditions and then be used as a fuel for transport. The ability of a metal cation to absorb hydrogen in quasimolecular form through charge polarization has been proved. These are currently being employed to dope porous materials with metal atoms to capture hydrogen in quasimolecular form. As they are reactive and are prone to alterations when assembled or maintained on a surface, most clusters investigated in the gas phase present a challenge. Producing magnetic particles from nonmagnetic materials is feasible by manipulating the material size and content. The size and form can be utilized to manage the "band gap" of nanoclusters effectively for their use in optical devices.

8.6 Problems with solutions

Q1. Why number of compounds of Si and Ge are limited as compared with carbon?

Answer: Due to high capacity for catenation, carbon forms lakhs of compounds. However, Si and Ge have much lower capacity for catenation. Therefore compounds of Si and Ge are limited.

Q2. What are EAN rule and 18-electron rule?

Answer: Sum of the electrons on the metal and the electrons donated by the ligands is called effective atomic number (EAN). When EAN for a compound is equal to 36 (Kr), 54 (Xe), or 86 (Rn), the EAN rule is said to be obeyed.

When the metal achieves an outer shell configuration of ns^2 $(n-1)d^{10}$ np^6, there will be 18 electrons in the valence shell. This is the stable configuration and termed as 18-electron rule and is given by Sidgwick in 1927. It is also known as the noble-gas rule as the metal center in an OM compound achieves the noble-gas configuration. Those OM compounds which follow this rule are considered stable compounds.

Q3. What do you mean by transmetallation?

Answer: In the transmetallation reaction one metal atom replaced the other metal center in an OM compound. In the electrochemical series, when the displaced metal is lower than the displacing metal such reactions are favorable. Mercury is replaced by the gallium metal.

Q4. Why CO_2 is a gas but SiO_2 is solid?

Answer: In CO_2, the double bond between carbon and oxygen (C = O) atoms is formed due to lateral overlapping of 2p (C) orbital and 2p (O) orbital. As the size and energy of both these orbitals are nearly same, they overlap effectively to form pi bonds and lead to the formation of monomeric O = C = O molecule. But in case of SiO_2, it requires the lateral overlapping of 3p (Si) and 2p (O) orbitals. Due to appreciable differences in the size and energy of these orbitals, they cannot overlap effectively to form pi bonds which hinders the formation of Si = O bond. SiO_2 forms giant molecules with −Si−O− single bonds.

8.7 Objective type questions

1. The chemical formula of Zeise's salt is
   a $(C_2H_5)Pb$
   b $(C_6H_6)_2Cr$
   c $(C_5H_5)_2Fe$
   d $K[PtCl_3(C_2H_4)]$
   Answer: (d)
2. Identify the molecule which follows the 18-electron rule.
   a $[Ni(H_2O)_6]^{2+}$
   b $Fe(CO)_5$
   c $V(CO)_6$
   d $(\eta^6 \text{-}C_6H_6)Ru$
   Answer: (b)
3. The complex which doesn't obey the 18-electron rule is:
   a $Ni(CO)_3(PPh_3)$
   b $Cr(C_5H_5)_2$
   c $Co_2(CO)_8$
   d $Mn(CO)_5Cl$
   Answer: (b)
4. Which of the following is a bridged organometallic compound?
   a Zeise salt
   b Grignard's reagent
   c Ferrocene
   d Dimeric trimethylaluminum
   Answer: (d)
5. The kappa convention is used to indicate:
   a The points of ligation
   b Bridging ligands
   c Unsaturated molecules
   d Anionic ligands
   Answer: (a)

6. Identify the Wilkinson's catalyst:
   a [RhCl(PPh$_3$)$_3$]
   b [Rh(CO)$_2$I$_2$]$^-$
   c [HCo(CO)$_4$]
   d [Rh(CO)(PPh$_3$)$_2$Cl]
   Answer: (a)
7. The catalyst used in hydroformylation process is:
   a Crabtree's catalyst
   b Zeigler-Natta catalyst
   c Octa carbonyl dicobalt
   d Wilkinson's catalyst
   Answer: (c)
8. The organometallic compound used as fuel additive:
   a Mg(C$_2$H$_5$)$_2$
   b (C$_2$H$_5$)$_4$Pb
   c (CH)$_3$SiCl
   d Ti(C$_5$H$_5$)$_2$Cl$_2$
   Answer: (b)
9. The organometallic compound [Co$_4$(CO)$_{12}$] have metal−metal bond:
   a Two
   b Four
   c Five
   d Six
   Answer: (d)
10. Zeigler-Natta catalyst is basically used for the preparation of:
    a Aldehyde
    b Polymers
    c Metallocenes
    d None of these
    Answer: (b)

References

[1] Mohapatra R.K. Engineering chemistry with laboratory experiments, PHI Delhi, 2016. ISBN: 9788120351585.
[2] Richard M. One hundred years of organosilicon chemistry. J Chem Educ 1965;42(1):41.
[3] Thomas NR, Stanley F. Kipping-pioneer in silicon chemistry: his life & legacy. Silicon 2010;2(4):187−93.
[4] Clemens W. Mittheilungen über des Germanium. Zweite Abhandlung. J Prak Chem 1887;36:177−209.
[5] Puri B.R., Sharma L.R., Kalia K.C. Principles of inorganic chemistry, milestone publishers & distributors, Delhi, 2013-14. ISBN: 9788192143330.
[6] Waltenburg HN, Yates JT. Chem Rev 1995;95:1589.
[7] Hamers RJ, Wang Y. Chem Rev 1996;96:1261.
[8] Campbell SA. The science and engineering of microelectronic fabrication. Oxford: *Oxford University Press*; 1996.

[9] Hayat H, Iqbal MA. Mod Tech Synth Organomet Compd Germanium 2018;.
[10] Bardeen J, Brattain WH. Phys Rev 1948;71:230.
[11] Whall TE, Parker EHC. J Mater Sci Mater Electron 1995;6:249.
[12] Kubby JA. J Surf Sci Rep 1996;26:249.
[13] Paul DJ. Adv Mater 1999;11:191.
[14] Buriak JM. Organometallic chemistry on silicon and germanium surfaces. Chem Rev 2002;102(5):1271−308.
[15] Zanatta JP, Ferret P, Duvaut P, Isselin S, Theret G, Rolland G, et al. J Cryst Growth 1998;185:1297.
[16] Ranke W. Surf Sci 1995;342:281.
[17] Rodriguez-Reyes JCF, Teplyakov AV. Chemistry of organometallic compounds on silicon: the first step in film growth. Chem Eur J 2007;13:9164−76.
[18] Canham LT. Appl Phys Lett 1990;57:1046.
[19] Rochow EG. The direct synthesis of organosilicon compounds. J Am Chem Soc 1945;67(6):963−5.
[20] Yao S, Xiong Y, Driess M. A new area in main-group chemistry: zerovalent monoatomic silicon compounds and their analogues. Acc Chem Res 2017;50(8):2026−37.
[21] Ochiai T, Szilvási T, Franz D, Irran E, Inoue S. Isolation and structure of germylenegermyliumylidenes stabilized by N-heterocyclic imines. Angew Chem 2016;128(38):11791−6.
[22] Katir N, Matioszek D, Ladeira S, Escudié J, Castel A. Stable N-heterocyclic carbene complexes of hypermetallyl germanium (II) and tin (II) compounds. Angew Chem Int Ed 2011;50(23):5352−5.
[23] Filippou AC, Stumpf KW, Chernov O, Schnakenburg G. Metal activation of a germylenoid, a new approach to metal−germanium triple bonds: Synthesis and reactions of the germylidyne complexes [Cp(CO)2MGe-C(SiMe3)3](M = Mo,W). Organometallics 2012;31(2):748−55.
[24] Seow C, Xi H-W, Li Y, So C-W. Synthesis of a germylidenide anion from the C−C bond activation of a bis (germylene). Organometallics 2016;35(8):1060−3.
[25] Leung W-P, Chiu W-K, Mak TC. Synthesis and structural characterization of metallogermylenes, Cp-substituted germylene, and a germanium (II)-borane adduct from pyridyl-1-azaallyl germanium (II) chloride. Organometallics 2012;31(19):6966−71.
[26] Ismail MLB, Liu F-Q, Yim W-L, Ganguly R, Li Y, So C-W. Reactivity of a base-stabilized germanium(I) dimer toward group 9 metal(I) chloride and dimanganese decacarbonyl. Inorg Chem 2017;56(9):5402−10.
[27] Deak N, Petrar PM, Mallet-Ladeira S, Silaghi-Dumitrescu L, Nemeş G, Madec D, et al. O-chelated metallylenes (Ge, Sn) as adjustable ligands for iron and tungsten complexes. Chem Eur J 2016;22(4):1349−54.
[28] Dhungana TP, Hashimoto H, Tobita H. An iron germylene complex having Fe-H and Ge-H bonds: synthesis, structure and reactivity. Dalton Trans 2017;46:8167−79.
[29] Xiong Y, Szilvási T, Yao S, Tan G, Driess M. Synthesis and unexpected reactivity of germyliumylidene hydride [:GeH]$^+$ stabilized by a bis (N-heterocyclic carbene) borate ligand. J Am Chem Soc 2014;136(32):11300−3.
[30] Ma Y, Lou S-J, Hou Z. Electron-deficient boron-based catalysts for C-H bond functionalization. Chem Soc Rev 2021;50:1945−67.
[31] Chardon A, Osi A, Mahaut D, Saida AB, Berionni G. Non-planar boron Lewis acids taking the next step: development of tunable Lewis acids, Lewis superacids and bifunctional catalysts. Synlett 2020;31:1639−48.

[32] Brook MA. New control over silicone synthesis using SiH chemistry: the Piers-Rubinsztajn reaction. Chem Eur J 2018;24:8458−69.

[33] Rao B, Kinjo R. Boron-based catalysts for C-C bond formation reactions. Chem Asian J 2018;13:1279−92.

[34] Li H, Liu H, Guo H. Recent advances in phosphonium salt catalysis. Adv Synth Catal 2021;363:2023−36.

[35] Waked AE, Chitnis SS, Stephan DW. P(V) dications: carbon-based Lewis acid initiators for hydrodefluorination. Chem Commun 2019;55:8971−4.

[36] Bayne JM, Stephan DW. C-F bond activation mediated by phosphorus compounds. Chem Eur J 2019;25:9350−7.

[37] Chitnis SS, Krischer F, Stephan DW. Catalytic hydrodefluorination of C-F bonds by an air-stable PIII Lewis acid. Chem Eur J 2018;24:6543−6.

[38] Liberman-Martin AL, Bergman RG, Don Tilley T. Lewis acidity of bis(perfluorocatecholato)silane: aldehyde hydrosilylation catalyzed by a neutral silicon compound. J Am Chem Soc 2015;137:5328.

[39] Panisch R, Bolte M, Müller T. Hydrogen- and fluorine-bridged disilyl cations and their use in catalytic C-F activation. J Am Chem Soc 2006;128:9676−82.

[40] Lühmann N, Panisch R, Müller T. A catalytic C-C bond-forming reaction between aliphatic fluorohydrocarbons and arylsilanes. Appl Organomet Chem 2010;24:533−7.

[41] Sinhababu S, Singh D, Sharma MK, Siwatch RK, Mahawar P, Nagendran S. Ge(II) cation catalyzed hydroboration of aldehydes and ketones. Dalton Trans 2019;48:4094−100.

[42] Roy MMD, Fujimoro S, Ferguson MJ, McDonald R, Tokitoh N, Rivard E. Neutral, cationic and hydride-substituted siloxygermylenes. Chem Eur J 2018;24:14392−9.

[43] Fritz-Langhals E, Gowans S. Process for metal-free hydrosilylation of unsaturated compounds using silicon(IV) cationic complexes as catalysts. WO 2017-EP83669. Chem Abstr 2019;171:1249217.

[44] Fritz-Langhals E, Gowans S. Noble-metal free hydrosilylatable mixture 2020; PCT/EP2020/060581.

[45] Jutzi P, Mix A, Rummel B, Schoeller WW, Neumann B, Stammler H-G. The (Me$_5$C$_5$)Si$^+$ cation: a stable derivative of HSi$^+$. Science 2004;305:849−51.

[46] Fritz-Langhals E, Werge S, Kneissl S, Piroutek P. Novel Si(II)$^+$ and Ge(II)$^+$ compounds as efficient catalysts in organosilicon chemistry: siloxane coupling reaction. Org Process Res Dev 2020;24:1484−95.

[47] Leszczynska K, Mix A, Berger RJF, Rummel B, Neumann B, Stammler H-G, et al. The Pentamethylcyclopentadienylsilicon(II) cation as a catalyst for the specific degradation of oligo(ethyleneglycol) diethers. Angew Chem Int Ed 2011;50:6843−6.

[48] Fritz-Langhals E. Main group catalysis: cationic Si(II) and Ge(II) compounds as catalysts in organosilicon chemistry. Reactions 2021;2:442−56. Available from: https://doi.org/10.3390/reactions2040028.

[49] Mudi SY, Usman MT, Ibrahim S. Clinical and industrial application of organometallic compounds and complexes: a review. Am J Chem Appl 2015;.

[50] Rima G, Satge J, Dagiral R, Lion C, Fatome M, Roman V, et al. Appl Organometal Chem 1999;13:583−94.

Chapter 9

Sensing application of organometallic compounds

Durga Prasad Mishra[1] and Ashish Kumar Sarangi[2]

[1]*Department of Pharmaceutical Chemistry, School of Pharmacy, Centurion University of Technology and Management, Balangir, Odisha, India,* [2]*Department of Chemistry, School of Applied Sciences, Centurion University of Technology and Management, Balangir, Odisha, India*

9.1 Introduction

Any substance belonging to the class of chemicals known as organometallic compounds should contain at least one metal-to-carbon bond in which the carbon is a component of an organic group. A very wide class of molecules known as organometallic compounds has contributed significantly to the growth of the field of chemistry. They are widely utilized in laboratories and the industry as catalysts (substances that speed up reactions without being consumed themselves) and as intermediates [1]. The detection of dangerous and harmful gases in industrial production, atmospheric environmental testing, and fire explosion accident prevention all benefited greatly from the use of sensors. However, there are currently clear inadequacies in selectivity, stability, and other areas of sensitive material development; as a result, the study and development of new high-performance sensors will have significant realistic significance and usefulness [2]. Chemical signals are typically produced following detection by chemical sensors. Photochemical and electrochemical sensors are the two primary categories of chemical sensors [3]. In the field of sensing applications, chemical sensors are widely distributed in real-time monitoring, energy conservation, security alarms, environmental pollution monitoring, resource detection, process analysis, meteorological observation, remote diagnosis, industrial automation, fresh food preservation, production, telemetry, and in medicine. Chemical sensors' fundamental workings are comparable to the olfactory and gustatory systems in humans. Additionally, these sensors are capable of detecting chemicals like H_2, CO, NH_3, and methylbenzene that human organs are unable to. They are therefore very useful tools for the differentiation of analytes [4]. Chemical sensors are devices that detect chemicals and transform chemical data (such as concentration, pressure, and particle activity) into an electrical signal to produce qualitative or quantitative time- and space-resolved information about particular chemicals [5]. There are so many complicated challenges involved in the determination of the chemical substances and it is the exclusive character will be examined under the basis of the application of the chemical sensor. The acknowledgment of the sensing property of a molecule depends upon the molecular structure and its reactivity. This phenomenon is known as selectivity. This is because there are so many known molecular compounds (>106), making this recognition aspect crucial. The amount or concentration of the substance being studied was connected to the sensitivity and limit of detection [6].

9.2 Fundamental principles

The prefix "organo-" defines the essential ideas behind organometallic compounds and admits all compounds that possess a bond between a metal and a carbon atom of an organyl group (for instance, organoplatinum compounds). In addition to the conventional metals such as s, d, and f-block elements, the p-block elements boron, silicon, arsenic, and selenium are thought to produce organometallic compounds (shown in Scheme 9.1) [7].

The following components are found in sensors for the detection of organometallic compounds, such as n-butyllithium:

a) Insoluble support, that is, inorganic, organic, or polymeric in the sample being studied.
b) An indicator physically trapped, adsorbed, absorbed, dissolved, or chemically linked in an electrostatic or covalent manner to the support that reversibly reacts with the organometallic compound and whose reaction product gives a characteristic absorption, reflection, or emission band in the range 150–15,000 nm, such as 1,10-phenanthroline and derivatives, triphenylmethane, or N-[2-(4-hydroxy benzyl)phenyl].

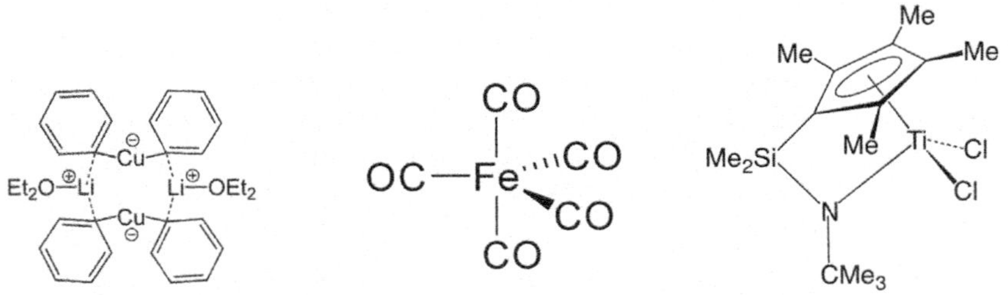

SCHEME 9.1 Different organometallic compounds (Organometallic Chemistry (June 23, 2022). In *Wikipedia*. https://en.wikipedia.org/wiki/Organometallic_chemistry).

c) An optical sensor that gauges the functionalized support's absorption, reflection, or emission at a particular wavelength and converts it to the amount of solution-bound organometallics. The process for determining the concentration of organometallic compounds by using an online sensor in anionic polymerizations is also explained [8].

9.3 Common methods of preparation

The synthetic procedure of 4,4′,4″-tris[trans-Pt (PEt3)2I(ethynyl phenyl)]amine shown in (Scheme 9.2). Preparation of tri[p-ethynylphenyl]amine, then the addition of tri(p-ethynyl phenyl)amine and trans-diiodobis(triethyl phosphine)platinum to a 100 mL round bottom Schlenk flask in the presence of 25 mL of dry nitrogen, degassed toluene and 10 mL of dry, diethyl amine were then added. Added 25 mg of CuI to the solution after continuous stirring for about 10 minutes at room temperature. Precipitation of a small amount of diethyl ammonium iodide occurs in the solution after keeping it for 16 hours at room temperature. After the removal of the solvent in presence of a vacuum then residue was obtained by using column chromatography on silica gel in the presence of a mixture of solvent hexane or benzene. The yellow residue was separated, then the final product was obtained with 64% yield (2:1). To yield the nitrated product (M) treating iodide derivative with silver nitrate. In an oil bath, the combined mixture was cooked overnight using the L (2.86 mg, 0.009 mmol) and M (12 mg, 0.006 mmol) solutions in methanol and acetone, at 50°C. After the addition of ether, 96% yield of product was obtained [9].

9.4 Mechanism involved

9.4.1 Ligand substitution

Mentioned Schemes 9.3 and 9.4 can be an associative, dissociative, or radical chain. The *trans*-effect is the kinetic effect of a ligand on the function of substitution at a position *trans* to itself in an octahedral or square complex (ground state weakening of bond). In Scheme 9.5, nucleophilic substitution occurs with the metal complex [10].

9.4.2 Oxidative addition

The reaction known as oxidative addition occurs when the metal center's oxidation state increases by two (Scheme 9.6). If the metal center of the complex doesn't have two units of accessible oxidation higher than the initial state, this reaction cannot take place. Lewis acid is recognized to function as the metal center of a complex. Under oxidative addition, it functions both as a Lewis acid and a Lewis base. When the metal center of the complexes accepts electrons from the sigma bond (σ) or donates electrons from the sigma antibonding (σ^*), it is a Lewis acid.

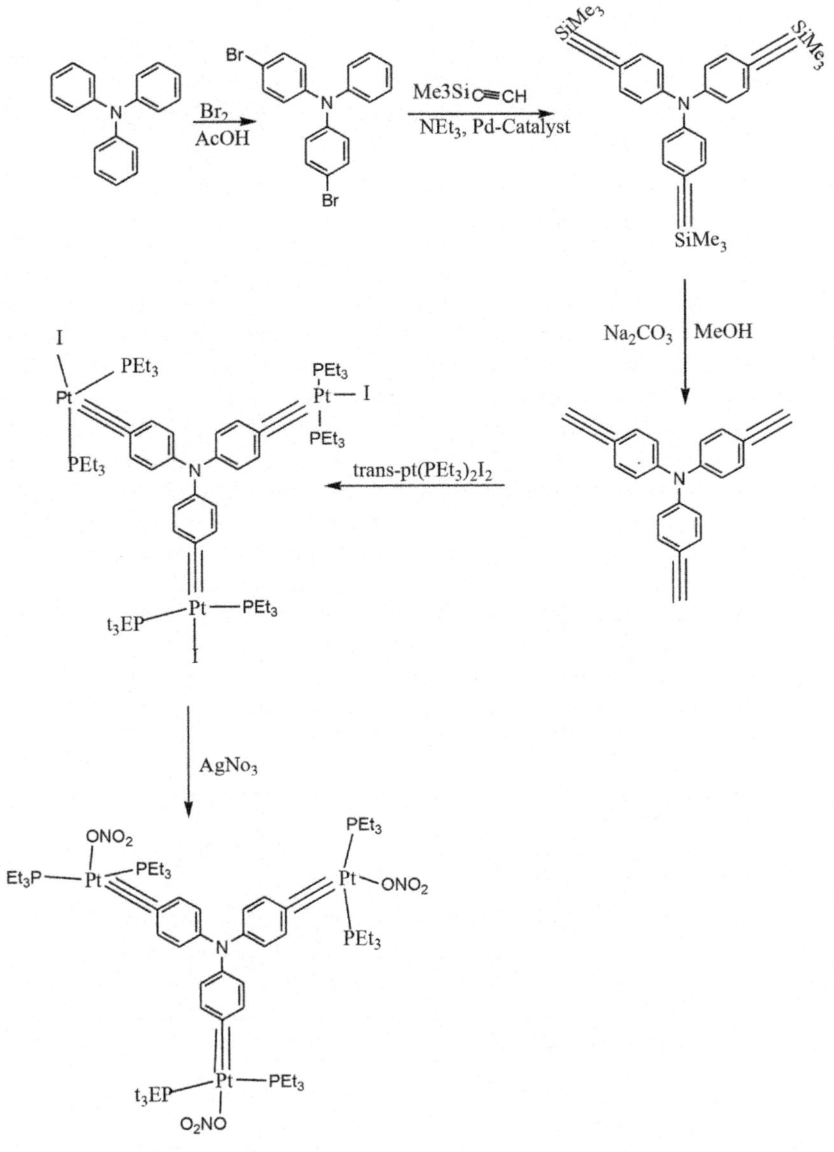

SCHEME 9.2 Synthesis of the organometallic Pt₃ acceptor M. *Procreated from Ghosh S, Mukherjee PS. Self-assembly of a nanoscopic prism via a new organometallic Pt3 acceptor and its fluorescent detection of nitroaromatics. Organometallics 2008;27(3):316–19.*

Sensing application of organometallic compounds Chapter | 9 **221**

$$M\text{-}L + L' \rightarrow M\text{-}L' + L$$

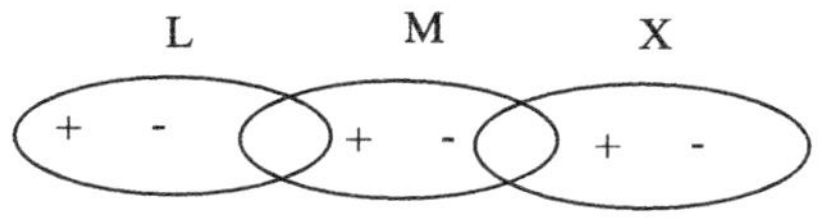

SCHEME 9.3 Substitution of ligand with metal.

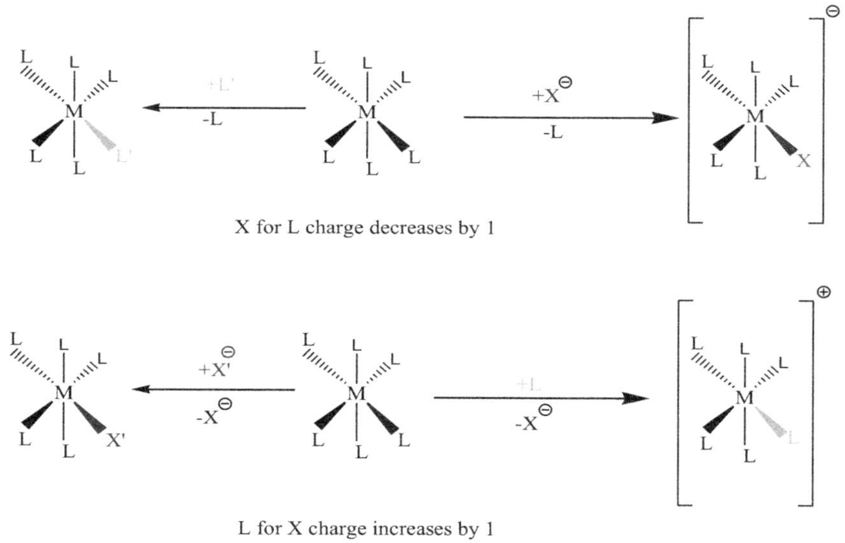

X for L charge decreases by 1

L for X charge increases by 1

SCHEME 9.4 Conservation of charge on ligand substitution.

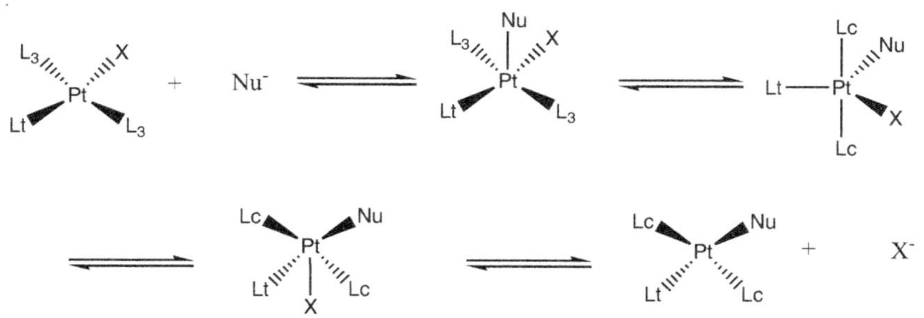

SCHEME 9.5 Nucleophilic substitution of metal and ligand derivative.

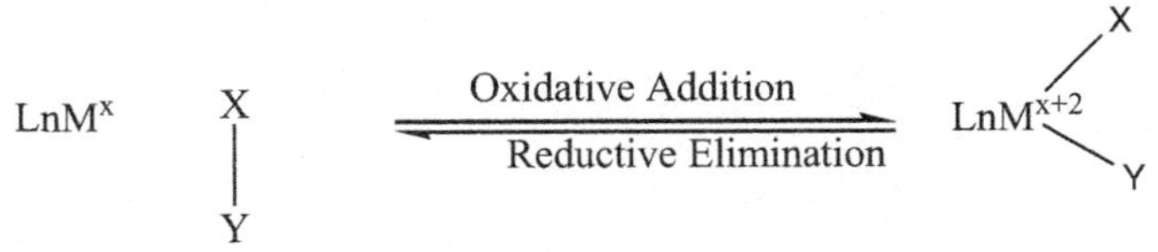

SCHEME 9.6 Oxidative addition and reductive elimination.

$$LnM^x + \begin{array}{c} X \\ | \\ Y \end{array} \xrightleftharpoons[\text{Reductive Elimination}]{\text{Oxidative Addition}} LnM^{x+2} \begin{array}{c} X \\ \diagup \\ \diagdown Y \end{array}$$

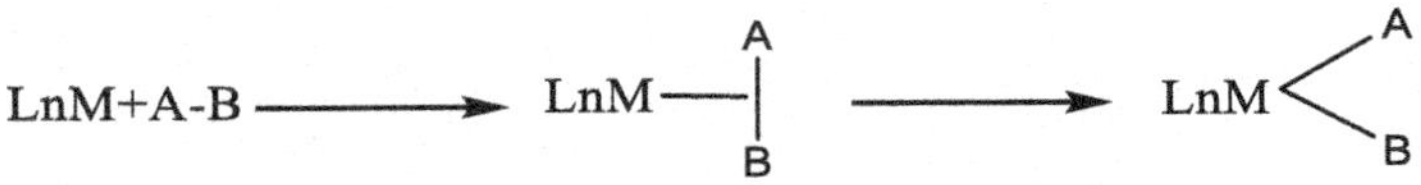

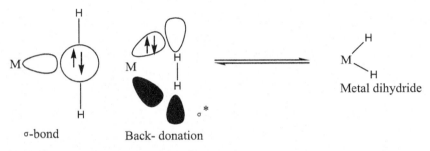

SCHEME 9.7 Donation of a pair of electrons to the interligand link's σ* orbital.

SCHEME 9.8 *cis* Conformation of ligand and metal.

9.4.3 Concerted track based

Hydrogen and hydrocarbons, which are nonpolar substrates, work together to carry out oxidative addition reactions. These substrates lack bonding due to the complicated utilization of three-cantered complexes. The creation of the oxidized complex is caused by the intramolecular cleavage of the ligand, which is most likely brought on by the donation of an electron pair into the σ* orbital of the interligand link in Scheme 9.7. The new ligands *cis* conformations will be reciprocal which is outlined in Scheme 9.8 [11].

Hydrogen like a homonuclear diatomic molecule is added to this chemical reaction. Many C—H activation events are coordinated by the creation of an M-complex (C—H). The relevant example for the interaction of hydrogen with Vaska's complex is *trans*-IrCl(CO)[P(C$_6$H$_5$)$_3$]$_2$. This change causes the iridium metal to transition from its regular oxidation state of +1 to +3 states. The generated product is typically bound using three single chloride and two hydride ligands. Following the steps in the procedure outlined in (Scheme 9.9), a metal complex with 16e and a coordination number of four is converted into a complex with six coordinates, 18e [12].

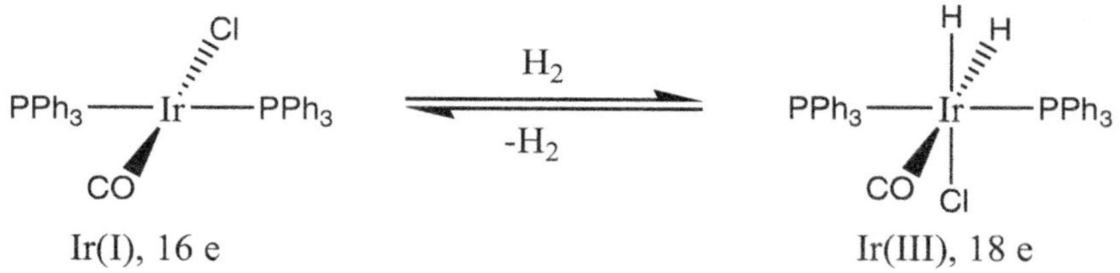

Sensing application of organometallic compounds Chapter | 9 **223**

SCHEME 9.9 Reaction mechanism of single chloride with two hydride ligands.

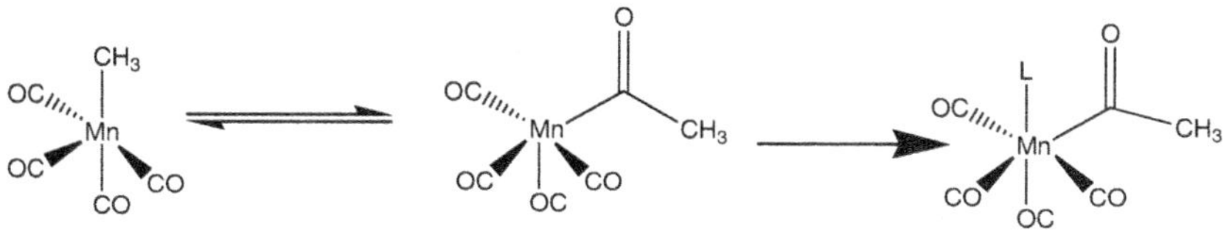

SCHEME 9.10 Insertion of carbonyl compound with metal complex.

The development of a trigonal bipyramidal dihydrogen intermediate is followed by the breakage of the H—H bond as a result of electron back donation into the H—H σ^* orbital [exhibit in Scheme 9.8]. The aforementioned reaction is also in chemical equilibrium, and in the metal center, the opposite reaction is ongoing as reduction by hydrogen gas removal takes place. Due to the electron in the H—H σ^* orbital that was donated via the rear side to split the H-H bond, metals rich in electrons are more favorable for the aforementioned process. *cis*-dihydride is the outcome. A migratory insertion of carbonyl, one of the most typical insertions is the introduction of carbonyl groups into alkyl groups like methyl groups. The carbonyl insertion results in the creation of an acyl group. A neutral ligand L, such as free CO, in which (see Scheme 9.10) fills the vacant site, can act as a catalyst for the process.

9.5 Recent developments with examples

Substances are an example that can be categorized in a number of ways. For instance, the ferrocene derived unit serves as both a structural component and a colorimetric reporter for ferrocene binding. The lability of the metal center is essential for hosts that utilize metals as structural elements. The host compound will bind as an anion group when the connections of metal-ligand are prolonged on a human time scale, just like, for instance, organic macrocyclic hosts. In Scheme 9.11, compound 1, one of the first hosts with a metal core, serves as a good example. The bond between its two tridentate terpyridyl ligands is defined by a ruthenium (II) center that is fairly inert. Long dicarboxylates, especially pimelate, are bound by the host. Since anions, cations, or both typically act as templates for the anion-metal-ligand aggregate, they fall under the category of

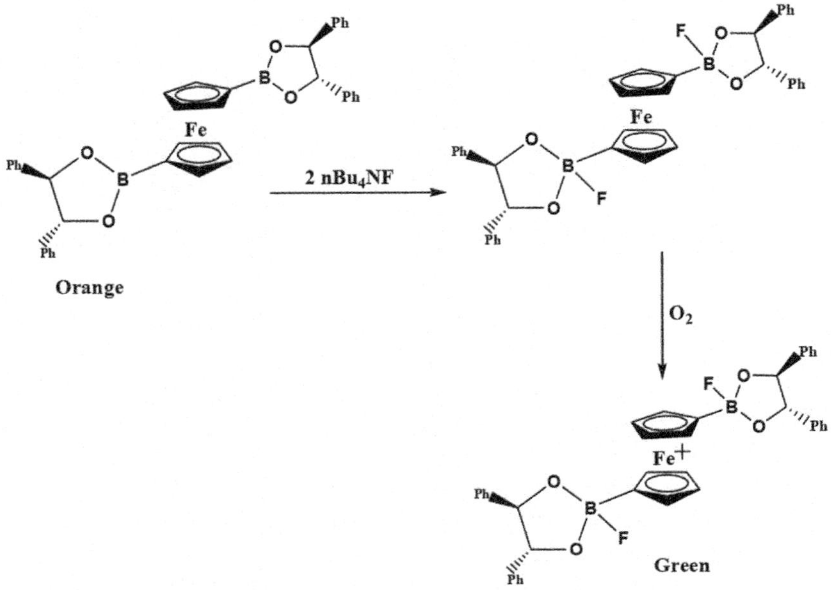

SCHEME 9.11 Proposed mode of binding between receptor 1 and fluoride.

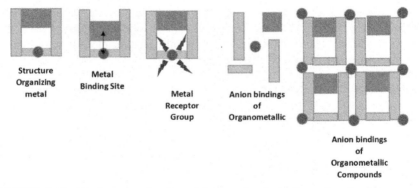

SCHEME 9.12 Development of variety of ferrocenyl and cobaltocene-based anion sensors. *Procreated from Steed JW. Coordination and organometallic compounds as anion receptors and sensors. Chem Soc Rev 2009;38(2):506–19.*

self-assembly. The ensemble can be viewed as a multicomponent self-assembled metal salt that includes ligands that can bind both cations and anions. A good illustration is aureidopyridyl ligand in compound 3, which has a single pyridyl activity and a binding group for an anion produced from urea. The complex self assembles when the labile Ag(I) cation and nitrate anion are present. The areneruthenium(II) derivatives 4 and 5 (in Scheme 9.12) show how labile and inert complexes differ from one another. The inert complexse hydrogen bonding interactions to attach a range of anions to the bispyrrole NH functions when a bidentate bispyrrole ligand is present. The monodentate pyridyl ligands are displaced over the course of hours by Cl, despite the fact that binds chloride, nitrate, and hydrogen sulfate with efficiency. This is because Cl coordinates directly to the semilabile metal center. Compound 5 uses both ruthenium (II) as a structural component and iron(II), a metal center, as a redox reporter group. Such ferrocenyl derivatives and cobalt ocenium analogs have frequently been used as redox sensing components in anion molecular sensors owing to the robustness and reproducibility of the Fe (II)/Fe(III) and Co(II)/Co(III) redox couples, particularly in work by Beer from the late 1980s onward. A good illustration is the 240-mV shift in the bis(amide) redox potential brought on by the interaction with H_2PO_4. Anion concentrations are increasingly being detected using colorimetric techniques or, more precisely, deviations in luminescence emission intensity. The group of lanthanideane-N_4 ("cyclen") derivatives, which skillfully integrate a metal ion as an anion binding and luminescent sensing element, includes a compound that is a luminous sensor selective for hydrogen carbonate. This complex also serves as a signaling particle by generating a metal-based signal that is sensitive to the presence of the complex. Even in the presence of competing anions like lactate, citrate, and phosphate, the concentration of solution hydrogen carbonate can be assessed by the ratio of the strength of the europium emission at up to three wavelengths. These illustrations demonstrate the adaptability and versatility of metal ions as essential elements in the formation of anion receptors. This tutorial will study the fundamentals and history of each type of metal-based receptor mentioned in Scheme 9.11, with a focus on the field's a rounding breadth as a result of recent developments with the process of biding in Scheme 9.13 [13].

The ferrocene and cobaltocene derivatives 1, 5, and 6 are particularly desirable redox-sensing components due to their powerful reversible redox chemistry. The field was reviewed in 2003 and 2005, and a variety of ferrocenyl- and cobaltocene-based anion sensors (Scheme 9.9) have been developed.

9.6 Applications in emerging fields

9.6.1 Chemical sensor

Chemical sensors are devices that convert chemical data into an analytically usable signal, such as the concentration of a particular sample component or a composition analysis of the entire sample. Chemical sensors normally consist of the detecting substance and the transducer as its two primary parts of this application. The target analyte is what the sensing material is supposed to contact with, and the result of this binding relationship is a change in a material property like mass or electrical conductivity. This change is subsequently converted by the transducer into a signal that can be read, typically

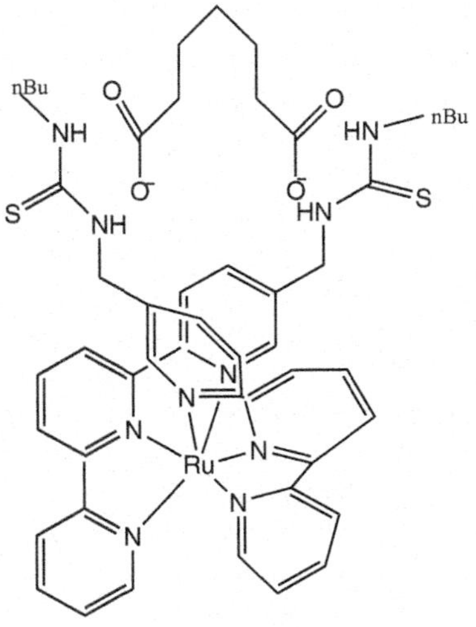

SCHEME 9.13 Mode of binding. *Procreated from Steed JW. Coordination and organometallic compounds as anion receptors and sensors. Chem Soc Rev 2009;38(2):506–19.*

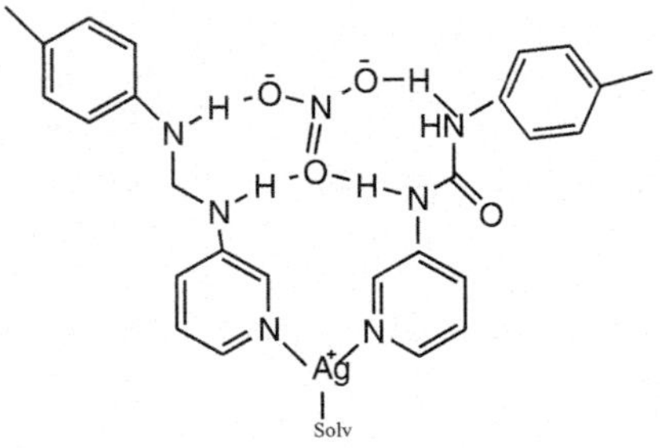

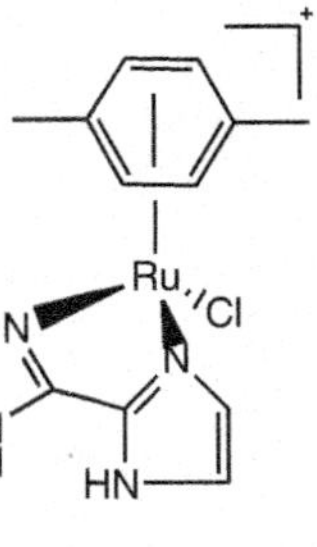

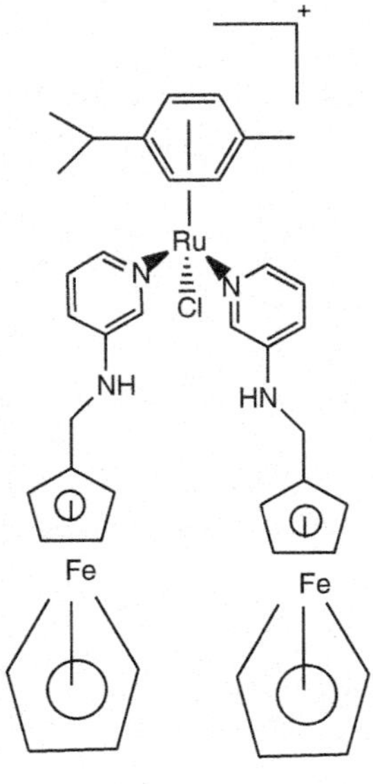

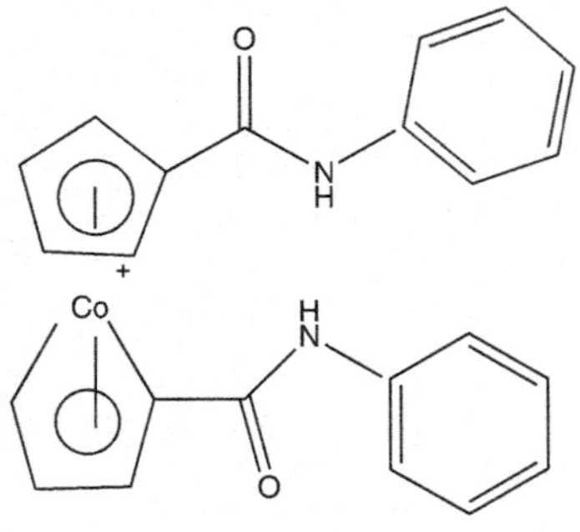

an electrical signal [14]. Given that they provide information on industrial production processes, food and beverage quality control and a variety of other purposes, chemical sensors are crucial for monitoring the environment in which we live [15].

Some of the expanding fields that make up the chemical sensor field include acoustics, membrane technology, electricity, optics, mechanics, semiconductor technology, microelectronics technology, chemistry, biology, and thermology [16]. Based on sensing object, chemical sensors can be classified into as follows (see Scheme 9.14).

Chemical sensors will continue to be used primarily for three purposes in the near future: disease detection and treatment, environmental monitoring, and continual quality of life improvement [17]. Chemical sensors are used to detect gases are based on multilayer systems that allow for the creation of three-phase limits with the gas. PbF_2, $PbSnF_4$, and particularly LaF_3 have undergone extensive research as fluoride-based solid electrolytes. LaF_3 is extremely reactive to oxygen at ambient temperature. Large LaF_3 single crystals were connected to porous Pt electrodes in the first configuration. The Pt electrode can be replaced for electrodes constructed of perovskite type oxides to boost the sensitivity of such cells but a more intriguing development occurs when the single crystal is changed to a thin layer of LaF_3

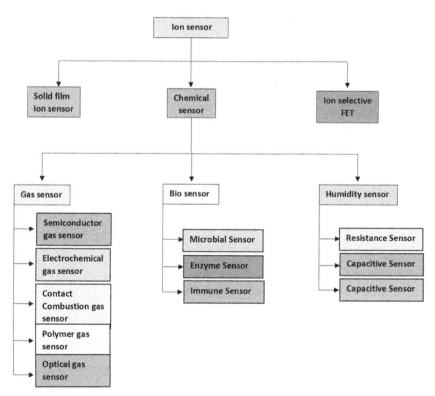

SCHEME 9.14 Classification of chemical sensors based on sensing objects. *Procreated from Wen W. 2016. https://doi.org/10.5772/64626.*

sprayed over an oxidized Si substrate. CaF_2 has also been the subject of a great deal of research in this field. Through surface treatment with Lewis acids like BF_3 or SbF_5, its ionic conductivity is improved. It is interesting to note that alkaline-earth difluorides can also act as a shield and antireflective layer over glass optics. Innovative spin-coating and sol-gel technologies have been used to manufacture nano-crystalline MgF_2 for use in films and composites. Reactive sputtering of Cr in an Ar/C_2F_6 plasma resulted in the creation of F/C/Cr thin films; electron diffraction analysis revealed that these films were crystalline. High reproducibility of film composition was achieved by careful control of response characteristics. Polycrystalline silicon films made from SiF_4/H_2 mixtures have been produced using a different plasma-based technique called plasma-enhanced chemical vapor deposition (PECVD) at temperatures as low as 100°C. However, combinations of CH_4 and CF_4 with various activation techniques have also been studied to synthesize atomic fluorine. A roughly identical process has been utilized to deposit diamond films using CF_4/H_2 mixtures. Successful experiments using WF_6 as a precursor for CVD-deposited tungsten films are closely related. The absence of contaminants brought on by the breakdown products of the organic ligands is one of their key benefits over organometallic precursors. The finding that introducing reducing agents GeH_4 or SiH_4 caused the deposition temperature to drop to 270°C led to further process development [18].

9.6.2 Gas sensor

A static test system was employed to conduct the gas sensing measurement. A sealed test chamber with a volume of 1 L was filled with the liquid using a microinjector to create different test gas concentrations. The gas-detecting gadget records a relative humidity of about 25 RH [19]. The industrial uses, gas sensing abilities, and optical properties of zinc oxide are particularly relevant. The substance $[Zn(C_6H_{11})_2]$ that exhibits these qualities [20]. The sensing elements based on perovskite are particularly appealing and exhibit a variety of benefits over the traditional metal-oxide sensing elements used for gas sensing (e.g., operational in high temperature and limited sensitivity). Organometallic perovskite films have excellent gas sensing properties. It has been discovered that $MAPbI_3$ films go through optical bleaching after being exposed to ammonia gas molecules. By manipulating their shape, content, and compatibility with other porous materials, chemically synthesized organometallic compounds allow for the improvement and regulation of the performance and applicability of humidity and gas sensing [21]. The complex blend of components used in common gas sensors typically includes catalysts. However, a little is known about the relationship between detecting mechanisms and microstructure, which is crucial for realizing the full potential of currently available materials or creating better sensor systems while simultaneously managing the stability of manufactured devices. In-depth characterization utilizing techniques like electron microscopy and surface analytical techniques is necessary to establish linkages between structure and property.

The ceramic sensor has not been the subject of a lot of modeling and simulation research. The difficulty stems from the need to construct and combine models at various wavelengths to adequately describe the broad variety of events that can take place in different kinds of sensors. Atomic-scale events are involved in fundamental chemical processes and electron transport across metal oxide surfaces and bulk regions. Even if the electrical characteristics of the device also consider behavior at the macroscopic level, conduction via the ceramic device's weakly sintered microstructure involves events at the granular or mesoscopic scale. It is necessary to combine effective models created at these various length scales into a unified framework that can be utilized to direct current research and testing targeted at creating sensors with the best potential features. The challenge comes from the requirement to build and combine models at multiple wavelengths to sufficiently represent the wide range of events that can occur in various types of sensors. Fundamental chemical reactions, electron transport through metal oxide surfaces, and bulk regions all include atomic-scale phenomena. Even if the device's electrical properties also take behavior at the macroscopic level into account, conduction through the poorly sintered microstructure of the ceramic device requires at the granular or mesoscopic scale. Effective models developed at these multiple-length scales must be combined into a single framework that can be used to guide ongoing research and testing aimed at developing sensors with the best possible features [22].

9.6.3 Optical sensor

Optical sensors are light-based biosensors that use the analyte's interaction with the bio-recognition component to alter the measurement of wavelength. Transducers for optical biosensors can be colorimetric, luminescent, interferometric, or fluorescent. A novel type of optical transducer-based biosensors called photonic crystal-based biosensors has recently emerged. Surface-enhanced Raman scattering biosensors, reflectometric interference spectroscopy biosensors, ellipsometric biosensors, and SPR-based sensors are a few examples of optical biosensors [23]. Plastics can be divided using optical sensors based on color or transparency. To sort bottles by color, use this method. These computer-controlled systems can quickly distinguish between the various shades of the plastic regrind since they were developed from the coffee bean separators that are used to remove immature green beans from the mixture. A charge-coupled device camera is used to instantly determine the color of each piece of plastic, which is then either allowed to flow downward or expelled into the reject or collection pile with a puff of air. Stake, for instance, makes equipment that is especially made to sort plastics according to hue, although their primary market is still the agriculture sector. In recycling applications, optical sensors are used to distinguish between the hues of HDPE and PET materials [24].

9.6.4 Biochemical sensor

The signal from the analyte is transmitted using a biological recognition process by a particular class of these sensor devices known as biosensors shown in Fig. 9.1. By serving as a bridge between the "soft" biological materials and the "hard" solid-state elements necessary to carry out this transduction, organometallic compounds which can be regarded as systems that convert chemical information into electrical output can moderate the information flow in biosensors [25]. The variability of biosensors influences both their operation and design since signal transduction takes place at the sensor surface. Both the biological recognition element and the organometallic element need to be immobilized on the surface since mass transport effects (diffusion and convection) are significant when the sensor is to be utilized continuously. The performance of the device may also vary significantly depending on which of the two phases is rate-limiting since the signal can be controlled by either a mass transport or reaction kinetics step [26].

This chapter discusses molecular probes that carry organometals, which are mostly utilized as reporter groups and that produce distinctive signals in response to biological targets. The most frequent bio organometal used for these reasons is Fc-receptor. Fc-receptor is a particularly ideal molecule for usage as

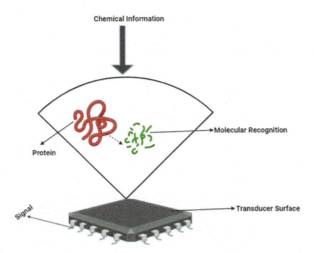

FIGURE 9.1 Sensor responses to biological target. *Procreated from Forrow NJ. Nicola 2004;15(1):137–144.*

reporter moiety in bioconjugate probes because of its stability in aqueous solution, flexibility to numerous chemical changes, and well-defined reversible electrochemical character. Fc-receptor with isostructural compound having a distinctive redox potential is cobaltocene. The distinct qualities of cyclometalated organoiridium complexes, including phosphorescence and catalytic activity, have been emphasized. A reporter molecule's phosphorescence is a very useful characteristic. When a time-resolved measurement approach is used, it is possible to get high-quality luminescence with little background. The HgC bond frequently remains stable in both air and moisture. Cytosine and mercury (II) combine to form an organometallic Janus nucleobase with double faces, which permits the coupling of opposing nucleobases. In other research, biorthogonal chemical procedures utilizing the soft atom Hg(II) were employed to activate the caged fluorophore on biomolecules. Additionally, platinum has the ability to create cyclometalated metals that are specifically attached to G4 DNA and exhibit sustained intrinsic phosphorescence. The extensive range of processes that metals can perform includes electrochemistry, catalysis, luminescence and pharmacology. The application of this flexibility in molecular biology as well as the quickly developing fields of bioanalytical and biomedical sciences would be extremely helpful and show great potential [27].

9.7 Future prospective and scope

Inorganic medicinal chemistry has been significantly improved by the recent development of metal-complex based sensing compounds, which helped in the synthesis of a small number of high metal-complex based on sensing. The majority of rapidly developed metal complexes have a range of sensing properties. The proposed study development, which covers the exploitation in chemical, biochemical, as well as medical and scientific sectors, highlights recent developments in medicinal inorganic chemistry. In the upcoming years inorganic species, particularly organometallic ones, should grow quickly in this area of imaging due to their potential for tuning. For future combinations of radio imaging, cell imaging, diagnostics, and therapeutics, they may even delve into multimodality. Organometallics are proven to be a very helpful tool for a range of sensing applications, especially for tiny molecules, aromatic compounds, gases, vapors, and explosives, according to studies. The exceptional sensitivity of this sort of detection is based on the fact that organometallics are label-free (without tagging biomolecules like antibodies, enzymes, and aptamers). Utilizing organometallic compounds for luminescence based molecular sensing is a workable alternative. For a wide range of sensing applications, especially for tiny molecules, aromatic compounds, gases, vapors, and explosives, organometallic compounds are proving to be a very effective tool. The great sensitivity label-free aspect of this method of detection is its main distinguishing feature (without tagging biomolecules like antibodies, enzymes, and aptamers).

9.8 Summary

The current overview of this chapter focuses on the use of metals as sensors and highlights some notable studies on new remedial chemicals. The investigation of sensing organometallic compounds in various rehabilitative domains is a growingly active and productive area of research. For monitoring intermediates and species in the solution state of the complexes, chemical and biological methods are applied. Finding and structurally characterizing the formed solid kind of complex benefits greatly from sensing characterization procedures. Our comprehension of the structural and molecular manifestations of the use of sensing metals is improved by these preclinical and clinical considerations. The sensing properties of organometallics explain their relevance in the chemical, electrochemical and biological domains. It would be very beneficial and have a lot of potentials to use this flexibility in molecular biology as well as the swiftly growing fields of bio analytical and biomedical sciences. Another aspect of sensing is signal regulation, which can be accomplished by either a mass transport or reaction kinetics step. Depending on which of the two phases is rate-limiting, the performance of the device may also vary significantly (Tables 9.1 and 9.2).

9.9 Problems with solutions

1. **Why is *cis*-platin not a substance that belongs in the class of organometallics?**

    Answer: *cis*-platin, also known as [Pt(NH$_3$)$_2$Cl$_2$], does not include a direct carbon-metal link, which is a prerequisite for a substance to qualify as an organometallic compound.

2. **How haptic must Cp be if the molecule [W(Cp)$_2$(CO)$_2$] adheres to the 18 e- rule?**

    Answer: We must determine the compound's valence electron concentration in order to forecast the hapticity of both (Cp) ligands (valence electron count; V.E.C.).

    Given: V.E.C. of the compound = 18

    Let us assume the hapticity of one Cp group be x and that of another Cp group be y.

TABLE 9.1 Electronic configuration of silicon and germanium.

Element	Symbol	Atomic number	Electronic configuration
Silicon	Si	14	[Ne] 3s^2 3p^2
Germanium	Ge	32	[Ar] 3d^{10} 4s^2 4p^2

TABLE 9.2 Physical properties of silicon and germanium.

Property	Silicon	Germanium
Atomic number	14	32
Molar mass (g/mol)	28.08	72.59
Density (g/cm^3)	2.49	5.35
Melting point (°C)	1410	937.5
Boiling point (°C)	2680	2830
Molar volume (cm^3/mol)	11.4	13.6
Atomic radius (A°)	1.11	1.22
Oxidation state (stable)	+4	+2, +4
Electronegativity	1.9	2.0
Ionization energy (kJ/mol)	786	761

Hence, V.E.C. of [W(Cp)$_2$(CO)$_2$]: $6 + x + y + 2(2) = 18$
V.E.C. of [W(Cp)$_2$(CO)$_2$]: $x + y = 8$
Hence, the hapticity of the two Cp groups must be a sum of 8.
Why is the 18 electron rule important for organometallic compounds?
Answer: To get 18 electrons in the metal's valence shell, the 18 electrons rule is applied. The ligands in organometallic compounds provide their valence electrons to the metal's valence shell. The valence shell of the metal may hold two electrons in its s-subshell, six in its p-subshell, and ten in its d-subshell. Consequently, a metal's valence shell can hold a total of 18 electrons.

3. **Illustrate the properties of the organometallic compounds.**
   Answer: The following are the organometallic compounds' five characteristics:
(i) Organometallic compounds are soluble in organic solvents like ether but insoluble in water. This is due to the nonpolar covalent bonds that hold the organometallic compounds together. (ii) Organometallic compounds possess at least one distinctive metal-carbon link and are particularly reactive, which is why they are stored in organic solvents. (iii) Organometallic compounds only dissolve in organic solvents, whereas nonpolar compounds only dissolve in nonpolar solvents. (iv) Organometallic compounds that contain metals from the main group or the transition metals exhibit this reactivity. (v) Organometallic compounds containing electropositive metals operate as reducing agents because of the polarity of the metal-carbon link. The highly electropositive metals that create flammable and easily combustible organometallic compounds.

9.10 Objective type questions

1. Which ligand among the following compounds only has one link to the metal?
   a) $W(CH_3)_6$
   b) $K[PtCl_3(C_2H_4)]$
   c) $(\eta^6\text{-}C_6H_6)_2Ru$
   d) $(\eta^5\text{-}C_5H_5)_2Fe$
   Answer: (a)

2. The organometallic compound with -bonds that has ethene as one of its constituents is
   a) Zeise's salt
   b) Ferrocene
   c) Dibenzene chromium
   d) Tetraethyl tin
   Answer: (a)

3. Which of the subsequent complexes has the highest metal oxidation state?
   a) $(\eta^6\text{-}C_6H_6)_2Cr$
   b) $Mn(CO)_5BR$
   c) $Na_2[Fe(CO)_4]$
   d) $K[Mn(Cu)_5]$
   Answer: (c)

4. Which organic ligand in which of the following complexes has a single bond with a metal?
   a) $W(CH_3)_6$
   b) $K[PtCl_3(C_2H_4)]$
   c) $(\eta^5\text{-}C_5H_5)_2Ru$
   d) $(\eta^5\text{-}C_6H_6)_2 Co$
   Answer: (a)

5. Which of the following neutral complexes adheres to the rule of 18 electrons?
   a) $(\eta^5\text{-}C_5H_5)Fe(Zn)_2$
   b) $(\eta^5\text{-}C_5H_5)2Mo(Cu)_3$
   c) $(\eta^5\text{-}C_5H_5)_2Co$
   d) $(\eta^5\text{-}C_5H_5)2Re(\eta^6\text{-}C_6H_6)$
   Answer: (d)

6. How many M—M bonds are present in $[Cp Mo(CO_3)]_2$?
   a) 9
   b) 8

c) 0
d) 5
   Answer: (c)

7. Which of the following statements about Zeise's salt is untrue?
   a) Diamagnetic salt is Zeise's salt.
   b) Pt in Zeis' salt has an oxidation state of +2.
   c) In Zeise's salt, the Pt—Cl bond lengths are all equal.
   d) The ethylene moiety in Zeise's salt has a longer C—C bond than a free ethylene molecule.
   Answer: (c)

8. Which of the following statements regarding the metal olefin complexes I [PtCl$_3$(C$_2$F$_4$)] and (ii) [PtCl3(C2H4)] is true?
   a) The length of the carbon-carbon bond is the same in both cases I and (ii)
   b) It is shorter in case I
   c) It is shorter in case (ii), and
   d) A metallacycle is created in each complex.
   Answer: (b)

9. Which of the subsequent complexes has the highest metal oxidation state?
   a) (η^6-C$_6$H$_6$)$_2$Cr
   b) [Cp Mo(CO$_3$)]$_2$
   c) Na$_2$[Fe(CO)$_4$]
   d) K[Mn(CO)$_5$]
   Answer: (c)

10. The π- bonded organometallic compound which has ethene as one of its components is
    a) Zeise's Salt
    b) Ferrocene
    c) Dibenzene chromium
    d) Tetraethyl tin
    Answer: (a)

11. Coordination compounds have great importance in biological systems. In this context which of the following statements is incorrect
    a) Cyanocobalamin is B$_{12}$ and contains cobalt
    b) Hemoglobin is the red pigment of blood contains iron
    c) Chlorophyll are green pigment in plant and contains calcium
    d) Carboxypepticase-A is an enzyme and contain zinc
    Answer: (c)

12. **Reactions of the CO ligands include**
    a) Nucleophilic attacks on the carbon atom and electrophilic attacks on the oxygen atom correct
    b) Nucleophilic attacks on the oxygen atom and electrophilic attacks on the carbon atom incorrect
    c) Nucleophilic attacks on both the oxygen and carbon atom
    d) Both b and c
    Answer: (a)

References

[1] Boudier A, Bromm LO, Lotz M, Knochel P. New applications of polyfunctional organometallic compounds in organic synthesis. Angew Chem Int Ed 2000;39(24):4414−35.

[2] Zhang X, Lin S, Liu S, Tan X, Dai Y, Xia F. Advances in organometallic/organic nanozymes and their applications. Coord Chem Rev 2021;429:213652.

[3] Angkawinitwong U, Williams GR. Electrospun materials for wearable sensor applications in healthcare. Electrospun Polym Compos 2021;405−32.

[4] Thomas S, Nguyen TA, Ahmadi M, Farmani A, Yasin G, editors. Nanosensors for smart manufacturing. Elsevier; 2021. p. 595−611.

[5] Abdul S, Judit T, Ilona F, Nikoletta M. Functional thin films and nanostructures for sensors. Fundam Nanopart 2018;485−519.

[6] National Research Council. Expanding the vision of sensor materials. Washington, DC: National Academies Press; 1995. p. 9−18.

[7] King RB. Applications of metal carbonyl anions in the synthesis of usual organometallic compounds. Acc Chem Res 1970;3(12):417−27.

[8] Trillo LM, Santa Quiteria VR, Moraleda GO, Franco AM, Dynasol Elastomeros SA. Sensors for the determination of organometallic compounds. US 6875615 (2005).

[9] Ghosh S, Mukherjee PS. Self-assembly of a nanoscopic prism via a new organometallic Pt_3 acceptor and its fluorescent detection of nitroaromatics. Organometallics 2008;27(3):316−19.

[10] Liu S, Gao S, Wang Z, Fei T, Zhang T. Oxygen vacancy modulation of commercial SnO_2 by an organometallic chemistry-assisted strategy for boosting acetone sensing performances. Sens Actuators B Chem 2019;290:493−502.

[11] Rataboul F, Nayral C, Casanove MJ, Maisonnat A, Chaudret B. Synthesis and characterization of monodisperse zinc and zinc oxide nanoparticles from the organometallic precursor [Zn(C6H11)$_2$]. J Organomet Chem 2002;643−644:307−12.

[12] Kymakis E, Panagiotopoulos A, Stylianakis MM, Petridis K. Organometallic hybrid perovskites for humidity and gas sensing applications. 2D Nanomater Energy Appl 2020;131−47.

[13] Steed JW. Coordination and organometallic compounds as anion receptors and sensors. Chem Soc Rev 2009;38(2):506−19.

[14] Mandoj F, Nardis S, Natale CD-I, Paolesse R. Porphyrinoid thin films for chemical sensing; 2018. p. 422−443.

[15] Abhilash M, Thomas D. Biopolymers for biocomposites and chemical sensor applications. Biopolymer composites in electronics. Elsevier; 2017. p. 405−35.

[16] Nasri A, Petrissans M, Fierro V, Celzard A. Gas sensing based on organic composite materials: Review of sensor types, progresses and challenges. Mater Sci Semicond Process 2021;128:105744.

[17] Luquin A, Bariáin C, Vergara E, Cerrada E, Garrido J, Matias IR, et al. New preparation of gold-silver complexes and optical fibre environmental sensors based on vapochromic $[Au_2Ag_2(C_6F_5)_4(phen)_2]_n$. Appl Organomet Chem 2005;19(12):1232–8.

[18] Milićev S, Lutar K, Žemva B, Ogrin T. Vibrational spectra of solid CrF_4 and of three new fluorochromate(IV) complexes with Xe fluorides, $XeF_2 \cdot CrF_4$, $(XeF_5{}^+CrF_5{}^-)_4 \cdot XeF_4$ and $XeF_5{}^+CrF_5{}^-$. J Mol Struct 1994;323:1–6.

[19] Renard L, Babot O, Saadaoui H, Fuess H, Brötz J, Gurlo A, et al. Nanoscaled tin dioxide films processed from organotin-based hybrid materials: an organometallic route toward metal oxide gas sensors. Nanoscale 2012;4(21):6806–13.

[20] Aubrecht J, Kalvoda L. Development of ammonia gas sensor using optimized organometallic reagent. J Sens 2016;2016:1–8.

[21] Wei BY, Hsu MC, Su PG, Lin HM, Wu RJ, Lai HJ. A novel SnO_2 gas sensor doped with carbon nanotubes operating at room temperature. Sens Actuators B: Chem 2004;101(1–2):81–9.

[22] Lee C, Dutta PK, Ramamoorthy R, Akbar SA. Mixed ionic and electronic conduction in Li_3PO_4 electrolyte for a CO_2 gas sensor. J Electrochem Soc 2005;153(1):H4–14.

[23] Vaz R, Frasco MF, Sales MG. Biosensors: concept and importance in point-of-care disease diagnosis. Biosens Based Adv Cancer Diagnostics 2022;59–84.

[24] Merrington A. Applied plastics engineering handbook: processing and materials; 2011. p. 177.

[25] Johnson CE, Eisenberg R. Stereoselective oxidative addition of hydrogen to iridium (I) complexes. Kinetic control based on ligand electronic effects. J Am Chem Soc 1985;107(11):3148–60.

[26] Cass AEG. Organometallic compounds in biosensing. Compr Organomet Chem III 2007;589–602.

[27] Ihara T. Organometallic complexes for biosensing. Adv Bioorganometalli 2019;277–303.

Chapter 10

Bioorganometallic chemistry: a new horizon on organometallic landscape

Mudasir Ahmad Hafiz, Moniza Qayoom, Tabee Jan, Mohd Mustafa, Tabasum Maqbool and Masood Ahmad Rizvi
Department of Chemistry, University of Kashmir, Hazratbal, Jammu and Kashmir, India

10.1 Introduction

The development of newer chemical systems open doors to explore the influence of novel chemical moieties in biological turf. While organic compounds mostly form the framework, inorganic constituents serve as functional components of the biosystems. Organometallics, which represent the ensemble of organic and metal components, have been restricted in biosystems except for coenzyme-B12. The life systems have not used organometallic compounds so often most probably due to the reactivity of metal−carbon bond toward aqueous aerobic conditions prevalent with biological systems; wherein metal−carbon bond undergoes hydrolysis with water and oxidation with molecular oxygen. In spite of these inherent drawbacks, a careful design of organometallic systems with relatively stabler metal−carbon bonds which survive biological conditions can bring remarkable advances, set newer medical regimes and address the unsettled challenges of therapeutics.

In its broader sense, "bioorganometallic chemistry" (BOC) refers to the chemistry of biomolecules with biologically active substances that have at least one carbon directly bound to a metal [1]. The field of BOC essentially involves the design and development of organometallic systems that are synchronous with the biological interface and selective for specific therapeutic goals [2]. BOC introduces a new domain in chemotherapeutics that aims to extend designed organometallic chemistry to bioanalytics and chemotherapeutics. In BOC, the function of the organometallic fragment can be multifaceted and frequently depends on the particular organometallic system. Generally speaking, pioneering organometallic species found in human beings included B_{12} coenzymes with a cobalt-carbon bond system

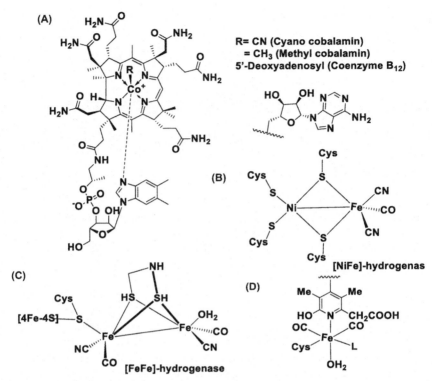

FIGURE 10.1 Some naturally occurring bioorganometallic moieties: (A) B_{12} coenzyme and (B–D) hydrogenase active sites.

(Fig. 10.1A). Recently, bioorganometallics have gained interest of synthetic chemists in perspective of biomimicking nature's efficient machinery to produce dihydrogen as the active sites of the hydrogenase enzymes contain carbon–metal bonds to iron center(s) [1]. The bioinspired design of novel organometallic-based systems to effectuate challenging reductions and provide solutions to the contemporary problems of energy, environment, and health mark the future of BOC. The synthetic bioorganometallic compounds mostly feature a carbon-based ligand connected by a coordinate bond to transition metal. The nature and properties of the bioorganometallic compound are predominantly ascribed to the ligand moiety, such as carbonyl-, aliphatic-, or arene-derived pi-bound ligands. The bioorganometallic domain broadly involves compounds that bind to biotargets, such as proteins and DNA for pharmacological action.

Bioorganometallics [2,3] also include metal based compounds with well-defined biomolecules, such as amino acids, carbohydrates, nucleic acids, steroids, or oligopeptides, as ligands with multipurpose applications in fields of medicinal chemistry, bioanalysis, and catalysis. Owing to their kinetic lability toward oxygen and water, the progression of BOC had been slow. The interest toward this applied domain of organometallics strengthened after exploration of the paths for kinetic stability of the metal−carbon bond under physiological conditions, along with the development of novel synthetic methodologies. In addition, with extensive spectroscopic advances in identification and characterization of chemico-biological interactions coupled with in silico modeling for lead identification and mechanistic insights of bioactivity, bioorganometallics is emerging as a distinctive field of synthetic drugs in medicine [1].

10.2 Shift from platinum metal based therapeutics to other bio-compatible less toxic nonplatinum complexes

Rosenberg et al. [4] in 1965, made an outstanding discovery of cisdiamminedichloroplatinum(II), known as cisplatin as a chemotherapeutic agent which marked the remarkable beginning of the chemotherapy [4]. Over the years the clinical cancer treatment mostly utilizes cisplatin or other platinum based derivatives(carboplatin, oxaliplatin, satraplatin, ormaplatin, aroplatin, enloplatin, zeniplatin, sebriplatin, miboplatin, picoplatin, and satraplatin). Serious side effects and significant limitations have been observed in case of platinum drugs [4]. Currently, their effectiveness has been restricted to solid tumors in case of ovarian, lung, and testicular malignancies, leaving a therapeutic vacuum for many other cancers of nonsurgical intervention and aggressive nature [4]. As a result of their nonspecific cytotoxic action, the platinum(II) anticancer compounds produce adverse effects of toxicity, nephrotoxicity, and neurotoxicity. Despite of all these limitations, it is practically impossible to deny the success of platinum drugs in chemotherapy [4].

However, for safe and secure chemotherapeutic treatments for aggressive cancers with nonsurgical intervention, novel drug molecules employing nonplatinum metal complexes are being explored as selective anticancer agents. The complexes of a variety of organic ligands with bioactive metals, such as selenium, iron, copper, ruthenium, rhodium, iridium, and gold, have been investigated so far with promising drug features for possible clinical trials [4].

10.2.1 Anticancerous properties of iron complexes

Iron as an essential element is utilized in several physiological processes. Some organic compounds, such as iron chelators, have been explored in therapy for colorectal cancer, although none of iron-containing compounds has been authorized for use in human clinical trials as an anticancer agent. Ferrocene, a biscyclopentadienideiron (II) with its metal center sandwiched between two cyclopentadienyl rings represents a pharmacophoric motif in bioactive iron compounds. To generate analogs of higher drug efficacy, functionalization of ferrocenyl cyclopentadienyl rings with appropriate substituents for tuned solubility, lipophilicity, acidity, and localization profiles is done. These tailored ferrocene derivatives have been used as antiproliferative agents. The antitumor characteristics of ferrocene derivatives were originally seen as ferrocene-linked tamoxifen derivates, which demonstrated good efficacy for some cancers considered incurable [4].

The redox characteristics of the ferrocenyl-ene-phenol moiety in ferrociphenols is known to produce reactive oxygen species (ROS) in cancer cells generating quinone methides that target thioredoxin reductase (TrxR) enzymes and induce cell death by apoptosis. Compounds **1** and **2** as ferrociphenols were synthesized and evaluated for anticancer activity against three ovarian epithelial cancer cell lines, one of which is resistant to the chemotherapy drug cisplatin (Fig. 10.2). Owing to drug resistance, the present pharmaceutical alternatives for treating epithelial ovarian cancer are among the worst because they are usually unsuccessful. Compound **1** as ferrociphenol based anticancer agent was equally potent against cisplatin-resistant cells and depicted nice cytotoxic propensity at nanomolar concentrations against the SK-OV-3, A2780, and A2780-*cis* cell lines. Although relatively less effective than compound **1**, compound **2** also showed cytotoxicity in nanomolar range. The exciting observation was compound **2** has double activity against resistant cells than it showed against nonresistant cells. Moreover, compound **1** caused a dose-dependent rise in the number of cells in the S phase, according to cell-cycle analysis, pointing to an apoptotic pathway. The exceptional activity of these heterocyclic ferrociphenols especially against resistant cell lines has rejuvenated interest in the ongoing biological studies [4].

Another important aspect of iron in bioinorganic chemistry is its use as NO carrier in the form of iron nitrosyl complex. NO is an important signaling molecule that modulates many physiological functions. NO is harmful for infections and tumor cells at higher concentrations (micromolar range), which is produced by macrophages. However, because NO has a short lifetime (2–5 seconds) under physiological conditions, it gets challenging to administer large levels of NO when needed. The body uses dinitrosyl iron

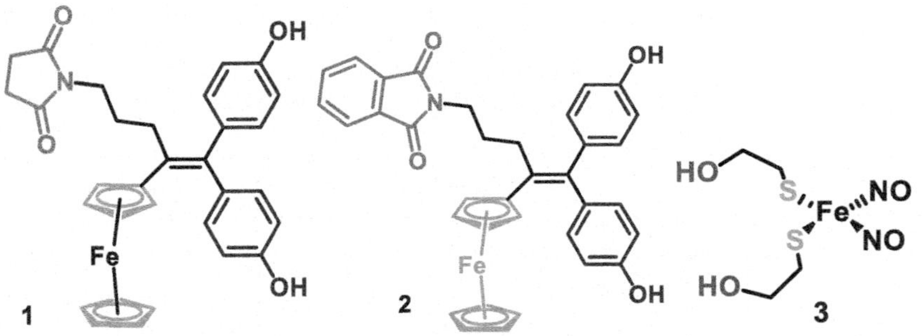

FIGURE 10.2 Iron based cytotoxic compounds: Ferrociphenol complexes **1** and **2** and NO-releasing compound **3**.

complex (DNIC) as an all-natural NO carrier. DNIC is used in the body as natural carriers of NO. Recent research on the biological effects of water-soluble DNIC compound **3**, a combination of biocompatible cysteamine-2 mercaptoethanol ligands that can control NO release from the iron center, was reported by Liaw and Wang [4]. The observed IC_{50} values for compound **3** for the studied cancer cell lines PC-3, SKBR-3, and CRL5866 ranged from 20 to 40 μM indicating lower cytotoxicity. Confocal imaging using a turn-on fluorescent activated xanthine dye that detects NO indicated that compound **3** is predominantly localized inside the mitochondria rather than the cytoplasm suggesting the onset of apoptosis. The treatment with compound **3** increases apoptosis was clearly observed from considerable differences between control cells and those incubated with it in annexin-V staining. Extended studies revealed that compound **3** provides double action by activating proteins linked to apoptosis while blocking proteins linked to survival in tumor cells [4].

In vivo studies of compound **3** on mice having subcutaneous PC-3 tumor xenografts were conducted. The mice were dosed for 21 days with compound **3** at 0.2 mg/kg every day. The treatment results indicated significantly slowed tumor growth compared to control tests, with 20 percent reduction in tumor size. These findings indicate that iron bearing compound **3** can be a suitable drug candidate for development as nonplatinum-based anticancer agent [4].

10.2.2 Anticancer properties of copper complexes

Copper is a vital component of many biological pathways and enzymatic reactions. Many peptides with cysteine amino acid residues easily bind to copper for its transport across cell membranes. It also plays a significant role in angiogenesis, which is crucial for the growth, invasion, and metastasis of tumors. Interest in utilizing copper complexes as anticancer agents has been stimulated by its importance in the functioning of normal cells and its potential to alter metabolism in cancer cells especially via ROS generation. The incorporation of copper into the doxorubicin drug motif resulted in anticancer medications with minimal side effects. Phase II clinical trials are currently investigating the synergistic effects of disulfiram and copper gluconate on recurrent glioblastoma. Plethora of copper complexes with a multitude of ligand architectures have been prepared and tested for anticancer activity. The bioorganometallic copper complexes with different geometries have been found to execute anticancer activity via DNA-binding under intercalative, groove binding modes or via inhibiting topoisomerase or proteasome enzymes. An example is copper II cyclen, compound **4**, which is observed to be cytotoxic being both effective inhibitor of DNA/RNA synthesis and DNA intercalator. Compound **4** on account of its low binding affinity, polymerase inhibition activity, and bulky groups in the minor groove satisfies the criteria for a transcription inhibitor.

Copper(II) complexes containing phenanthroline and indomethacin ligands have been synthesized by Suntharalingam group [4]. Indomethacin is an NSAID and a powerful inhibitor of the COX isoenzymes COX-1 and COX-2. The synthesized copper(II) compounds were tested against breast cancer stem cells (CSCs), that is, cells that are prevalent in higher proportions in breast tumors of the lowest life expectancies. The breast CSCs are very important for metastatic progression of breast cancer and they have higher propensity to survive cancer treatment regime. The cytotoxic studies of copper complexes on two human mammary epithelial cell lines, HMLER and HMLER-shEcad, the latter of which exhibits a sizable CSC-like population, were attempted. Compound **5** demonstrated selective potency for HMLER-shEcad over HMLER among the tested compounds, with submicromolar activity and a selectivity index of 3.3. Both the reduction in size and viability of breast CSC mammospheres were observed postaddition of compound **5**. Mechanistic studies revealed that compound **5** (Fig. 10.3) induces DNA damage and apoptosis in breast CSCs by inhibiting COX-2 activity and producing more ROS in CSCs than in non-CSCs.

Copper complexes are also interesting for cancer treatment in photodynamic therapy domain. The use of copper(II)-boron-dipyrromethene(BODIPY) conjugates as photoactive anticancer compounds was reported by Chakravarty group [4]. Copper(II) compound **6** was localized in the mitochondria of HeLa and MCF-7 cancer cells where, upon exposure to visible light, it induced apoptosis through the generation of ROS. This was accomplished by utilizing the mitochondria-targeting and singlet oxygen photosensitization ability of BODIPY species. When exposed to light, compound **6** demonstrated a superior

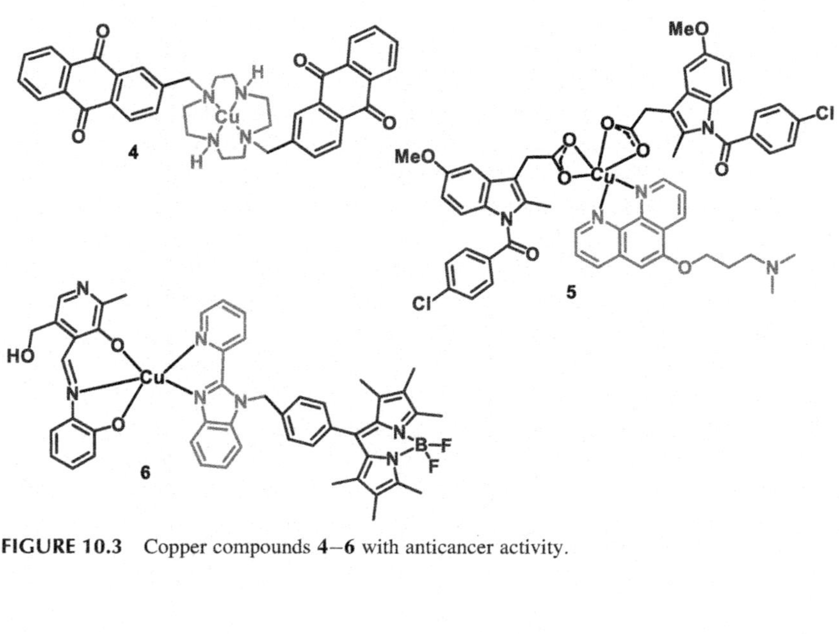

FIGURE 10.3 Copper compounds 4–6 with anticancer activity.

photoselectivity index that was far superior to that of the clinically used photodynamic therapy drug, Photofrin, with nanomolar IC_{50} values [4].

10.2.3 Anticancer properties of ruthenium complexes

Unlike square planar platinum(II) complexes, ruthenium(II) as d^6 systems have tendency to form low spin inert octahedral complexes, that present the possibility to investigate as newer metallodrugs. The most recent method for designing ruthenium-based drugs is the creation of ruthenium organic directing molecules. Under this approach, an enzyme's active site gets bound to the organic part of molecule, and nearby amino acid residues of enzyme bind to ruthenium ion part. The benefit of this strategy is that a compound has a known biological target that can be used to perform targeted actions like enzyme inhibition rates. The lower toxicity of ruthenium complexes can be explained by their capacity to mimic iron binding serum proteins, which enhances their concentration in cancer cells than in plasma. Cancer cells with transferrin receptors have a high affinity for plasma-bound ruthenium complexes. This capability explains the pharmacodynamic differences between cancerous and healthy cells, and forms the basis for higher cytotoxicity of trans-[tetrachlorobis(1H-indazole)ruthenate (III)] KP1019 compared to imdazolium-trans-tetrachloro(dimethyl sulfoxide) imidazole ruthenate (III)NAMI-A compounds 7 and 8 (Fig. 10.4).

Passive diffusion as well as transferrin-dependent transport are both parts of the proposed cellular uptake mechanism for Ru(III). Importantly, these ruthenium(III) anticancer compounds get activation by reduction to ruthenium(II) under hypoxic conditions of cancer cells using suitable reduction potential intracellular reducing agents like glutathione. It is reported that with an increase in Ru(III) reduction potential, antiproliferative activity of corresponding Ru(III) complexes also increases [4].

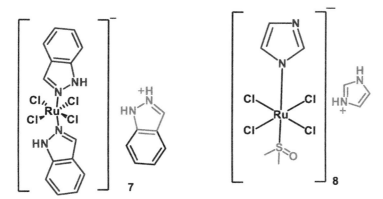

FIGURE 10.4 The structure of ruthenium(III) anticancer agents KP1019 (compound 7) and NAMI-A (compound 8) under clinical trials.

Ru$^{(II)}$-arene complexes, like Ru(η^6-p-cymene)-(PTA)Cl$_2$ (RAPTA-C), and Ru(η^6-toluene)-(PTA)Cl$_2$, (RAPTA-T) are reported to be cytotoxic. These Ru$^{(II)}$ complexes have an amphiphilic 1,3,5-triaza-7-phosphaadamantane (PTA) ligands (Fig. 10.5). Selective binding of RAPTA-C to histone protein of chromatin, activates chloride ligands for aquation and induces apoptosis that mildly inhibits the growth of primary tumors in vivo. RAPTA-C (Fig. 10.5) produces effective tumor growth inhibition at very low doses without producing toxic side effects as in case of other anticancer medications [5].

10.2.3.1 Substitution modulated anticancer activity of Half-sandwich ruthenium (II) complexes with heterocyclic ancillary ligands

In the search for novel metallo-drugs that can substitute cisplatin as an anticancer agent, ruthenium-based complexes have been at the center of attention. Two such ruthenium based anticancer complexes NAMI-A and KP1019 (Fig. 10.5) are currently under human clinical trial, for their antimetastatic and anticancer activities respectively. The change of Ru$^{(III)}$ metal center to its Ru$^{(II)}$ active form within the cancer cells prompts medicinal chemists toward synthesizing Ru$^{(II)}$ based complexes as anticancer complexes. In the bio-organometallic domain, half-sandwich Ru$^{(II)}$-arene type complexes are most attractive owing to their transferrin glycol protein mediated transport inside the cancer cells. For example, the RM175 half-sandwich complex is transported inside the cells via binding with His-242 and His-576 of the transferrin enzyme. Moreover, by tuning the structural components of RM175 such as increasing the hydrophobicity of arene group, increases its anticancer propensity [6]. Another example includes that of RAPTA-C which is an organoruthenium half-sandwich compound having arene group, chlorides, and 1,3,5-triaza-7-phosphaadamantane (PTA) connected to ruthenium (II) center. This compound is reported to have good antimetastatic and antiangiogenic activity. Compared to highly cytotoxic platinum compounds,

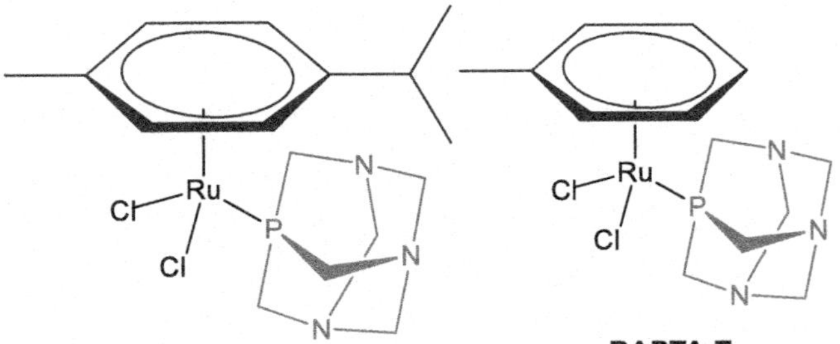

RAPTA-C   **9**   **RAPTA-T**   **10**

FIGURE 10.5 Ru(II)-arene based anticancer complexes RAPTA-C (compound **9**) and RAPTA-T (compound **10**).

NAMI-A is much less cytotoxic and with a different mechanism of cytotoxic action. Being comparatively less cytotoxic, the interaction of RAPTA-C with other cellular targets like cathepsin-B, cyclin-D kinase, TrxR, topoisomerase Iiα, etc., for broader mechanistic observations [6].

The ancillary ligands in half-sandwich Ru$^{(II)}$ organometallic compounds modulate their anticancer properties. Ru$^{(II)}$ organometallic compounds with heterocyclic compounds letrozole, mebendazole or chloroquine bound as ancillary ligands have shown significant anticancer propensities. The case study of anticancer activity of half-sandwich ruthenium(II) compounds was reported by Samuelson et al. [6] in which a library of more than 10 half-sandwich bioorganometallic ruthenium(II) compounds with mercaptobenzothiazole, mercaptobenzoxazole, mercaptobenzimidazole as heterocyclic ancillary ligands were synthesized in dichloromethane using dinuclear [(η^6-cymene)-RuCl$_2$]$_2$ and the corresponding ancillary ligands in a stoichiometric ratio of 1:2 (Fig. 10.6) [6].

Structural analysis of the synthesized compounds depicted a pseudo-octahedral geometry wherein the heterocyclic ligand and Cl ligands produce a three-legged piano stool configuration capped by the hexadentate cymene (η^6). The prominent supramolecular interactions providing major stability to the half-sandwich framework in these compounds was strong hydrogen-bonding between the Cl ligands and the NH-group of the heterocyclic ligand. The structure activity response on the cytotoxic propensity of the synthesized half-sandwich ruthenium(II) compounds was investigated using growth inhibition (GI50) against A2780 (ovarian carcinoma), Ovkar-3 (ovarian carcinoma) and KB (nasopharyngeal) cell lines. Among synthesized derivatives, mercaptobenzothiazole ligand framework depicted the highest cytotoxicity. Only fluorine as a halogen substituent on the aromatic ring of the ancillary ligand produced good cytotoxicity, all other substituents reduced the cytotoxic propensity of the half-sandwich compounds on the observed cell lines. Bromine atom or a carboxylic acid group addition on the aromatic ring of the ancillary ligand caused a significant reduction in anticancer activity.

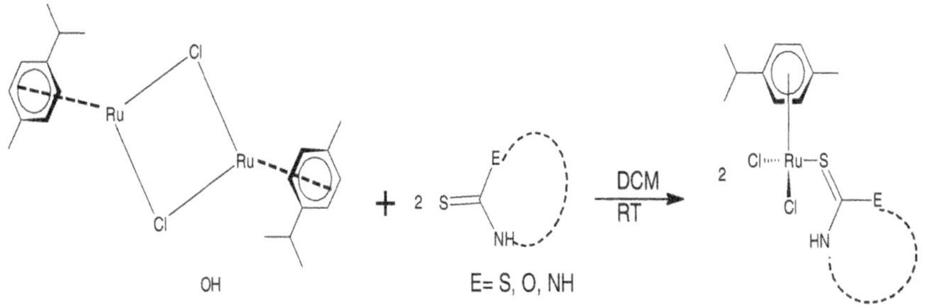

FIGURE 10.6 Synthetic routes for the preparation of half-sandwich Ru$^{(II)}$ organometallic compounds.

To explore the possibility of DNA cleavage as the probable mechanism of cytotoxicity, interaction of synthesized half-sandwich Ru$^{(II)}$ compounds with calf thymus DNA was carried out using Ethidium Bromide as the site marker. The observed binding constants of compounds with DNA could not distinguish their relative anticancer propensities which were latter examined from the thermal denaturation behavior of CT-DNA. The melting behavior of CT-DNA was examined in presence synthesized Ru$^{(II)}$ compounds under a CT-DNA and Ru$^{(II)}$ compound ratio of 0.5 to 1.0 depicted diverse and case subjective responses. From CT-DNA studies, it can be concluded that dinuclear ruthenium complexes affect the stability and melting behavior of CT-DNA evident from biphasic curve. However, the observed result does not unequivocally suggest that DNA damage as the mechanism of the anticancer activities of these compounds. Bio-organometallic compounds are also known to show anticancer activities via enzyme inhibition if not by DNA interactions. As a part of mechanistic exploration of synthesized Ru$^{(II)}$ compounds, Samuelson et al. [6] analyzed topoisomerase and TrxR activities in the presence of these compounds. All studied Ru$^{(II)}$ compounds showed TrxR inhibition around 10 μm concentration. However the extent of TrxR inhibition by Ru$^{(II)}$ compounds of greater cytotoxicity was not significantly higher than that of those derivatives with the lower cytotoxicity which leads to conclude that anticancer activity of these compounds cannot be exclusively due to TrxR inhibition [6].

10.2.4 Anticancer properties of gold complexes

Gold(III) complexes can be an additional cisplatin substitute. Gold(III) bioorganometallic compounds with varied carbon-based ligands have been synthesized with proven extraordinary cytotoxic properties. The development of anticancer compounds of lower toxicity and selective binding to the active site of biological targets is the primary attribute of designing gold complexes, for instance, Gold(III) complexes have the potential to be selective toward thiol-containing enzymes like TrxR. On cisplatin-resistant cell lines, the cytotoxic gold(III) complexes have been observed to be active. Recently, cytotoxicity studies of an organometallic titanocene-gold compound [(η^5-C$_5$H$_5$)$_2$Ti{OC(O)CH$_2$PPh$_2$AuCl}$_2$] against prostate and renal cell lines was evaluated [5].

10.2.5 Anticancer properties of rhodium and iridium complexes

Compared to platinum, gold, and ruthenium based bioorganometallic anticancer compounds, iridium and rhodium based organometallic derivatives were relatively less developed with none of the derivatives in the major clinical trials so far. The discovery of a rhodium(III) compound that inhibits Wee1 in TP53-mutated triple-negative breast cancer, a subtype of breast cancer that is

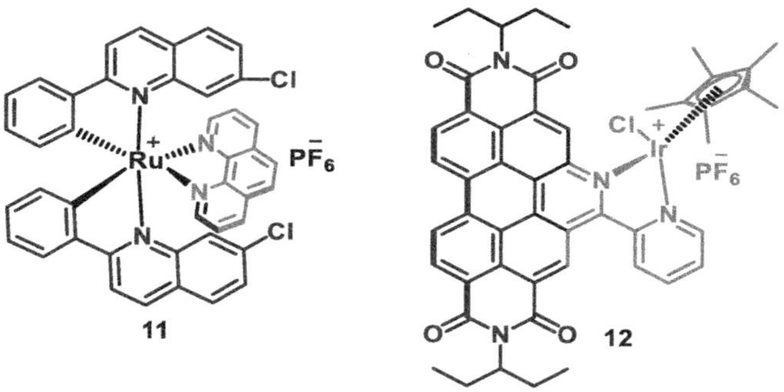

FIGURE 10.7 Rhodium(III) and iridium(III) complexes as photosensitizers for photodynamic therapy.

associated with the worst prognosis was reported by the Leung group [4] (Fig. 10.7). In TP53-mutated triple-negative breast cancer cells, which solely rely on the G2/M checkpoint to repair damaged DNA, Wee1 is a tyrosine kinase that activates the G2/M checkpoint in response to dsDNA breaks. As a Wee1 inhibitor, MK-1775, is being tested in clinical trials; however, its use has hematopoietic disorder, nausea, vomiting, and fatigue as major side effects. A single-site semiquantitative immunoassay with compound **11** at 3 μM revealed significant inhibition of Wee1(88.1%) greater than MK-1775 as Wee1 inhibitor in clinical trials [4].

For MDA-MB-231 human breast cancer cell line compound **11** inhibitedWee1 with an IC$_{50}$ of 11.2 ± 1.8 nM. From structure−activity view point, position of the methyl substituents on the phenanthroline ligand was vital, as phenanthroline ligand by itself displayed a milder inhibitory activity. Compound **11** was applied to a panel of different breast cancer cell lines, each of which contained varying levels of Wee1, and the IC$_{50}$ values were found to be negatively correlated with the levels of Wee1. This suggests that compound **11** cytotoxic activity is caused by Wee1 activity inhibition. Cell death is likely caused by the inhibition of Wee1 by compound **11** which results in G2/M defects, premature mitosis, and DNA damage accumulation. Iridium and ruthenium complexes with azabenzannulated perylene bisimide ligands were recently created, and their biological effectiveness as photosensitizers for photodynamic therapy was evaluated (Fig. 10.7). The ruthenium compound had a relatively low singlet oxygen quantum yield compared to its corresponding iridium compound **12** with a quantum yield of approximately 80% in acetonitrile. A number of cell lines, including cervical cancer HeLa, ovarian epithelial cancer A2780 (cisplatin sensitive), and A2780R (cisplatin resistant), as well as noncancerous MRC-5 human fibroblasts, were used in light and dark cytotoxicity studies using the two compounds. When exposed with 420 nm light, the iridium complex displayed nanomolar IC$_{50}$ values against the cancer cells, while its activity was hardly detectable in the dark which corresponds to photoselectivity indices greater than 20. While as, the ruthenium compound when exposed to radiation showed very low micromolar activity with photoselectivity index values of around 2 only. The research efforts are aimed to make iridium compounds more soluble and to shift photodynamic therapy (PDT) wavelength further into the red region.

The Sadler group is working on developing iridium (III) complexes of the type IrCp(N^N)Cl, where Cp* stands for pentamethylcyclopentadienyl [6]. One or more phenyl rings could be added to the Cp* to increase its anticancer properties, or the N^N chelating ligand could be swapped out for an N^C cyclometallating chelator. While complexes with phenyl extended Cp* ligands were also capable of binding DNA via intercalation, hydrolysis studies demonstrated that the chloride ligand was quickly replaced with water in aqueous systems. The same research team also demonstrated that these Ir$^{(III)}$ complexes can take up hydrides directly from coenzyme nicotinamide adenine dinucleotide (reduced form; NADH) to participate in the redox mechanism in cells. As a result, cancer cells produce more nicotinamide-adenine dinucleotide (NAD$^+$), which leads to oxidative stress and induces apoptosis [4].

10.3 Organoselenium compounds as potent chemotherapeutic agents

The escalating catchment of mortality associated with cancer incidents continue to challenge human life, making it an immediate need to look for more effective therapies with lesser burdens on patients. Due to their lower toxicity, greater selectivity and effectiveness among metallodrugs currently in use of antitherapies, the development of Se based cytotoxic agents can be an extraordinary approach of potentiating cancer chemotherapy [7]. Compared to selenium compounds, in inorganic form (selenite, selenate) the organoselenium compounds as symmetric aromatic diselenides/selenides, have been observed to be safer, potent drug moieties with higher biocompatibility. The methylseleninic acid, selenoesters, 1,2-benzisoselenazole-3[2H]-one and selenophene-based derivatives and also selenoamino acids as synthetic organoselenium compounds are also being analyzed in medicinal domain [7].

The redox behavior under physiological conditions makes selenium compounds more interesting. These get metabolized to methylselenol (e.g., methylseleninic acid, selenoesters) and hydrogen selenide (e.g., selenite) type redox-active metabolites which exhibit a potent mechanism of cytotoxicity under comparable drug doses to conventional metallodrugs used. Additionally, strong reactivity of these metabolites is due to their enhanced nucleophilic nature, which ensures their good anticancer efficiency. In addition to their direct cytotoxic behavior, studies show that selenium based compounds as chemo preventives as well. For Se-containing compounds, major mechanism of action against cancer is due to induction of safer cell death via apoptosis; however, studies also report these alter gene expression, DNA damage and repair, as well as the angiogenesis process or metastasis [7].

Like a two in one mechanistic machine, selenium compounds depending on dose can have antioxidant or prooxidant properties. If taken as a nutritional dose, the redox active selenium compounds have only antioxidant activity. In supranutritional doses, these show good prooxidant and anticancer properties. So, Se compounds under specific chemical conditions may activate both internal and exterior pathways to cause cell death other than apoptosis. Additionally, nonapoptotic processes, such as cell cycle arrest; autophagy, necrosis, ferroptosis, anoikis, entosis, and NETosis may also take place. Dixon and Stockwell first recognized ferroptosis in 2012, which appears to be one of the more intriguing cell death processes brought on by Se compounds. Lipid peroxides build up in the membrane as a result of nonapoptotic, iron-dependent cell death that is controlled and accompanied by an increase in ROS. Its fundamental mechanism is caused by an imbalance between cellular antioxidant systems and the generation of oxidizing substances (ROS), which results in oxidative stress and an elevated quantity of ROS inside the cell [7].

Iron, which builds up inside the cell and follows the Fenton reaction, results in the creation of free radicals, which turns out to be the catalyst for this particular sort of cell death. The main enzyme that prevents the development of ferroptosis is phospholipid hydroperoxide glutathione peroxidase (PHGPx) and Se-dependent glutathione peroxidase (GPX). Thus the body's amounts of this trace element determine its expression and activity (too low Se concentration limits their activity). This is owing to the fact that the selenocysteine residue in the active site of GPX4 and glutathione (GSH) is responsible for reducing lipid peroxides to the respective alcohols, which reduces the production and buildup of ROS. Without the component that regenerates it, GPX4 cannot exist on its own. Glutathione (GSH) is in charge of reducing oxidized GPX4. Thus they are both very important to prevent ferroptosis and antioxidant protection of the cell. In summary, ferroptosis occurs when the amount of the oxidant in the cells exceeds (excessive formation of ROS/lipid peroxides by high iron levels and other reactions leading to their formation) or when the antioxidant system is weakened (decrease in expression/inactivation/depletion of GPX4 or depletion/decrease in GSH Level). Fig. 10.8 illustrates many types of cell death that may be brought on by compounds containing selenium [7].

The type of biological activity and mechanisms of cytotoxic behavior of the synthetic organo-selenium compounds depends on their chemical architecture; some of the cases are as under:

10.3.1 Inorganic selenium compounds

The selenite (SeO_3^{2-}) and its sodium salt, disodium selenite, Na_2SeO_3, as inorganic selenium compounds were among the first Se-containing compounds tested for anticancer properties. As a result of selenite metabolism, higher production of

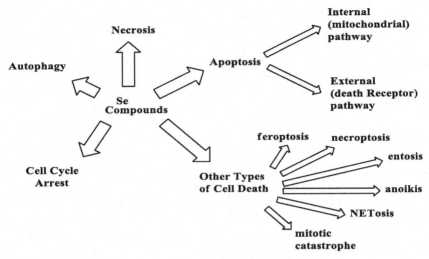

FIGURE 10.8 Cytotoxicity routes of Se containing compounds.

ROS was observed resulting in DNA damage, phosphorylation and p53 activation. This ultimately leads to enhanced Bax protein transcription, which can cause cell death by piercing the outer membrane of mitochondria. The mitochondrial membrane gets destabilized and when cytochrome-c is released into the cytosol, caspases are activated, which ultimately results in death of cancer cells. Other nonapoptotic forms of cell death, such as necrotic or autophagic phenotype, have also been noted under selenite treatment in addition to the apoptotic phenotype. Miranda et al. [7], observed that selenite induces necrosis in MCF-7 breast cancer cells, whereas Soukupová et al. [7] observed necrosis and autophagy in human bladder cancer RT-112/D21 cells (chemoresistant). Subburayan et al. found that sodium selenite (Na_2SeO_3) induces ferroptosis in human cancer cells, such as prostate (PC-3), breast (MCF-7), and glioma (U-87MG) [7] (Fig. 10.9).

10.3.2 Organic selenium compounds

On account of nonodorous nature, kinetic stability, lower reducibility and relatively better bioavailability, organic derivatives are more desired as synthetic derivatives over inorganic selenium compounds. Among the synthetic organic selenium derivatives, the symmetric and mixed diselenides are very common. A group of compounds known as symmetric diselenides having a general chemical formula R_2Se_2 with the Se—Se bond in their structure. A compound known as dimethyldiselenide (**17**) has antioxidant effects and strongly activates NADPH quinone oxidoreductase, whereas dipropyl diselenide(**18**) and dibutyl diselenide(**19**) even at low doses, are well-known for their prooxidant effects [7] (Fig. 10.10).

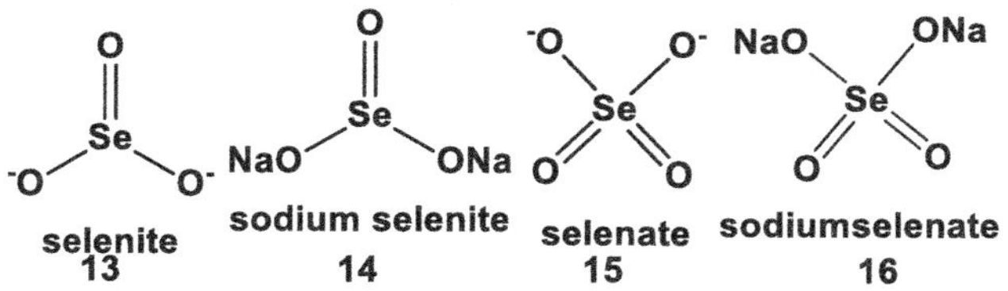

FIGURE 10.9 Chemical structures of some bioactive inorganic selenium compounds.

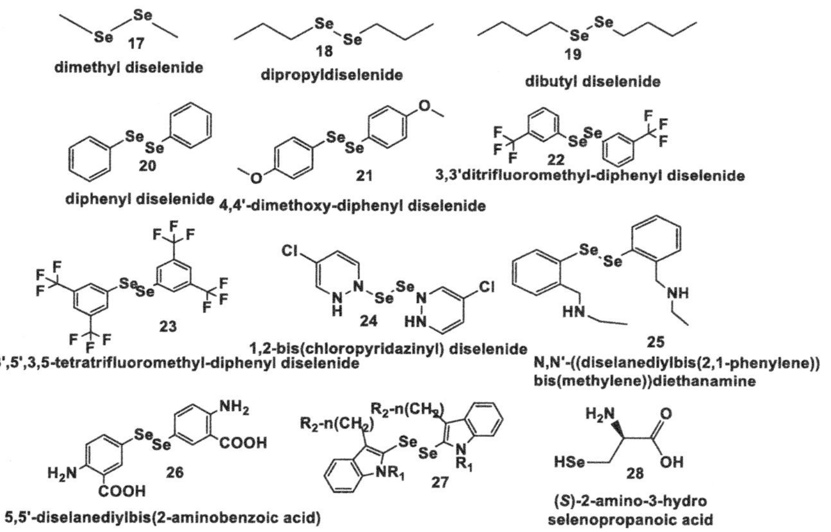

FIGURE 10.10 Chemical structures of bioactive symmetric diselenide derivatives (**17–27**) 7.

10.3.3 Selenocysteine

Selenocysteine (Sec) a vital selenoamino acid for humans like L-selenocysteine (**28**) is present in the active sites of enzymes with antioxidant characteristics, making it crucial for the body's proper operation. There have been many reviews on Sec's physiological function. It is also a component of Se-enriched yeasts and is not hazardous to living things. By causing apoptosis and a cell cycle stop in S-phase, Sec slows the development of cancer cells. Cell cycle arrest was found to be associated with down regulated cyclin-dependent kinase-2 (CDK-2) and cyclin-A. In addition, p53 regulation and ROS production contribute to Sec's anticancer activity [7].

10.3.4 Selol

Another intriguing substance is selol, which has selenium in the +4 oxidation state in its structure. Selol is not harmful and does not cause mutations.

However, the method of administration affects how hazardous it is to the body. Parenterally delivered selol was not poisonous, but when taken orally, it became significantly toxic, which may imply the development of more damaging chemicals during digestion. It should be regarded as solely being supplied parenterally as a result. Additionally, the Se content of a Selol determines its action. The researchers found that although 2% of Selol exerted antioxidant activity by stimulating enzymes of the second phase of detoxification, while 7% of Selol exhibited cytotoxic properties on human colorectal adenocarcinoma Caco-2 cells, demonstrating its chemopreventive properties [7].

It is also important to note that its high bioavailability and antioxidant activity are a result of selenoproteins incorporating selenium from selol. Prostate cancer and leukemia were two of the cancer cell lines on which selol were tested. Selol decreased cell proliferation and promoted apoptosis in HL-60 and leukemia cells in vitro, with resistant lines being more severely affected. This compound did not exhibit this effect against normal cells, but it decreased cell proliferation in the cancer cells of androgen-dependent prostate cancers and likely steered the cells by the apoptotic pathway. Additionally, Sochacka et al. identified 5% Selol as a TrxR inhibitor and a prooxidant (which results in increased ROS generation) [7] (Fig. 10.11).

10.3.5 Seleninic acids

Methylselenic acid (also known as methaneseleninic acid, MSA, Fig. 10.12) is one of the Se-containing compounds with anticancer characteristics that have been the subject of the most research. It belongs to the family of oxoacids and is formed via the transamination reaction from methylselenocysteine (MSC). As a result of the presence of lyase in tissues, the difference between MSA and MSC's activity in vivo and in vitro is less pronounced. In contrast to selenite, MSA is not hazardous to healthy body tissues and is nongenotoxic.

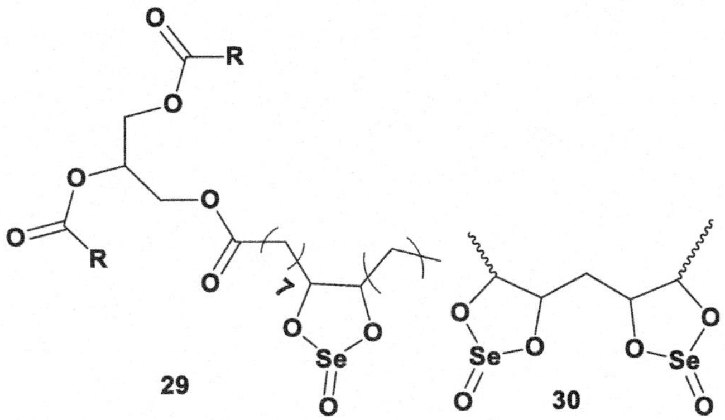

FIGURE 10.11 The chemical structure of Selol compound **29**-seleol(single dioxaselenolane rings) compound **30**-selol(double dioxaselenolane rings), R-polyunsaturated fatty acid moiety.

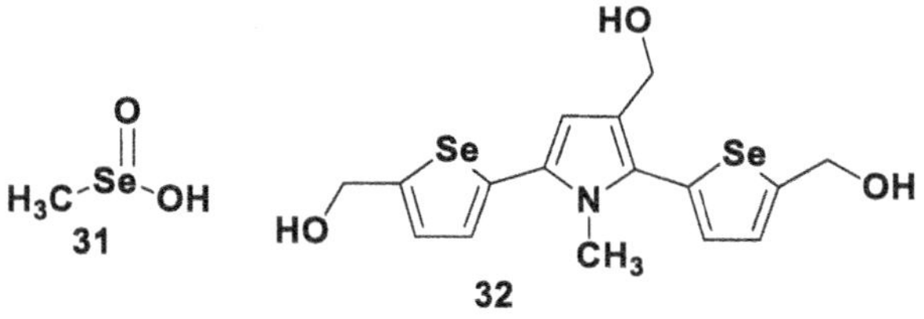

FIGURE 10.12 Chemical structures of methylseleninic acid **31** and selenophene derivative (D501036) **32**.

Colon, prostate, lung, breast, pancreatic, and liver cancer cell lines were used to study its effects on cancer cells both in vivo and in vitro. Through a variety of mechanisms, including cell cycle arrest in phase G1 and its antiangiogenic effect, MSA prevents the growth of cancer cells. The increased production of ROS is also one of the potential pathways for MSA [7].

10.3.6 Selenophene-based derivatives

2,5-bis(5-hydroxymethyl-2-selenienyl)-3-hydroxymethyl-*N*-methylpyrrole (D-501036) (Fig. 10.12) is one of the chemical compounds in this class that has sparked the most interest because of its diverse spectrum of antihuman cancer cell line activities. The dose and exposure duration employed determine its effect. It has been proven that D-501036 exposure increased caspase activity. Additionally, this substance selectively causes breaks in DNA double-strand and apoptosis in cancer cells (DSBs). Its remarkable efficacy against cells displaying multidrug resistance (MDR), which is characterized by glycoprotein P overexpression, raises hopes for an improvement in therapeutic effectiveness in cancers that are resistant to treatment [7].

10.3.7 1,2-Benzisoselenazole-3[2*H*]-one derivatives

Ebselen, also known as PZ 51, Dr3305, SPI-1005, and 2-phenyl-1,2-benzisoselenazol-3(2H)-one, is a Se-bearing heterocyclic molecule that exhibits chemo-preventive characteristics via antioxidant and antiinflammatory activity. GPx-like activity, but with weaker effects. Because it scavenges ROS, this substance protects against DNA mutations and the oxidation of cell components, which are all symptoms of cellular oxidative stress [7]. Yang et al. [7] in their work showed that the effects of hydrogen peroxide on lipid peroxidation, DNA damage and growth inhibition in HepG2 cells were diminished when the cells were treated with 25 μM ebselen. There are numerous reports of ebselen's potential proliferation-inhibiting actions, and a wide range of the antioxidant capabilities. The apoptosis process was accelerated in HepG2 cells when this compound was used at doses of 50−75 μM. This is due to the fact that Thiols within the cell underwent oxidative stress, which was the root cause of this activity [7]. Ethaselen/(BBSKE) (**34**) is another derivative compound belonging to the 1,2-benzisoselenazole-3[2H]-one class. It is a promising chemical compound with possible anticancer characteristics. The C-terminal active site of the enzyme TrxR is the target of this compound, it has been found to be a mixed-type mammalian inhibitor of TrxR. Increase of ROS and Ethaselen within the cell is caused by the inhibition of the reduction of oxidized Trx by attaching to the redox couple Sec-Cys. On several cell lines, including cancer types like the lung, stomach, tongue, liver, prostate, colon, nasopharynx, cervix and leukemia, ethaselen's suppressive effect has been demonstrated on the cancer cell proliferation both in vivo and in vitro conditions. In addition to being beneficial when used alone, ethaselen has also been shown to be more successful when combined with radiation or other cytostatic medications. Patients with nonsmall-cell lung cancer (NSCLC) are currently receiving this medication as clinical trials of phase 1c in China (NCT number: NCT02166242) at doses of 600 mg per day (tolerated dose: 1200 mg per day) [7] (Fig. 10.13).

10.3.8 Analysis of anticancer property of selenium-bearing 4-anilinoquinazoline compounds

Tubulin polymerization inhibition is one of the key strategies in the fight against cancer. Numerous novel 4-anilinoquinazoline hybrids containing selenium have been synthesized and tested as inhibitors for tubulin polymerization. It has been confirmed from antiproliferative activities assay that very low nanomolar doses of the compound inhibit human sensitive cancer cells. The mechanistic study showed that this is the most effective compound which altered microtubule dynamics, decreased mitochondrial membrane potential, and finally stopped Hela cells in the G2/M phase, which ultimately led to cellular apoptosis [8].

10.3.8.1 Anticancer agent

To check the antiproliferative activities, selenium-containing 4-anilinoquinazoline hybrids were tested against a panel of six human tumor cell lines, including human gastric cancer cell line, human colon cancer cell line, human hepatoma carcinoma cell line, human breast adenocarcinoma, human epithelial cervical cancer cell line, and human colon cancer cell line, RKO, HEPG2, MCF-7, MGC803, HeLa, and HCT116. The results indicated that each and every target species possesses excellent antiproliferative activity

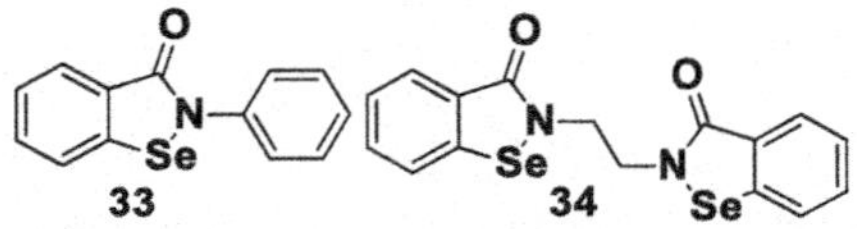

FIGURE 10.13 Chemical structure of 1,2-benzoselenozole-3[2H]-one derivatives.

with an IC$_{50}$ value of nM level with the 2-chloro-*N*-methyl-*N*-(4-selenocyanatophenyl)quinazolin-4-amine **34** exhibiting the best efficiency with the IC$_{50}$ value range from 2.09 to 9.98 nM [8].

10.3.9 Evaluation of indole chalcone and diarylketone derivatives containing selenium as tubulin polymerization inhibitory agents

Sixteen novel diarylketone and indole chalcone compounds containing selenium were created and tested as inhibitors for tubulin polymerization. Among them compound **36** demonstrated the strongest antiproliferative effects with IC$_{50}$ values against six human cancer cell lines ranging from 0.004 to 0.022 μM. Immunofluorescence experiment and microtubule dynamics assay both showed that compound **36** diarylketone selenium derivative could successfully inhibit tubulin polymerization (IC$_{50}$ = 2.1 0.27 μM). The G2/M phase arrest caused by compound **36** was discovered by additional cellular mechanism research, and the decline in mitochondrial membrane potential provided additional support (MMP) [9] (Fig. 10.14).

10.3.9.1 Effect of diarylketone selenium derivative(compound-36) on the advancement of cell apoptosis

Based on structure activity observations, a hypothesis can be made that the compound **36** treatment would cause the apoptosis. To assess this hypothesis cell cycle study using flow cytometry of HeLa cells involving fluorescent dyes was attempted. The protein annexin-V (V-FITC) underwent immunolabeling. The cells of HeLa were collected; then stained by Annexin V-FITC/PI, then subjected to varied doses of compound **36** for 48 or 72 hours. The results were evaluated by flow cytometry. The percentages of initial and late apoptotic results in HeLa cells were 5.89%(initial), 13.47%, and 23.19%, after 48 hours of exposure to compound **36** at dosages of 1, 5, and 10 nM. When the incubation period was raised to 72 hours (Fig. 10.15B), the initial and late apoptotic results rose to 18.13%(initial), 37.99%, and 71.33%. This result showed that compound **36** triggered time and concentration dependent cell apoptosis [9].

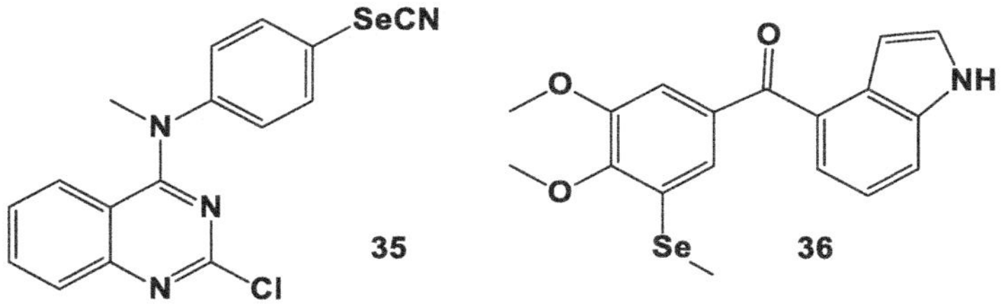

FIGURE 10.14 Chemical structure of 2-chloro-N-methyl-N-(4-selenocyanatophenyl)quinazolin-4-amine **35** and diarylketone selenium derivative compound **36**.

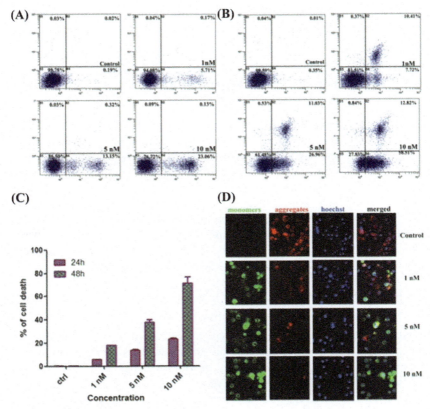

FIGURE 10.15 Effect of compound **36** on HeLa cells mitochondrial membrane potential and progression of cell apoptosis. Treatment with compound **36** causes the cultivated HeLa cells to undergo apoptosis. The apoptotic HeLa (Annexin-V positive) cells were treated with compound **36** (1, 5 &10 nM) for 24 h (A) and 48 h (B) and shown as a dot plot of Annexin-V-FITC-flourescence (x-axis) vs Pl-flouresence (y-axis). The HeLa cells were treated for 48 hours with compound **36** at various concentrations (1, 5, 10 nM) or DMSO (0.01%), and then incubated for 30 minutes with the fluorescent probe JC-1. The cells were then examined using fluorescence microscopy (D) 9.

10.3.10 Chemopreventive applications of isoselenocyanate compounds

Isothiocyanates (ITCs), have shown significant chemopreventive and anticarcinogenic activities in humans. Isoselenocyanates (ISCs) are more cytotoxic to cancer cells than their respective sulfur analogs. Research in the last ten years has demonstrated the potent anticancer therapeutic benefits of different ISCs. Greater cytotoxicity of ISCs compared to their corresponding ITCs has been demonstrated by both in vitro and in vivo models toward cancer cells. In a number of malignancies, including lung cancer, liver cancer, melanoma, etc., this extraordinary therapeutic activity has been established [10].

The usage of ISCs as anticancer drugs is particularly intriguing given the dearth of potent, long-lasting treatments as well as the anticipated increase in cancer incidence and fatalities.

In vivo and in vitro growth inhibitory potency of phenylalkyl ISCs was improved by lengthening the alkyl chain that bridges the ISC moiety and phenyl ring. Importantly, in ISC investigations, the phenylhexyl selenocyanate showed the enhanced growth inhibitory effects brought on synergistically by a combination of the ISC structure and Se. Additionally, it was hypothesized that the lengthening of alkyl chain would make it easier for the ITC or ISC groups to access and attach to crucial residues required for enzymatic activity [10].

The anticancer activity of the organofluorine ISC analogs of sulforaphane (SFN) was seen to be enhanced by the oxidation of the sulfur atom if there is a methylene group positioned between the sulfoxide group and the 4-fluorophenyl ring. Additionally, a decline in the compound's ability to selectively target cancer cells was observed when the length of alkyl chain between the ISC moiety and sulfur-based functionality was increased (from 4 to 5 carbons). The observations lead to the conclusion that the length of alkyl chain, ring substitutions, and oxidation state all affect the effectiveness and efficacy of the ISCs [10].

10.4 Other biological activities of organoselenium compounds

10.4.1 GPx activity

10.4.1.1 Catalytic GPx-like activity of 2,7-dialkoxy-substituted naphthalene-1,8-peri diselenides

Several 2,7-dialkoxy-substituted naphthalene-1,8-peri-diselenides were synthesized and examined for antioxidant catalytic action for the reduction of hydrogen peroxide. They show enhanced catalytic activity under acidic conditions, whereas suppressed activity under basic conditions. These compounds act like the antioxidant selenoenzyme glutathione peroxidase mimics. The mechanism for the catalytic activity shows that the hydrogen peroxide slowly oxidizes diselenides to the selenolseleninates, formation of which takes place via hydroxyselenonium intermediates, then the parent diselenides are formed back by rapid reduction by benzyl thiol Fig. 10.16. Mesomeric effects, coordination effects, and in some cases, anomeric effects are the factors which determine the relative reaction rates of the diselenides [11].

10.4.1.2 pH-sensitive organoselenium compounds as pH-responsive glutathione peroxidase mimics

The diaminodiselenide compound **37** and diaminoselenide compound **38** show pH-responsive glutathione peroxidase (GPx) mimic activity via

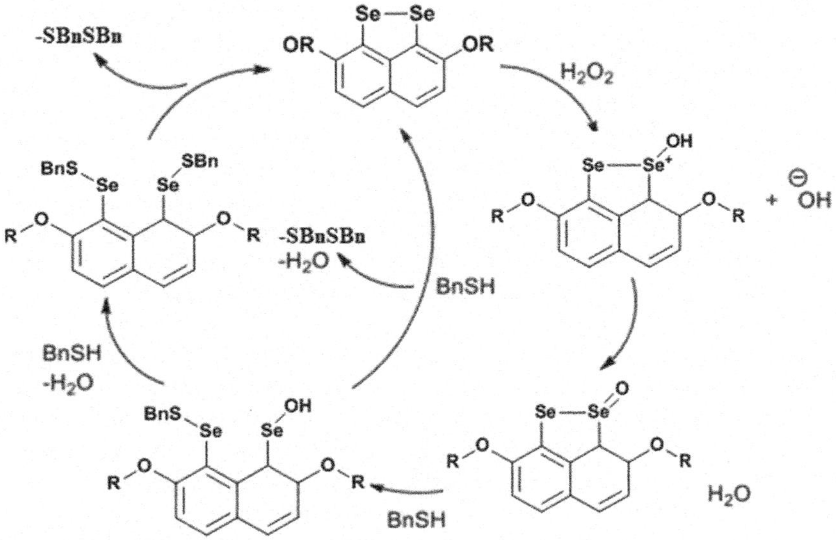

FIGURE 10.16 Catalytic pathway showing the hydrogen peroxide reduction with benzyl thiol and naphthalene peri-diselenides.

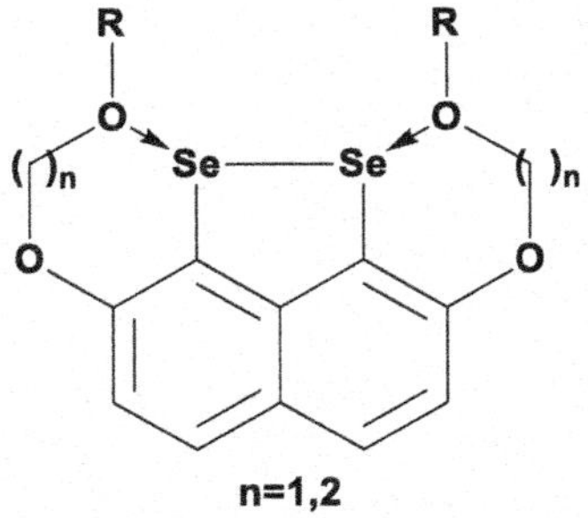

FIGURE 10.17 Possible coordination of O....Se in naphthalene peri-diselenide compounds.

different catalytic pathways. Compound **37** shows a ping-pong type mechanism and the catalytic mechanism of compound **38** is a sequential one. The catalytic activity is switched under different pH. Moreover, for compound **37,** the switching of catalytic activity was reversed at pH range of 7 and 9 or for compound **38** between 7 and 10 pH. The development of such smart pH-sensitive GPx mimics will open up new routes for the creation of artificial intelligent enzymes [12] (Fig. 10.17).

H₂N～～Se—Se～～NH₂   H₂N～～Se～～NH₂

37   38

10.4.1.2.1 Mechanistic study

Enzyme kinetic analysis helps us to fully understand the variation in GPx activity of compounds **37** and **38** using reduced GSH reductase NADPH linked test. GSH and H_2O_2 were used as two substrates in the test. Compound **37** or **38** GPx activities were tested at various concentrations of GSH while the concentration of H_2O_2 was fixed. The reciprocal plots of GPx activity versus GSH concentration were used to create the double-reciprocal plots, GSH concentration was kept constant while under various concentrations of H_2O_2, GPx action of compound **37** or compound **38** was assessed. Thus ping-pong type mechanism of compound **37** was confirmed by having parallel line in plots [12] (Fig. 10.18).

When compared to compound **37**, which produced parallel double-reciprocal plots for both substrates, compound **38** produced a sequence of intersecting linear plots for both substrates, indicating that its catalytic mechanism was sequential. The single product (GSSG), according to unavoidably released in the final stage of the sequential process [12]. Both of the substrates joined the enzyme to form a complex between enzyme and substrate before the product was released [12]. After careful examination of the catalytic process of selenide containing organoselenium compounds, it was discovered that the compounds were in their oxidized state (RSe(OH)₂R) in the catalytic cycle. The combined research of Carsol and Engman [12] suggests that the mechanism shown below (Fig. 10.19) might be used to catalyze the reaction of compound **38** [12].

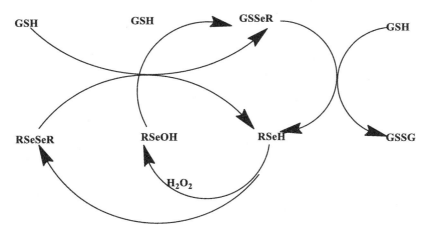

FIGURE 10.18 Proposed catalytic mechanism of compound **37**.

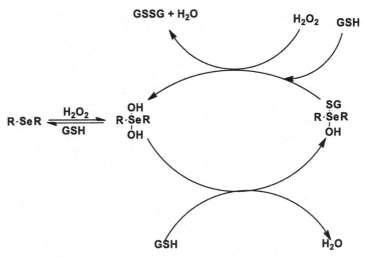

FIGURE 10.19 Proposed catalytic mechanism of compound **38**.

10.4.1.3 Enhanced activity as glutathione peroxidase mimics when benzo[d]isoselenazol-3-ones are combined with sterically hindered alicyclic amines and nitroxides

The GPx activity of several closely related benzisoselenazolones demonstrates how the synergistic interaction between the pyroline and selenium substituents increases the antioxidant activity of these substances. Because amino and selenium functional groups work together synergistically, changing the ebselen unit by adding pyroline redox-active groups dramatically improves the GPx activity of ebselen. The findings support a novel hypothesis that selenium may work in concert with sterically hindered amino groups close to the GPx active site to reduce hydroperoxides [13].

10.4.2 Anti-Alzheimer activity

10.4.2.1 Selenyl-dihydrofurans as anti-Alzheimer agents

A novel approach of the electrosynthetic intramolecular oxidative oxyselenylation is being used to create allylnaphthol and allylphenol derivatives using diselenides in an environment friendly manner. In this reaction using an electrolyte of 0.2 equiv. of nBu_4NClO_4 and working Pt electrodes in an undivided cell, results in the excellent yield of selenyl-dihydrofurans. Many of the synthesized selenylated compounds possess inhibition property against the acetylcholinesterase (AchE) enzyme like galantamine (commercially-available drug), this shows they can act as therapeutic agents to cure Alzheimer's disease due to AChE inhibition [14].

10.4.2.2 Selenodihydropyrimidinones as potential multi-targeted treatments for Alzheimer's disease

These compounds exhibit great effects as AchE inhibitors and are just as potent as the standard drugs. These compounds also exhibit excellent antioxidant activity via various modes of action [15]. On account of their ability to combat oxidative stress and also direct AchE inhibitory activity, these are considered to be potential and multifaced drug compounds for the treatment of Alzheimer's disease (Fig. 10.20) [15].

10.4.2.3 AChE inhibitory activity

Many derivatives of novel seleno-DHPMs (seleno-dihydropyrimidinones) were evaluated for AChE inhibitory activity. A modification of metal approach was used to measure the enzymatic activity. Plotting the inhibition against the concentrations of the sample solution allowed us to identify the (IC$_{50}$) value that inhibited substrate hydrolysis by 50%. All of the novel seleno-DHPMs derivatives exhibited significant AchE inhibition. The outcomes show that almost every derivative was just as effective as the common alkaloid galantamine, which serves as the main ingredient in reminyl, a medication in commercial use to treat Alzheimer's disease. The novel seleno-DHPMs derivatives evaluated on the comparison of the electron

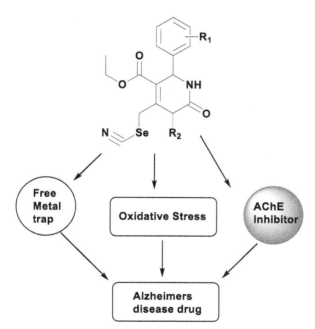

FIGURE 10.20 Seleno-dihydropyrimidinones as multi-targeted therapeutics for Alzheimer's diseases.

releasing or withdrawing groups at various places of the aromatic ring did not show a straightforward structure-activity link. The most potent member of the sequence is the compound bearing the simple phenyl substituent (Fig. 10.20) [15].

10.4.3 Cytoprotection

10.4.3.1 Organoselenium antioxidant compounds with cytoprotective effects

Novel N-methylated ebselenamine with antioxidant properties were synthesized from the respective diselenides and iodomethane. They showed outstanding chain-breaking and also mimicked GPx-like activity with efficiency double than the pure ebselen in the coupled reductase assay. They could more efficiently inhibit lipid peroxidation than α-tocopherol. Furthermore, they were observed to scavenge the ROS generated at a lower concentration (10 μM) with very less toxicity. These substances may be utilized to treat a number of ailments brought on by autoxidation, protecting animals from cell death [16]. The proposed mechanistic paths [16] of antioxidant property of N-methylated ebselenamine is shown in Fig. 10.21.

On the basis of experimental and also theoretical studies, a novel mechanism involving proton-coupled electron transport was proposed (Fig. 10.21, cycle A). Initially, an O-atom is transferred to Se atom from peroxyl radical (LOO•), creating an alkoxyl radical (LO•), and then an H atom is added in situ from the nearby —NHMe group, resulting in the formation of an aminyl radical selenoxide. Along with the creation of dehydroascorbate (DHA) from ascorbic acid AscOH [16].

In conclusion, they are discovered to be superior to α-tocopherol at quenching peroxyl radicals and in the presence of aqueous AscOH.

10.4.3.2 Bromo substituted diselenide compounds potential to scavenge ROS

The overproduction of reactive oxygen and nitrogen species (ROS and RNS) is a hallmark of diseases such as inflammation, atherosclerosis, and stroke. These reactive species were scavenged by bromo substituted diselenide catalyst as demonstrated below in a more mechanistic context [17] (Fig. 10.22).

Bromo substituted diselenide was used in ^{77}Se NMR investigations in deuterated DMSO in order to take the mechanistic details of its GPx-activity. The diselenide peak totally vanished after the addition of one equivalent of GSH, which was dissolved in 200 μL of water, and a signal at = 511 ppm, corresponding to the structure of selenosulfide, was seen instead (Fig. 10.22). One more equivalent of GSH was added to the mixture; however, this did not result in any discernible spectral changes, and there was no peak of selenol observed [17].

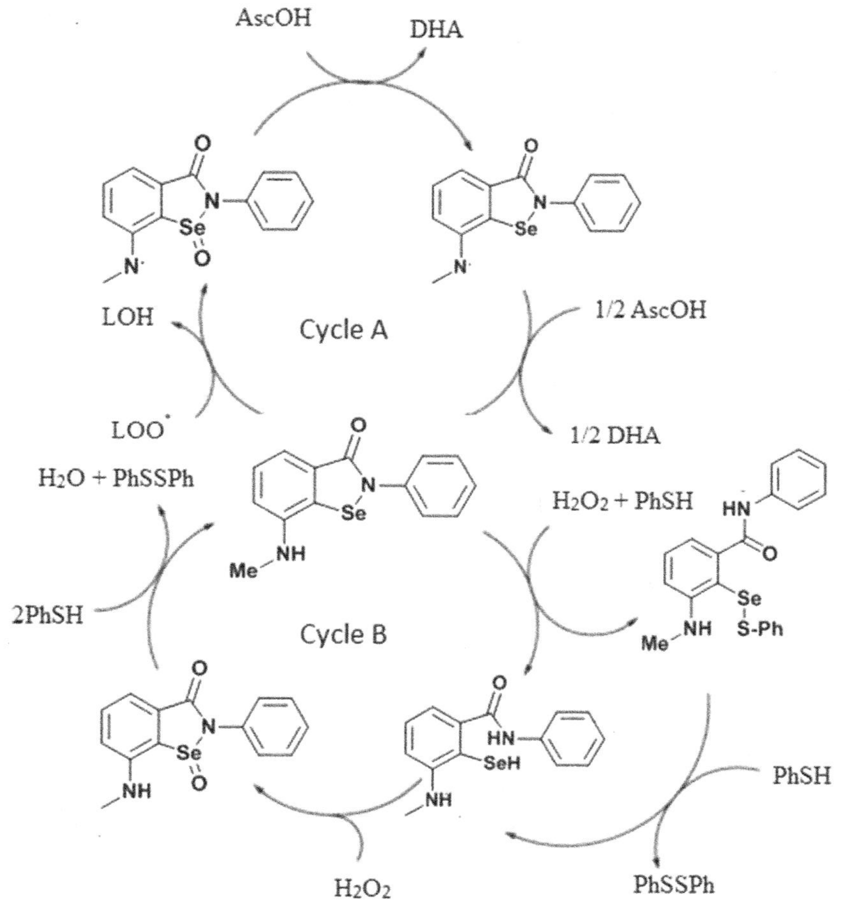

FIGURE 10.21 Proposed radical quenching mechanism (cycle A) and hydrogen peroxide decomposition (cycle B).

However, it could be possible to get indirect evidence of selenol production. The peak corresponding to butyl pyridyl selenide at = 342 ppm was obtained when GSH was added to the solution of Bromo substituted diselenide and iodobutane. A broad peak at $\delta = 142$ ppm was observed due to selenol when Bromo substituted diselenide was allowed to react with an excess of GSH. Bromo substituted diselenide was reformed when one equivalent of hydrogen peroxide was added to the mixture of selenosulfide/selenol produced from diselenide plus two equivalents of GSH. A peak at $\delta = 1215$ ppm due to selenic acid appeared upon further addition of GSH. Based on all these experiments, the mechanism for the reduction of hydrogen peroxide could be pictured as shown in Fig. 10.22 [17].

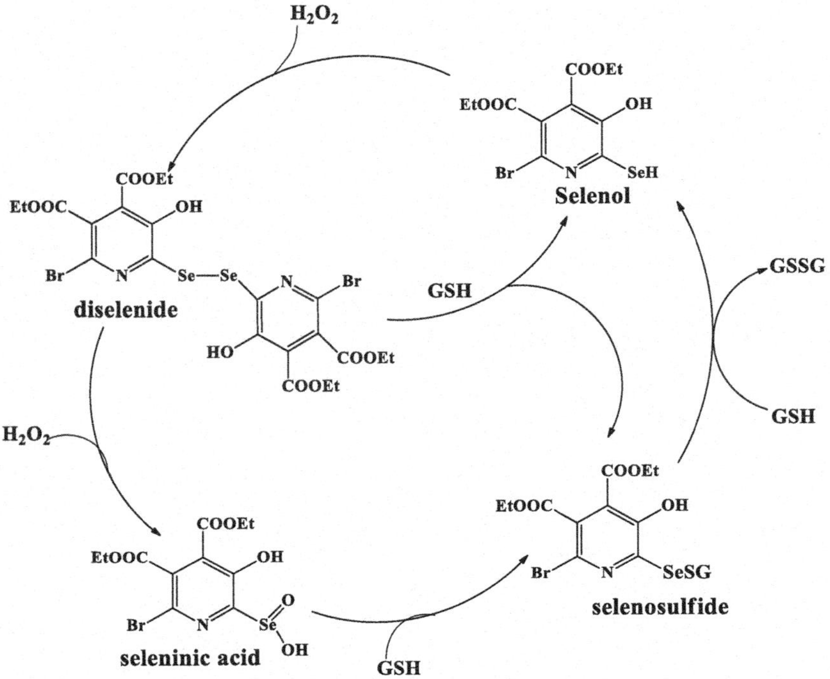

FIGURE 10.22 Preposed catalytic cycle for hydrogen peroxide decomposition in the presence of catalyst bromo substituted diselenide and GSH.

10.4.4 Insecticidal activity

10.4.4.1 Selenoethers and their insecticidal activities

Many of the synthetic selenoether compounds (**39–41**) show significant insecticidal properties. Using the leaf disk approach, the target compounds insecticidal activities were assessed under a broad screening against Tetranychus cinnabarinus and Plutella xylostella organisms. The studies revealed compound **40** to be an effective insecticide with 50% and 100% mortality rates against Plutella xylostella and Tetranychus cinnabarinus respectively at concentration of 600 mg/L [18].

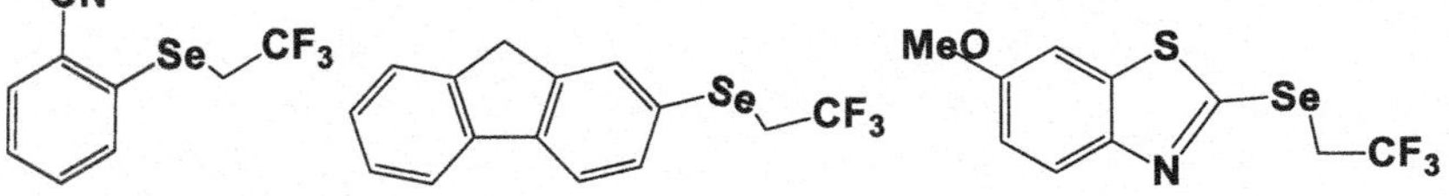

10.4.4.2 2,2-Difluoro-1,3-benzoxathioles (selenoles) with insecticidal activities

The reaction of trifluoromethanethiolates (selenolates) with o-bromophenols is an effective synthetic path for the production of 2,2-difluoro-1,3-benzoxathioles/selenoles These compounds exhibit excellent insecticidal activities. Using the leaf disk approach, the target compounds' insecticidal activities were assessed under a broad screening. Among the studied compounds insecticidal activities of compound **44** with a fluorine group at the C-5 position were observed to be strongest against the insects Myzuspersicae, Leucania separate, and Plumella xylostella. The selenium effect on the insecticidal activity was established from the comparative study of compounds **43** and **47** which differ in just sulfur and selenium atoms whereas sulfur analog **43** had little or no activity the corresponding selenium analog **47** showed great insecticidal activity. These findings suggested that the selenium atom might be crucial to the insecticidal effects [19].

10.4.4.3 Selenium-containing imidazolium ionic liquids with antibacterial activity

Imidazolium based ionic liquids functionalized with selenium in the side chain were efficiently synthesized via a simple two step methodology using PhSeSePh as selenium precursor, The observed activity of selenium modified ionic liquids is observed to be influenced by the substituents attached to the selenium moiety and by the nature of the counterion. These compounds showed a range of antimicrobial activities and were particularly effective against algae, according to the minimum inhibitory concentration data. A panel of microorganisms was tested for antimicrobial resistance including bacteria (Staphylococcus aureus ATCC 25923, Escherichia coli ATCC 25922, Pseudomonas aeruginosa ATCC 27853), yeast-like filamentous fungi (Aspergillus fumigatus ATCC 204305), fungi (Candida albicans ATCC 24433), and algae (Prototheca z).

The minimum inhibitory concentration (MIC) revealed that the activity of these compounds is modulated by structural modifications in the aryl group attached to selenium as well as by the counterion associated with the cationic part [20].

10.4.4.4 Cyclization of Z-selenoenynes by iron(III)-PhSeSePh results in selenophenes with antidepressant-like activity

A group of 2,5-disubstituted 3-(organoseleno)-selenophenes was produced using a unique synthetic method, the intramolecular cyclization of (Z)-chalcogenoenynes by $FeCl_3$-diorganyl dichalcogenide. Good yields of the cyclized products were achieved. The findings revealed by mouse forced-swimming test that these compounds induced an antidepressant-like action. The experiments unambiguously demonstrate that the organoselenium group at position three and the phenyl group at position two of the selenophene ring are essential for the antidepressant-activity. Further analysis revealed from the results is that the fluorophenyl element in the organoselenium group is essential for the antidepressant effects of this class of organochalcogens [21].

The 3-(4-fluorophenylselenyl)-2,5-diphenylselenophene was used as an archetype molecule for the structure-activity relationship study, and a number of chemical modifications were made. The mouse forced-swimming test was used to examine the possible antidepressant activity of synthesized 3-chalcogen selenophenes. The development of passive behavior, which disengages the animal from active means stressful inputs with copying, or persevere failure in escape-directed action following prolonged stress are regarded to be the two possible explanations for immobility. This immobility in animals is known as behavioral despair and is thought to replicate a state like human depression and diminishes by a number of therapeutic treatments for this ailment [21].

10.5 Different plausible mechanisms of action for bioorganometallic complexes

Due to their unique electronic and structural characteristics, which include multiple oxidation states of variable stabilities, coordination geometries, type and ligand numbers, metal-based therapeutics stand out and are exceptional in their performance. They offer novel reactivity routes of lowered energy barriers, such as different types of ligand substitution, metal and ligand-based redox processes, and amazing catalytic loops. The metal based compounds in a way represent 'prodrugs' which can undergo activation on the way to or at the target site, which is an altogether different feature from organic drugs. This typically involves the dissociation of one or more labile ligands, chelate ring-opening, type reactions or a change in the oxidation state (or spin state) of the metal in the complex. Additionally, metal complexes can be selectively activated at the target site by an external stimulus, such as pH, concentration gradient, and heat/mechanical energy/radiation (photodynamic therapy) [22].

10.5.1 Activation by means of hydrolysis

In drug design, the inertness of transition metals (TMs) often becomes a limiting factor. The "inertness" of metal ions can primarily be changed by selecting the appropriate ligands. For TM based drugs, hydrolysis involves a frequent activation mechanism via displacement of donor ligands (that are only weakly bound) with water molecules under the influence of mostly pH, concentration gradient, or strongly binding donor site on the amino acid residue at the active site [22].

10.5.1.1 Square-planar Pt^{II} complexes

One example of activation by hydrolysis is in case of cisplatin (cis-$[Pt^{II}Cl_2(NH_3)_2]$). Hydrolysis is inhibited in extracellular environments due to higher $[Cl]^-$ ion concentration (>100 mM). While as $[Cl]^-$ concentration being relatively lower inside cells (around 23 mM in the cytoplasm and 4 mM in the nucleus), aquation happens more readily. This produces more reactive mono and di-aquated ions $[Pt^{II}(OH_2)Cl(NH_3)_2]^+$, $[Pt^{II}(NH_3)_2(OH_2)_2]^{2+}$, which readily bind to DNA bases Guanine and Adenine. Only about 1% of the DNA-reacting activity of intracellular cisplatin results in apoptosis and cell cycle arrest. Strong interactions exist between "soft" Pt^{II} and proteins that have cysteine thiolate groups. Pt^{II} can be detoxified by tripeptide L-glutathione (GSH, 2–10 mM in cells), particularly in cancers that are treatment-resistant. These kinds of reactions may also have unintended side effects. A chelated dicarboxylate was added to the second-generation drug carboplatin, which exhibits reduced nephrotoxicity and is very slow to hydrolysis, as hydrolysis rates depend on the ligands [22].

10.5.1.2 Ti^{IV} dichloride complexes

$[Ti^{IV}Cp_2Cl_2]$ (Titanocene dichloride) underwent clinical anticancer trials for about ten years, but its entry into phase II clinical trials was hampered by its high propensity to hydrolyze and pH-dependent interactions with DNA. Not only are the Cl^- ligands labile but also Cp ring is displaced in reactions with serum transferrin, where Ti^{IV} binds to the Fe^{III} sites, providing a delivery system for Ti^{IV} into cancer cells. To reduce hydrolysis rates, later titanocene derivatives included the addition of bulky aryl substituents, tethering Cp rings together, and even tethering Cp rings to labile groups to reduce ligand displacement [22].

10.5.1.3 Ru^{II} and Os^{II} half-sandwich complexes

Half-sandwich pseudo-octahedral Ru^{II} and Os^{II} η^6-arene diamine anticancer [Ru^{II}/Os^{II}(η^6-arene)(N,N)Cl]$^+$ complexes, such as RM175, hydrolyze and bind to DNA in a similar way to cisplatin, but only monofunctionally because they have a single labile monodentate chloride ligand. The rate of hydrolysis and reactivity are significantly influenced by the ligands present. The hydrolysis rate of Ru^{II} decreases with the monodentate ligand in the following order: Cl$^-$ > Br$^-$ > I$^-$ > N$_3^-$. The ability of the arene to donate electrons tends to increase the rate of hydrolysis, and the bidentate ligand significantly affects hydrolysis. Switching from neutral N- to anionic O-donors increases the rate of hydrolysis significantly. Compared to aqua ligands, hydroxyl ligands were observed to be more labile and likely to produce bridging species which happened in culture media for [Os^{II}(η^6-p-cym)(acac)Cl] leading to hydroxo-bridged species that are inert to cancer cells.

Strong-donor ligands(such as en, acac, and pico) typically promote rapid hydrolysis and the formation of basic aqua adducts. Acceptor ligands promote slower hydrolysis. The Os-I bond in complex FY26, [Os^{II}(η^6-p-cym)(azpy-NMe$_2$)I]PF$_6$, undergoes very little hydrolysis in aqueous environments over the time period of 24 hours. Studies using ^{131}I-radiolabelling demonstrated that cancer cells promote rapid Os-I link breakage. GSH appears to be attacking the azo-bond during the process (Fig. 10.23) [22].

10.5.1.4 Hydrolysis of Rh^{III} and Ir^{III} Cp complexes

For water molecules, the first coordination shell of aqua Rh^{III} and Ir^{III} has very long settle times (10^8 to 10^{10}s). The analog of NAMI-Ir^{III} A, (**Ir1**), has poor antiproliferative activity and is inert to hydrolysis. However, [Ir^{III}(H$_2$O)$_6$]$^{3+}$ hydrolysis rates dramatically increase by 14 orders of magnitude when Cp* is added. Thus in Rh^{III} and Ir^{III} anticancer complexes, a variety of N,N, N,O, O,O, and C,N-coordinated ligands hydrolyze quickly

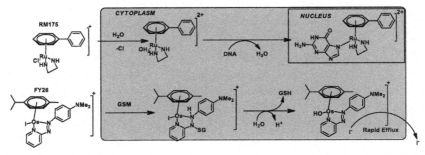

FIGURE 10.23 Hydrolytic activation of two types of half-sandwich complex; RM175(Ru1), bearing a σ-donor bidentate ligand (en), for which hydrolysis is activated by reduced [Cl$^-$] levels in cells; FY26(Os4), bearing a strong π-acceptor ligand (azpy), for which hydrolysis is activated by GSH attack on the azo-bond. *Copyrighted from Royal Society of Chemistry.*

(minutes). Similar to Ru^{II} and Os^{II} arene complexes, Rh^{III} exhibits higher pKa values for its aqua species than its heavier congener Ir^{III}, just like Ru^{II} and Os^{II} arene complexes. Many Ir^{III} Cp* chlorido complexes with bipy, phpy, en, pico, and acac bidentate ligands hydrolyze quickly and have no anticancer effects. Extremely potent Ir^{III} complexes with a Cp* ligand bound to a "labile" pyridine ligand, which is a solvent moiety used in biological assays, are known to be activated via chelate ring opening [22].

Ir1

10.5.2 Redox activation

Although essential for the production of energy in eukaryotic organisms, oxygen can occasionally be fatal if it produces ROS. Modulating redox balance, which causes cancer cells to become hypoxic, is an effective anticancer strategy. Through reduction or oxidation at metal or ligand centers, or indirectly through interaction with biomolecules in redox pathways, metal complexes can directly or indirectly change the cellular redox balance. The ideal outcome of redox activation of metal complexes is the selective formation of excess cytotoxic species in cancer cells, which minimizes adverse effects to the noncancerous cells [22].

10.5.2.1 Metal reduction

In tumors, hypoxia and low oxygen levels can encourage the selective activation and reduction of inert metal complexes like Pt^{IV}, Ru^{III}, and $Co^{III.}$ These prodrugs are reduced in healthy, oxygenated cells but are swiftly reoxidized and rendered inactive, while re-oxidation in hypoxic cancer cells occurs gradually. Bioreductants like GSH, ascorbate, NAD(P)H, and cysteine-containing proteins can reduce octahedral low-spin $5d^6$ Pt^{IV} complexes to release active ligands and DNA-binding Pt^{II} species. Compared to traditional Pt^{II} agents, these complexes are more stable in biological media. The four Pt^{IV} complexes that have entered clinical trials are tetraplatin (**Pt1**), iproplatin (**Pt2**), satraplatin (**Pt3**), and **LA-12** bis(acetato)(1-adamantylamine)amminedichloroplatinum (IV) (**Pt4**); however, none of them have been approved due to side effects and subpar efficacy. Through Pt^{IV} reduction, a Ru^{II}-Pt^{IV} conjugate (**Ru-Pt1**) bridged by 1,6-diaminohexane releases cytotoxic Pt^{II}, the histone deacetylase inhibitor phenylbutyrate, and photosensitive Ru^{II} (480 or

595 nm irradiation), which has a variety of targets and actions. The NAMI-A (**Ru4**) and KP1019 (**Ru5**) Ru^{III} complexes underwent successful phase I clinical trials (NAMI-A was terminated in phase II due to low activity). The Ru^{II} complexes produced by reduction have an octahedral structure and can bind to DNA and proteins [22].

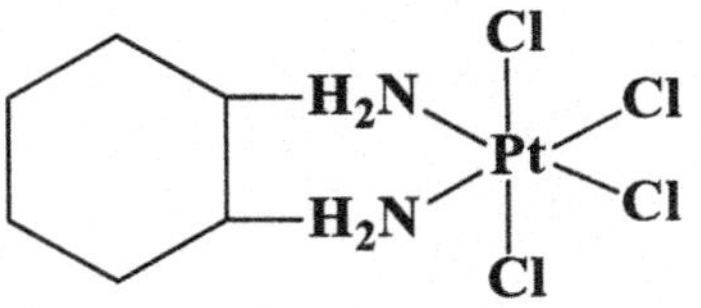

tetraplatin(Pt1)

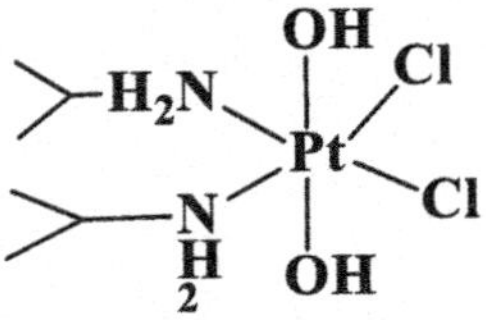

iproplatin(Pt2)

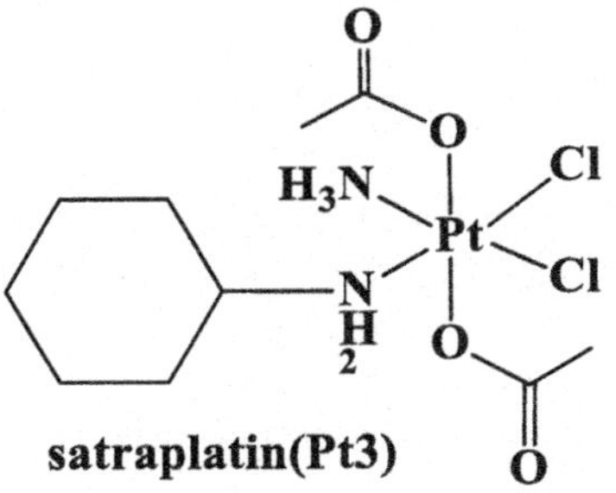

satraplatin(Pt3)

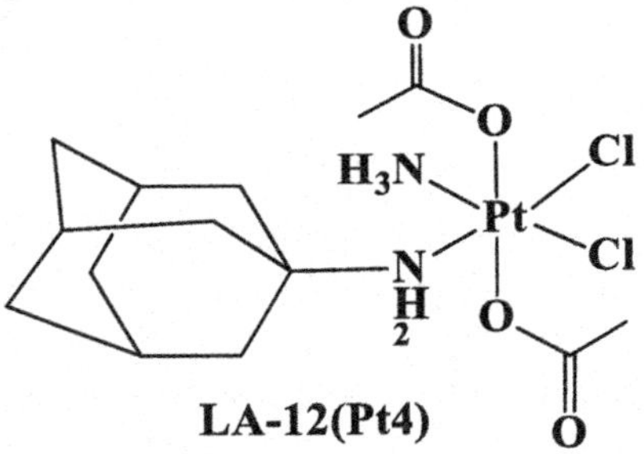

LA-12(Pt4)

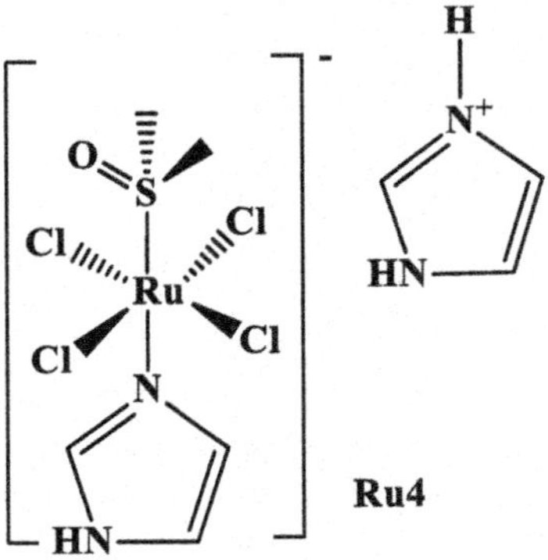

Ru4

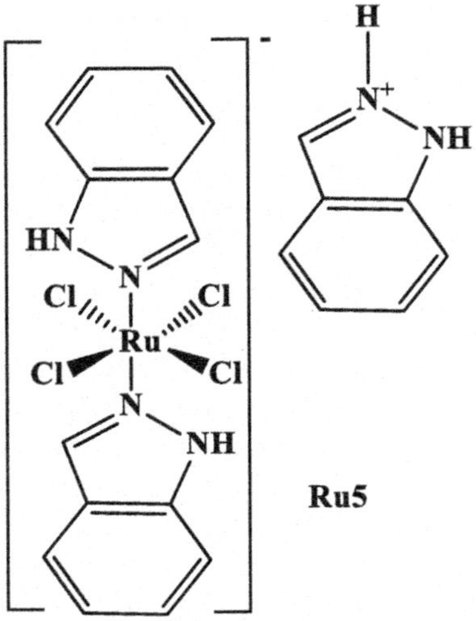

Ru5

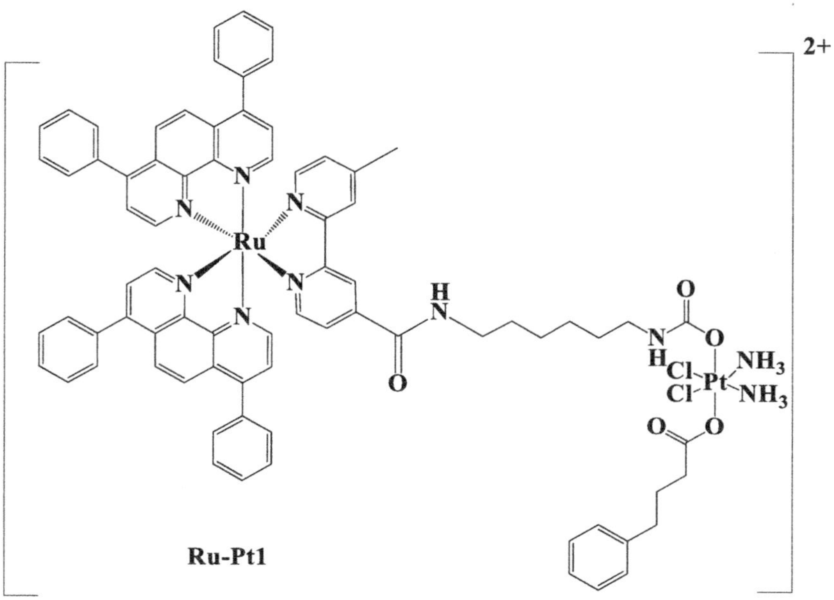

Ru-Pt1

10.5.2.2 Ligand reduction

The activation of Ru^{II} arene complexes [Ru(η^6-bip)(azpy-NMe$_2$)X]$^+$ (X = I, OH) (**Ru6**) by the reductive addition of GSH results in the production of GSSG, which increases the ROS levels in cancer cells. Through nucleophilic aromatic substitution with thiols, the meso-carbon in the porphyrin ring of **Au1** can change cysteine thiols in cancer-related proteins [22].

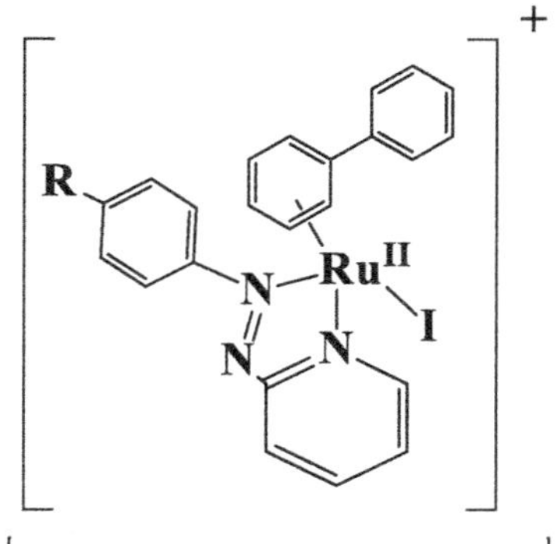

Ru6

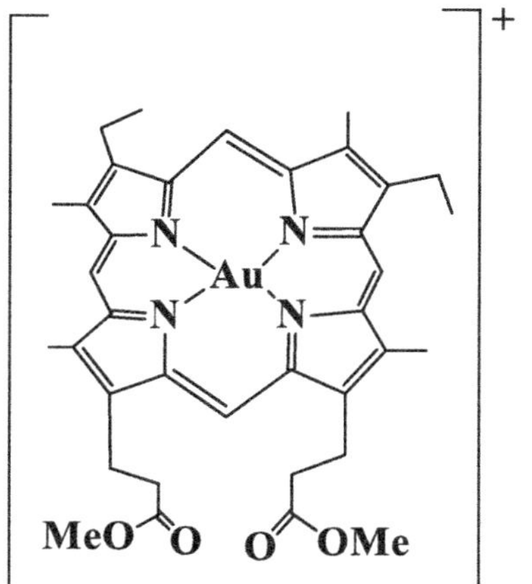

Au1

10.5.2.3 Metal oxidation

Because cancer cells produce an excessive amount of H_2O_2, the derivatives of ferrocene can undergo Fenton-type reactions in cells and generate HO• radicals that can cleave DNA. Ferrocifen (**Fe1**), a highly effective antibreast cancer drug that is not dependent on hormone expression, is a ferrocenyl analog of hydroxytamoxifen. **Fe1** easily loses $2e^-$ and $2H^+$ with the help of the Fc^+/Fc redox couple to create a stable and cytotoxic quinone methide intermediate. However, cells exposed to hydroxytamoxifen derivatives do not produce ROS. Cells exposed to ferrocifens do. Antimalarial medication Ferroquine (**Fe2**) also causes cytotoxicity by generating Fe^{III} species and HO• radicals. The $Fc^{+/0}$ reduction potentials are correlated with the cytotoxicity of ferrocene N-heterocycle-linked Ru^{II}p-cym complexes, such as **Ru-Fe1**, which is consistent with ferrocene's simple oxidation to ferrocenium and the subsequent generation of ROS [22].

10.5.2.4 Ligand oxidation

The hydrolysis product $[Ru^{II}(\eta^6\text{-bip})(en)H_2O]^{2+}$ of the anticancer complex $[Ru^{II}(\eta^6\text{-bip})(en)Cl]^+$ reacts initially with GSH to form $[Ru^{II}(\eta^6\text{-bip})(en)(SG)]^+$. However, even under hypoxic conditions, oxidation to the more labile sulfenato species, [Ru^{II}(η^6-bip)(en)(OSG)]$^+$, can take place, which allows the sulfenate to be easily replaced by N7-cGMP. Another enzyme that can experience this Ru-induced thiol oxidation (to sulfonate) is glutathione S-transferase, which is overexpressed in solid tumors. Cellular bioreductants, such as GSH to produce GSSG, can attack the iodide ligand in photoactive trans-[Pt^{IV}(en)(OH)$_2$I$_2$] and liberate iodine, resulting in the reduction of Pt^{IV} to Pt^{II} [22].

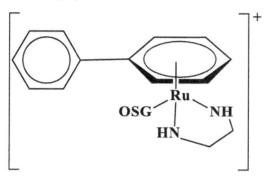

[Ru^{II}(η^6-bip)(en)(OSG)]$^+$

10.5.3 Photoactivation

In the context of (i) photodynamic therapy (PDT), (ii) photothermal therapy (PTT), and (iii) photoactivated chemotherapy, photoactivation selectively activates metallodrugs in cancer cells. As a minimally invasive technique with clinical approval, photodynamic therapy uses a photosensitizer (PS) and controlled light (typically red light between 600 and 800 nm, the "therapeutic window") to kill cancer cells in a catalytic manner under oxygen presence. Longer wavelength light penetrates further and is less damaging to cells. An electron is moved from its singlet ground state to its singlet excited state by the absorption of a photon. From there, the electron goes through the intersystem crossing and nonradiative decays into a low-lying triplet state. An electron can be transferred from this state to biological substrates to produce radicals, which then interact with oxygen to form ROS and oxidative stress, which eventually results in cell death (type 1) [22].

In addition, there is a chance of direct energy transfer from the PS excited triplet state to the ground state triplet oxygen (^{3}O$_2$), resulting in the production of highly reactive singlet oxygen (^{1}O$_2$) with a lifetime of three seconds in cell nuclei and a diffusion distance of approximately 300 nm (type II). Tissues primarily utilize type II processes, and a ruthenium-based PS TLD-1433 (**Ru8**) that relies on these processes heavily is presently undergoing clinical trials [22]. In photothermal therapy, the metallodrugs are used to convert photons into heat, which selectively kills cancer cells. PTT typically uses nanomaterials, such as carbon nanotubes, graphene oxide sheets, Cu^{II} nanocrystals, Bi^{III} nanorods, and Au^0 nanoparticles, as its agents (AuNPs). AuroLase in clinical trials for prostate cancer is currently being used as a photothermal therapeutic with NCT04240639. When AuNPs are exposed to near-infrared light (between 750 and 1400 nm), localized surface plasmon resonance develops. This resonance loses energy through radiative and nonradiative processes and results in hyperthermic cell death. Although effective ablation requires temperatures over 50°C, tumor damage begins to occur at temperatures over 41°C [22].

The PACT is responsible for using light to activate an inert prodrug, creating photoproducts that support the therapeutic effects. Pt^{IV} azido complexes $[Pt^{IV}L_1L_2(N_3)_2(OH)_2]$ (L_1, L_2 = am(m)ine ligands) L_1, L_2 = amine ligands-trans-Pt(**Pt5**) and cis-Pt(**Pt6**) structures, exposed to visible light with the additional release of azidyl radicals are reduced to cytotoxic Pt^{II} species. A dithienylcyclopentene-Pt^{II} complex Pt-Pt exhibits increased cytotoxicity toward melanoma and colorectal cancer cells after photoswitching from its open to a closed form. PACT agents have an advantage over PDT in hypoxic tumor environments because of their oxygen-independent mechanism. Due to this, prodrugs that combine PDT and PACT have been developed, such as chlorine conjugated Pt^{IV} micelles loaded on nanoparticles that can generate oxygen upon photodecomposition and then can be used for PDT. Clinical trials for any metal-based PACT agents have not yet begun [22].

10.5.4 Ionizing radiation, sonodynamic and thermal activation

The potency of a drug can be activated by X-rays and other ionizing radiation sources. Ionizing radiation and radiosensitizers together produce a synergistic biological response that is greater than each of their individual effects. This collaboration was noted early on in the clinical use of platinum drugs, which are now frequently combined with external beam radiotherapy. In preclinical studies, a wide range of TM-based radiosensitizers, mainly from groups 6−8, have been investigated. A cyclometallated Ir^{III}

compound (**Ir2**) was recently found to be localized in the mitochondria of cancer cells and to increase radio sensitization when compared to cisplatin. This was explained by increased ROS production upon X-ray irradiation [22].

A form of external beam radiation called synchrotron stereotactic radiotherapy (SSR) uses monochromatic high-fluence X-ray beams to treat cancer. The presence of heavy elements may enhance the therapeutic effect if the X-ray beam is tuned to the energy of the K-electrons of the sensitizer. Despite the fact that conventional linear accelerators also had a similar effect, the outcomes for brain tumors in rats treated with SSR and cisplatin were noticeably better [22]. The PS must either be directly excited by X-rays or indirectly by the X-ray excitation of a luminescent substance (scintillator), which can then activate the PS, in order to perform X-ray PDT (XPDT). For the direct approach, metalloporphyrins have been studied, and for the indirect approach, inorganic-based nanoparticles have been studied. The main advantage of XPDT is its ability to overcome the shallow penetration depth of visible light [22].

PDT is being substituted with sonodynamic therapy that uses a mechanical wave approach to deepen tissue penetration. Sonoluminescence, which is produced by intensely concentrating ultrasound energy, can activate PS's. Porphyrins and metalloporphyrins demonstrated promising in vitro results when metal-based nanoparticles and metal-organic frameworks were investigated [22]. Heat can be used to selectively activate metallodrugs, for instance by conjugating them to thermo activatable fragments. Heating, for instance, increases both the solubility and cytotoxicity of RAPTA-derivatives with perfluorinated ligands (**Ru9**). Metaloenediynes, which are sources of the enediyne function that are thermally activated include **Pt7**, spontaneously to produce extremely potent 1,4-benzenoid diradical species [22].

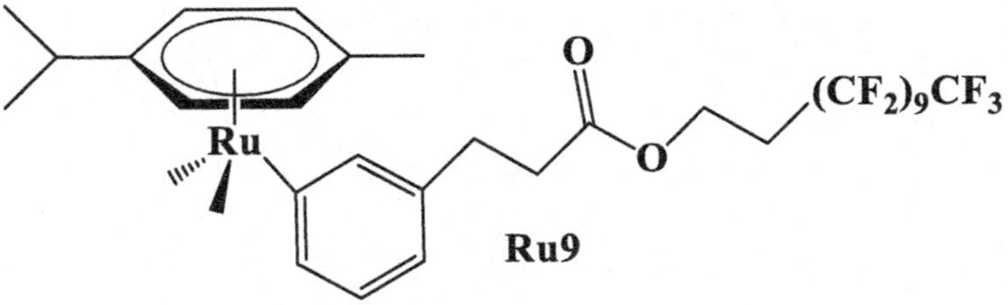

Ru9

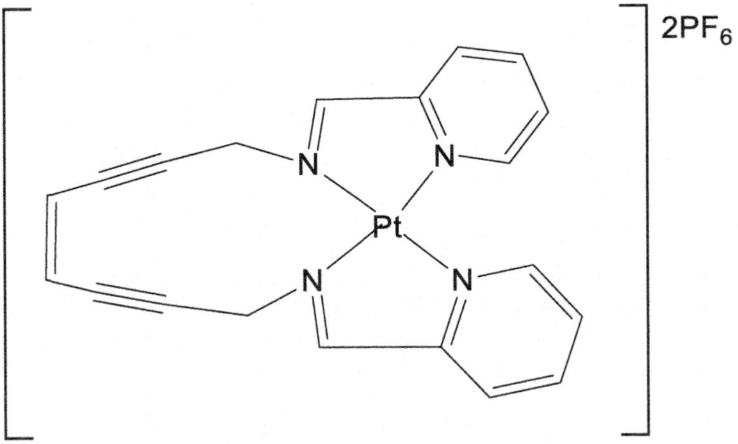

Pt7 (2PF$_6$)

10.5.5 Catalytic metallodrugs

The TM catalysts can deliver low-dose substances that can carry out biocatalytic reactions on either endogenous or exogenous substrate. Among the possible catalytic reactions are transfer hydrogenation, C−C bond cross-coupling, bond cleavage by hydrolysis or oxidation, azide-alkyne cycloaddition, and allyl carbamate cleavage. For a reaction to work in living systems, it needs to have high conversion rates under physiological conditions, good enantioselectivity, tolerance to aqueous environments, and resistance to nucleophilic poisoning [22].

10.5.5.1 Reduction enroute transfer hydrogenation

The biomolecules in cancer cells reduced by external catalysts have the potential to disrupt the metabolic cellular processes required for cell survival. Numerous cellular processes involving redox homeostasis require the cofactors nicotinamide adenine dinucleotide (NAD$^+$) and reduced

NADH. The NAD$^+$/NADH ratio in the cytosol of cancer cells is always lower than that of healthy cells. The "piano-stool" metal complexes Ir^{III} Cp*, Rh^{III} Cp*, and Ru^{II} arene have been used in cells to achieve transfer hydrogenation reactions of NAD$^+$/NADH. Formate can act as a hydride source at nontoxic concentrations. The catalytic mechanism includes the initial formation of M-formate adducts hydride transfer to the metal, release of CO$_2$, and hydride transfer to NAD$^+$. Noyori-type Ru^{II} arene complexes with sulfonamido ethylenediamines as ligands exhibit a 50-fold increase in cytotoxicity against human ovarian cancer cells when formate is present. Nonapoptotic cell death results from this type of catalysis, which also increases the reductant pool [22]. Lactate dehydrogenase and NADH convert pyruvate to lactate in hypoxic cancer cells. The 16-electron chiral catalytic Os^{II} complex (**Os1**) can convert pyruvate to synthetic D-lactate enantioselectively and preferentially for cancer cells in the presence of formate [22].

Os1

10.5.5.2 Oxidation of NADH and thiols

Metal complexes can also act as catalysts in the oxidation of NADH to NAD$^+$. For instance, half-sandwich Rh^{III} and Ir^{III} Cp* complexes reduce ketones when NADH is present as a cofactor. Ir^{III} complexes can produce H$_2$O$_2$ by transferring hydride to molecular oxygen. The Ir^{III} photocatalyst [Ir(ttpy)(pq)Cl]PF$_6$ (**Ir3**) can oxidize NADH and generate NAD radicals when exposed to light. Ir-Cl bond in **Ir6** is highly stable and resistant to hydrolysis under 463 nm radiations. Additionally, the equilibrium between reduced GSH and oxidized glutathione (GSSG), which varies from 30:1 to 100:1, is essential for mediating cellular redox processes. The first one-electron step in the catalysis of GSH oxidation to GSSG by arene Ru^{II}azpy complexes is the reduction of the azo-bond by GSH. At micromolar concentrations, the Ru^{II} catalyst converts millimolar GSH to GSSG. **Ir4** complex [Ir^{III}(Cp^{xbiph})(phpy)(py)]$^+$ with hydrosulfide ligands convert GSH to GSSG without the creation of Ir-SG adducts or the need for hydrolytic intermediaries.

10.5.5.3 Degradation and cleavage of biomacromolecules

Biomolecules can be broken down by strong Lewis acidic metal complexes both hydrolytically and oxidatively. Metal complexes with tetra-N-methylated cyclam (TMC) ligands can split at specific amino acid residues in protein, Aβ peptides associated with Alzheimer's disease can be catalytically split by metals $Co^{II} > Zn^{II} > Cu^{II} > Ni^{II}$ is its efficiency order. One of the amino-terminal copper/nickel (ATCUN) peptide binding motifs is CuII-GGHK-R (**Cu1**), where R is a target recognition sequence. This can catalyze the cleavage of viral RNA and G-quadruplex telomeric DNA, in which the redox-active Cu^{II}/Cu^{III} generates ROS to cleave bonds [22].

10.6 Summary

This chapter was aimed to introduce the less explored potential of organometallic compounds in the medicinal domain as synthetic drugs. Their inherent kinetic lability toward water, oxygen and temperature restricted their development as bioactive compounds. The careful design and development of organometallic compounds synergize the effect of desirable properties of metals with the corresponding carbon based bioactive scaffolds. Bio-organometallic compounds can have a vast diversity based on possible combinations of metal and carbon moieties. This peculiar property allows modulation in their bioactivities and potentiates the scope for their utilization as a replacement to platinum based anticancer compounds, chemo-preventive compounds or multitargeted diagnostic and therapeutic agents. Bioorganometallic compounds on account of novel reactivity routes and amazing catalytic type properties can register newer mechanisms of drug action with desired drug efficacies. Organometallic compounds can have distinctive features, such as in situ activation under physiological conditions or at the target site, selectivity in drug action stimulated by pH, concentration gradient, change in oxidation state, heat /mechanical, radiation a spin state change of the metal prompting a prospective pharmacological response favoring drug action. This chapter focuses on case studies of organo iron compounds as cytotoxic agents, organoselenium compounds as chemo-preventive and half sandwich ruthenium compounds as scaffolds of structure activity based modulation of cytotoxicity. A segment describing the different plausible mechanisms of metallo-drug action is also presented. The development of bioorganometallic compounds as multipronged drugs toward neurodegenerative disorders was described using organoselenium compounds as AchE inhibitors.

10.7 Problems with solutions

Q1. Which of the following statements regarding the Michaelis-Menten equation is incorrect?

a. The relationship between velocity and enzyme concentration is inverse.
b. Velocity is directly proportional to the turnover number.
c. As substrate concentration increases, velocity increases as well.
d. With increase in enzyme concentration, velocity also increases.

Answer: (a) The relationship between velocity and enzyme concentration is inverse.

Explanation: The Michaelis-Menten equation can be expressed as:
$V = kcat[E0][S]/Km + [S]$

As a result, the velocity is not inversely proportional to the enzyme concentration E0.

Q2. If adding 2 M of substrate results in a reaction rate of 10 mol/s, what is the highest reaction rate? (Km = 2 M)

a. 7.5 mol/s
b. 15 mol/s
c. 20 mol/s

d. Cannot be determined from the given information

Answer: (c) 20 mol/s

Explanation: The key to answering this question is seeing that the substrate concentration and Km are same. Remember that Km is the substrate concentration needed to achieve a reaction rate that is half to that of the maximum rate. Given that the substrate concentration and Km are both given as being the same, we can infer that the reaction rate of 10 mol/s is half of the maximum reaction rate, which is 20 mol/s.

Q3. What is the assumption of Michaelis-Menten model?

a. There might not be a rate-limiting step.
b. The enzyme-substrate complex breaks down into the product alone.
c. None of the reactions are catalyzed by enzymes.
d. Enzyme, substrate, and enzyme−substrate complex are in equilibrium.

Answer: (d) Enzyme, substrate, and enzyme−substrate complex are in equilibrium.

Explanation: In this model, an intermediate is formed when the substrate binds to an enzyme. The intermediate decomposes to an enzyme and a product, not just the product alone. The model requires enzyme-catalyzed reactions that include a rate-limiting step of the enzyme−substrate complex breaking down into the enzyme and product.

Q4. The ratio of Vo/Vmax when [S] = 4KM, is?

a. 1/2
b. 1/3
c. 4/5
d. 5/6

Answer: (d) 4/5

Explanation: The Michaelis-Menten equation is used to provide the following explanation:
Vo = Vmax[S]/[S] + KM
To find the desired ratio, rearrange the equation.
Vo/Vmax = [S]/[S] + KM
Substitute 4KM for [S] to simplify.
Vo/Vmax = 4/5

Q5. Assuming an enzyme with a KM of 1 mM, at what substrate concentration will the enzyme's speed reach 1/4 of its maximum speed? Vmax = 400 mmol/s

a. 0.12 mM
b. 0.5 mM

c. 0.33 mM
d. 0.20 mM

Answer: (c) 0.33 mM
Explanation: To solve this, we need the solution for [S] in the Michaelis-Menten equation:
Vo = Vmax[S]/Km + [S]
We know the following information:
KM = 1 mM
Vmax = 400mmols
Vo = 1/4Vmax = 100 mmol/s
Plug in these numbers and solve for substrate concentration.
100 = 400[S]/1 + [S]
[S] ≈ 0.33 mM

10.8 Objective type questions

i. Reason behind shifting from platinum based drugs to nonplatinum-based drugs is:
   a. Nonspecific cytotoxic action of platinum drugs.
   b. Ototoxicity
   c. Nephrotoxicity
   d. All of the above
   Answer: (d)

ii. The bioorganometallic copper complexes execute anticancer activity via:
   a. DNA binding
   b. Inhibiting topoisomerase
   c. Inhibiting proteasome enzyme
   d. All of the above
   Answer: (d)

iii. Possible cellular uptake mechanism of $Ru^{(III)}$ is:
   a. Passive diffusion
   b. Transferrin dependent transport
   c. Both a & b
   d. None above
   Answer: (c)

iv. Which is the bioactive form of Ruthenium:
   a. $Ru^{(III)}$
   b. $Ru^{(II)}$
   c. Both a and b
   d. None of them
   Answer: (b)

v. Mechanism of anticancer action of Se-containing compounds is:
   a. Apoptosis
   b. Morph gene expression
   c. Angiogenesis
   d. All of the above
   Answer: (d)

vi. Organoselenium compounds are desired over inorganic selenium compounds because:
   a. Kinetic stability
   b. Nonodorous nature
   c. Better bio availability
   d. All of the above
   Answer: (d)

vii. Which of the following complexes shows mechanism of action via hydrolysis:
   a. Square-planar Pt^{II} complexes
   b. Ti^{IV} dichloride complexes
   c. Ru^{II} and Os^{II} half-sandwich complexes
   d. All of the above
   Answer: (d)

viii. What is the challenging aspect of organometallic compounds in biosystems:
   a. Hydrolysis of metal-carbon bond
   b. Oxidation of metal-carbon bond
   c. Both a and b
   d. None of them
   Answer: (b)

ix. The first organometallic compound found in living system is:
   a. Coenzyme B_{12}
   b. Cisplatin
   c. Oxaliplatin
   d. None of them
   Answer: (a)

x. The metal present in Coenzyme B_{12}:
   a. Co
   b. Fe
   c. Ni
   d. Cu
   Answer: (a)

References

[1] Monney A, Albrecht M. Transition metal bioconjugates with an organometallic link between the metal and the biomolecular scaffold. Coord Chem Rev 2013;257(17−18):2420−33.

[2] Fish RH, Jaouen G. Bioorganometallic chemistry: structural diversity of organometallic complexes with bioligands and molecular recognition studies of several supramolecular hosts with biomolecules, alkali-metal ions, and organometallic pharmaceuticals. Organometallics 2003;22(11):2166−77.

[3] Charles R. Bioorganometallic chemistry of cobalt and nickel. Compr Coord Chem II 2004;8:677−713.

[4] (a) Rosenberg B, Vancamp L, Krigas T. Inhibition of cell division in Escherichia coli by electrolysis products from a platinum electrode. Nature 1965;205(1):698−9.
(b) Simpson PV, Desai NM, Casari I, Massi M, Falasca M. Metal-based antitumor compounds: beyond cisplatin. Future Med Chem 2019;11(2):119−35.
(c) Wu S-C, Lu C-Y, Chen Y-L, et al. Water-soluble dinitrosyl iron complex (DNIC): a nitric oxide vehicle triggering cancer cell death via apoptosis. Inorg Chem 2016;55(18):9383−92.
(d) Boodram JN, McGregor IJ, Bruno PM, Cressey PB, Hemann MT, Suntharalingam K. Breast cancer stem cell potent copper(II)−non-steroidal anti-inflammatory drug complexes. Angew Chem Int Ed 2016;55(8):2845−50.
(e) Mukherjee N, Podder S, Mitra K, Majumdar S, Nandi D, Chakravarty AR. Targeted photodynamic therapy in visible light using BODIPY-appended copper(II) complexes of a vitamin B6 Schiff base. Dalton Trans 2018;47(3):823−35.
(f) Zhang P, Sadler PJ. Advances in the design of organometallic anticancer complexes. J. Organomet. Chem. 2017;839:5−14.

[5] Ndagi U, Mhlongo N, Soliman ME. Metal complexes in cancer therapy - an update from drug design perspective. Drug Des Devel Ther 2017;11:599−616.

[6] (a) Yan YK, Melchart M, Habtemariam A, Sadler PJ. Chem Commun 2005;4764−76.
(b) Raja M, Samuelson AG. Substitution-modulated anticancer activity of *half-sandwich* ruthenium(II) complexes with heterocyclic ancillary ligands. Eur J Inorg Chem 2014;22:3536−46.

[7] Radomska D, Czarnomysy R, Radomski D, Bielawski K. Selenium compounds as novel potential anticancer agents. Int J Mol Sci 2021;22(3):1009.

[8] Baijiao A, Zhang S, Jinhui H, Tingting P, Huang L, Tang JC, et al. The design, synthesis and evaluation of seleniumcontaining 4-anilinoquinazoline hybrids as anticancer agents and a study of their mechanism. Org Biomol Chem 2018;16(25):4701−14.

[9] Zhang S, Baijiao A, Jiayan L, Jinhui H, Huang L, Xingshu L, et al. Synthesis and evaluation of selenium-containing indole chalcone and diarylketone derivatives as tubulin polymerization inhibition agents. Org Biomol Chem 2017;15(35):7404−10.

[10] Frieben EE, Amin S, Sharma AK. Development of isoselenocyanate compounds' syntheses and biological applications. J Med Chem 2019;62(11):5261−75.

[11] Doig AI, Tuck TA, LeBlanc B, Back TG. Synthesis, catalytic GPx-like activity, and SET reactions of conformationally constrained 2,7-dialkoxy-substituted naphthalene-1,8-peri-diselenides. ACS Omega 2022;7(31):27312−23.

[12] Li J, Jia W, Ma G, Zhang X, An S, Wang T, et al. Construction of pH sensitive smart glutathione peroxidase (GPx) mimics based on pH responsive pseudorotaxanes. Org Biomol Chem 2020;18(16):3125−34.

[13] Kalai T, Mugesh G, Roy G, Sies H, Berente Z, Hideg K. Combining benzo[d]isoselenazol-3-ones with sterically hindered alicyclic amines and nitroxides: enhanced activity as glutathione peroxidase mimics. Org Biomol Chem 2005;3(19):3564–9.

[14] Scheide MR, Schneider AR, Jardim GAM, Martins GM, Durigon DC, Saba S, et al. Electrochemical synthesis of selenyl-dihydrofurans via anodic selenofunctionalization of allyl-naphthol/phenol derivatives and their anti-alzheimer activity. Org Biomol Chem 2020;18(26):4916–21.

[15] Canto RF, Barbosa FA, Nascimento V, de Oliveira AS, Brighente IM, Braga AL. Design, synthesis and evaluation of seleno-dihydropyrimidinones as potential multi-targeted therapeutics for Alzheimer's disease. Org Biomol Chem 2014;12(21):3470–7.

[16] Kumar M, Chhillar B, Yadav M, Sagar P, Singhal NK, Gates PJ, et al. Catalytic and highly regenerable aminic organoselenium antioxidants with cytoprotective effects. Org Biomol Chem 2021;19(9):2015–22.

[17] Singh VP, Poon JF, Butcher RJ, Lu X, Mestres G, Ott MK, et al. Effect of a bromo substituent on the glutathione peroxidase activity of a pyridoxine-like diselenide. J Org Chem 2015;80(15):7385–95.

[18] Dong J, Li Z, Weng Z. Copper-catalyzed synthesis of 2,2,2-trifluoroethyl selenoethers and their insecticidal activities. Org Biomol Chem 2018;16(47):9269–73.

[19] Zhang M, Lu J, Weng W. Copper-catalyzed synthesis of 2,2-difluoro-1,3- benzoxathioles (selenoles) and their insecticidal activities. Org Biomol Chem 2018;16(24):4558–62.

[20] Alberto EE, Rossato LL, Alves SH, Alvesc D, Braga AL. Imidazolium ionic liquids containing selenium: synthesis and antimicrobial activity. Org Biomol Chem 2011;9(4):1001–3.

[21] Gai BM, Stein AL, Roehrs JA, Bilheri FN, Nogueira CW, Zeni G. Synthesis and antidepressant-like activity of selenophenes obtained via iron(III)−PhSeSePh-mediated cyclization of Z-selenoenynes. Org Biomol Chem 2012;10(4):798–807.

[22] Anthony EJ, Bolitho EM, Bridgewater HE, Carter OWL, Imberti C, Donnelly JM, et al. Metallodrugs are unique: opportunities and challenges of discovery and development. Chem Sci 2020;11(48):12888–917.

Chapter 11

Azole-based organometallic compounds as bioactive agents

Krishna[1], Deepak Yadav[2], Sunil Kumar[3], Meenakshi[3], Aman Kumar[4] and Vinod Kumar[4]

[1]*Department of Chemistry, University of Delhi, New Delhi, Delhi, India,* [2]*Department of Chemistry, Gurugram University, Gurugram, Haryana, India,* [3]*Department of Chemistry, Government P.G. College, Hisar, Haryana, India,* [4]*Department of Chemistry, Central University of Haryana, Mahendergarh, Haryana, India*

11.1 Introduction

Azoles are a group of compounds that contain a five-membered ring with at least one nitrogen atom. They possess a wide range of biomedical activities and are particularly known as antifungal agents which are commonly prescribed in clinical practice [1]. The azole ring systems may be of different types like oxadiazole, thiadiazole, triazole, pyrazole, imidazole, isoxazole, and other related rings [2]. Bacteria and fungi, due to their high rate of development and adaptation, develop resistance towards drugs used for their treatment and create health issues at the global level [3–10]. Azole-based compounds have been widely used for the treatment of fungal diseases for many years due to their properties, such as chemical stability, broad spectrum of action, and oral bioavailability [11]. As azole drugs are orally administered and have few side effects they have been used widely.

Azole-type compounds (e.g., fluconazole) are known for sterol biosynthesis inhibitors which prevent the growth of certain parasites and fungi by targeting the cytochrome P-450 dependent enzyme, sterol 14α-demethylase cytochrome P-450, preventing the synthesis of ergosterol, a major component of fungal plasma membranes [12]. The mechanism of action of azoles against fungi is based on the inhibition of ergosterol [11]. Ergosterol plays a hormone-like role in fungal cells, which stimulates growth, the net effect of azoles is the inhibition of the fungal growth. Azoles also show antibacterial activity by inhibiting the enoyl acyl carrier proteins reductase [13]. In the last few years, fungi have become a great global threat to forest, crops, and animals [14,15]. Every year, fungi kill more than 120 million tons of potatoes, maize, rice, wheat and soybean [15]. Azole-based antifungal drugs used to treat human mycoses have serious side effects [16] and use of azole-based agricultural pesticides has also been associated with the increased number of azole-resistant fungal species that affect humans [14]. In view of the mentioned problems, there is urgent need to develop new and potent antifungal compounds with alternative mechanisms of action.

The combination of azoles and metals is a promising strategy to develop new efficient drugs, even against drug-resistant pathogens [17−20]. The development of these new drugs is related to the complexation of azole derivatives with transition metals that show moderate toxicity and are biocompatible [18,21,22]. According to earlier investigations, the resulting complexes may have greater antibacterial activity than the corresponding free azoles ligands [23,24]. The pursuit of synergism by coupling a metal ion-containing moiety and a bioactive ligand is a recurring idea in bioinorganic chemistry. It has been suggested that binding of the metal ion, even when the ligand's active site appears to be blocked, can have a positive impact on biological activity through a variety of mechanisms including different modes of action, higher solubility, increased absorption, and even changed biological activity [25,26]. In this chapter, we have discussed different types of organometallic complexes containing azole moiety. These complexes facilitated comparative research to investigate how various metal−ligand ratios and antifungal agents may influence the biological properties of the compounds.

11.2 Azoles as antifungal drugs

In general, azole antifungal drugs inhibit cytochrome P-450 dependent enzymes involved in the biosynthesis of cell membrane and causing the production of toxic byproducts that are fatal to fungi. The azoles based antifungal drugs are divided into two subgroups: triazoles and imidazoles. Triazoles and imidazoles contain three and two nitrogen atoms in the five-membered heterocyclic ring, respectively.

i. The first report of antifungal activity of an azole compound, that is, benzimidazole, was described in 1944 by Woolley [27], but not employed in clinical setting until 1959 with the release of topical agents, such as ointments, Chlormidazole [28]. Imidazole-based drugs in the market are represented by Bifonazole (**1**), Econazole (**2**), Miconazole (**3**), Clotrimazole (**4**), Fenticonazole (**5**), Isoconazole (**6**), Sertaconazole (**7**), Tioconazole (**8**), Butoconazole (**9**), Luliconazole (**10**), Oxiconazole (**11**), Sulconazole (**12**), Omoconazole (**13**), and Ketoconazole (**14**) (Fig. 11.1). In the last 1960s, four most popular imidazoles compounds that well-established azoles as antifungal agents are Clotrimazole (**CTZ**), Tioconazole (**TCZ**), Ketoconazole (**KTZ**) and Miconazole (**MCZ**)

Azole-based organometallic compounds as bioactive agents **Chapter | 11** **289**

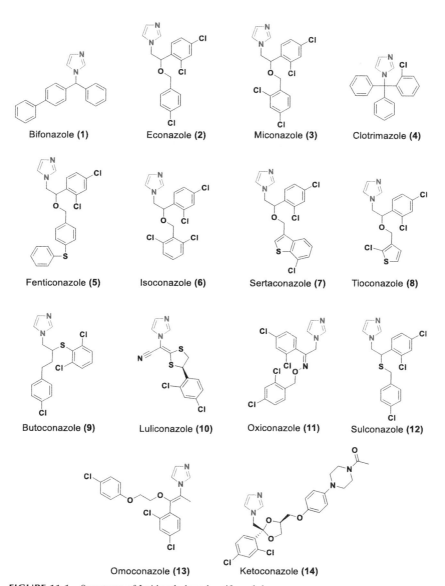

FIGURE 11.1 Structures of Imidazole-based antifungal drugs.

(Fig. 11.1). The first imidazole antifungal drug which hit the market was Clotrimazole and developed by Bayer AG (Germany). Antifungal drugs, Miconazole and Econazole, both are developed by Janseen Pharmaceutica (Beerse, Belgium) [28]. Ketoconazole (KTZ) was the first orally administrated imidazole fungicidal agents with abroad spectrum of activity against the treatment of systemic and superficial fungal infection.

It can be administered both orally and topically, especially in case of infections caused by *Histoplasma capsulatum and Blastomyces dermatitidis*, for which it is frequently used in nonimmuno compromised patients. Additionally, it is effective against dermatophyte infections, cutaneous candidiasis, and pityriasis versicolor [29]. It is also effective against mucosal candidiasis. As adverse gastrointestinal side effects more regularly noted, triazole was introduced to take its place.

ii. A single representative of triazole antifungals, abafungin is also known. When compared to imadazoles, triazoles generally have a wider range of antifungal efficacy and less toxicity. The triazole based antifungal drugs are Voriconazole (**VRC**) (**15**), Fluconazole (**FCZ**) (**16**), Ravuconazole (**RAV**) (**17**), Posaconazole (**POS**) (**18**), Itraconazole (**ITC**) (**19**), and Terconazole (**TER**) (**20**) (Fig. 11.2). Isavuconazole (**IAC**) (**21**) and Albaconazole (**ABC**) (**22**) are two novel antifungal compounds under investigation. Terconazole is a new broad-spectrum antifungal agent for the treatment of dermatomycoses and vaginal candidiasis and was the first triazole-based antifungal agent that specifically used to improve the

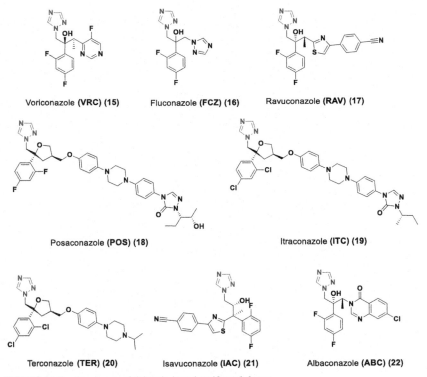

FIGURE 11.2 Structures of Triazole-based antifungal drugs.

antifungal activity for human use. For treating different types of systemic mycoses, the triazoles (Fluconazole and Itraconazole) have superseded amphotericin B as the gold standard for azoles. Fluconazole is an orally active triazole agent, that extremely popular for use as a single dose oral therapy for vulvovaginal candidiasis as well as widespread use as a prophylaxis and treatment for superficial and invasive *Candida* infection as well as *Cryptococcus* infections and several forms of coccidioidomycosis, as well as candidemia in nonneutropenic hosts [30]. Histoplasmosis, blastomycosis, coccidioidomycosis, consolidation therapy for cryptococcosis, and several aspergillosis types have also been successfully treated with Itraconazole. New triazoles, such as Voriconazole (VRC) and Posaconazole (POS) have also been approved for human use [31,32]. The emergence of high-level resistance is accompanied by widespread use of azole agents. Due to its high water solubility and poor affinity for plasma proteins, Fluconazole has a better bioavailability than other azoles. Fluconazole resistance in emerging *Candida species*, including global health threat *Candida auris*, has been reported extensively and is attributed to this drug being the most widely used azole [33]. Furthermore, Fluconazole-resistant clinical isolates of *Histoplasma capsulatum* and *Cryptococcus neoformans*, and Itraconazole-resistant clinical isolates of *Aspergillus fumigatus* have also been reported [34–36].

11.3 Mechanistic approach of azole-based organometallic complexes for infection

The primary targets for designing and producing antiinfection (antimicrobial) drugs are often enzymes that are involved in the biosynthesis pathways of microbial cell walls. Azole derivatives are used to create drugs because they have been found to be effective inhibitors of the glucosamine-6-phosphate synthase enzyme (GlcN-6-P synthase). In essence, it is necessary for the initial stage of the synthesis of N-acetyl-glucosamine, an important amino sugar that serves as a crucial building block for the cell walls of bacteria and fungi. The GlcN-6-P synthase enzyme is the one of the best targets for the development of antifungal and antibacterial drugs [37].

Similarly, ergosterol is the main component of fungal cell membranes which is responsible for maintaining membrane fluidity, structural stability, cell viability and acts as a permeability barrier. Antifungal substances generated from azoles target the ergosterol biosynthesis pathway and successfully inhibit the cytochrome P450 dependent sterol 14α-demethylase enzyme, which is necessary for the conversion of lanosterol to ergosterol. As a result, the decrease in ergosterol affects the consistency and systematic functions of fungal membranes, which inhibiting fungal cell growth [38,39].

11.4 Azole based metal-complexes and their use in medicinal chemistry

Several metals have an utmost priority in the human body. Metallodrugs can be used for both diagnosis and treatment where essential elements are involved in therapy in terms of homeostasis [40]. Probably the most popular and effective metal-based substance is the anticancer drug, cisplatin (cis-[PtCl$_2$(NH$_3$)$_2$]) by Barnett Rosenberg in 1960, which is able to bind with DNA or other specific biological targets and cause cancer cells to undergo apoptosis [41–45]. The biological actions of azoles can be further enhanced by formation of metal complexes. Azole based metal complexes have wide range of biological activities. Over the past few decades, there has been a significant increase in the use of metal complexes in a variety of medical domains, including in the area of neglected tropical diseases (NTDs) [46]. Multitarget drugs are recommended for NTDs, in this paradigm, metal-containing compounds have several advantages over carbon-based molecules. In actual, metal complexes may exert their effects *via* a variety of different routes by combining the pharmacological effects of the metal and azole-based ligands. Antitumor complexes that can bind and/or intercalate DNA have been among the first metal complexes tested for their antiparasitic activity due to the similarities between the metabolic pathways of trypanosomes and rapidly dividing tumor cells (e.g., high rate of aerobic glycolysis with key glycolytic enzyme, such as catalase, thioredoxin reductases, pyruvic kinase, hexokinase, phosphofructokinase, and need of nucleic acid biosynthesis) [46–49].

11.5 Recent advances of azole-based organometallic compounds

Many scientists consider as "organometallic" a compound that simply containing both an "organic moiety and a metal." However, following the IUPAC's criteria, a compound is considered to be an organometallic compound if it contains at least one bond between a metal atom and a carbon atom [50]. Why should organometallic molecules be so unique? The metal center is often low-valent, making it scarcely harmful as oxidant. Additionally, organometallic frequently exhibits kinetic inertness and maintains its integrity in solution. Due to this, organometallic may function as a whole compound, instead of releasing the metal ion and the organic ligand like coordination complexes. Finally, the relatively strong metal-carbon bond makes possible easy transformation of the organic moiety without modification of the structure of the organometallic core. Prodrugs, or chemicals that require chemical conversion by metabolic processes before becoming active pharmacological agents, are one specific applications for organometallic substances. It is exceedingly challenging to use CO safely as a gas, despite the fact that it has shown to have very important protective effects against inflammation, apoptosis, and endothelial oxidative damage. Then, transition metal carbonyls have proven to be versatile experimental CO releasing molecules in a controlled way, and are actively studied for clinical applications [51].

If organic components coupled with the metal is of some biological importance, the organometallic complex is considered as "bioorganometallic" sub-family. Whereas the term "medicinal bioorganometallic chemistry" refers to the biological assessment of organometallic for diagnostic or therapeutic purposes. Since 1985, when the phrase "bioorganometallic chemistry" first appeared in a publication, a specialized class of molecules have shown an rapid development [52–55]. Sanofi-Aventis worked on medicines for Malaria Venture to create the antimalarial compound ferroquine. Ferroquine entered in phase I in 2003 and it was able to conduct a phase II combination study (ferroquine + artefenomel) in 2015 [56]. This chapter will focus on the recent development in the use of azole-based transition metal organometallic compounds as bioactive agents.

11.5.1 Azole derivatives-based ruthenium organometallic complexes

The biological properties of organoruthenium complexes of antibacterial agents of the quinolone family were studied and discovered that in some cases the antibacterial potency of the complexes is not significantly altered. Some of the complexes possessed discrete toxicity against specific cancer cell lines and inhibited the activity of cathepsins which are involved in a variety of pathological processes, including cancer [57–59]. The R.A. Sanchez-Delgado group recently developed two series of novel ruthenium complexes bearing azole antifungal drugs Clotrimazole [60] and Ketoconazole [61]. Further their potential as antiparasitic agents was also assessed. It is known that ruthenium complexes are active against a variety of microorganisms that cause tropical diseases [62,63]. The organoruthenium complex with ofloxacin showed a minor amount of activity in leishmaniasis and Chagas diseases [59]. R. A. Sanchez-Delgado discovered that out of all the complexes, the organoruthenium species containing the aromatic cymene ligand were the most efficient against the two parasites while exhibiting no appreciable damage to human cells. The action against the parasites was lowest in nonorganometallic complexes, and it was marginally higher when the two chlorido ligands in organometallic compounds were replaced by bidentate ligands, such as bipy (2,2′-bipyridine), acac (acetylacetonato), or en (ethylenediamine).

Nine organoruthenium complexes conjugated with azole antifungal agents (L = Clotrimazole (CTZ), Miconazole (MCZ) and Tioconazole (TCZ)) (Fig. 11.3) with the general formula [η^6-(p-cymene)-RuCl$_2$(L)] (**23**), [η^6-(p-cymene)-RuCl(L)$_2$]Cl (**24**), and [η^6-(p-cymene)-Ru(L)$_3$](PF$_6$)$_2$ (**25**) (Fig. 11.3) were evaluated for their antifungal and antiparasitic activity by Kljun *et al.* in

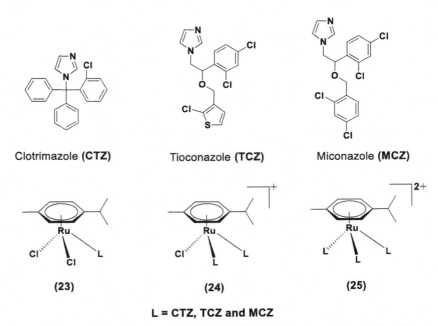

FIGURE 11.3 Mono-, *bis*-, and *tris*-imidazole based organoruthenium complexes for antifungal activity.

2014. Among these compounds, the complex [η^6-(*p*-cymene)-RuCl$_2$(CTZ)] was described by Sanchez-Delgado et al [60]. and complex [η^6-(*p*-cymene)-RuCl$_2$(MCZ)] by the Kljun group [64]. NMR spectroscopy was used to assess the stability of these metal-azole complexes in deuterated DMSO. The monoazole complexes showed partial DMSO-mediated ligands dissociation, resulting in a combination monoazole-ruthenium and DMSO-ruthenium species. However, it was found that bis and tris-azole were stable in both DMSO and aqueous solution, with only a small amount of decomposition being visible after 12 hours. At low millimolar concentrations, these mono-, bis-, and tris-azoles complexes prevented the growth of *Culvularialunata*; however, the effectiveness diminished with an increase in the number of coordinating ligands. The scientists hypothesized that these complexes must work in distinct ways because in azole ruthenium complexes N3 atom unable to bind to the heme iron within the active site of lanosterol 14α-demethylase [21].

Recently, Castonguay *et al.* revealed the antifungal properties of three Ru-cyclopentadienyl complexes, including two cationic compound [Ru-ACN]$^+$ (**27**) and [Ru-ATZ]$^+$ (**28**) and one neutral Ru-Cl (**26**) compound (Fig. 11.4) against five different *Candida* strains and one kind of *Cryptococcus species*. An aromatase inhibitor that contains triazoles is called Anastrozole (ATZ) that acts by blocking the production of estrogen in the treatment and prevention of breast cancer in woman. In contrast, [Ru-ACN]$^+$

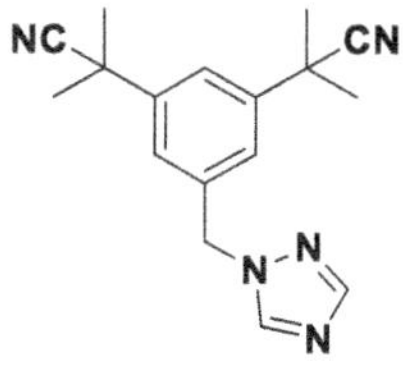

Azole-based organometallic compounds as bioactive agents Chapter | 11 295

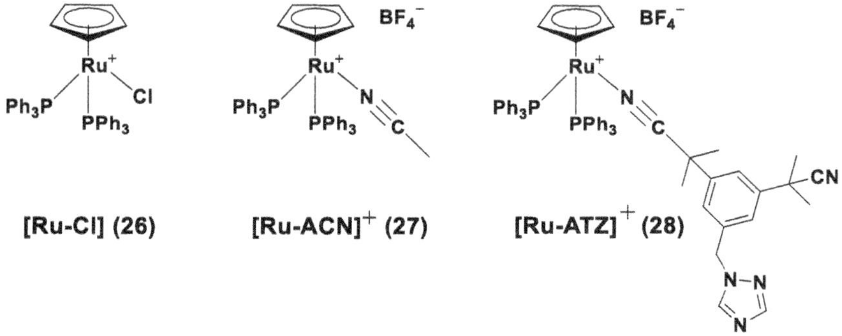

FIGURE 11.4 Ruthenium-cyclopentadienyl-based organometallic complexes containing **ATZ** Imidazole for antifungal activity.

(**27**) and [Ru-ATZ]$^+$ (**28**) complexes showed a significant fungal growth inhibition with MIC values ranging from 2.4 to 9.3 mM. At 20 mM, neutral Ru-Cl (**26**) complex, ATZ, and NaBH$_4$ revealed negligible antifungal activity. Both cationic species have shown lower MIC values then FCZ against the FCZ-resistant strains *C. glabrata* and *C. krusei*. It was discovered that [Ru-ATZ]$^+$ (**28**) was taken up by *C. glabrata* cells in a dose-dependent manner and produced sizable levels of intracellular ROS when it was applied to the cells. The Ru-Cl (**26**) compound has ROS (Reactive Oxygen Species) levels similar to untreated cells [65].

11.5.2 Manganese based organometallic complexes with azole ligands

Five manganese(I) tricarbonyl complexes with the general formula [Mn(CO)$_3$(N)(azole)] (**29**) (Fig. 11.5) incorporating a chelating 2,2'-bipyridine as well as a monodentate azole ligand were synthesized. The azole ligands were chosen from a variety of well-established antifungal drugs, such as Ketoconazole, Miconazole, and Clotrimazole, to investigate the affect of metal coordination on biological activity in the context of antibacterial and antiparasitic treatments. Their antiparasitic activity was tested against

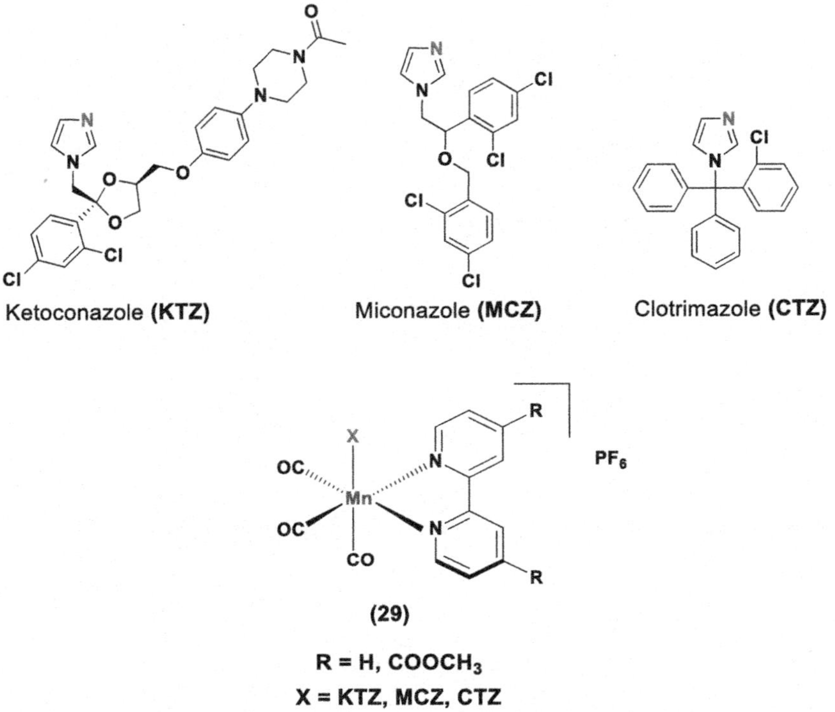

Ketoconazole (KTZ)   **Miconazole (MCZ)**   **Clotrimazole (CTZ)**

(29)

R = H, COOCH$_3$
X = KTZ, MCZ, CTZ

FIGURE 11.5 Manganese(I) tricarbonyl complexes with **KTZ**, **MCZ**, and **CTZ** azole for antitrypanosomal and antibacterial activity.

Trypanosoma brucei (*T. brucei*) and *Leishmania major* (*L. major*). Eight different bacterial strains, both Gram-positive and Gram-negative, were tested for antibacterial activity. While there was no effect on the latter microorganisms, however, CTZ complex exhibited submicromolar activity against *Staphylococcus aureus* and *S. epidermidis*, both of which had MIC values of 0.625 M. The tested compounds were completely inert against Gram-negative bacterial strains but exhibited low to submicromolar MIC values on Gram-positive bacteria, in particular on *S. aureus* and *S. epidermidis*. Metal complexes showed better activity (low micromolar IC$_{50}$ values) than the parent organic antifungals even though their SI in human embryonal kidney 293 T cells or murine macrophages J774.1 were not so high, especially in the case of *L. major* [66].

11.5.3 Copper-, gold-, and platinum-based organometallic complexes with azole derivatives

Navarro *et al.* reported a series of metal-based KTZ and CTZ derivatives (M = Au, Cu, and Pt). Excellent activity was shown by these metal complexes against *Trypanosoma cruzi*'s epimastigote form [67–69]. Further, the inhibitory effects of three previously reported metal-azole complexes, such as [Pt(CTZ)$_2$Cl$_2$] (**32b**), [Au(PPh$_3$)(KTZ)]PF$_6$ (**33a**), and [Au(PPh$_3$)(CTZ)]PF$_6$ (**33b**) and five newly reported metal-azole complexes, namely [Cu(PPh$_3$)$_2$(KTZ)$_2$]NO$_3$ (**30a**), [Cu(PPh$_3$)$_2$(CTZ)$_2$]NO$_3$ (**30b**), [Au(KTZ)$_2$]Cl (**31a**), [Au(CTZ)$_2$]Cl (**31b**), and [Pt(KTZ)$_2$]Cl$_2$ (**32a**) more investigated against *Sporothrix* strains, such as *S. schenckii*, *S. brasiliensis*, and *S. globose* (Fig. 11.6). The coordination geometries of these complexes (**30a-b**, **31a-b**, and **32a**) were deduced *via* use of a range of techniques. Complexes **30a/b** and **33a/b**, which contained phosphine as an auxiliary ligand, shown good activity with MIC values less than 2 nM against all tested isolates and were found more potent than parent drugs KTZ and CTZ. Complexes **33a** and **33b**, even able to kill *Sporothrix* spp. at lower concentration than the reference antifungal drugs (CTZ, KTZ). Scanning electron microscopy (SEM) was used to analyze the effect of four most active complexes (**30a**, **30b**, **33a**, and **33b**) on the morphology of *Sporothrix* species. Furthermore, the cytotoxicity tests on human red blood cells and mouse fibroblasts revealed that these metal complexes were more selective for fungus than mammalian cells. Metal complexes **30a**, **30b**, and **33b** proved to be promising fungistatic drug candidates, while complex **33a** was excluded due to its hemolytic effect [70].

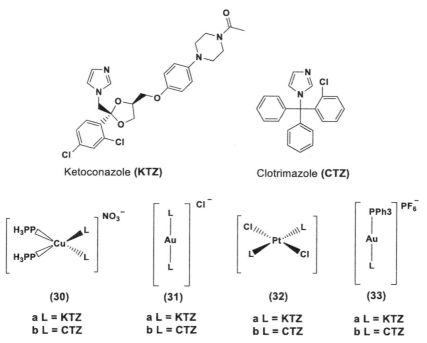

FIGURE 11.6 KTZ- and CTZ-based metal (Cu, Au and Pt) complexes for antifungal activity.

11.5.4 Copper- and zinc-based organometallic complexes with azole derivatives

Navarro et al. reported another series of Cu(II) and Zn(II) KTZ- and CTZ-containing complexes, including [Zn(KTZ)$_2$(Ac)$_2$].H$_2$O (**34a**), [Zn(KTZ)$_2$(Cl)$_2$]. (0.4)CH$_3$OH (**35a**), [Zn(KTZ)$_2$(H$_2$O)(NO$_3$)]NO$_3$ (**36a**), [Cu(KTZ)$_2$(Ac)$_2$].H$_2$O (**37a**), [Cu(KTZ)$_2$(Cl)$_2$].(3.2)H$_2$O (**38a**), [Cu(KTZ)$_2$(H$_2$O)(NO$_3$)](NO$_3$).H$_2$O (**39a**), [Zn(CTZ)$_2$(Ac)$_2$]0.4H$_2$O (**34b**), [Zn(CTZ)$_2$Cl$_2$] (**35b**), [Zn(CTZ)$_2$(H$_2$O)(NO$_3$)](NO$_3$)0.4H$_2$O (**36b**), [Cu(CTZ)$_2$(Ac)$_2$].H$_2$O (**37b**), [Cu(CTZ)$_2$Cl$_2$]0.2H$_2$O (**38b**), and [Cu(CTZ)$_2$(H$_2$O)(NO$_3$)](NO$_3$)0.2H$_2$O (**39b**) (Fig. 11.7). These metal complexes were tested for their antifungal properties against the fungi *Candida albicans*, *Candida neoformans*, and *Sclerotium brasiliense*. Against one or more species, all the metal complexes were discovered to be more effective than the parent drugs. Amongst all tested compounds, complexes **34a** and **36a**, which had MIC values ranging from 0.03−0.25 mM, were found the most effective against the three fungus species. These numbers were less than KTZ's (MIC = 0.125−0.5 mM) values. SEM was used to describe the surface-level changes in *S. brasiliensis* morphology that occurred after exposure to sublethal doses of complexes **34a** and **36a**. When compared to untreated cells, treatment with complexes **34a** and **36a** produced dramatic alterations, including enlarged cells, damaged cell walls, and loss of cellular structure. Cytotoxicity studies showed that compounds **34a** and **36a** were more toxic than KTZ against mammalian cells. However, when compared to KTZ, compound **34a** and **36a** showed identical SI values toward fungus over LLC-MK2 cells [71].

11.5.5 Metallocenyl derivatives with azole-based ligands

Iron is a critical micronutrient for the most living organisms as it is involved in several metabolic processes, such as the synthesis of proteins and iron-sulfur cluster-containing proteins, amino acids, DNA, lipids, and sterols [72,73]. Recently it was also shown that *Candida albicans* cell walls can be changed due to high iron concentrations which decreases the survival of fungal cells when exposed to antifungal drugs [74]. The discovery of ferrocene, a sandwich compound with two cyclopentadienyl (Cp) rings and one iron atom, launched the study of organometallic chemistry. The exceptional physicochemical characteristics of ferrocene makes it popular substituent in medicinal chemistry for a number of reasons: (1) the general lack of toxicity of ferrocene, which permits it to be injected, inhaled, or consumed orally without creating serious health issues; (2) Ferrocene's stability in aqueous and aerobic environments [75]; (3) the robustness of ferrocene, allowing for a straightforward accessibility of a large variety of derivatives; and (4) the lipophilicity of the ferrocene moiety makes it easier for ferrocenyl derivatives to cross cell membranes [76]. Biot et al. were able to produce ferroquine (FQ) by attaching a ferrocenyl moiety into the carbon chain of

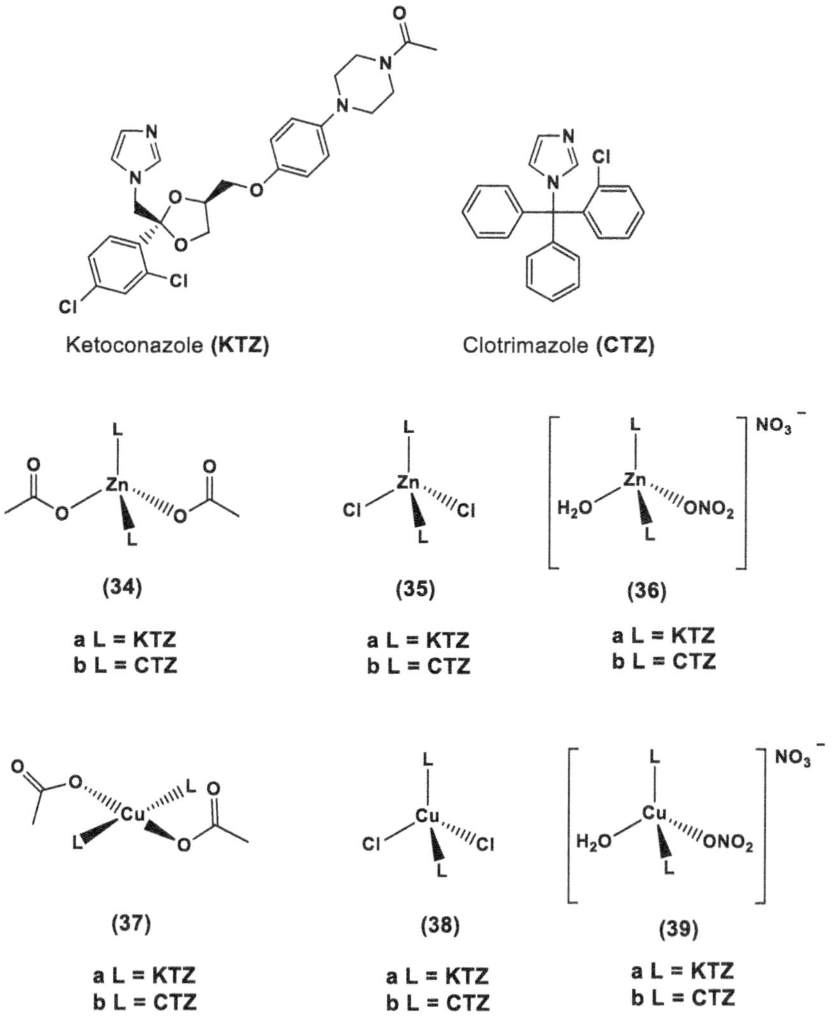

Azole-based organometallic compounds as bioactive agents Chapter | 11 299

FIGURE 11.7 KTZ- and CTZ-based metal (Cu and Zn) complexes for antifungal activity.

chloroquine (CQ). FQ was found to be active against both CQ-susceptible, CQ-resistant, culture-adapted strains of *Plasmodium falciparum* lineages by a proliferation test. Efficacy of an artemisinin-based combination therapy combining artesunate and FQ in malaria patients is now being investigated in phase IIb clinical trials using FQ [77].

The first example of implementation of this strategy in antifungal research was reported by Biot *et al.* in 2000. They created the first FCZ derivative (**40**) with ferrocene by swapping the 2,4-difluorophenyl ring of FCZ with a ferrocenyl moiety [78]. Fang and coworkers synthesized numerous ferrocene compounds with 1,2,4-triazole, thiazole, including ferrocenyl-substituted vinyl 1,2,4-triazole derivatives (**41Z, 41E**) [79], ferrocene-triadimefon analogs (**42**) [80], ferrocene-triadimentol analogs (**43**) [81], 1-ferrocenyl-3-aryl-2-(1H-1,2,4-triazol-1-yl)-prop-2-en-1-one derivatives (**44**) [82], and N-substituted benzylidene-4-ferrocenyl-5-(1H-1,2,4-triazol-1-yl)-1,3-thiazol-2-amine derivatives (**45**, Fig. 11.8) [83]. The *E*-isomer showed more inhibitory action than the *Z*-isomers for all drugs. There was no positive control provided for comparison, although the *E*-isomers of **41b-e**, **41g**, **41j**, and **41l** exhibited a 20%−40% increase in efficacy compared to the *Z*-isomer against brown rusts. It was reported that the *E*-isomer may have a better chance of binding to the receptor. The antifungal efficacy of two series of ferrocene-triadimefon and triadimenol derivatives were studied against powdery mildews and brown rusts. It was found that all the examined compounds exhibited lesser antifungal activity than the parent compounds, triadimefon or triadimenol. The substituted aryl groups were spatially attracted to and turned toward the neighboring triazole group due to the presence of the bulky ferrocene moiety as in compound **42f**.

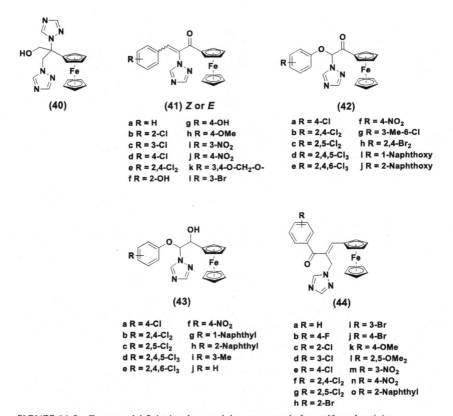

FIGURE 11.8 Ferrocenyl 1,2,4-triazole-containing compounds for antifungal activity.

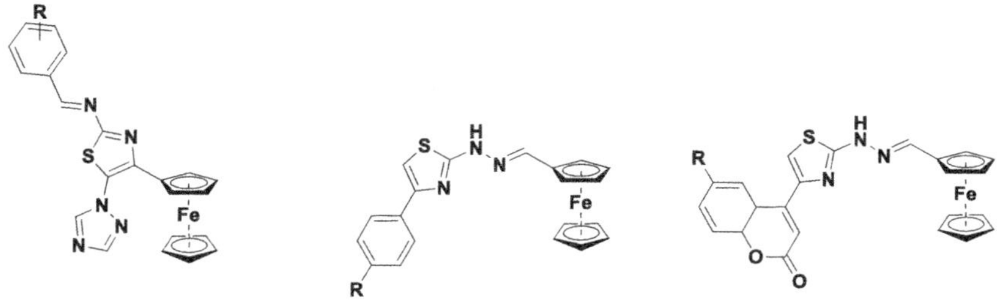

(45)
a R = H   j R = 2,4-di-OMe
b R = 4-F   k R = 4-Me
c R = 3-F   l R = 2-OH
d R = 2-Cl   m R = 2-NO$_2$
e R = 2-F   n R = 3-OH
f R = 4-Cl   o R = 4-Br
g R = 2,4-di-Cl p R = 4-OH
h R = 2,4-di-Me q R = 4-OMe
i R = 2-OMe

(46)
a R = H
b R = CH$_3$
c R = F
d R = Cl
e R = Br
f R = NO$_2$

(47)
a R = H
b R = Cl
c R = Br

FIGURE 11.9 Ferrocenyltriazole and thiazole derivatives for antifungal activity.

It was suggested that the reduced biological activity was caused by the large ferrocene near the triazole cycle.

Chandak et al. examined the antifungal activity against A. niger and A. flavus strains ferrocene-containing thiazole compounds **46** and **47** having phenyl and coumarin analogs, respectively (Fig. 11.9). It was reported that compounds **46e** exhibited a >55% reduction of antifungal activity against both strains. In contrast, FCZ (Fluconazole) demonstrated inhibition 75.3% and 74.6% against A. niger and A. flavus, respectively. Overall, it was shown that the phenyl analogs have greater antifungal activity than the coumarin analogs against both fungi [84].

11.6 Summary

The pharmacological effects of azole-based organometallic complexes have been investigated in this chapter. To examine the role of metals and chelating agents in coordination chemistry, a comparison of the biological activities of ligands and complexes was made. The prevalence of diseases is increasing along with the level of antibiotic resistance. In order to combat resistant antimicrobial strains with effective outcomes and minimal side effects, new drugs must be developed. Finally, metal-based drugs have been developed to address the problem of widespread resistance. They have greater biological activity when compared to Schiff base ligands and commercial organic-based drugs, such as fluconazole, streptomycin, and cisplatin.

This chapter highlights that ruthenium, manganese, copper, gold, platinum, zinc and iron are preferable for designing therapeutic drugs. The literature showed that chemists and pharmacists have been intrigued by azole derivatives and their metal complexes to use them as drugs. Additionally, it has been discovered that in some cases, cationic species, such as cationic Ru-cyclopentadienyl complexes, are more efficient than their neutral counterparts. Significant antifungal action was also seen in complexes with metal ions including manganese, copper, gold, platinum, zinc and iron, as well as ligands like azole based moiety.

11.7 Objective type questions

i What factors are responsible for biological activity of complexes?
  a Oxidation state of metal
  b Coordinated ligand
  c Lipophilicity of complexes
  d All of them
ii Which type of the complexes have antiparasitic potential?
  a [Mn(CO)$_3$(N − N)(azole)]
  b [Ru^{II}(η^6-p-cymene)Cl$_2$(CTZ)]
  c [cation]$^+$[Os^{IV}Cl$_5$(H azole)]$^-$
  d None of them
iii Most preferred coordination mode of azole to metal center is through ————- atom (s).
  a Nitrogen
  b Carbon
  c Both
  d None of them
iv What are the applications of azole-based organometallic compounds?
  a Catalytic
  b Organic light emitting diode
  c Both a and b
  d Neither a nor b
v Which mode of coordination is not preferred by azole motif?
  a η^5
  b η^6
  c η^2
  d η^1
vi What is the most preferred oxidation state of Ruthenium in azole-based organometallic compounds?
  a Ru(II)
  b Ru(I)
  c Ru(III)
  d Ru(IV)
vii What is the prerequisite for a complex to be organometallic?
  a M−C bond
  b M−N bond

c M—O bond
   d None of them
viii Some of the common reactions of organometallic compounds include
   a Oxidative addition
   b Reductive elimination
   c Substitution reactions
   d All of them

References

[1] Allen D, Wilson D, Drew R, Perfect J. Azole antifungals: 35 years of invasive fungal infection management. Expert Rev Anti Infect Ther 2015;13(6):787–98.

[2] Jain AK, Sharma S, Vaidya A, Ravichandran V, Agrawal RK. 1,3,4-Thiadiazole and its derivatives: a review on recent progress in biological activities. Chem Biol Drug Des 2013;81(5):557–76.

[3] Crump JA, Ramadhani HO, Morrissey AB, Saganda W, Mwako MS, Yang LY, et al. Invasive bacterial and fungal infections among hospitalized HIV-infected and HIV-uninfected adults and adolescents in northern tanzania. Clin Infect Dis 2011;52(3):341–8.

[4] Richardson MD, Warnock DW. Fungal infection: diagnosis and management. West Sussex, UK: John Wiley & Sons; 2012. p. 40–5.

[5] Browning DF, Busby SJ. Local and global regulation of transcription initiation in bacteria. Nat Rev Microbiol 2016;14(10):638–50.

[6] Kontoyiannis DP, Vaziri I, Hanna HA, Boktour M, Thornby J, Hachem R, et al. Raad II. Risk factors for candida tropicalis fungemia in patients with cancer. Clin Infect Dis 2001;33(10):1676–81.

[7] Kothavade RJ, Kura MM, Valand AG, Panthaki MH. Candida tropicalis: its prevalence, pathogenicity and increasing resistance to fluconazole. J Med Microbiol 2010;59(8):873–80.

[8] Silva S, Negri M, Henriques M, Oliveira R, Williams DW, Azeredo J. Candida glabrata, candida parapsilosis and candida tropicalis: biology, epidemiology, pathogenicity and antifungal resistance. FEMS Microbiol Rev 2012;36(2):288–305.

[9] Bihan CL, Zahar JR, Timsit JF. Staphylococcus aureus transmission in the intensive care unit: the potential role of the healthcare worker carriage. Ann Infect 2017;1:1–5.

[10] Price JR, Cole K, Bexley A, Kostiou V, Eyre DW, Golubchik T, et al. Transmission of staphylococcus aureus between health-care workers, the environment, and patients in an intensive care unit: a longitudinal cohort study based on whole-genome sequencing. Lancet Infect Dis 2017;17(2):207–14.

[11] Kathiravan MK, Salake AB, Chothe AS, Dudhe PB, Watode RP, Mukta MS, et al. The biology and chemistry of antifungal agents: a review. Bioorg Med Chem 2012;20(19):5678–98.

[12] Sheehan DJ, Hitchcock CA, Sibley CM. Current and emerging azole antifungal agents. Clin Microbiol Rev 1999;12(1):40–79.

[13] Joshi SD, Dixit SR, More UA, Aminabhavi T, Kulkarni VH, Gadad AK. Enoyl-ACP reductase as effective target for the synthesized novel antitubercular drugs: a-state-of-the-art. Mini Rev Med Chem 2014;14(8):678–93.

[14] Fisher MC, Henk D, Briggs CJ, Brownstein JS, Madoff LC, McCraw SL, et al. Emerging fungal threats to animal, plant and ecosystem health. Nature 2012;484(7393):186–94.

[15] Kupferschmidt K. Attack of the clones. Science 2012;337(6095):636–8.

[16] Horvat S, McWhir J, Rozman D. Defects in cholesterol synthesis genes in mouse and in humans: lessons for drug development and safer treatments. Drug Metab Rev 2011; 43(1):69−90.

[17] El-Gammal OA, Bekheit MM, Tahoon M. Synthesis, characterization and biological activity of 2-acetylpyridine-α-naphthoxyacetylhydrazone its metal complexes. Spectrochim Acta A Mol Biomol Spectrosc 2015;135:597−607.

[18] Murcia RA, Leal SM, Roa MV, Nagles E, Muñoz-Castro A, Hurtado JJ. Development of antibacterial and antifungal triazole chromium (III) and cobalt (II) complexes: synthesis and biological activity evaluations. Molecules 2018;23(8):2013.

[19] Alaghaz AN, Ammar RA. New dimeric cyclodiphosph (V) azane complexes of Cr (III), Co (II), Ni (II), Cu (II), and Zn (II): preparation, characterization and biological activity studies. Eur J Med Chem 2010;45(4):1314−22.

[20] Tarafder MT, Ali MA, Wee DJ, Azahari K, Silong S, Crouse KA. Complexes of a tridentate ONS schiff base. synthesis and biological properties. Transit Met Chem 2000;25(4):456−60.

[21] Kljun J, Scott AJ, LanišnikRižner T, Keiser J, Turel I. Synthesis and biological evaluation of organoruthenium complexes with azole antifungal agents. first crystal structure of a tioconazole metal complex. Organometallics 2014;33(7):1594−601.

[22] Bello-Vieda NJ, Pastrana HF, Garavito MF, Ávila AG, Celis AM, Muñoz-Castro A, et al. Antibacterial activities of azole complexes combined with silver nanoparticles. Molecules 2018;23(2):361.

[23] Pahontu E, Fala V, Gulea A, Poirier D, Tapcov V, Rosu T. Synthesis and characterization of some new Cu (II), Ni (II) and Zn (II) complexes with salicylidene thiosemicarbazones: antibacterial, antifungal and in vitro antileukemia activity. Molecules 2013;18(8):8812−36.

[24] Yousef TA, El-Reash GA, Al-Jahdali M, El-Rakhawy EB. Synthesis, spectral characterization and biological evaluation of Mn (II), Co (II), Ni (II), Cu (II), Zn (II) and Cd (II) complexes with thiosemicarbazone ending by pyrazole and pyridyl rings. Spectrochim Acta A Mol Biomol Spectrosc 2014;129:163−72.

[25] Turel I, Kljun J. Interactions of metal ions with DNA, its constituents and derivatives, which may be relevant for anticancer research. Curr Top Med Chem 2011;11(21):2661−87.

[26] Turel I. The interactions of metal ions with quinolone antibacterial agents. Coord Chem Rev 2002;232(1−2):27−47.

[27] Woolley DW. Some biological effects produced by benzimidazole and their reversal by purines. J Biol Chem 1944;152(2):225−32.

[28] Fromtling RA. Overview of medically important antifungal azole derivatives. Clin Microbiol Rev 1988;1(2):187−217.

[29] Field MC, Horn D, Fairlamb AH, Ferguson MA, Gray DW, Read KD, et al. Antitrypanosomatid drug discovery: an ongoing challenge and a continuing need. Nat Rev Microbiol 2017;15(4):217−31.

[30] Groll AH, Piscitelli SC, Walsh TJ. Clinical pharmacology of systemic antifungal agents: a comprehensive review of agents in clinical use, current investigational compounds, and putative targets for antifungal drug development. Adv Pharmacol 1998;44:343−500.

[31] Pascual A, Calandra T, Bolay S, Buclin T, Bille J, Marchetti O. Voriconazole. Therapeutic drug monitoring in patients with invasive mycoses improves efficacy and safety outcomes. Clin Infect Dis 2008;46(2):201−11.

[32] Schiller DS, Fung HB. Posaconazole: an extended-spectrum triazole antifungal agent. Clin Ther 2007;29(9):1862−86.

[33] Rex JH, Rinaldi MG, Pfaller MA. Resistance of Candida species to fluconazole. Antimicrob Agents Chemother 1995;39(1):1−8.

[34] Venkateswarlu K, Taylor M, Manning NJ, Rinaldi MG, Kelly SL. Fluconazole tolerance in clinical isolates of cryptococcus neoformans. Antimicrob Agents Chemother 1997; 41(4):748−51.
[35] Wheat J, Marichal P, Vanden Bossche H, Le Monte A, Connolly P. Hypothesis on the mechanism of resistance to fluconazole in histoplasma capsulatum. Antimicrob Agents Chemother 1997;41(2):410−14.
[36] Denning DW, Venkateswarlu K, Oakley KL, Anderson MJ, Manning NJ, Stevens DA, et al. Itraconazole resistance in aspergillus fumigatus. Antimicrob Agents Chemother 1997;41(6):1364−8.
[37] Aouad MR, Mayaba MM, Naqvi A, Bardaweel SK, Al-Blewi FF, Messali M, et al. Design, synthesis, in silico and in vitro antimicrobial screenings of novel 1, 2, 4-triazoles carrying 1, 2, 3-triazole scaffold with lipophilic side chain tether. Chem Cent J 2017; 11(1):1−3.
[38] Tatsumi Y, Nagashima M, Shibanushi T, Iwata A, Kangawa Y, Inui F, et al. Mechanism of action of efinaconazole, a novel triazole antifungal agent. Antimicrob Agents Chemother 2013;57(5):2405−9.
[39] Nath PS, Ashish P, Rupesh M. Triazole: a potential bioactive agent (synthesis and biological activity). Int J Res Ayurveda Pharm 2011;2:1490−4.
[40] Barry NP, Sadler PJ. Exploration of the medical periodic table: towards new targets. Chem Comm 2013;49(45):5106−31.
[41] Ghosh S. Cisplatin: the first metal based anticancer drug. Bioorg Chem 2019;88:102925.
[42] Komeda S, Casini A. Next-generation anticancer metallodrugs. Curr Top Med Chem 2012;12(3):219−35.
[43] De Almeida A, Oliveira BL, Correia JD, Soveral G, Casini A. Emerging protein targets for metal-based pharmaceutical agents: an update. Coord Chem Rev 2013;257(19−20):2689−704.
[44] Mjos KD, Orvig C. Metallodrugs in medicinal inorganic chemistry. Chem Rev 2014; 114(8):4540−63.
[45] Medici S, Peana M, Nurchi VM, Lachowicz JI, Crisponi G, Zoroddu MA. Noble metals in medicine: latest advances. Coord Chem Rev 2015;284:329−50.
[46] Sánchez-Delgado RA, Anzellotti A. Metal complexes as chemotherapeutic agents against tropical diseases: trypanosomiasis, malaria and leishmaniasis. Mini Rev Med Chem 2004;4(1):23−30.
[47] Williamson J, Scott-Finnigan TJ. Trypanocidal activity of antitumor antibiotics and other metabolic inhibitors. Antimicrob Agents Chemother 1978;13(5):735−44.
[48] Kinnamon KE, Steck EA, Rane DS. Activity of antitumor drugs against African trypanosomes. Antimicrob Agents Chemother 1979;15(2):157−60.
[49] Farrell NP, Williamson J, McLaren DJ. Trypanocidal and antitumour activity of platinum-metal and platinum-metal-drug dual-function complexes. Biochem pharmacol 1984; 33(7):961−71.
[50] Connelly NG, Damhus T, Hartshorn RM, Hutton AT. Nomenclature of inorganic chemistry: IUPAC recommendations 2005. Cambridge, UK: *Royal Society of Chemistry Publishing/IUPAC*; 2005.
[51] Schatzschneider U. Novel lead structures and activation mechanisms for CO-releasing molecules (CORMs). Br J Pharmacol 2015;172(6):1638−50.
[52] Hartinger CG, Dyson PJ. Bioorganometallic chemistry-from teaching paradigms to medicinal applications. Chem Soc Rev 2009;38(2):391−401.
[53] Hillard EA, Jaouen G. Bioorganometallics: future trends in drug discovery, analytical chemistry, and catalysis. Organometallics 2011;30(1):20−7.

[54] Gasser G, Metzler-Nolte N. The potential of organometallic complexes in medicinal chemistry. Curr Opin Chem Biol 2012;16(1−2):84−91.
[55] Arrais A, Gabano E, Ravera M, Osella D. Transition metal carbonyl clusters in biology: a futile or niche research area? Inorganica Chim Acta 2018;470:3−10.
[56] Held J, Jeyaraj S, Kreidenweiss A. Antimalarial compounds in phase II clinical development. Expert Opin Investig Drugs 2015;24(3):363−82.
[57] Hudej R, Kljun J, Kandioller W, Repnik U, Turk B, Hartinger CG, et al. Synthesis and biological evaluation of the thionated antibacterial agent nalidixic acid and its organoruthenium (II) complex. Organometallics 2012;31(16):5867−74.
[58] Kljun J, Bytzek AK, Kandioller W, Bartel C, Jakupec MA, Hartinger CG, et al. Physicochemical studies and anticancer potency of ruthenium η6-p-cymene complexes containing antibacterial quinolones. Organometallics 2011;30(9):2506−12.
[59] Turel I, Kljun J, Perdih F, Morozova E, Bakulev V, Kasyanenko N, et al. First ruthenium organometallic complex of antibacterial agent ofloxacin. crystal structure and interactions with DNA. Inorg chem 2010;49(23):10750−2.
[60] Martínez A, Carreon T, Iniguez E, Anzellotti A, Sánchez A, Tyan M, et al. Searching for new chemotherapies for tropical diseases: ruthenium−clotrimazole complexes display high in vitro activity against leishmania major and trypanosoma cruzi and low toxicity toward normal mammalian cells. J Med Chem 2012;55(8):3867−77.
[61] Iniguez E, Sánchez A, Vasquez MA, Martínez A, Olivas J, Sattler A, et al. Metal−drug synergy: new ruthenium (II) complexes of ketoconazole are highly active against leishmania major and trypanosoma cruzi and nontoxic to human or murine normal cells. J Biol Inorg Chem 2013;18(7):779−90.
[62] Demoro B, Sarniguet C, Sánchez-Delgado R, Rossi M, Liebowitz D, Caruso F, et al. New organoruthenium complexes with bioactive thiosemicarbazones as co-ligands: potential anti-trypanosomal agents. Dalton Trans 2012;41(5):1534−43.
[63] Glans L, Ehnbom A, de Kock C, Martínez A, Estrada J, Smith PJ, et al. Ruthenium (II) arene complexes with chelating chloroquine analogue ligands: synthesis, characterization and in vitro antimalarial activity. Dalton Trans 2012;41(9):2764−73.
[64] Patra M, Joshi T, Pierroz V, Ingram K, Kaiser M, Ferrari S, et al. DMSO-mediated ligand dissociation: renaissance for biological activity of N-heterocyclic-[Ru (η6-arene) Cl₂] drug candidates. Eur J Chem 2013;19(44):14768−72.
[65] Golbaghi G, Groleau MC, Lopez de los Santos Y, Doucet N, Déziel E, Castonguay A. Cationic Ru (II) cyclopentadienyl complexes with antifungal activity against several candida species. Chembiochem 2020;21(21):3112−19.
[66] Simpson PV, Nagel C, Bruhn H, Schatzschneider U. Antibacterial and antiparasitic activity of manganese (I) tricarbonyl complexes with ketoconazole, miconazole, and clotrimazole ligands. Organometallics 2015;34(15):3809−15.
[67] Navarro M, Lehmann T, Cisneros-Fajardo EJ, Fuentes A, Sánchez-Delgado RA, Silva P, et al. Toward a novel metal-based chemotherapy against tropical diseases. Part 5. synthesis and characterization of new Ru (II) and Ru (III) clotrimazole and ketoconazole complexes and evaluation of their activity against trypanosoma cruzi. Polyhedron 2000; 19(22−23):2319−25.
[68] Navarro M, Cisneros-Fajardo EJ, Lehmann T, Sánchez-Delgado RA, Atencio R, Silva P, et al. Toward a novel metal-based chemotherapy against tropical diseases. 6. synthesis and characterization of new copper (II) and gold (I) clotrimazole and ketoconazole complexes and evaluation of their activity against trypanosoma cruzi. Inorg Chem 2001; 40(27):6879−84.

[69] Sánchez-Delgado RA, Navarro M, Lazardi K, Atencio R, Capparelli M, Vargas F, et al. Toward a novel metal based chemotherapy against tropical diseases 4. synthesis and characterization of new metal-clotrimazole complexes and evaluation of their activity against trypanosoma cruzi. Inorganica Chim Acta 1998;275:528–40.

[70] Gagini T, Colina-Vegas L, Villarreal W, Borba-Santos LP, de Souza Pereira C, Batista AA, et al. Metal−azole fungistatic drug complexes as anti-sporothrix spp. agents. N J Chem 2018;42(16):13641–50.

[71] de Azevedo-França JA, Borba-Santos LP, de Almeida Pimentel G, Franco CH, Souza C, de Almeida Celestino J, et al. Antifungal promising agents of zinc (II) and copper (II) derivatives based on azole drug. J Inorg Biochem 2021;219:111401.

[72] Misslinger M, Lechner BE, Bacher K, Haas H. Iron-sensing is governed by mitochondrial, not by cytosolic iron−sulfur cluster biogenesis in aspergillus fumigatus. Metallomics 2018;10(11):1687–700.

[73] Schaible UE, Kaufmann SH. Iron and microbial infection. Nat Rev Microbiol 2004;2(12):946–53.

[74] Tripathi A, Liverani E, Tsygankov AY, Puri S. Iron alters the cell wall composition and intracellular lactate to affect candida albicans susceptibility to antifungals and host immune response. J Biol Chem 2020;295(29):10032–44.

[75] Van Staveren DR, Metzler-Nolte N. Bioorganometallic chemistry of ferrocene. Chem Rev 2004;104(12):5931–86.

[76] Hartinger CG, Metzler-Nolte N, Dyson PJ. Challenges and opportunities in the development of organometallic anticancer drugs. Organometallics 2012;31(16):5677–85.

[77] Biot C, Glorian G, Maciejewski LA, Brocard JS, Domarle O, Blampain G, et al. Synthesis and antimalarial activity in vitro and in vivo of a new ferrocene-chloroquine analogue. J Med Chem 1997;40(23):3715–18.

[78] Biot C, François N, Maciejewski L, Brocard J, Poulain D. Synthesis and antifungal activity of a ferrocene-fluconazole analogue. Bioorg Med Chem Lett 2000;10(8):839–41.

[79] Fang JX, Jin Z, Li ZM, Liu W. Preparation, characterization and biological activities of novel ferrocenyl-substituted azaheterocycle compounds. Appl Organomet Chem 2003;17(3):145–53.

[80] Jin Z, Hu Y, Huo A, Tao W, Shao L, Liu J, et al. Synthesis, characterization, and biological evaluation of novel ferrocene-triadimefon analogues. J Organomet Chem 2006;691(11):2340–5.

[81] Fang J, Jin Z, Hu Y, Tao W, Shao L. Synthesis and evaluation of novel ferrocene-substituted triadimenol analogues. Appl Organomet Chem 2006;20(12):813–18.

[82] Liu J, Liu T, Dai H, Jin Z, Fang J. Synthesis, structure and biological activity studies of 2-[(1H-1, 2, 4-triazol-1-yl) methyl]-1-aryl-3-ferrocenyl prop-2-en-1-one derivatives. Appl Organomet Chem 2006;20(10):610–14.

[83] Yu H, Shao L, Fang J. Synthesis and biological activity research of novel ferrocenyl-containing thiazole imine derivatives. J Organomet Chem 2007;692(5):991–6.

[84] Chandak N, Kumar P, Sharma C, Aneja KR, K Sharma P. Synthesis and biological evaluation of some novel thiazolylhydrazinomethylideneferrocenes as antimicrobial agents. Lett Drug Des Discov 2012;9(1):63–8.

Chapter 12

Hybrid organometallic compounds as potent antimalarial agents

Preeti Singh[1], Yadav Preeti[2], Badri Parshad[3], Deepak Yadav[4], Sushmita[5] and Manjeet Kumar[6]

[1]*Department of Chemistry, Swami Vivekanand Subharti University, Meerut, Uttar Pradesh, India,* [2]*Department of Chemistry, Maitreyi College, University of Delhi, New Delhi, Delhi, India,* [3]*Wellman Center for Photomedicine, Massachusetts General Hospital, Harvard Medical School, Boston, MA, United States,* [4]*Department of Chemistry, Gurugram University, Gurugram, Haryana, India,* [5]*Department of Chemistry, Netaji Subhas University of Technology, Dwarka, Delhi, India,* [6]*Department of Chemistry, Central University of Haryana, Mahendergarh, Haryana, India*

12.1 Introduction

The biological evolution of the malaria parasite makes it a highly coordinated and well-adapted species that has maintained resistance to therapy over hundreds of years [1]. The term "Malaria" originated from the Italian words; mala means bad, and aria means air [2]. The malaria parasites were evidenced in blood for the first time by Alphonse Laveran and Patrick Mansion (1844–1922) and the role of mosquitoes as a vector in the transmission of the malady was revealed, and their investigations were further continued by Ronald Ross (1857–1932) [3–5]. Ronald Ross received Nobel Prize in Medicine in 1902 for his research on malaria parasite and its life cycle in mosquitoes. In the same decades, three scientists named Giovanni Batista Grassi (1854–1925), Amico Bignami (1862–1919), and Giuseppi Bastianelli (1862–1959) investigated the development stages of the malaria parasite in vector [6,7].

In the 17th century, cinchona bark was commonly used by Jesuit priests for the treatment of ague (fever) in Europe [8]. Quinine, first isolated in 1820 from the cinchona bark, had been used for a longer time as an herbal remedy and it was the first successful drug that was frequently used to treat malaria [9,10]. From the early 19th to 20th century, herbal treatment of malaria was substituted by drug therapy [11]. In 1891 Paul Ehrlich identified methylene blue as a possible drug against malaria (Fig. 12.1) and it continued to be used as antimalarial drug until world war II, when two major side effects, that is, blue sclera and turning the urine green or blue were observed in the soldiers [12]. The malaria parasite was considered to be the first pathogenic microbe that could develop drug resistance capability for several decades [13]. Later in the 19−20th century, when malaria was very common in several parts of North America and Europe, the malaria disease played a considerable role to oppose the expansion of the European nation into the tropical world [14].

Malaria, a mosquito-borne infectious disease, is widely distributed in tropical and subtropical regions around the world [15]. It is one of the deadliest diseases along with tuberculosis, cancer, and heart diseases. Malaria has been found to infect human population for over 50,000 years. The symptoms usually associated with malaria include fever, headache, fatigue and vomiting and in severe case it can also cause yellow skin, seizures or even death. As per the recent World malaria report, there were 227 million cases of malaria in 2019 which increased to 241 million in 2020 and caused estimated 627,000 deaths in 2020 with an increase of ∼69000 deaths that were reported in the previous year. Around two-thirds of these increased deaths, that is, ∼47,000 were caused by disruptions in providing the malaria diagnosis, prevention, and treatment facilities during the COVID-19 pandemic [16].

Open water collection served as the breeding grounds for female *Anopheles* mosquitoes which assists as the vector for malaria [17]. Malaria mainly affects pregnant women and children and an average a child dies every minute because of this disease [18]. African sub-Saharan countries continue to be highest malaria affected regions, accounting for ∼95% of all cases and ∼96% of all malaria deaths in 2020. Approximately 80% of deaths in these regions are of children under five years of age. Mainly the four *Plasmodium* species viz *Plasmodium falciparum*, *Plasmodium ovale*, *Plasmodium vivax*, and *Plasmodium malariae* were found to cause malaria in humans [19]. There is also fifth species of *Plasmodium*, that is, *Plasmodium knowlesi* (common in South East Asia) which is also responsible for malaria and frequently diagnosed as *P. malariae* [20]. Worldwide, more than 50% of cases of malaria are

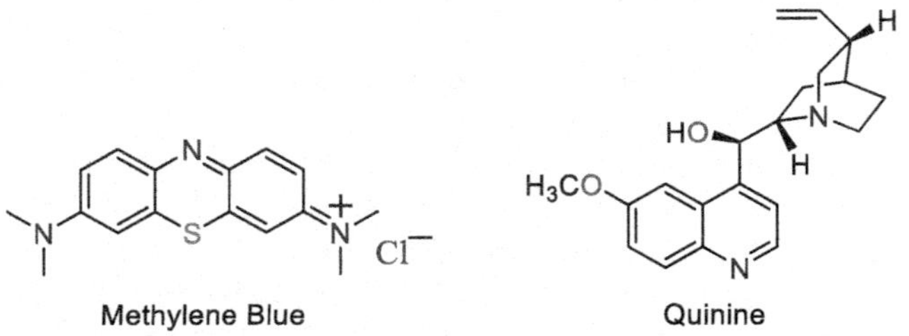

FIGURE 12.1 Molecular structures of methylene blue and quinine.

caused by *P. falciparum* and it is considered to be the deadliest among all the four species whereas the second most effective is *P. vivax* [21,22].

12.2 Life cycle of malaria

It was observed that the malaria involves the cyclical infection of female *Anopheles* mosquitoes and humans. The malaria parasites grow in human body which multiply in the liver cells and then in the red blood cells (RBCs) (Fig. 12.2). After that, the successive broods of the malaria parasite develop in the RBCs which destroys them and release the daughter parasite termed as "merozoites" that continues the malaria cycle by invading other red cells [23].

In the malaria life cycle, the symptoms of malaria are caused by blood stage parasites. Certain forms of the blood stage parasites such as gametocytes are ingested by the female *Anopheles* mosquito during blood feeding, then they begin the cycle of growth, and the multiplication of parasites takes place. After that in 10–18 days, sporozoite which is one of the forms of the malaria parasite migrates into the salivary glands of the mosquito. Then, the female *Anopheles* takes the blood from the other human body, the anticoagulant saliva is injected together with sporozoites that migrates into the liver and begins a new malaria cycle. In this way, an infected *Anopheles* mosquito transfers the disease from one to another human by acting as a "vector" [24]. In this process, unlike the human host, the mosquito vector was found not to be suffered from the presence of parasites.

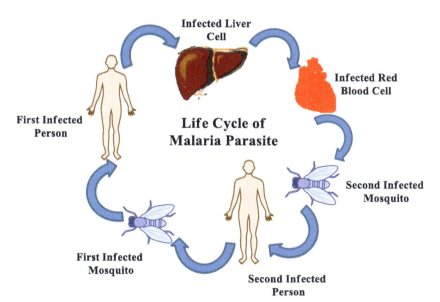

FIGURE 12.2 Life cycle of the malaria parasite.

This reflects that the life cycle of the malaria parasite combines two phases: one is sexual (mosquito) and another is asexual (human). When the sporozoite parasite infects the liver cell and matures into the schizonts it results in the rupture which then releases the merozoites [25]. Later on, there is asexual multiplication of the malaria parasite in the erythrocytes. The merozoites are responsible for the infection of RBCs. Then the trophozoites matured into schizonts and rupture the merozoites. Several malaria parasites are differentiated into the sexual erythrocytic stages which are known as gametocytes. These gametocytes are further divided into male, that is, microgametocytes, and female, that is, macrogametocytes. During a blood meal, these gametocytes are then ingested by the *Anopheles* mosquito. This is said to be a sporogonic cycle when there is a multiplication of parasites in the mosquito. The microgametes penetrate the stomach of the mosquito which generates zygotes. These developed zygotes become motile and penetrate the midgut wall of the Anopheles where they grow up into the oocysts. These oocysts grow, rupture, and then release sporozoites, which make their way to the salivary glands of the mosquito. The penetration of the sporozoites into the new human host continues the malaria life cycle [26].

12.3 Some common examples of antimalarial drugs

Recently, a number of studies have shown that people suffering from the deadliest diseases such as malaria, tuberculosis, cancer, and multiresistant infections have sharply hiked, which leaves the researchers without any alternative but to look for new treatment methodologies as well as strategies. For the treatment of malaria, chloroquine and its derivative hydroxychloroquine are among the oldest antimalarial clinical drugs. These drugs along with quinine also act as reference molecules and played a major role in the development of other antimalarial compounds such as mefloquine, for the treatment of malaria with better efficacy (Fig. 12.3).

During the mechanism of action, the uncharged form of chloroquine freely diffuses and accumulates in the food vacuole of the parasite and subsequently gets protonated because of the acidic pH of the food vacuole. Afterward, the protonated species binds with the specific receptor (e.g., heme) in the food vacuole [27–29]. One of the major roles of chloroquine is to inhibit the conversion

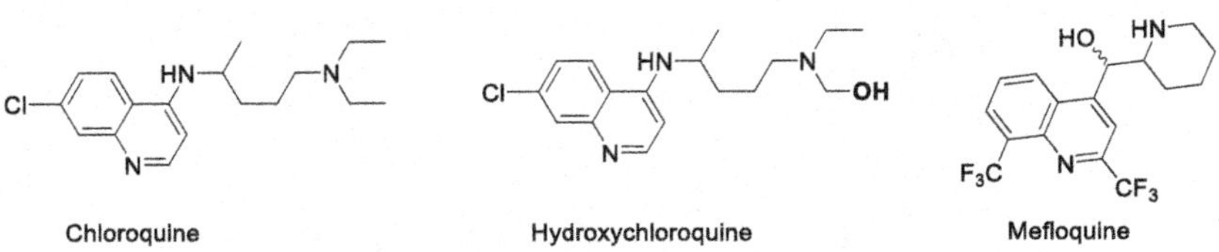

FIGURE 12.3 Molecular structures of some important antimalarial drugs; chloroquine, hydroxychloroquine and mefloquine.

Hybrid organometallic compounds as potent antimalarial agents Chapter | 12   313

of free heme (released by the digestion of hemoglobin) to hemozoin. The presence of free heme causes the lyses of the membrane which ultimately leads to the death of the parasite.

It was observed that the chloroquine resistance would be associated with the decreased accumulation of the chloroquine owing to an increased level of drug efflux in the food vacuole. Chloroquine was used frequently till the 1960s, but the rise of *P. falciparum* resistance afterward leads the drug discovery to a standstill [30,31]. This parasite is among the first microbial pathogens that displayed the behavior of drug resistance for several years and its resistance against chloroquine is nearly all over the world where *P. falciparum* is transmitted. Later, *P. falciparum* was found to develop resistance against almost all the other currently available drugs such as mefloquine, halofantrine, sulfadoxine/pyrimethamine, and quinine.

A breakthrough in antimalarial drugs after the failure of chloroquine was the introduction of artemisinin, a natural product isolated from *Artemisia annua* that exhibit very low toxicity and very high antiplasmodial potency against *P. falciparum*[32,33]. *Artemisia annua* has been used in traditional Chinese medicines for hundreds of years owing to its potent medicinal values. Tu Youyou for the first time isolated the artemisinin from *Artemisia annua* in 1970s and studied its effect on malaria and she was awarded with 2015 Nobel Prize in Physiology or Medicine for his work. Artemisinin is a sesquiterpene lactone containing an endoperoxide moiety. The antimalarial activity of artemisinin toward cerebral malaria is because of the presence of this endoperoxide moiety (Fig. 12.4).

The mechanism of action of artemisinin is always under debate. The parasite's food vacuole that is rich in hemoglobin Fe^{2+} and exogenous Fe^{2+} ions is one of the most considerable pieces of evidence to activate the mechanism of action of artemisinin in which the heme can act as a primary activator as shown in Fig. 12.5 [34].

In this mechanism, Fe(II) ion can either attack on O1 or O2 of the artemisinin unit which then produces two different types of highly reactive intermediates known as oxy radicals, which are responsible for the deterioration of the parasite but the exactly targeted moiety is still on the dispute [35,36]. According to previous theories, to generate the reactive oxy radical artemisinin interact with

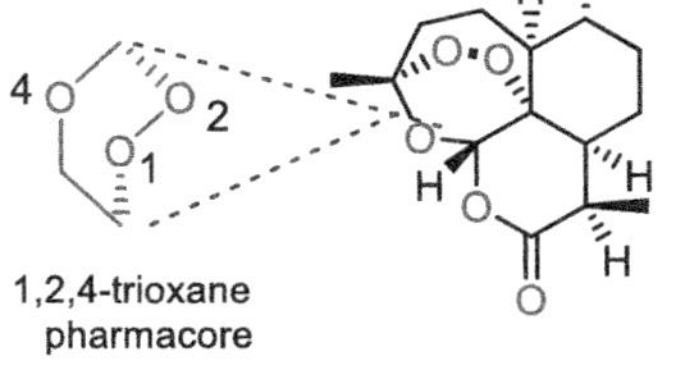

FIGURE 12.4   1,2,4-Trioxane scaffold of artemisinin.

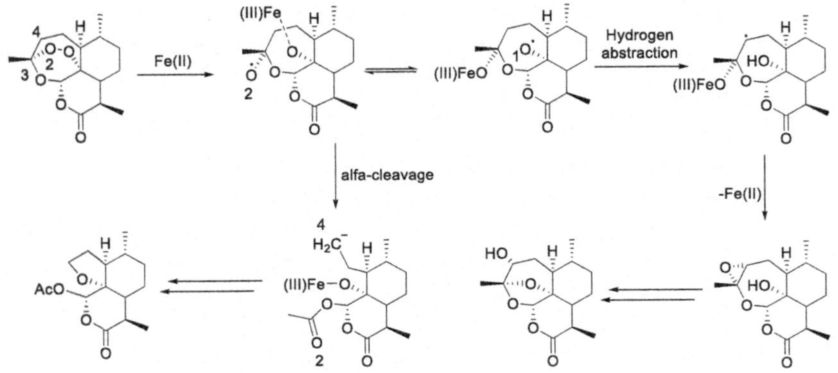

FIGURE 12.5 Mechanism of action of artemisinin.

iron-rich biomolecules such as heme in the food vacuole of the parasite which fazed the growth of the parasite by inhibiting PfATP6.

Hemoglobin (Fe^{2+}) is observed to be degraded by various successive actions of two enzymes namely protease and plasmepsin in the parasite's digestive vacuole which resulted in oxidized hematin (Fe^{3+}) as an end product [37,38], which is then responsible for the elimination of proteins and amino acids. The formation of the end product is highly toxic for the parasite; hence, parasites exhibit a new mode of mechanism. In this mechanism, hematin gets changed to an insoluble nontoxic hemozoin (Fe^{3+}) which is termed as malaria pigment by the biomineralization method. Recently resistance toward artemisinin has emerged in some parts of Southeast Asia, limiting the efficacy of this important antimalarial drug [39].

12.4 An approach to hybrid antimalarial drugs

The emergence of resistance in *P. falciparum* species against treatment has always been a challenging situation. To overcome the issue, various approaches, like covalent biotherapy [40], and artemisinin combination therapy [41], were introduced from time to time. Presently, artemisinin combination therapy is the first-line treatment in most malaria endemic countries. As artemisinin combination therapy is the last arsenal to malaria; hence, there is an urgent need to find new and more potent antimalarial approach.

Recently a highly acknowledged and successful approach known as "hybrid drug approach," where two or more active pharmacophores are introduced in a single target molecule has been introduced [42]. While incorporating properties of all the combined molecules the hybrid compound usually display significantly improved properties and offer extra advantages of more controlled pharmacokinetics. Fig. 12.6 represents some of the important organic hybrid molecules which behave as potential antimalarial candidates.

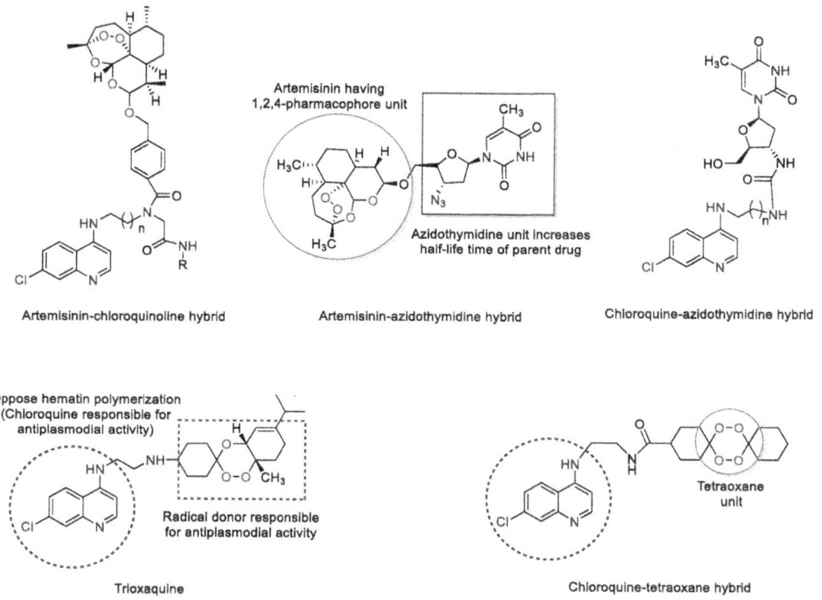

FIGURE 12.6 Molecular structures of some important antimalarial drug candidates.

As demonstrated by researchers, the hybrid approach based on drugs with independent targets offers the best way to lower the resistance of the malaria parasite [40,43]. Nowadays, scientists are focusing more on hybrid molecules for drug discovery because of their excellent potency toward biological profile, for example, they exhibit specific selectivity in addition to the advanced mode of action, that is, dual/multiple mode of action. The advantages of the hybrid drugs also include improved activity, selective toxicity, and good bioavailability. Several advantages are noted as dosage compliance, cheap preclinical evaluation, lower toxicity, and the ability to provide more and better combinations of drugs that are responsible for delaying resistant development in the parasite.

Hybrid drugs provide enormous modifications toward the design and synthesis of various drugs along with excellent potency against endemic malaria. Hybrids of artemisinin, that is, artemisinin-quinine [44], trioxane such as trioxaquine [45,46], and acridine such as acridine-endoperoxide [47] revealed improved in vitro activities in micro to nanomolar range with lower cytotoxicity as compared to their precursors.

12.5 Hybrid organometallic antimalarial compounds

Because of the widespread resistance to most of the available drugs, now it has been considered a major setback for treating malaria diseases more effectively. This situation led to the introduction of organometallic compounds as antimalarial candidates. A large number of organometallic compounds have shown considerable potential in the drug discovery of hybrid antimalarial drugs for the treatment of malaria. Ferrocene, an organometallic compound which is formed by the combination of two cyclopentadienyl ions with an iron (Fe^{2+}), found to display physicochemical properties that exhibit favorable effects on the living organism. However, the poor bioavailability of ferrocene and its related compounds pushed up for the strategy for the discovery of new drug candidates as antimalarial agents.

Among them, one of the successful strategies in drug discovery has been the development of hybrid organometallic molecules which involves the hybridization of organometallic moiety with other useful or active pharmacophore. This hybrid strategy has proven to display the tendency to work on multiresistant diseases more effectively. If one considers about organometallic species, there are a number of studies available on combination of iron, rhodium, and ruthenium-based organometallic compounds with chloroquine. These types of combined drugs showed enhanced antimalarial activity against the various strains of *P. falciparum* especially against chloroquine-resistant strains.

Ferroquine is a combined analog of ferrocene and chloroquine and is considered one of the first and most successful organometallic derivatives of chloroquine as an antimaterial agent (Fig. 12.7). It comprises a ferrocenyl group which is covalently linked to a 4-aminoquinoline and a basic alkylamine moiety. The envisaged structure of ferroquine was created to act similarly to chloroquine to inhibit the formation of hemozoin by residing in the food vacuole of the parasite [48,49]. Hence, it was considered to behave similarly to chloroquine in the treatment of malaria, as it adopts a similar mechanism. Ferroquine is found to be a promising drug for the treatment of malaria as the parasite has not developed any resistance to this antimalarial drug yet.

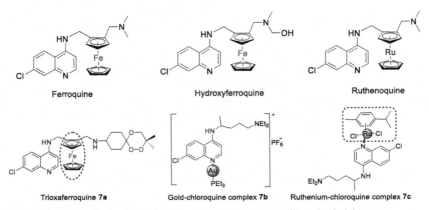

FIGURE 12.7 Molecular structures of some hybrid organometallic antimalarial agents.

Currently, ferroquine is under phase II clinical trial in humans against the uncomplicated *P. falciparum* [50–52]. Not only ferroquine but its derivatives have also been reported to show promising antimalarial potential [53].

Several analogs of ferroquine such as hydroxyferroquine and ferrocenophanes are also reported to display enhanced antimalarial activity against various strains of the *P. falciparum* parasite as compared to that of parent drug candidates [54]. Furthermore, the drug candidates containing ruthenium in place of iron in ferroquine known as ruthenoquine were also found to exhibit better activity than that of the ferroquine against some of the chloroquine-resistant plasmodium parasite strains. A number of gold-chloroquine complexes have also been developed and evaluated against different strains of *P. falciparum* in vitro [55]. Gold-chloroquine complex **7b** showed potent activity against chloroquine-resistant FcB1 strains which is five times higher than the control chloroquine diphosphate.

Due to the inherent biological properties, versatile nature of metal chelating properties and, the possibility to undergo further modification, that hybrid organometallic compounds are emerging as a new class of potential antimalarial agents.

12.6 Recent developments with examples: some representative hybrid organometallic antimalarial agents

12.6.1 Trioxaferroquine containing hybrid organometallic drug

Dual functional hybrid drug ferroquine consisting of ferrocene and chloroquine units was introduced by Sanofi et al. [56] Later on, Bellot et al. extended the concept of hybrid molecules and subsequently designed another hybrid molecule trioxoferroquine **8a** which consists of ferroquine and trioxane motifs and was found to be even more active than ferroquine [57]. Fig. 12.8 represents the structures of trioxaferroquine **8a** and trioxaferrocene **8b** hybrid organometallic compounds.

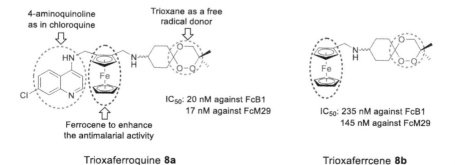

FIGURE 12.8 Structures of trioxaferroquine and trioxaferrocene hybrid organometallic compounds.

In the representative structure (Fig. 12.8), ferrocene and trioxane moieties are connected through the secondary amine (–NH–) linkage in the molecule. This linkage is further responsible for the enhanced biological activity of the compound. The desired molecule can form cis and trans stereoisomers and its biological activity can be tuned by restricting the ratio of cis to trans stereoisomers [58]. Trioxaferroquine 8a displayed excellent antiplasmodial activity with IC_{50} values of 20 and 17 nM against the two different strains of *P. falciparum*, that is, FcB1 and FcM29, respectively. While trixaferrocene **8b** derivative without a quinoline moiety exhibit lower antimalarial activity against (IC_{50}: FcB1–235 nM and FcM29–145 nM). The superior *in vitro* activity of the trioxaferroquine hybrid over trioxaferrocene demonstrate the importance of quinoline moiety which also plays a crucial role in enhancing the antimalarial activity even against chloroquine-resistant strains.

12.6.2 Ferrocene-quinoline containing hybrid organometallic compounds

Ferroquine mimic ferrocene-quinoline containing hybrid organometallic compounds were synthesized by Minić et al. and screened for antiplasmodial activity against the chloroquine-resistant and chloroquine-sensitive strains of *P. falciparum* [59]. The *in vitro* results displayed that three of the synthesized hybrid compounds **9a-c** exhibited antiplasmodial activity in the low micromolar range against chloroquine-sensitive NF54 strain. In addition to this, compounds **9b** and **9c** demonstrated potential antiplasmodial behavior against chloroquine-resistance strains as well with IC_{50} values of 0.77 and 0.82 μM, respectively. Fig. 12.9 represents the ferrocene-quinoline containing hybrid organometallic complexes.

12.6.3 Organoruthenium aminoquinoline–trioxane hybrid

Ruthenium-based organometallic compounds have been developed and studied recently in various biological applications [60,61]. Their low toxicity and

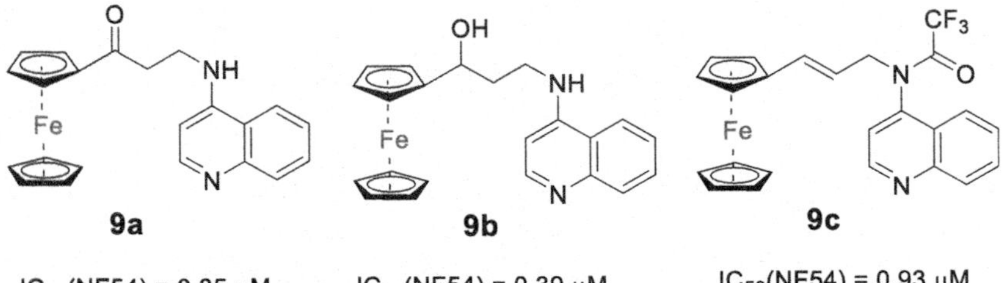

9a

$IC_{50}(NF54) = 0.85\ \mu M$
$IC_{50}(K1) = 3.91\ \mu M$
$IC_{50}(CHO) = 251.68\ \mu M$

9b

$IC_{50}(NF54) = 0.39\ \mu M$
$IC_{50}(K1) = 0.77\ \mu M$
$IC_{50}(CHO) = 23.14\ \mu M$

9c

$IC_{50}(NF54) = 0.93\ \mu M$
$IC_{50}(K1) = 0.82\ \mu M$
$IC_{50}(CHO) = 19.90\ \mu M$

FIGURE 12.9 Ferrocene-quinoline containing hybrid organometallic compounds with antiplasmodial activity.

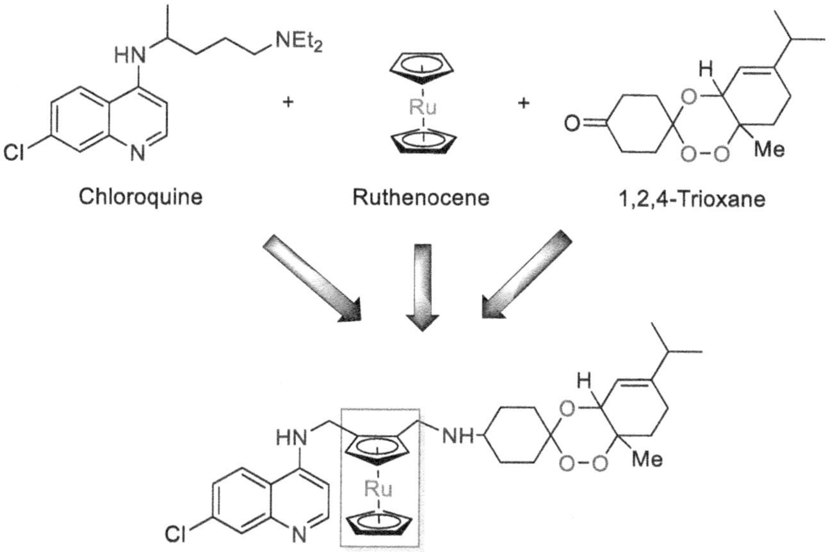

FIGURE 12.10 Molecular structure of ruthenocene containing hybrid organometallic compound.

good biocompatibility make them a promising candidate for the further medicinal applications against malaria. A ruthenium-based organometallic hybrid compound (**10**) containing 4-aminoquinoline, ruthenocene, and 1,2,4-trioxane molecule in a single molecule was synthesized and evaluated for antimalarial activity by Martínez et al. (Fig. 12.10) [62].

In order to ascertain the importance of each entity incorporated in hybrid 10, its antimalarial activity was evaluated against different chloroquine-resistant and chloroquine-sensitive strains of *P. falciparum* and compared with the active drugs such as artemisinin and chloroquine and with those of precursors; trioxane ketone and ruthenocene-trioxane without any aminoquinoline moiety. The antimalarial activity of hybrid 10 was found to be better than the parent trioxane, chloroquine, and ruthenocene-trioxane compounds. It exhibits potential antimalarial activity with low nanomolar IC_{50} value and low toxicity toward healthy mammalian cells. In this context, the results validate the proposed hypothesis that the hybrid compound exhibits enhanced activity.

12.6.4 Organoiridium based antimalarial hybrid complexes

Iridium-based organometallic compounds are among the most important platinum group coordination compounds for their excellent pharmacological activity. As the antimalarial, the quinoline-based organoiridium compounds

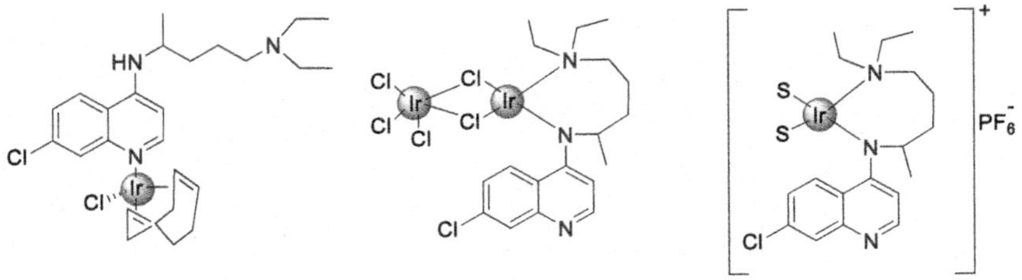

FIGURE 12.11 Chemical structures of iridium-chloroquine complexes (**11a-c**).

were brought into the limelight in 2009 by Navarro and coworkers. They incorporated the iridium into the quinolinyl scaffold to accomplish newer antiplasmodial iridium complexes **11a–c** (Fig. 12.11) [63]. These modified compounds displayed improved efficacy when compared to the parent drug candidate against the *P. berghei* strain of malaria.

Because of the increased lipophilicity and higher stability, the incorporation of iridium metal into quinoline-based compounds is favored. It was observed that the complexation of cyclooctadiene-based iridium compound with chloroquine gave rise to overall enhanced lipophilicity which led to increased retention within the active site of the parasite, that is, the digestive vacuole and subsequently improved activity. Comparison of IC_{50} values indicated that the complex **11a** showed higher activity than chloroquine diphosphate while the complex **11c** showed lower activity. Moreover, compound **11b** did not show any change in activity, that is, it displayed activity similar to that of chloroquine diphosphate.

12.6.5 Organorhodium-based antimalarial complexes

Organorhodium complexes, especially when rhodium is present in +3 oxidation state, have been often overlooked in biological applications because of the perception that they might exhibit limited biological activities as they are chemically inert. However, recent studies have shown that rhodium complexes have displayed growing interest among biologists as well as chemists to further expand the antimalarial drug arsenal. Rh(I)-chloroquine conjugate was one of the first examples of an antimalarial hybrid rhodium complex which was formed by coordinating the RhCl(COD) unit where COD is 1,5-cyclooctadiene linking through the N-atom of the quinoline nucleus. It exhibited the enhanced antimalarial activity when compared with the control drug in vivo against *P. berghei* strain.

Various heteronuclear ferrocenyl azine complexes **12a–c** were reported with the rhodium based organometallic motif linked through the bidentate NO-coordination [64]. The compound without rhodium found to exhibit weak to moderate activity against chloroquine-sensitive NF54 and

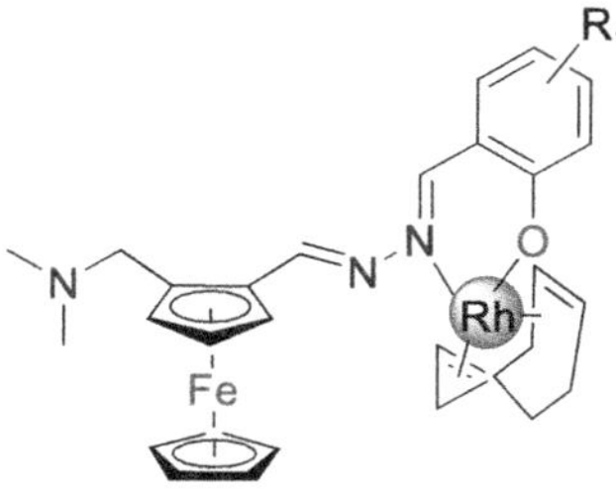

12a; R₁ = 5-Cl
12b; R₁ = 3-OMe
12c; R₁ = H

FIGURE 12.12 Structural representation of Rh(I)COD motif via bidentate N̂O-coordination.

chloroquine-resistant K1 strains of *P. falciparum* while the rhodium complexation increased the activity against both strains (Fig. 12.12) [65]. The synthesized compounds showed antimalarial activities in the low micromolar range against both strains of *P. falciparum* and low resistance indices for **12a–c** (RI = 0.22–0.85). While the haemozoin formation inhibition was considered the possible action mechanism of these compounds, the exact target has not been reported yet.

12.6.6 Organoosmium-based antimalarial complexes

Osmium is considered one of the interesting candidates which exhibit several biological applications. Osmium complexes exist in various oxidation states such as +2, +3, and +4 or higher. The tendency to exist in various oxidation states enables the better tuning of redox properties of the organoosmium complexes which is accountable for the regulation of redox-dependent biological processes like the generation of reactive oxygen species in the cell of the targeted pathogen [66]. The versatility of the organoosmium complexes has become grown with great promise for utilization in cancer treatment [67,68].

Despite having the considerable potential applications of osmium complexes, the antimalarial potential of these molecules has been studied very less. Till now, only a few studies demonstrating the antiplasmodial activity of osmium complexes are reported. In one of the reports, Nordlander and coworkers evaluated the antimalarial potential of organoosmium complexes **13a–c** and their organoruthenium analog complexes **13d–f** (Fig. 12.13) in vitro against the chloroquine-sensitive(NF54 and D10) and chloroquine-resistant (Dd2) *P. falciparum* strains [69]. Coordination of osmium and ruthenium arene moieties to the parent ligands resulted in decreased antiplasmodial activities as compared to the parent ligand while the resistance index was found to be better for ruthenium complexes as compared to chloroquine [70]. Overall, the ruthenium complexes displayed better antimalarial activity than their corresponding osmium analogs. Based on these results, it can be concluded that the osmium-based organometallic complexes need further attention in the field of malaria treatment.

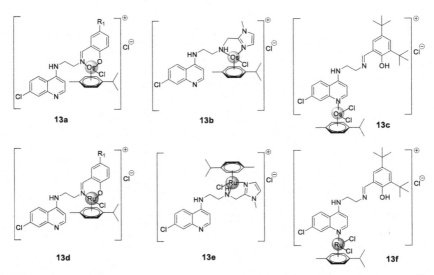

FIGURE 12.13 Representative structures of the ruthenium- and osmium-based organometallic compounds.

12.6.7 Organoplatinum-based antimalarial complexes

Platinum is another commonly used metal employed in synthesizing organometallic complexes for various biological applications. The anticancer activity of the platinum complex *cis*-diamminedichloroplatinum(II) commonly known as cisplatin was discovered by Rosenberg and his team long back in the 1960s [71]. The next-generation platinum(II) derivatives, such as carboplatin, oxaliplatin, and nedaplatin, were also introduced to address the challenges faced by cisplatin, such as the development of clinical resistance, limited selectivity, and acute toxicity [72,73]. Various organoplatinum complexes have been reported in recent years exhibiting efficient antiplasmodial activity [74,75]. Fig. 12.14 represents some of the organometallic platinum complexes (**14a−f**) which were screened for antimalarial activity and their mode of action was studied [76,77]. It was found that the platinum coordination usually enhances the antimalarial activity of the formed complex as compared to its free parent ligand.

Chellan and coworkers synthesized tridentate CNS cycloplatinated complexes and evaluated them for antiplasmodial activity against chloroquine-sensitive D10 and chloroquine-resistant Dd2 strains (Fig. 12.14) [78]. Cycloplatinum complex **14b** was obtained by the reaction 3,4-dichloroacetophenone thiosemicarbazone ligand with $K_2[PtCl_4]$. Further modification of this tetrameric complex was performed by taking mono-phosphino, that is, triphenylphosphine (PPh_3) or 1,3,5-triaza-7-phosphaadamantane (PTA), and bi-phosphino, that is, bis(diphenylphosphino)ferrocene or trans-bis(diphenylphosphino)ethylene ligands, which resulted into the formation of

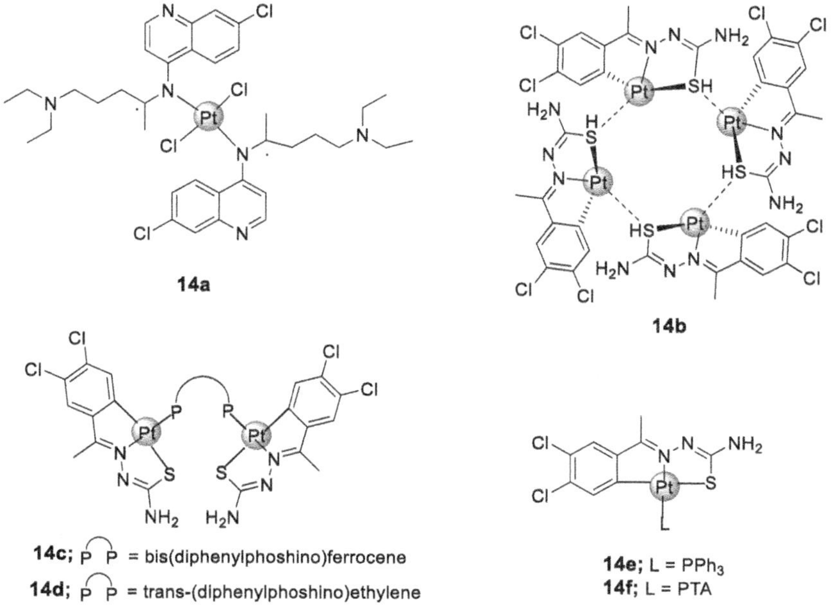

14c; ⌒P P = bis(diphenylphoshino)ferrocene
14d; ⌒P P = trans-(diphenylphoshino)ethylene

14e; L = PPh3
14f; L = PTA

FIGURE 12.14 Representative structures of platinum-based organometallic complexes.

corresponding mono-(**14e–f**) as well as binuclear cycloplatinated thiosemicarbazone products (**14c–d**). It was observed that tetranuclear and mononuclear candidates (**14b** and **14e–f**) exhibited antimalarial activity *in vitro* with IC$_{50}$ values in the range of 19.93–32.29 μM, while the binuclear counterparts 14c-d did not show any activity at the highest tested concentration. Moreover, complex 14e containing PPh3 ligand showed better activity as compared to compound **14f** containing PTA ligand against both the tested strains. Moreover, organoplatinum complex **14e** exhibited better selectivity against the chloroquine-resistant Dd2strain (IC$_{50}$ = 14.47 ± 1.98 μM) as compared with the chloroquine sensitive D10 variant (19.93 ± 3.74 μM).

12.7 Future prospective and scope

Over the past few decades, hybrid organometallic compounds have emerged as new scaffolds against various diseases and have made a significant contribution to the biomedical field. The design of potent hybrid scaffolds can be achieved by combining an organometallic compound with another active moiety with a different target or mechanism of action. Moreover, the optimization of hybrid compounds and their integration with the biological system in vivo is still challenging. Further studies should be focused on the design and development of new advanced hybrid organometallic compounds and examine their in vivo mechanism of action and pharmacokinetic and pharmacodynamic behavior. In the case of hybrid organometallic compounds, it is important to explore their stability, solubility, and target specificity and to optimize their nature through the insertion of different types of metal with different oxidation states.

12.8 Summary

The resistance of malaria parasites to the existing drugs continues to grow and progressively limits our ability to control this fatal disease. However, this is reassuring that many new approaches to antimalarial drug discovery are under evaluation. Antimalarial drugs are firmly established in combination therapies to treat drug-resistant malaria. The discovery of organometallic-based hybrid antimalarial drugs led to the establishment of the fact that it comprises a combination of two or more molecules along with the metal into at least one drug molecule to prevent the development of resistant strains of the disease. The reason behind the incorporation of organometallic pharmacophores into drug is to introduce bioactive functionalities in molecules with poor levels of uptake or bioactivity. For the eradication of malaria disease, organometallic hybrid drugs play an important role as these hybrid candidates have the potential to increase the biopharmaceutical efficacy, improved dosage compliance, reduced cost, and potential to overcome the development of resistant problems with minimum toxicity. Based on the literature discussed in this chapter, newer targeted therapeutic complex molecules can be generated for specific and selective use to treat drug-resistant malaria which is a global threat.

12.9 Objective type questions

i Which of the following drugs is used in the treatment of malaria?
   a Chloroquine
   b INH
   c Diazepam
   d Pyrazinamide

ii All the given drugs are 4-aminoquinoline derivatives except——.
   a Hydroxychloroquine
   b Chloroquine
   c Amodiaquine
   d Primaquine

iii Malaria is caused due to———.
   a *P. vivax*
   b *P. falciparum*
   c *P. malariae*
   d All of the them

iv Which of the followings is the disease from tropical and subtropical region?
  a Tuberculosis
  b Malaria
  c Gonorrhea
  d Syphilis

v Malaria is transmitted from———.
  a Bite of bee
  b Bite of female Anopheles mosquito
  c Bite of dog
  d Bite of snake

vi Chloroquine acts by——.
  a Inhibiting mycolic acid synthesis
  b Inhibiting hemozoin formation
  c Inhibiting benzoin formation
  d DHFR inhibition

vii The sexual life cycle of the malarial parasite is known as ———.
  a Sporogony
  b Schizogony
  c Zygomany
  d Aagami

viii Why does only female anopheles mosquito cause malaria?
  a Blood is the only diet to suck blood
  b Males find it difficult to suck blood
  c Because females need blood to nourish eggs
  d None of them

ix Which of the following drugs is suitable for the treatment of malaria during pregnancy?
  a Quinine
  b Chloroquine
  c Pyrimethamine
  d (d)Primaquine

x During the mechanism of action of artemisinin, hemoglobin is present in which oxidation state?
  a Fe^{2+}
  b Fe^{3+}
  c Both Fe^{2+} and Fe^{3+}
  d None of them

References

[1] Carter R, Mendis KN. Evolutionary and historical aspects of the burden of malaria. Clin Microbiol Rev 2002;15(4):564–94.

[2] Vos T, Allen C, Arora M, Barber RM, Bhutta ZA, Brown A, et al. Global, regional, and national incidence, prevalence, and years lived with disability for 310 diseases and injuries, 1990–2015: a systematic analysis for the Global Burden of Disease Study 2015. Lancet 2016;388(10053):1545–602.

[3] Milleliri JM. La médecine tropicale en images. Une mémoire pour l'histoire des sciences. *Images & Mémoires − Bulletin no 27.* Hiver 2010–1;10–15.

[4] Cook GC, Zumla A. Manson's tropical diseases. Lancet Infect Dis 2009;9(7):407–8.

[5] Dutta A. Where Ronald Ross (1857–1932) worked: the discovery of malarial transmission and the Plasmodium life cycle. J Med Biogr 2009;17(2):120–2.

[6] Snowden FM. The conquest of malaria: Italy, 1900–1962. Italy: Yale University Press; 2020.

[7] Yoeli M. Sir Ronald Ross and the evolution of malaria research. Bull N Y Acad Med 1973;49(8):722–35.

[8] Keeble TW. A cure for the ague: the contribution of Robert Talbor (1642-81). J R Soc Med 1997;90(5):285–90.

[9] Belay WY, Gurmu AE, Wubneh ZB. Antimalarial ACTIVITY OF STEM BARK OF PERIPLOCALINEARIFOLIA DURING EARLY AND ESTABLISHED PLASMODIUM INFECTION IN Mice. Evid Based Complement Altern Med 2018;2018:4169397.

[10] Achan J, Talisuna AO, Erhart A, Yeka A, Tibenderana JK, Baliraine FN, et al. Quinine, an old anti-malarial drug in a modern world: role in the treatment of malaria. Malar J 2011;10(1):144.

[11] Jamshidi-Kia F, Lorigooini Z, Amini-Khoei H. Medicinal plants: past history and future perspective. J Herbmed Pharm 2018;7:1–7.

[12] Schirmer RH, Adler H, Pickhardt M, Mandelkow E. Lest we forget you—methylene blue. Neurobiol Aging 2011;32(12):2325 e7–16.

[13] Rocamora F, Zhu L, Liong KY, Dondorp A, Miotto O, Mok S, et al. Oxidative stress and protein damage responses mediate artemisinin resistance in malaria parasites. PLoS Pathog 2018;14(3):e1006930.

[14] Caminade C, Kovats S, Rocklov J, Tompkins AM, Morse AP, ColónGonzález FJ, et al. Impact of climate change on global malaria distribution. Proc Natl Acad Sci 2014;111(9):3286–91.

[15] Pang YP, Ekstrom F, Polsinelli GA, Gao Y, Rana S, Hua DH, et al. Selective and irreversible inhibitors of mosquito acetylcholinesterases for controlling malaria and other mosquito-bornediseases. PLoS One 2009;4(8):e6851.

[16] World Health Organization. World Malaria Report 2022. Geneva: World Health Organization. <https://www.who.int/publications/i/item/9789240064898>, 2022.

[17] Slack RD, Jacobine AM, Posner GH. Antimalarial peroxides: advances in drug discovery and design. Med Chem Comm 2012;3(3):281–97.

[18] Black RE, Allen LH, Bhutta ZA, Caulfield LE, de Onis M, Ezzati M, et al. Maternal and child undernutrition: global and regional exposures and health consequences. Lancet 2008;371(9608):243–60.

[19] Cox-Singh J, Davis TME, Lee KS, Shamsul SSG, Matusop A, Ratnam S, et al. Plasmodium knowlesi malaria in humans is widely distributed and potentially life threatening. Clin Infect Dis 2008;46(2):165–71.

[20] Sabbatani S, Fiorino S, Manfredi R. The emerging of the fifth malaria parasite (Plasmodium knowlesi). A public health concern? Braz J Infect Dis 2010;14(3):299–309.

[21] Dhingra N, Jha P, Sharma VP, Cohen AA, Jotkar RM, Rodriguez PS, et al. Adult and child malaria mortality in India: a nationally representative mortality survey. Lancet 2010;376(9754):1768−74.

[22] Loha E, Lindtjørn B. Model variations in predicting incidence of Plasmodium falciparum malaria using 1998−2007 morbidity and meteorological data from south Ethiopia. Malar J 2010;9(1):166.

[23] Favuzza P, Ruiz M, Cowman AF. Dual plasmepsin-targeting antimalarial agents disrupt multiple stages of the malaria parasite life cycle. Cell Host Microbe 2020;27:642−58.

[24] Yang T, Ottilie S, Istvan ES, Godinez-Macias KP, Lukens AK, Baragaña B, et al. Accelerating malaria drug discovery. Trends Parasitol 2021;37(6):493−507.

[25] Tuteja R. Malaria − an overview. FEBS J 2007;274(18):4670−9.

[26] Habtewold T, Sharma AA, Wyer CAS, Masters EKG, Windbichler N, Christophides GK. Plasmodium oocysts respond with dormancy to crowding and nutritional stress. Sci Rep 2021;11:3090.

[27] Chinappi M, Via A, Marcatili P, Tramontano A. On the mechanism of chloroquine resistance in Plasmodium falciparum. PLoS One 2010;5(11):e14064.

[28] Klayman DL. Qinghaosu (artemisinin): an antimalarial drug from China. Science 1985;228(4703):1049−55.

[29] Brown GD. Artemisinin and a new generation of antimalarial drugs. Educ Chem 2006;43 (4):97−9.

[30] Mueller I, Galinski MR, Baird JK, Carlton JM, Kochar DK, Alonso PL, et al. Key gaps in the knowledge of Plasmodium vivax, a neglected human malaria parasite. Lancet Infect Dis 2009;9:555−66.

[31] Nwaka S, Ridley RG. Virtual drug discovery and development for neglected diseases through publiceprivate partnerships. Nat Rev Drug Discov 2003;2(11):919−28.

[32] Wang J, Xu C, Wong YK, Li Y, Liao F, Jiang T. Tu Y. Artemisinin, the magic drug discovered from traditional Chinese medicine. Engineering 2019;5(1):32−9.

[33] Lin AJ, Klayman DL, Milhous WK. Antimalarial activity of new water soluble dihydroartemisinin derivatives. J Med Chem 1987;30(11):2147−50.

[34] Zhan S, Gerhard GS. Heme activates artemisinin more efficiently than hemin, inorganic iron, or hemoglobin. Bioorg Med Chem 2008;16(16):7853−61.

[35] O'neill PM, Barton VE, Ward SA. The molecular mechanism of action of artemisinin-the debate continues. Molecules 2010;15(3):1705−21.

[36] Olliaro PL, Haynes RK, Meunier B, Yuthavong Y. Possible modes of action of the artemisinin-type compounds. Trends Parasitol 2001;17(3):122−6.

[37] Coghi P, Basilico N, Taramelli D, Chan WC, Haynes RK, Monti D. Interaction of artemisinins with oxyhemoglobin Hb-FeII, Hb-FeII, carboxyHb-FeII, Heme-FeII, and carboxyhemeFeII: significance for mode of action and implications for therapy of cerebral malaria. Chem Med Chem 2009;4(12):2045−53.

[38] Egan TJ. Recent advances in understanding the mechanism of hemozoin (malaria pigment) formation. J Inorg Biochem 2008;102(5−6):1288−99.

[39] Müller O, Sié A, Meissner P, Schirmer RH, Kouyaté B. Artemisinin resistance on the Thai−Cambodian border. Lancet 2009;374(9699):1419.

[40] Muregi FW, Ishih A. Next-generation antimalarial drugs: hybrid molecules as a new strategy in drug design. Drug Dev Res 2010;71(1):20−32.

[41] Nosten F., White N.J. Artemisinin-based combination treatment of falciparum malaria. In: Breman JG, Alilio MS, White NJ, editors. Defining and Defeating the Intolerable Burden of Malaria III: Progress and Perspectives: Supplement to Volume 77 (6) of American

Journal of Tropical Medicine and Hygiene; 2007. https://www.ncbi.nlm.nih.gov/books/NBK1713/.
[42] Agarwal D, Gupta RD, Awasthi SK. Are antimalarial hybrid molecules a close reality or a distant dream? Antimicrob Agents Chemother 2017;61(5) e00249−17.
[43] Meunier B. Hybrid molecules with a dual mode of action: dream or reality? Acc Chem Res 2008;41(1):69−77.
[44] Walsh JJ, Coughlan D, Heneghan N, Gaynor C, Bell A. A novel artemisinin−quinine hybrid with potent antimalarial activity. Bioorg Med Chem Lett 2007;17(13):3599−602.
[45] Dechy-Cabaret O, Benoit-Vical F, Robert A, Meunier B. Preparation and antimalarial activities of "trioxaquines," new modular molecules with a trioxane skeleton linked to a 4-aminoquinoline. Chembiochem 2000;1(4):281−3.
[46] Coslédan F, Fraisse L, Pellet A, Guillou F, Mordmüller B, Kremsner PG, et al. Selection of a trioxaquine as an antimalarial drug candidate. Proc Natl Acad Sci 2008;105 (45):17579−84.
[47] Rudrapal M, Chetia D. Endoperoxide antimalarials: DEVELOPMENT, structural diversity and pharmacodynamic aspects with reference to 1, 2, 4-trioxane-based structural scaffold. Drug Des Devel Ther 2016;10:3575−90.
[48] Wani WA, Jameel E, Baig U, Mumtazuddin S, Hun LT. Ferroquine and its derivatives: new generation of antimalarial agents. Eur J Med Chem 2015;101:534−51.
[49] Dubar F, KhalifeJ, Brocard J, Dive D, Biot C. Ferroquine, an ingenious antimalarial drug-thoughts on the mechanism of action. Molecules 2008;13(11):2900−7.
[50] Mombo-Ngoma G, Supan C, Dal-Bianco MP, Missinou MA, Matsiegui PB, Salazar CLO, et al. Phase I randomized dose-ascending placebocontrolled trials of ferroquine - a candidate anti-malarial drug - in adults with asymptomatic Plasmodium falciparum infection. Malar J 2011;10:53.
[51] Supan C, Mombo-Ngoma G, Dal-Bianco MP, Salazar CLO, Issifou S, MazuirF, et al. Pharmacokinetics of ferroquine, a novel 4- aminoquinoline, in asymptomatic carriers of Plasmodium falciparum infections. Antimicrob Agents Chemother 2012;56(6):3165−317.
[52] National Library of Medicine (US). Study to investigate the clinical and parasiticidal activity and pharmacokinetics of different doses of artefenomel and ferroquine in patients with uncomplicated plasmodium falciparum malaria. Identifier NCT03660839. <https://clinicaltrials.gov/ct2/show/NCT03660839>.
[53] Chavain N, Davioud-Charvet E, Trivelli X, Mbeki L, Rottmann M, Brun R, et al. Antimalarial activities of ferroquine conjugates with either glutathione reductase inhibitors or glutathione depletors via a hydrolyzable amide linker. Bioorg Med Chem 2009;17(23):8048−59.
[54] Salas PF, Herrmann C, Cawthray JF, Nimphius C, Kenkel A, Chen J, et al. Structural characteristics of chloroquine-bridged ferrocenophane analogues of ferroquine may obviate malaria drug-resistance mechanisms. J Med Chem 2013;56(4):1596−613.
[55] Navarro M, Vasquez F, Sánchez-Delgado RA, Pérez H, Sinou V, Schrével J. Toward a novel metal-based chemotherapy against tropical diseases. 7. Synthesis and in vitro antimalarial activity of new gold − chloroquine complexes. J Med Chem 2004;47(21):5204−9.
[56] Fraisse L., Struxiano A., Sanofi S.A. Use of ferroquine in the treatment or prevention of malaria. US20120270851A1 (2012).
[57] Bellot FO, Cosledan F, Vendier L, Brocard J, Meunier B, Robert A. Triox-aferroquines as new hybrid antimalarial drugs. J Med Chem 2010;53(10):4103−9.
[58] Boissier J, Cosledan F, Robert A, Meunier B. In vitro activities of trioxaquines against Schistosoma mansoni. Antimicrob Agents Chemother 2009;53(11):4903−6.

[59] Minić A, Van de Walle T, Van Hecke K, Combrinck J, Smith PJ, Chibale K, et al. Design and synthesis of novel ferrocene-quinoline conjugates and evaluation of their electrochemical and antiplasmodium properties. Eur J Med Chem 2020;187:111963.

[60] Bergamo A, Gaiddon C, Schellens JHM, Beijnen JH, Sava G. Approaching tumour therapy beyond platinum drugs: status of the art and perspectives of ruthenium drug candidates. J Inorg Biochem 2012;106(1):90–9.

[61] Bergamo A, Sava G. Ruthenium anticancer compounds: myths and realities of the emerging metal-based drugs. Dalton Trans 2011;40:7817–23.

[62] Martínez A, Deregnaucourt C, Sinou V, Latour C, Roy D, Schrével J, et al. Synthesis of an organo-ruthenium aminoquinoline-trioxane hybrid and evaluation of its activity against Plasmodium falciparum and its toxicity toward normal mammalian cells. Med Chem Res 2017;26(2):473–83.

[63] Navarro M, Pekerar S, Pérez HA. Synthesis, characterization, and antimalarial activity of new iridium–chloroquine complexes. Polyhedron 2007;26(26):2420–4.

[64] Stringer T, Guzgay H, Combrinck JM, Hopper M, Hendricks DT, Smith PJ, et al. J Organomet Chem 2015;788:1–8.

[65] Ma DL, Wang M, Mao Z, Yang C, Ng CT, Leung CH. Rhodium complexes as therapeutic agents. Dalton Trans 2016;45:2762–71.

[66] Zhang P, Huang H. Future potential of osmium complexes as anticancer drug candidates, photosensitizers and organelle-targeted probes. Dalton Trans 2018;47:14841–54.

[67] Lazic S, Kaspler P, Shi G, Monro S, Sainuddin T, Forward S, et al. Novel osmium-based coordination complexes as photosensitizers for panchromatic photodynamic therapy. Photochem Photobiol 2017;93(5):1248–58.

[68] Zhang P, Wang Y, Qiu K, Zhao Z, Hu R, He C, et al. NIR phosphorescent osmium (II) complex as a lysosome tracking reagent and photodynamic therapeutic agent. Chem Commun 2017;53:12341–4.

[69] Ekengard E, Glans L, Cassells I, Fogeron T, Govender P, Stringer T, et al. Antimalarial activity of ruthenium (II) and osmium (II) arene complexes with mono-and bidentate chloroquine analogue ligands. Dalton Trans 2015;44:19314–29.

[70] Bozec HL, Touchard D, Dixneuf PH. Organometallic chemistry of arene ruthenium and osmium complexes. Adv Organomet Chem 1989;29:163–247.

[71] Rosenberg B, Vancamp L, Trosko JE, Mansour VH. Platinum compounds: a new class of potent antitumour agents. Nature 1969;222(5191):385–6.

[72] Johnstone TC, Suntharalingam K, Lippard SJ. The next generation of platinum drugs: Targeted Pt (II) agents, nanoparticle delivery, and Pt (IV) prodrugs. Chem Rev 2016;116 (5):3436–86.

[73] Dilruba S, Kalayda GV. Platinum-based drugs: past, present and future. Cancer Chemother Pharmacol 2016;77(6):1103–24.

[74] Baartzes N, Stringer T, Smith GS. Targeting sensitive-strain and resistant-strain malaria parasites through a metal-based approach. In: Hirao T, Moriuchi T, editors. Advances in bioorganometallic chemistry. Amsterdam, The Netherlands: Elsevier; 2019. p. 193–213.

[75] Sekhon BS, Bimal N. Transition metal-based anti-malarial. Indian J Pharm Educ Res 2012;20112:52–63.

[76] Macedo TS, Villarreal W, Couto CC, Moreira DR, Navarro M, Machado M, et al. Platinum (ii)–chloroquine complexes are antimalarial agents against blood and liver stages by impairing mitochondrial function. Metallomics 2017;9(11):1548–61.

[77] Navarro M, Castro W, Madamet M, Amalvict R, Benoit N, Pradines B. Metal-chloroquine derivatives as possible anti-malarial drugs: evaluation of anti-malarial activity and mode of action. Malar J 2014;13:471.

[78] Chellan P, Land KM, Shokar A, Au A, An SH, Clavel CM, et al. Exploring the versatility of cycloplatinated thiosemicarbazones as antitumor and antiparasitic agents. Organometallics 2012;31(16):5791−9.

Chapter 13

Therapeutic approach of polynuclear organometallic complexes

Ashish Kumar Sarangi
Department of Chemistry, School of Applied Sciences, Centurion University of Technology and Management, Balangir, Odisha, India

13.1 Introduction

Almost all drugs applied today are entirely organic compounds based drugs. Particularly after the tremendous success of the cisplatin drug (Fig. 13.1) in the treatment of tumor cell, involvement of metal complexes has developed [1]. Synthetic organometallic complexes are in general studied to be harmful or incompatible with biological systems. Contempt this sensing, the therapeutic properties of polynuclear organometallic compounds, in specific organotransition metal complexes have been examined for a long period of time and in the last few years the area has developed substantially. D-block metals have a significant place within medicinal biochemistry [2]. These elements have partly filled d-shells in any of their normally happening oxidation state. Transition metal complex or coordination compound is a structure comprising of a central metal atom, the mode of bonding to a surrounding layout of ligands (i.e., molecules or ions), which donate electron pair to the central metal atom. In recent years, there has been a major advancement in the use of transition metal complexes as medicines to treat a variety of human ailments, including antiinflammatory, lymphomas, cancer, infection control, psychiatric disorders, and diabetes. All d-block elements can interact with ligand molecules and display different oxidation states. This behavior of transition metals has started the development of metal-based medications with anticipated pharmacological application and may increase special therapeutic opportunities.

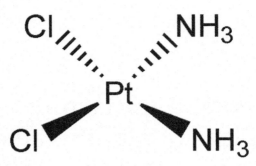

FIGURE 13.1 cis-diamminedichloroplatinum(II) (Cisplatin).

13.2 Fundamental principles

The prefix "organo-" (e.g., organoplatinum compounds) designates compounds that contain metal and carbon bond in an organyl group, which includes all conventional metals such as s, p, d block metals and post transition metals.3 In addition, f-block, semimetals, lanthanides, actinides, and the p-block elements arsenic, selenium, All metal carbonyls, including ferrocene, iron penta, and tetracarbonyl nickel, among others. Transition metals are present in the majority of organometallic compounds. Et2Zn, Bu3SnH, Et3B, and Me3Al are more examples of organometallic compounds.

A naturally extracted organometallic complex is Vitamin B_{12}, that is, methylcobalamin, that carries a bond between cobalt-methyl.

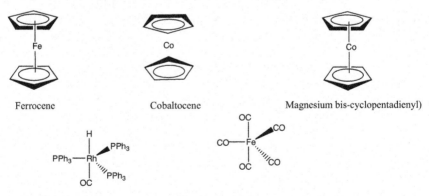

Tris(triphenylphosphine)rhodium carbonyl hydride Iron(0) pentacarbonyl

13.3 Common methods of preparation

Organometallic metal cluster complexes were initially entirely created by chance, and even now, deliberate syntheses of metal clusters from lower nuclearity precursors are still infrequently carried out. In the simplest and most common syntheses, the carbonyl ligand of a cluster bearing the appropriate metal composition must be replaced with a hydrocarbon, and then the hydrocarbon unit must be adjusted to claim that the metal-cluster centre has not been altered [1].

13.3.1 Substitution of ligand on metal carbonyls

The primary method of thermal activation, which removes carbonyl ligands from metal clusters like $M_3(CO)_{12}$ (M = Ru, Fe, and Os), was used to prepare substituted ligands for the preparation of organometallic cluster compounds. The highest activation energy necessary for CO loss mandated that most stable organometallic cluster complexes yield could be separated and less stable intermediates were insensible. For example, low-temperature pathways were not found during the pyrolysis of $Os_3(CO)_{12}$ at 120°C in the presence of ethylene, which produced $H_2Os_3(3-2-C = CH_2)(CO)_9$ as opposed to the intermediates, $Os_3(CO)_{11}(CH_2 = CH_2)$ and $HOs_3(2-2-CH = CH_2)(CO)_{10}$, which can be synthesized. When the end products are thermally reactive, light-integrated processes like photolysis can be more productive. Equations only provide a few examples of these kinds of reactions (13.1–13.3) [1–4].

$$Os_3(CO)_{12} + C_6H_6 \rightarrow H_2O_3\left(\mu_3 - \eta^2 - C_6H_4\right)(CO)_9 \quad (13.1)$$

$$Os_3(CO)_{12} + PPh_3 \xrightarrow{n=1-3} Os_3(CO)_{12} - n(PPh_3)_n \quad (13.2)$$

$$Ru_3(CO)_{12} + H_2 = CHCH = CH_2 \rightarrow (\mu - H)Ru_3\left(\mu_3 - \eta^3 - HCCHCMe\right)(CO)_9 \quad (13.3)$$

The displacement of CO ligands has opened up the potential of planned syntheses. The three most efficient techniques are: [5] the process of changing a bound CO ligand into a CO_2 molecule using trimethylamine N-oxide; [6] Applying one electron to create high sensitive metal cluster radical intermediates; and [7] By applying halide nucleophilesions to encourage CO dissociation. The trimethylamineN-oxide has the potential to chemically eliminate carbonyls as CO_2; Eq. (13.4). Applying the monosubstituted product necessitates by adding of one equivalent of trimethylamine N-oxide. The next step in the mechanism is the oxide's nucleophilic attack on a carbonyl carbon. This reaction occurs quickly but requires of an electrophilic carbonyl ligand. As a consequence, the chemical agent is ineffective for highly substituted metal clusters and anionic metal carbonyls.

$$Os_3(CO)_{12} + ONMe_3 + PPh_3 \rightarrow Os_3(CO)_{11}(PPh_3) + CO_2 + NMe_3 \quad (13.4)$$

13.3.2 Template method of preparation

The metal clusters can be gathered across the main group of atoms, such as sulphur or phosphorous, that forms a extremely strong bonds to metals, that is adequate to form a bridge bonding [8]. An example is given in equation (v) [9].

$$2Ru_3(CO)_9(\mu_3 - PhCCH)(\mu_3 - S) \rightarrow Ru_6(CO)_{16}(\mu_3 - PhCCH)_2(\mu_4 - S)_2 + 2CO \quad (13.5)$$

13.3.3 Metal fragments condensation

By condensing unsaturated metal complexes with metal donors atoms, the majority of refined syntheses of mixed-metal clusters are now being done [10]. Consider the synthesis of cluster complexes $H_2Os_3W(3-CR)(CO)_{12}$ as a result of the condensation of the alkylidyne complex $CpW(CO)_2(CR)$ with the unsaturated metal cluster $H_2Os_3(CO)_{10}$. Metal-metal, carbon-metal, or metal-metal multiple connections may be the sites where metal pieces condense. Eq. (13.6) give a few other examples (13.7).

Ru(cyclooctatriene)(cyclooctadiene)

$$+ CpW(CO)_2(CR) \rightarrow RuW_2(\mu_3- RCCR)(CO)_7Cp_2 + RuW_3(\mu- CR)(CO)_4Cp_3 \quad (13.6)$$

$$(H)Ru_3(COMe)(CO)_{10} + CpW(CO)_3(CCPh) \rightarrow CpWRu_3(COMe)(CO)_9(C = CHPh) \quad (13.7)$$

13.3.4 Oxidative addition reactions

In addition to H-C, H-Si, H-Sn, and Cl-C, other -bonded atom pairs have also been used to add group 14 elements ligands donor to the clusters' metal centers. Gains in the H-C, H-Si, and H-Sn bonds indicate that there will be an unsaturated metal core. The mechanism analyses are consistent with one that adds a single metal atom through a synchronous, three-centre process; this is similar to what usually happens for single metal complexes [11]. As shown in Eqs. (13.8) and (13.9), respectively, a few pertinent examples of H-Si addition to unsaturated Os3 metal clusters are proposed. Inactive C-H bonds are occasionally added through oxidation. The selective addition of terminal C-H type bonds to $H_3Ru_3(3-S)(Cp^*)_3$ results in the formation of $H_2Ru_3(3-S)(3-CR)(Cp^*)_3$ (Scheme 13.1) [12].

$$H_2Os_3(CO)_{10} + HSiEt_3 \rightarrow H_3Os_3(CO)_{10}(SiEt_3) \quad (13.8)$$

$$Os_3(CO)_{11}(NCMe) + HSi(OEt)_3 \rightarrow HOs_3(CO)_{11}[Si(OEt)_3] + NCMe \quad (13.9)$$

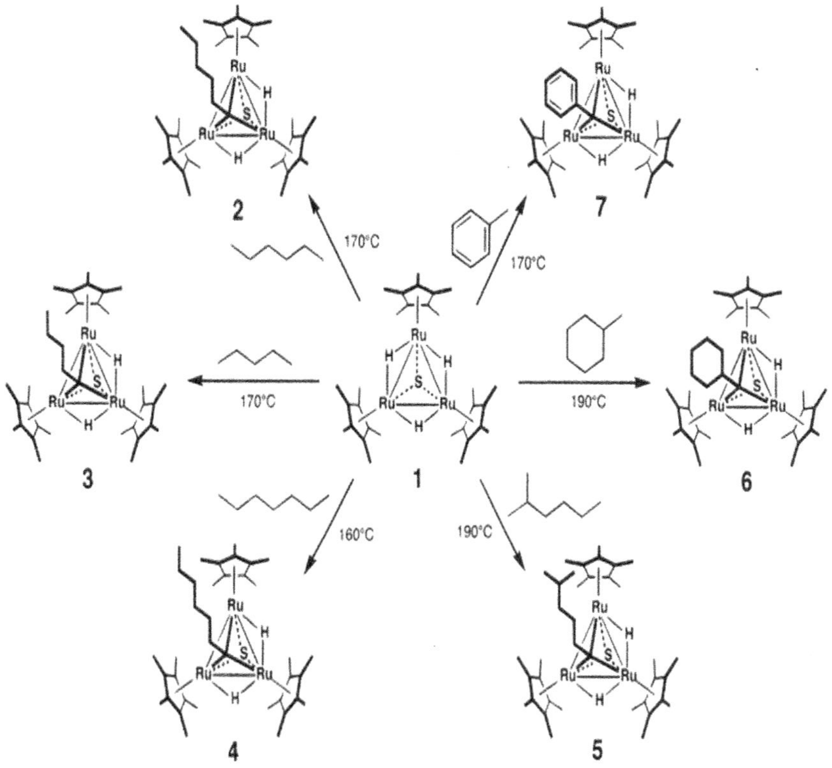

Therapeutic approach of polynuclear organometallic complexes Chapter | 13 335

SCHEME 13.1 Addition of terminal C-H type bonds to $H_3Ru_3(3-S)(Cp^*)_3$. Procreated from K. Matsubara, A. Inagaki, M. Tanaka, and H. Suzuki, J. Am. Chem. Soc., *1999*, *121*, *7421*.

13.4 Mechanism involved

13.4.1 Ligand substitution

$$M-L + L' \rightarrow M-L' + L$$

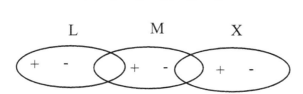

The aforementioned response can be radical chain, dissociative, or associative. A ligand's kinetic influence on the function of substitution at a position trans to itself in a square or octahedral complex is known as the transeffect (ground-state weakening of bond). Negative charge is repelled to transposition by L M.

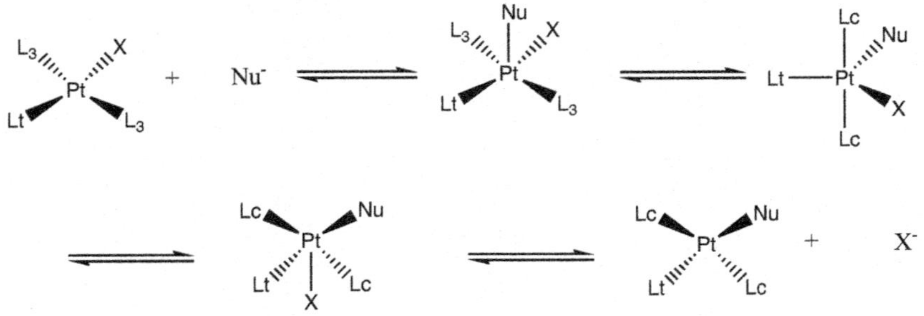

13.4.2 Oxidative addition

Numerous tracts that depend on the position of the metal and the movement of the substrates are traversed by the oxidative additions. Here are some of the crucial tracks:

13.4.2.1 Concerted track based

Hydrogen and hydrocarbons are examples of nonpolar substrates that seem to coordinately undertake oxidative addition processes. These substrates lack -bonds, necessitating the use of a three-cantered complex. The oxidized complex is then produced by the intramolecular cleavage of the ligand, which is most likely brought on by the donation of an electron pair into the sigma* orbital of the interligand link. The ensuing ligands will be reciprocally cis [13].

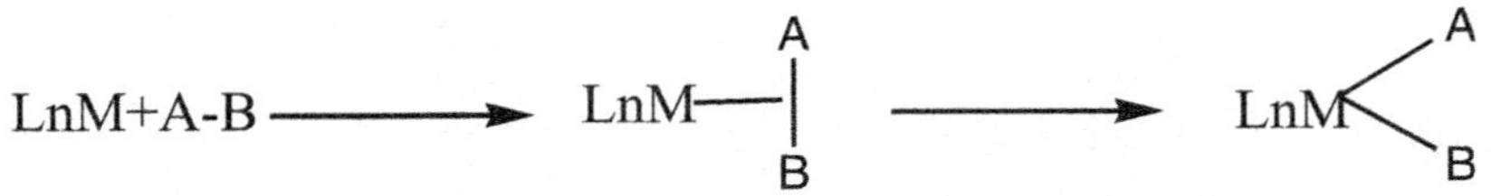

$$LnM + A-B \longrightarrow LnM{-}{\overset{A}{\underset{B}{|}}} \longrightarrow LnM\overset{A}{\underset{B}{\diagdown}}$$

In this chemical process, homonuclear diatomic molecules, such the H_2 molecule, are added. Many C-H activation events in the agnostic complex also follow a coordinated mechanism through the creation of an M- (C−H) [10]. Trans-IrCl(CO)[P(C$_6$H$_5$)$_3$] The appropriate illustration of how hydrogen interacts with Vaska's complex is shown in Fig. 13.2. This change causes the iridium metal to transition from its regular oxidation state of +1 to +3 states. The generated product is typically bound using three anions—a single chloride and two hydride ligands. The following reaction results in an 18e-based complex with six coordinates from an initial metal complex having 16e- and a coordination number of 4.

The back donation of electron into the H-H *-orbital, the establishment of a trigonal bipyramidal dihydrogen intermediate is adopted by the breaking of

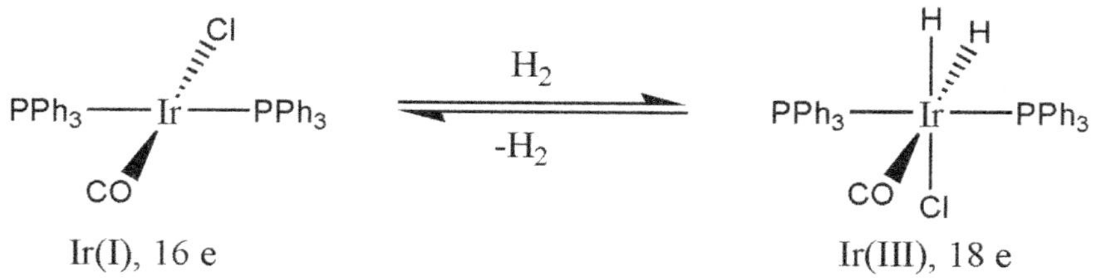

PPh₃—Ir(Cl)(CO)—PPh₃ + H₂ ⇌ PPh₃—Ir(H)(H)(Cl)(CO)—PPh₃ (−H₂)

Ir(I), 16 e   Ir(III), 18 e

FIGURE 13.2 Hydrogen interacts with Vaska's complex.

the H-H bond. The aforementioned reaction is also in chemical equilibrium, and the metal centre experiences a reduction while the opposite reaction continues by removing hydrogen gas [14].

Electron-rich metals favor the aforementioned reaction due to the electron in the H-H *-orbital that was donated via the rear side to split the H-H bond [14]. Finally the product is a cis dihydride.

13.4.2.2 Bimolecular nucleophilic substitution reactions-based

In organic synthesis, a few well-known bimolecular nucleophilic substitution procedures continue with oxidative additions. When a nucleophile comes into contact with a metal that has a less electronegative atom than the substrate, the R-X bond is broken, resulting in the production of a [M-R] + species. In this stage, the anion quickly coordinates with the cationic metal centre. The complex with square plans, for instance. Reaction of methyl iodide

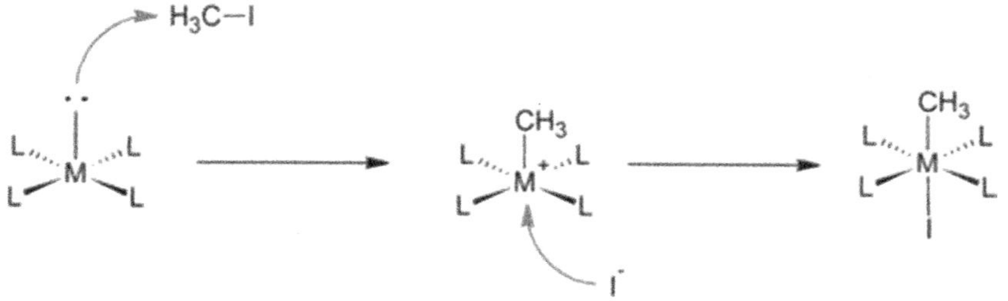

13.4.2.3 Ionic-based

Both the ionic-based mechanism of oxidative addition and the bimolecular nucleophilic substitution processes type involve the sequential addition of two distinct ligand fragments. Ionic processes require substrates, but there is a substantial distinction in that they also require dissociation in solution before they can interact with metal ions. Ionic oxidative addition is used in the instance of adding hydrochloric acid [13].

13.4.2.4 Radical-based

Alkyl halides and nearly identical substrates can add to a metal ion centre via a radical mechanism process in addition to bimolecular nucleophilic substitution processes [13].

Lednor and his coworkers offered a similar one example [15].

Step −I Initiation reaction

$$[(CH_3)_2C(CN)N]_2 \rightarrow 2(CH_3)_2(CN)C^\bullet + N_2$$

$$(CH_3)_2(CN)C^\bullet + PhBr \rightarrow (CH_3)_2(CN)CBr + Ph^\bullet$$

Step-II Propagation reaction

$$Ph^\bullet + [Pt(PPh_3)_2] \rightarrow [Pt(PPh_3)_2Ph]^\bullet$$

$$[Pt(PPh_3)_2Ph]^\bullet + PhBr \rightarrow [Pt(PPh_3)_2PhBr] + Ph^\bullet$$

13.4.3 Reductive elimination

In reductive elimination, a new covalent connection between two ligands is created while the metal center's oxidation state is reduced. This microscopic activity, which is the stage of many catalytic processes where products are formed, then transforms into an oxidative addition reaction. Despite the fact that oxidative addition and reductive elimination are the opposite reactions, the same principles apply to both. The product's equilibrium depends on both thermodynamic forces [16,17].

During the reductive elimination process, it also manifests in higher oxidation levels. This requires either a two-electron switch at a single metal (mononuclear) or a one-electron switch at each of two metals (binuclear, dinuclear, or bimetallic) [16,17].

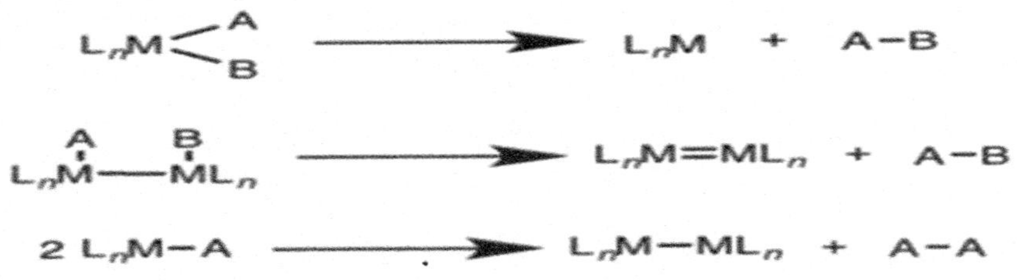

$$L_nM\begin{pmatrix}A\\B\end{pmatrix} \longrightarrow L_nM + A-B$$

$$L_nM-ML_n \text{ (with A, B substituents)} \longrightarrow L_nM=ML_n + A-B$$

$$2\,L_nM-A \longrightarrow L_nM-ML_n + A-A$$

The metal ions' oxidation status is decreased by two in the mononuclear reductive elimination process, while their number of d-electron counts increases by two. This orbital system's d8 orbital tract is comparable to that of metals like Ni(II), Pd(II), and Au (III). Similar to this, the d6 orbital system containing the metals Pt(IV), Pd(IV), Ir(III), and Rh integrates (III). The groups that make up the mononuclear reductive elimination also require that they be cis to one another on the metal ion [18].

$$\text{[Ph}_2\text{P-Pd(CH}_3)_2\text{-PPh}_2\text{]} \xrightarrow{80\,°\text{C}, \text{ DMSO}} \text{H}_3\text{C-CH}_3 + \text{Pd}^0$$

$$\text{[Ph}_2\text{P-Pd(CH}_3)\text{(H}_3\text{C)-PPh}_2\text{ phenanthrene]} \xrightarrow{80\,°\text{C}, \text{ DMSO}} \text{no reaction}$$

Each metal's oxidation state decreased by one throughout the binuclear reductive elimination process, whereas each metal's d orbital electron count increased by one. With first row metals, a similar level of reactivity is typically observed. Both second row and third row transition metals have been found to exhibit the oxidation state in this reaction, which prefers a one-unit change in transition metals [19].

$$\text{Os}_2(\text{CO})_8(\text{CH}_3)(\text{H}) \xrightarrow{\Delta \text{ or } h\nu} \text{Os}_2(\text{CO})_8(=\text{Os})\text{-CO} + \text{CH}_4$$

13.4.4 Migratory insertion

The migratory insertion reaction in organometallic chemistry is a reaction in which two ligands join with a metal complex.

$$\text{R-CH}_2\text{CH}_2\text{-Zr(Cp)}_2\text{Cp} \text{ (16e}^-\text{)} \longrightarrow [\text{O}\equiv\text{C-Zr(Cp)}_2\text{Cl CH}_2\text{CH}_2\text{R}]\text{ (18e}^-\text{)} \longrightarrow \text{R-CH}_2\text{CH}_2\text{-C(=O)-Zr(Cp)}_2\text{Cl} \text{ (16e}^-\text{)}$$

13.4.5 Hydrometalation / β-hydride elimination

A chemical component having a hydrogen to metal bond (M-H, or metal hydride), combines with a compound with an unsaturated bond, such as an alkene

(RC = CR), to create a novel product with a carbon to metal connection during the hydrometalation or hydrometallation process (RHC-CRM) [20]. Since hydrogen is more electronegative than metals, the reverse reaction is the removal of hydrides.

$$\underset{R}{\overset{R}{\diagup}}=\underset{R}{\overset{R}{\diagdown}} \quad \xrightarrow{H-MR_n} \quad H-\underset{R}{\overset{R}{\underset{|}{C}}}-\underset{R}{\overset{R}{\underset{|}{C}}}-MR_n \quad \xrightarrow{\text{electrophile E}} \quad H-\underset{R}{\overset{R}{\underset{|}{C}}}-\underset{R}{\overset{R}{\underset{|}{C}}}-E$$

13.4.6 Transmetalation

An organometallic reaction known as transmetalation or transmetallation takes the shift of ligands from single metal to another.

$$R\text{-}M + M'\text{-}X \longrightarrow R\text{-}M' + M\text{-}X$$

$$R\text{-}M + M' - X \rightarrow R\text{-}M' + M - X$$

13.4.7 Recent developments with examples

An innovative palladium (II) anticancer agent: Synthesis, characterization, DFT viewpoint, CT-DNA and BSA interaction studies by in-vitro and in silico methods [21]. Recently, in the realm of metal complexes, the 1,10-phenanthroline ligand acac and the Pd(II) complex of the formula [Pd(phen)(acac)]NO$_3$ were found. This compound was made by combining sodium salt of acetylacetone and [Pd(phen)(H$_2$O)$_2$](NO$_3$)$_2$ in a 1:1 molar ratio. Structure could be characterized with the aid of spectral methods. The geometry optimization of the prepared complex at the level of density functional theory places the palladium metal atom in a square-planar shape. The complex's efficiency against K562 cancer cells has also been researched. The fundamental interactions between the palladium metal complex, CTDNA, a direct molecule for anticancer medicines, and BSA, a contain protein, were investigated using a variety of techniques. Through hydrophobic and H-bonding interactions as well as van der Waals forces in the UV-Vis and fluorescence emission bands, the palladium metal complex communicates with EB + CT-DNA and BSA. The fluorescence quenching mechanism worked for both macromolecules. Circular dichroism and Förster resonance energy transfer assays were performed to ascertain the viscosity of BSA, while gel electrophoresis methods were used to further demonstrate the interaction for CT-DNA. Based on the partition coefficient data, the results show that the [Pd(phen)(acac)]NO$_3$ complex is more lipophilic than well-known cisplatin. Cu(I), Ag(I), and Au(I) trinuclear macrocyclic HC/urea complexes have antibacterial and anticancer properties [22]. Newly created macrocyclic tricarbene/urea N-heterocyclic carbene (NHC) metal complexes of Cu(I), Ag(I), and Au(I) are said to have fluorescence properties that are comparable to those of the corresponding tetracarbene metal complexes. Ag(I) and Au(I), two recently found compounds, emit light with maximum emission at 481 and 334 nm, respectively. Additionally, produced complexes are tested for their ability to limit the growth of bacteria strains and cancer cell lines (HeLa and MCF7) (*Staphylococcus aureus* and *Escherichia coli*). The metal was observed to have made a significant influence. The Cu(I) and Au(I) complexes are inactive, in contrast to bacteria. The final state of a control action is described by the Ag(I) complex (MIC 30M). The Cu(I) and Ag(I) complexes exhibit antiproliferative effects in both cancer cell types. The results showed values between 25.1 and 3.03M (Scheme 13.2).

NHeterocyclic Carbene Silver Complexes in Novel Fast-Acting Pyrazole/Pyridine-Functionalized Nanoparticles Show Improved Safety and Efficacy as Anticancer Therapeutics [23]. Four multinuclear silver complexes with newly developed N-heterocyclic carbene (NHC) ligands that were functionalized with pyrazole or pyridine. X-ray methods verified the crystal structures of the silver-NHC complexes. According to in vitro testing, the silver-NHC complexes strongly

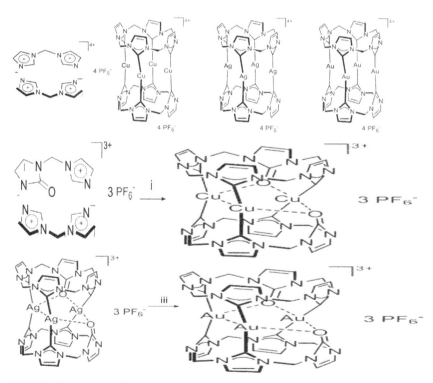

SCHEME 13.2 Previously reported Cu(I), Ag(I) and Au(I)tetracarbene complexes [21,22]. *Procreated from Christian H.G. Jakob, Angela Weigert Muñoz, Jonas F. Schlagintweit, Vanessa Weiß, Robert M. Reich, Stephan A. Sieber, João D.G. Correia, Fritz E. Kühn, J. Organometallic Chem., 2021, 932, 121643.*

destroy a range of cancer cells after a brief drug exposure, acting as quick-acting cytotoxic agents. In the instance of the cisplatin-resistant A549 cancer cells, none of the silver-based complexes were cross-resistant to the medication cisplatin. With reduced systemic drug toxicities proportionate to cisplatin, the nanotherapeutics with silver complexes demonstrated a maximum safety margin in animal models. The xenograft model of human colorectal cancer and the organization of the nanotherapeutics resulted in a notable suppression of tumor development. DNA-Binding Properties and Anticancer Activities of Ruthenium(II) Cymene Complexes with (Poly)cyclic Aromatic Diamine Ligands [24]. Four multi-nuclear silver complexes with pyrazole or pyridine functionalities were linked with newly synthesized N-heterocyclic carbene (NHC) ligands. X-ray methods verified the crystal structures of silver-NHC complexes. According to in vitro experiments, the silver-NHC complexes effectively kill a wide range of cancer cells after only a short exposure to the medicine. Every last bit of silver-based Recently synthesized Ru(II) arene complexes with the general formula [RuCl(6-p-cymene)(diamine)] were tested experimentally for their chloro/aqua exchange properties. PF6 (diamine = 1,2-diaminobenzene, 2,3-diaminonaphthalene, 9,10-diaminophenanthrene, 2,3-diaminophenazine, and 1,2-diaminobenzene) Computational studies show that the use of guanine in place of water is possible for every given complex, and that the bonds formed between Ru and Nguanine are remarkably robust. To establish even more potent chemical affinity for DNA, aromatic systems can adopt angles out of the plane that reveal crucial covalent interaction with the molecule. There is reason for optimism, as the ruthenium arenes complexes utilized as examples in this study have been shown to have anticancer characteristics. Half-maximal inhibitory concentration (IC50) values against three cancer cell lines are comparable to or better than cisplatin. Multinucleated ruthenium organometallic complexes with 1,3,5-triazine ligands: synthesis, DNA interaction, and biological activity [25]. Recently produced novel ruthenium complexes are an attractive option to platinum-based anticancer drugs. Researchers may have access to novel complexes with more potent biological effects if they are able to build multinuclear counterparts. This is now used to synthesize two out of three trinuclear complexes consisting of 2,4,6-tris(di- 2-pyridylamino)-1,3,5-triazine as the core and (arene)RuCl as the organometallic ruthenium fragment. (Arene = benzene [6], p-cymene [5], or hexamethylbenzene) [7]. The structural interaction of the resulting complexes with DNA was extensively analyzed using a variety of biophysical probes and a molecular docking approach. The complexes have been found to bind to DNA with apparent binding constants between 2.20 and 4.79 104 M1. We did not have a definitive answer for how to bind to the nucleic acid. All of the complexes showed considerable antibacterial activity against the organisms tested, in addition to anticancer activity at high (> 100 M) concentrations. Synthesis and in vitro evaluation of new polynuclear organometallic Ru(II), Rh(III), and Ir(III) pyridyl ester complexes for use against parasites and cancer [26]. Polynuclear organometallic Platinum and its Metal (PGM) complexes containing di- and tripyridyl ester ligands have been synthesized and characterized using a wide range of analytical and spectroscopic techniques.

After combining these polypyridyl ester ligands with [Ru(p-cymene)Cl2], [Rh (C5Me5)Cl2], or [Ir(C5Me5)Cl2]2, di- or trinuclear organometallic complexes were obtained. This coordinating mechanism was ultimately supported after the molecular structures of two of the dinuclear complexes were elucidated. Each of the metal centers is served by a monodentate donor provided by the polyaromatic ester ligand. The di- and trinuclear metal complexes were investigated for their capacity to inhibit the growth of Trichomonas vaginalis strain G3, Plasmodium falciparum strain NF54 (chloroquine sensitive), and human ovarian cancer cell lines A2780 (cisplatin-sensitive) and A2780cisR (cisplatin-resistant). All of the complexes exhibited moderate to strong antiplasmodial activity, and the strongest compounds were examined for their ability to inhibit the manufacture of synthetic hemozoin in a cell-free setting. In vitro anticancer testing revealed that trinuclear pyridyl ester complexes had low levels of activity against the two tumor cell lines, but were safer for use in model healthy cells.

13.5 Applications in emerging fields

13.5.1 In the field of medicine

New Indol-3-Acetic Acid-Based Zinc(II), Palladium(II), Platinum(II), and Silver(I) Complexes; Solid and Solution Interaction with DNA and Anticancer DNA interaction and the prevention of cancer: New Complexes of Silver, Palladium, Platinum, and Indol-3-Acetic Acid Based on Solid and Solution Studies [27]. For zinc(II), palladium(II), platinum(II), and silver, novel complexes of N,N- (2,2'-bipyridyl; bpy) or P- (triphenylphosphine; PPh3) as a secondary chelate and indol-3-acetic acid (HIAA) as a primary chelate have allegedly been developed (I). The spectral data from IR, NMR (1 H, 13 C, 31 P), UV-Vis, EI, and maldi-mass measurements, elemental analysis, molar conductivity, and TGA are used to explore the molecular structures of these complexes. The two O-atoms of -COO, deprotonated O $=$ C-O, or carbonyl O-atom (HO-C $=$ O), respectively, are used by HIAA to coordinate the metal ions in a mononegative bidentate, mononegative monodentate, or neutral monodentate manner. As a mononegative bidentate ligand, HIAA binds two Ag(I) ions through carbonyl O- and deprotonated -O- atoms in the case of the binuclear Ag(I) complex. The molar ratio and Job's methods could be used to determine the dissociation constant (pKa) of HIAA and the stoichiometry of the complexes in solution. Examined was how well some of the complexes inhibited the growth of VC-8-BRCA and VC-8 cell lines, which represent human ovarian cancer. UV-vis spectroscopy was used to examine the water-soluble complexes' DNA-binding capabilities. Due to the complexes' hypochromism, intercalative CT-DNA binding may occur. The complexes comprising CT-DNA have high to moderate binding capacities as determined by their binding constants (Kb). Tertamyldithiocarbamate and DACH ligands have recently been used to produce anticancer Pd and Pt complexes against the HT29 and Panc1 cell lines. Two new platinum and palladium complexes with the formula [M(DACH)(tertamyl.dtc)]NO3 were developed and characterized by spectroscopic techniques to study the reduction of side effects of commercial anticancer medications like cisplatin, according to research proposed by Moghadam et al. in 2021. The IC50 values of [Pt(DACH)(tertamyl.dtc)]NO3 and [Pd(DACH)(tertamyl.dtc)]NO3 against Panc1 cell line were 263.1 and 198.7 M, and against HT29 cell line were 241.9 and 258.2 M, respectively, according to in vitro cytotoxicity tests on HT29 and Panc1 cell lines. They were close to or below the oxaliplatin value and below the cisplatin value. Both metal complexes' data on circular dichroism and temperature stability point to the possibility that electrostatic bonding may be how they bind to DNA. The alignment of B-DNA to C-DNA was altered via electrostatic interaction, but groove interaction may now be more common. A molecular docking simulation revealed that the [Pd(DACH)(tertamyl.dtc)]NO$_3$ complex had a higher negative docking energy and a higher propensity for DNA contact. Two new Pt and Pd compounds have been tested for their cytotoxicity in vitro against the HT29 and Panc 1 cell lines, and the results indicate that they are nearly as toxic as oxaliplatin but less toxic than cisplatin. Both compounds bind to DNA via electrostatic interactions, as shown by CD findings and thermal stability experiments. According to molecular docking, the [Pd(DACH)(tertamyl.dtc)]NO$_3$ complex has a higher negative docking energy and a higher probability of contact. Phenanthridinium Derivatives Synthesized via Gold-Catalyzed Artificial Metallocenzyme Synthesis for Prodrug Activation [28]. A novel approach in the realm of targeted drug delivery is the creation of a biotic metal-triggered prodrug mechanism that can control the release of bioactive medicines. Nowadays, prodrugs that use a biotic metal as a trigger heavily rely on encaging methods. Here, we outline a technique based on a prodrug's physiological hydroamination following activation by phenanthridinium and gold. To make the prodrug approach biocompatible, a gold artificial metalloenzyme (ArM) based on human serum albumin was utilized as a trigger for prodrug activation as opposed to the free gold metal complex. In vitro, the albumin-based gold ArM preserved the bound gold metal's catalytic activity even in the presence of up to 1 mM glutathione. The drug developed using the gold ArM demonstrated a therapeutic effect in cell-based testing, indicating the gold ArM's potential for use in anticancer applications.

13.5.2 In the field of dye sensitized solar cell

The extraordinary [Ru(L1)2(L2)]n(BF4)2n metal-organic polymer dye is described in this article. L1 stands for 2,2′-bipyridyl 4,4′-dicarboxylic acid, while L2 stands for N-Methyl Morpholine. The name of this dye is RuMOP-NMM1. The material was examined using the N-Methy UV-Vis absorption, Elemental Analyzer (CHNS/O), Raman, Thermo Gravimetric/Differential Scanning Calorimetry (TG/DSC), Nuclear Magnetic Resonance (NMR), Fluorescence, Infrared (FT-IR), and Cyclic Voltametry methods. The long-established organic basis at the centre of this color (CV). At longer wavelengths, the polynuclear molecule displays an absorption band with a high molar extinction coefficient. A brilliant light is produced by the complex of supramolecular molecules that is mediated by metals.

Therefore, compared to mononuclear complexes, these metal-organic polymers (RuMOP-NMM1) are more thermally and chemically stable. The synthesized metal-organic polymer is kinetically and energetically suitable to act as sensitizers in energy-relevant applications, specifically in dye-sensitized solar cells, as shown by these photoluminescence and electrochemical characterization results (DSSC). Ruthenium polypyridyl complexes with L-proline coordination and their photo reactivity in relation to steric bulk and solvent 29 If ruthenium polypyridyl complexes can successfully photo substitute one of their ligands while staying stable in the dark, they are attractive candidates for photo activated chemotherapy (PACT). Here, the ruthenium-based PACT compounds [Ru(bpy)2(L-prol)]PF6 ([1a]PF6; bpy = 2,2'-bipyridine and L-prol = L-proline), [Ru(bpy)(dmbpy)(L-prol), and [Ru(bpy)(dmbpy)(L-prol)]PF6 are used to investigate the use of the ([3a]PF6). Only two of the four potential regioisomers, [2a]PF6 and [2b]PF6, were produced as a result of the synthesis of the tris-heteroleptic complex containing the dissymmetric proline ligand. Spectroscopy and calculations based on the density functional theory were used to separate both isomers and provide comprehensive descriptions of them. The photo reactivity of all four complexes—[1a]PF6, [2a]PF6, [2b]PF6, and [3a]PF6—was investigated in acetonitrile (MeCN) and water using mass spectrometry, 1 H NMR spectroscopy, circular dichroism, UV-vis, and circular dichroism spectroscopy. H2O did not undergo photo substitution when exposed to visible light in the presence of oxygen; instead, the amine in compound [1a]PF6 was photo oxidized to an imine. Contrary to predictions, the ligand was not replaced by light in phosphate-buffered saline when the steric strain was increased by adding two ([2b]PF6) or four ([3a]PF6) methyl substituent's (PBS). However, the ability to transfer electrons by the methyl substituent's probably contributed to its resistance to photo oxidation. Despite the fact that [2b]PF6 was photo stable in PBS, [2a]PF6 quantitatively isomerized to [2b]PF6 when exposed to light. The non-strained molecule [1a]PF6 was photo stable in pure MeCN, in contrast to [2a]PF6 and [3a]PF6, which displayed non-selective photo substitution of the L-proline and dmbpy ligands. The selective photo substitution of L-proline by [3a]PF6 in H2O/MeCN solutions serves as evidence that the solvent has a significant impact on the photo reactivity of this family of complexes. It is explained how the polarity and coordination characteristics of the solvent affect the photochemical characteristics of polypyridyl complexes (Scheme 13.3).

13.6 Future prospective and scope

Inorganic medicinal chemistry has advanced significantly as a result of the current development of metal-complex based medicines, which helps to synthesis a small number of high metal-complex based symptomatic or therapeutic molecules. The rapid expansion following the notable success of the most well-known medication, cisplatin. The planned research will take new developments in medicinal inorganic chemistry into account, such as the use of the metal ions' nuclear and electronic characteristics in vivo via fluorescence, electronic properties, positron emission, or gamma emission.

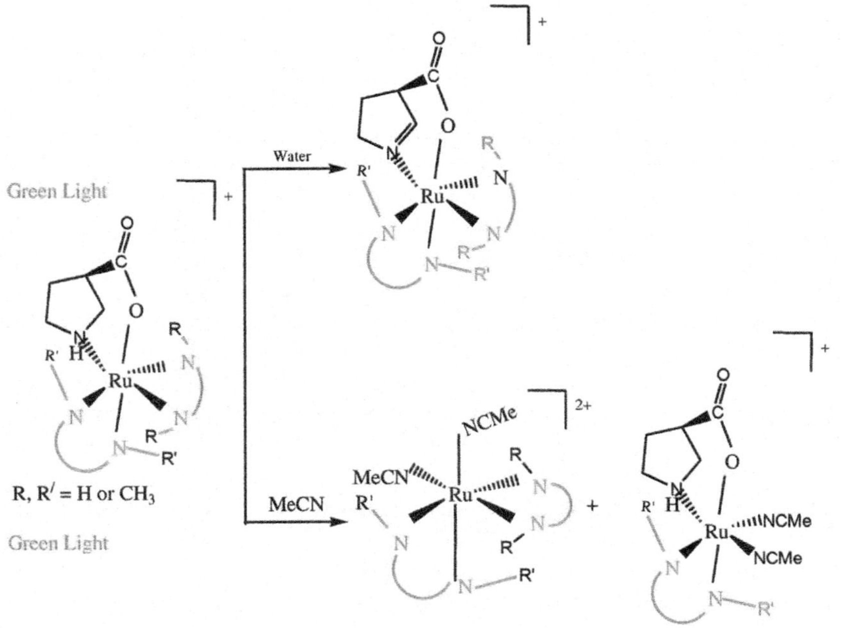

SCHEME 13.3 Photochemical characteristics of polypyridyl complexes. *Procreated from Jordi-AmatCuello-Garibo, Elena Pérez-Gallent, Lennard van der Boon, Maxime A. Siegler and Sylvestre Bonnet,* J. Inorg. Chem., *2017, 56, 4818–4828.*

13.7 Summary

The current summary of this chapter was based on a few noteworthy investigations on novel remedial compounds and the therapeutic applications of polynuclear organometallic complexes. An increasingly busy and productive area of research is the exploration of polynuclear organometallic compounds in a number of restorative fields. Potentiometric techniques and optical spectroscopy are used for monitoring intermediates and species in the solution state of the complexes, while spectroscopic characterization methods such as time-resolved infrared spectra are applied for discovering and structurally characterizing the prepared solid type of complex. Additionally, these preclinical and clinical considerations have a significant impact on how well we understand the composition and underlying mechanisms of therapeutic polynuclear organometallic complexes.

Problems with solutions

1. One of the ligands in the chemical compound $FeC_9H_7O_3Cl$ is a substituted cyclopentadienyl group. It is difficult to dissolve in hydrocarbon liquids since it is an ionic molecule. When the chemical is reacted with

AgNO₃ solution, a white precipitate results. If the 18 electron criteria are met, describe the structure of the object. Solution:
Solution:

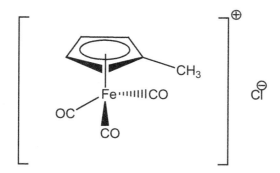

2. The synthesis of Mn(CO)₄(PPh₃)[C(O)CH₃] starting from Manganeseacetate, Mn(OAc)₂, should be described.
Solution:

$$2Mn(OAc)_2 + 4Na + 10CO \xrightarrow[\text{High Prussure}]{\text{High Temp.}} Mn_2(CO)_{10} + 4NaOAc$$

$$Mn_2(CO)_{10} + 2Na \rightarrow 2NaMn(CO)_5$$

$$2NaMn(CO)_5 + CH_3I \rightarrow CH_3Mn(CO)_5$$

$$CH_3C(O)Mn(CO)_5 \xrightarrow{E} CH_3C(O)Mn(CO)_5 PPh_3$$

Or at step 3 direct reactions with acyl chloride instead of MeI. Step 1 other reducing agents e.g., AlEt₃ can also be used.

3. Which of the following substances won't be subject to the oxidative addition of methyl iodide? Give arguments
(a) Ir(PPh₃)₂(CO)Cl
(b) [RhI₂(CO)₂]⁻
(c) η⁵-Cp₂Ti(Me)Cl
Solution:
(c) As titanium is already in its highest oxidation state of +4 and therefore no d electrons present for further oxidation

4. Mark against each statement the appropriate type of reaction or reactions from the list (oxidative addition, reductive elimination, migratory insertion, and β-H transfer)
(a) *cis* orientation of the participating ligands is a must.
(b) This reaction does not occur for d^0 metal complexes.
(c) This reaction is enthalpy favored and entropy prohibited.
(d) A vacant coordination site on the metal center is a prerequisite.
(e) The more electron rich the metal center, the more facile is the reaction.

(f) There is an increase in the electron count of the metal complex by two units during this reaction.

Solution:
(a) Red elimination and migratory insertion
(b) Oxidative addition
(c) Migratory insertion
(d) Oxidative addition and β-H transfer
(e) Oxidative addition
(f) Oxidative addition and β-H transfer

5. The values of n in the following complexes can be determined using the 18 electron rule as a reference.
   (a) Na$_2$Fe(CO)$_n$
   (b) MnBr(CO)$_n$
   (c) W(η^6-C$_6$H$_6$)(CO)$_n$
   (d) Rh(η^5-C$_5$H$_5$)(CO)$_n$
   (e) Cr(η^3-C$_5$H$_5$)(CO)$_n$(CH$_3$)
   (f) IrBr$_2$(CO)$_n$(PPh$_3$)$_2$(CH$_3$)

Solution:
(a) Na$_2$Fe(CO)$_n$ n = 4
(b) MnBr(CO)$_n$ n = 5
(c) W(η^6-C$_6$H$_6$)(CO)$_n$ n = 3
(d) Cr(η^3-C$_5$H$_5$)(CO)$_n$(CH$_3$) n = 4
(e) Rh(η^5-C$_5$H$_5$)(CO)$_n$ n = 2
(f) IrBr$_2$(CO)(PPh$_3$)$_2$(CH$_3$) n = 1

Objective type questions

Q1. Which metal comprises the Ziegler-Natta catalyst, an organometallic complex?

(a) Iron
(b) Zirconium
(c) Rhodium
(d) **Titanium**

Q2. Wilkinson's catalyst is used in

(a) Polymerization
(b) Condensation
(c) Halogenation
(d) **Hydrogenation**

Q3. Which of the following is an organo-metallic compound?

(a) Lithium ethoxide
(b) **Ethyl lithium**

(c) Lithium acetate
(d) Lithium carbide

Q4. Coordination molecules are crucial in biological systems. Which of the following claims is false in the present situation?

(a) Cyanocobalamin is B_{12} and contains cobalt.
(b) Hemoglobin is the red pigment of blood and contains iron
(c) Chlorophylls are green pigments in plants and contain calcium.
(d) Carboxypepticase-A is an enzyme and contains zinc.

Q5. The ethene-containing organometallic compound with -bonding is known as

(a) Ferrocene
(b) Dibenzene chromium
(c) Tetraethyl tin.
(d) Zeise's salt

Q6. Metals are coordinated by the amino acid histidine in the active sites of several enzymes. What element of histidine supplies the electrons required for the formation of the coordination bond?

(a) Oxygen
(b) Nitrogen
(c) Sulfur
(d) Carbon

Q7. What is the name of the following anticancer drug?

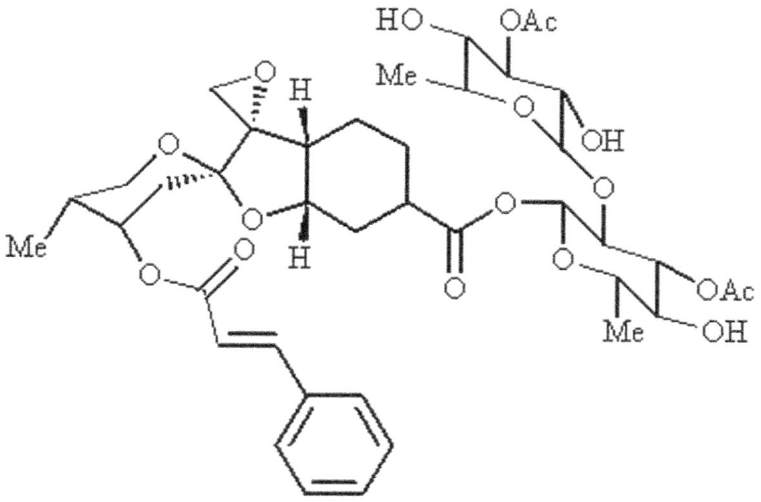

(a) Spongistatin 1
(b) **Cryptophycin 1**
(c) Phyllanthoside
(d) Maytansine 1

Q8. What kind of therapy includes giving a patient an antibody that is connected to an enzyme, causing the enzyme to activate a prodrug?

(a) ADAPT
(b) GDAPT
(c) **ADEPT**
(d) GDEPT

Q9. What exactly does synthetic biology mean?

(a) The synthesis of naturally occurring compounds.
(b) **The genetic modifications of microbial cells such that they produce compounds that they would not normally produce.**
(c) Researching the biosynthesis of naturally existing substances in living organisms
(d) The creation and synthesis of enzymes capable of performing a specific reaction during the synthesis of organic compounds.

Q10. Which of the following statements about the metal olefin complexes I [PtCl3(C2F4)] and (II) [PtCl3(C2H4)] is true?

(a) a metallacycle is formed in each complex
(b) carbon-carbon bond length in (ii) is smaller
(c) **carbon-carbon bond length in (i) is smaller**
(d) carbon-carbon bond length is same both in (i) and (ii)

References

[1] Sappa E. In: Abel EW, Stone FGA, Wilkinson G, Shriver DF, Bruce MI, editors. Comprehensive organometallic chemistry II, 7. Oxford: Pergamon Press; 1995, p. 803.
[2] Shriver DF, Bruce MI, Pergamon Press, Oxford, 1995, Vol. 7, Chap. 12, p. 683.
[3] Henkel G, Weissgraber S. In: Braunstein P, Oro LA, Raithby PR, editors. Metal clusters in chemistry, 1. Wienheim: Wiley-VCH Verlag GmbH; 1999, p. 63.
[4] Deeming A.J. Comprehensive organometallic chemistry II', eds. in-chief Abel EW., Stone FGA, Wilkinson G. 1982–1994.
[5] Abu-Surrah AS, Kettunen M, Leskelä M, Al-Abed Y. Platinum and palladium complexes bearing new (1R,2R)-(-)-1,2-Diaminocyclohexane (DACH)-based nitrogen ligands: evaluation of the complexes against L1210 leukemia. Z für anorganische und allgemeine Chem 2008;634 (14):2655–8. Available from: https://doi.org/10.1002/zaac.200800281 (October 2008).
[6] Rafique S, Idrees M, Nasim A, Akbar H, Athar A. Transition metal complexes as potential therapeutic agents. Biotechnol Mol Biol Rev 2010;5(2):38–45 (April 2010).
[7] IUPAC—Goldbook: organometallic compounds. *goldbook.iupac.org*. Retrieved 2 January 2021.

[8] Adams RD. In: Shriver DF, Kaesz HD, Adams RD, editors. The chemistry of metal cluster complexes. New York: VCH; 1990, p. 121. Chap. 3.
[9] Adams RD. In: Abel EW, Stone FGA, Wilkinson G, Adams RD, editors. Comprehensive organometallic chemistry II, 10. Oxford: Pergamon Press; 1995, p. 1. Chap. 1.
[10] Hall RJ, Serguievski P, Keister JB. Organometallics 2000;19:4499 and references therein.
[11] Matsubara K, Inagaki A, Tanaka M, Suzuki H. J AmChem Soc 1999;121:7421.
[12] 10. Crabtree, Robert. The organometallic chemistry of the transition metals. Wiley-Interscience; 2005, p. 159–80. ISBN 0–471-66256-9.
[13] Johnson C, Eisenberg R. Stereoselective oxidative addition of hydrogen to Iridium(I) complexes. Kinetic control based on ligand electronic effects. J Am Chem Soc 1985;107(11):3148–60.
[14] Hall TL, Lappert MF, Lednor PW. Mechanistic studies of some oxidative-addition reactions: free-radical pathways in the Pt^0-RX, Pt^0-PhBr, and Pt^{II}-R'SO$_2$X Reactions (R = alkyl, R' = aryl, X = halide) and in the related rhodium(I) or iridium(I) Systems. J Chem Soc, Dalton Trans 1980;8:1448–56. Available from: https://doi.org/10.1039/DT9800001448.
[15] The Organometallic Chemistry of the Transition Metals (6 ed.). Wiley. p. 173. ISBN 978-1-118–13807-6.
[16] Hartwig JF. Organotransition metal chemistry − from bonding to catalysis. University Science Books; 2010, p. 321. ISBN 978-1-891389-53-5.
[17] Gillie A, Stille JK. Mechanisms of 1,1-reductive elimination from palladium. J Am Chem Soc 1980;102(15):4933–41. Available from: https://doi.org/10.1021/ja00535a018.
[18] Okrasinski SJ, Nortom JR. Mechanism of reductive elimination. 2. Control of dinuclear vs. mononuclear elimination of methane from cis-hydridomethyltetracarbonylosmium. J Am Chem Soc 1977;99:295–7. Available from: https://doi.org/10.1021/ja00443a076.
[19] Elschenbroich C. Organometallics. Weinheim: Wiley-VCH; 2006. ISBN 3–527-29390-6.
[20] Feizi-Dehnayebi M, Dehghanian E, Mansouri-Torshizi H. A novel palladium(II) antitumor agent: synthesis, characterization, DFT perspective, CT-DNA and BSA interaction studies via in-vitro and in-silico approaches. Spectrochim Acta Part A Mol BiomolSpectrosc 2021;249:119215.
[21] Jakob CHG, Muñoz AW, Schlagintweit JF, Weiß V, Reich RM, Sieber SA, et al. Anticancer and antibacterial properties of trinuclear Cu(I), Ag(I) and Au(I) macrocyclic NHC/urea complexes. J Organomet Chem 2021;932:121643.
[22] Chen C, Zhou L, Xie B, Wang Y, Ren L, Chen X, et al. Dalton Trans 2020. Available from: https://doi.org/10.1039/C9DT04751D.
[23] Alsaeedi MS, Babgi BA, Abdellattif MH, Jedidi A, Humphrey MG, Hussien MA. DNA-binding capabilities and anticancer activities of Ruthenium(II) cymene complexes with (poly)cyclic aromatic diamine ligands. Molecules 2021;26:76. Available from: https://doi.org/10.3390/molecules26010076.
[24] Beckford FA, Niece MB, Lassiter BP, et al. Polynuclear ruthenium organometallic complexes containing a 1,3,5-triazine ligand: synthesis, DNA interaction, and biological activity. J Biol Inorg Chem 2018;23:1205–17. Available from: https://doi.org/10.1007/s00775-018-1599-8.
[25] Smith G, Chellan P, Land K, Shokar A, Au A, An S, et al. Synthesis and evaluation of new polynuclear organometallic 1 Ru(II), Rh(III) and Ir(III) pyridyl ester complexes as *in vitro* antiparasitic and antitumor agents. Dalton Trans 2013. Available from: https://doi.org/10.1039/C3DT52090K.

[26] Ismail AM, Butler IS, Abou El Maaty WM, Mostafa SI. Anticancer and DNA Interaction of New Zinc(II), Palladium(II), Platinum(II) and Silver(I) Complexes Based on Indol-3-Acetic Acid; Solid and Solution Studies. Polycycl Aromat Compd 2021. Available from: https://doi.org/10.1080/10406638.2021.1892779.

[27] Chang TC, Vong K, Yamamoto T, Tanaka K. Prodrug activation by gold artificial metalloenzyme-catalyzed synthesis of phenanthridinium derivatives via hydroamination. Angew Chem 2021;60(22):12446−54.

[28] Cuello-Garibo JA, Pérez-Gallent E, Van der Boon L, Siegler MA, Bonnet S. Influence of the steric bulk and solvent on the photoreactivity of ruthenium polypyridyl complexes coordinated to l-Proline. J Inorg Chem 2017;56:4818−28.

CHAPTER

14

Role of organometallic compounds in neglected tropical diseases

Deepak Yadav[1], Sushmita[2], Shramila Yadav[3], Sunil Kumar[4], Manjeet Kumar[5] and Vinod Kumar[5]

[1]Department of Chemistry, Gurugram University, Gurugram, Haryana, India, [2]Department of Chemistry, Netaji Subhas University of Technology, Dwarka, Delhi, India, [3]Department of Chemistry, Rajdhani College, University of Delhi, New Delhi, Delhi, India, [4]Department of Chemistry, Government P.G. College, Hisar, Haryana, India, [5]Department of Chemistry, Central University of Haryana, Mahendergarh, Haryana, India

14.1 Introduction

The World Health Organisation (WHO) considered a group of several communicable and poverty-related diseases as "Neglected Diseases" due to the restricted development of their diagnostic method and therapeutic treatment. The neglected diseases disproportionately affect the people living in tropical areas and are generally called neglected tropical diseases (NTDs). The WHO currently prioritized 20 NTDs, namely, Buruli ulcer, Chagas disease, dengue and chikungunya, dracunculiasis (Guinea-worm disease), echinococcosis, foodborne trematodiases, human African trypanosomiasis (sleeping sickness), leishmaniasis, leprosy (Hansen's disease) lymphatic filariasis, mycetoma, chrornoblastomycosis and other deep mycoses, onchocerciasis (river blindness), rabies, scabies and other ectoparasitoses, senistosomiasis, soil-transmitted helminthiases, snakebite envenoming, tenias/cysticercosis, trachoma, and yaws and other endemic treponematoses [1]. These diseases are caused by different pathogens, such as bacteria, parasitic worms, protozoa, and viruses and mainly thrive among the poorest people who live in rural, urban slum or conflict areas where access to healthcare, clean water and adequate sanitation is limited. The NTDs are present in 149 countries, affect over 1 billion people and impose a devastating economic and social burden worldwide. The unavailability of treatment and timely care leaves millions of people severely disabled, disfigured or debilitated. Unfortunately, the pharmaceutical industries either ignored or have less interest in developing novel drugs for NTDs because most patients with such diseases lives in less developed countries with low socio-economic status and they are not able to pay for expensive drugs.

In 2011 the World Health Organisation made a roadmap with a target to eliminate several of NTDs by 2020 and intensified control of the remaining diseases [2]. Although, after the adoption of the roadmap in 2012, significant progress has been made to overcome the global impact of NTDs (2012−2020), however, many of the targets were not achieved by 2020. WHO has revised its NTDs roadmap (2021−2030) to control, prevent, eliminate, or eradicate such diseases [3].

Different types of drugs are available in the market for the treatment of various NTDs but their side effects, unavailability and lack of selectivity, limit the use of such drugs and there is a need to modify the existing drugs or develop new drugs for the treatment of NTDs. Organometallic compounds, containing at least one carbon-metal bond, are much more known for catalytic activities rather than their medicinal applications. The specific and unique characteristics of organometallic compounds, such as structural diversity, redox properties, easy synthesis and the high possibility of ligand exchange, always attracted scientists to explore their applications in medicine [4,5]. Over the past few years a several drug candidates have been developed, mostly derived by attaching an organometallic moiety to the existing drugs to increase the activity of these drugs or to overcome the side effects of existing drugs. In this chapter, various organometallic compounds which are effective against various NTDs have been presented and discussed.

14.2 Diseases

14.2.1 Chagas disease

Chagas disease, also known as American trypanosomiasis is caused by the protozoan parasite *Trypanosoma cruzi* [6]. It spreads through the feces of bloodsucking triatomine bugs, also called kissing bugs. The triatomine bug gets infected by parasites during sucking the blood of an infected person and transfers the parasites to a noninfected person by deposing feces on them. These parasites penetrate through the soft skin of the eyes or the broken skin and enter to the human body. Alternatively, Chagas disease also spread by the consumption of contaminated drinks and food. The common symptoms of this disease include headache, skin lesions, fever, and swelling of tissues and lymph nodes. Over the last few decades, considerable progress has been made in the treatment and control of the Chagas disease [7], still it is one of the major health problems affecting many people worldwide. According to WHO, it is estimated that about 6−7 million people worldwide, mostly from Latin America get infected with Chagas disease. Currently, nitro compounds like benznidazole and nifurtimox are recommended drugs for the treatment of Chagas disease (Fig. 14.1) [8].

The several toxic side effects, low efficacy, and inability or less effective to treat different stages of the disease limit the use of these drugs and further developments are still required. Due to metal-drug synergism, various transition metal-drug complexes have been developed and tested for the treatment of Chagas disease [9]. In the next section, we will discuss about the organometallic compounds that exhibit antitrypanosomal activity.

14.2.1.1 Ruthenium-azole complexes

The sterol biosynthesis inhibitors, such as ketoconazole (KTZ) **1** and clotrimazole (CZT) **2**, exhibit trypanocidal activity but their toxic effects and low efficacy leave them far from ideal (Fig. 14.2) [10]. In addition to trypanocidal activity, these azole compounds show binding affinity toward transition metals *via* their imidazole moieties. Based on this characteristic numerous metal-azole complexes have been synthesized. The first inorganic complex of CZT was with ruthenium $RuCl_2(CZT)_2$ **3**, which displayed anti-*T. cruzi* activity on the proliferation of its epimastigoat form. The activity of $RuCl_2(CZT)_2$ is higher than free clotrimazole and ketoconazole [11]. Also, $RuCl_2(CZT)_2$ was found to be nontoxic to the mammalian cells.

Inside the cell, the $RuCl_2(CZT)_2$ undergoes hydrolysis, releasing one CZT fragment that exerts sterol biosynthesis inhibition, and the remaining fragment (Ru-CZT) causes parasitic nuclear damage. Therefore the higher activity of $RuCl_2(CTZ)_2$ complex is due to its multitargeting properties. Later on, various M-CTZ and M-KTZ (M = Ru, Rh, Au and Cu) complexes have been developed to enhance the activity of CTZ and KTZ [12,13]. Unfortunately, due to the low solubility of these complexes in an aqueous

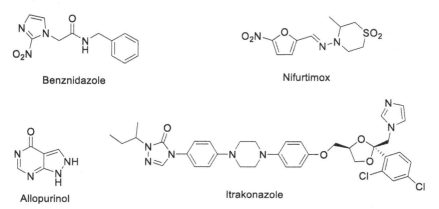

FIGURE 14.1 Chemical structures of selected drugs which are effective against Chagas disease.

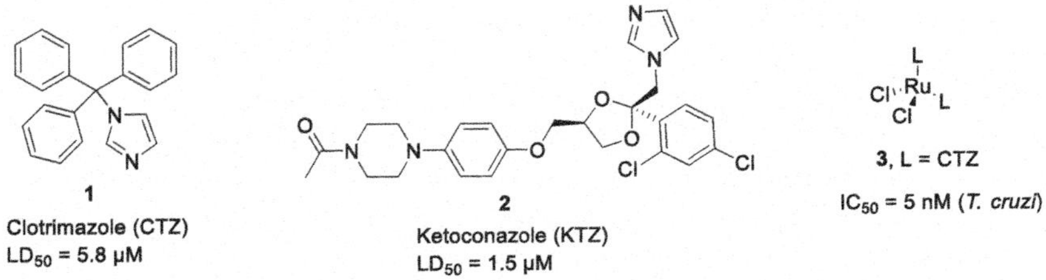

1
Clotrimazole (CTZ)
LD$_{50}$ = 5.8 μM

2
Ketoconazole (KTZ)
LD$_{50}$ = 1.5 μM

3, L = CTZ
IC$_{50}$ = 5 nM (*T. cruzi*)

FIGURE 14.2 Structures of clotriamzole (CTZ), ketoconazole (KTZ) and inorganic complex of RuCl$_2$(CTZ)$_2$.

medium, their in vivo applications are restricted. The shifting from CZT-metal inorganic complexes to organometallic complexes, such as CZT or KTZ-Ru complexes (**4–7**), containing *p*-cymene ligand (Fig. 14.3) showed improvement in activity against *T. cruzi* epimastigotes, with higher water solubility and low toxicity to human osteoblasts and fibroblasts [14,15]. In the clotrimazole complexes, **7a** was found most active (LD$_{50}$ = 0.1 μM) and exhibited 58 times higher activity than free clotrimazole and six-fold than RuCl$_2$(CTZ)$_2$ complex. The Ru-KTZ complexes showed lower activity compared with Ru-CTZ complexes. Inside the cell, Cl ligands are substituted by water, generating Ru-KTZ cationic complex which could release KTZ and inhibits sterol biosynthesis. The remaining cationic ruthenium species might causes nuclear damage by interaction with parasitic DNA.

Another organoruthenium complex of CZT with cyclopentadienyl ligand [RuCp(PPh$_3$)$_2$(CTZ)][(CF$_3$SO$_3$)] **8a** (Fig. 14.3) was found to be cytotoxic on *T. cruzi* epimastigotes (IC$_{50}$ = 0.25 μM), showing a six-fold higher activity than free CTZ and 30 times higher activity compared with the reference drug nifurtimox (IC$_{50}$ = 8.0 μM) [16]. The mode of action of the ruthenium-cyclopentadienyl complex is different from the ruthenium-*p*-cymene complexes. It does not damage the parasitic DNA but it inhibits the biosynthetic pathway of transformation of squalene to squalene oxide in *T. cruzi*.

14.2.1.2 Ruthenium-hydrazone complexes

It has been reported that aryl(4-oxothiazolyl)-hydrazone (AZT) **9** derivatives act as cysteine protease cruzain (TCC) enzyme inhibitors but overall effect of AZT increases on binding with Ruthenium metal [17,18]. The RuCl$_2$(AZT)(COD) complexes **10a-h** (Fig. 14.4) were found to be active against the epimastigote (IC$_{50}$ = 1.8–87.4 μM) and trypomastigote form (IC$_{50}$ = 3.3–27.2 μM) of *T. cruzi*.

The cytotoxicity of these organometallic complexes was tested using BALB/c mice splenocytes. Even though, many of these complexes displayed higher activity than the corresponding free AZT but all these complexes except **10h** showed undesired cytotoxicity to splenocytes.

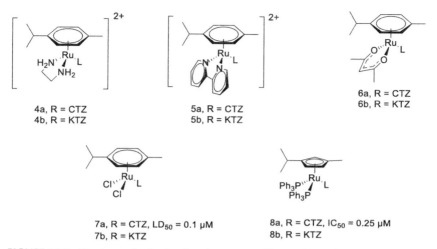

FIGURE 14.3 Structures of ruthenium-based organometallic complexes of CTZ and KTZ.

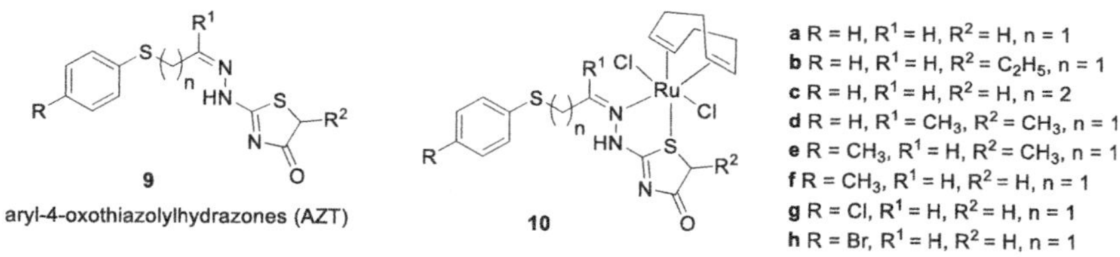

FIGURE 14.4 Structure of aryl-4-oxothiazolylhyrazone (ATZ) and Ru-ATZ complex [RuCl$_2$(ATZ)(COD)].

14.2.1.3 Ruthenium-thiosemicarbazone complexes

5-Nitrofuran based compound Nifurtimox (Nfx) is an effective drug for the treatment of Chagas disease. Continuing the approach of attachment of metal ions with bioactive ligands to improve activity of existing drugs several studies have been conducted. It has been reported that the organoruthenium complexes with 5-nitrofuryl-containing thiosemicarbazone derivatives and a *p*-cymene ligand i.e. [Ru$_2$(p-cymene)$_2$L$_2$]X$_2$ where, X = Cl or PF$_6$ and L = 5-Nitrofuryl-containing thiosemicarbazones) **11a−11c** showed very low in vitro activity on epimastigote form but higher activity on trypmastigoate form of *T. cruzi* (Fig. 14.5) [19]. Out of these Ru complexes, **11b** was found most active against trypmastigoate form of *T. cruzi* and its activity was higher as compared to free Nifurtimox and thiosemicarbazone.

The mechanism of antiparasitic action involves the generation of reactive organic species by the reduction of the nitro group of 5-nitrofuryl moiety and/or inhibition of cruzipain by thiosemicarbazone pharmacophore.

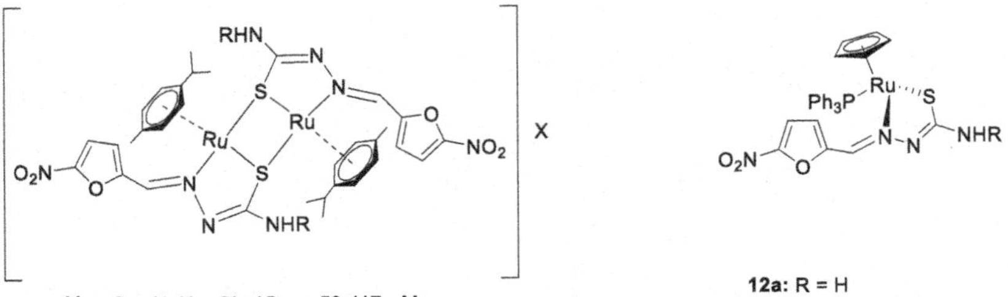

11a: R = H, X = Cl₂ IC₅₀ = 59-117 µM
11b: R = Ph, X = Cl₂ IC₅₀ = 8.68-11.69 µM
11c: R = Me, X = (PF₆)₂, IC₅₀ = 193-231 µM

12a: R = H
12b: R = Me IC₅₀ = 0.41 µM
12c: R = Et

FIGURE 14.5 Structures of ruthenium-thiosemicarbazone organometallic complexes.

Also, Ru-arene could cause nuclear damage through interactions with parasitic DNA. The other organometallic complexes with formula RuCp(PPh₃)(L), L = 5-nitrofuryl-TSC **12a−12c** were also found active against *T. cruzi* trypomastigote and were observed more effective on *T. cruzi* than that of [Ru₂(p-cymene)₂L₂]X₂ complexes [20]. Compound **12b** was found 49 times more active (IC₅₀ = 0.41 µM) than Nifurtimox (IC₅₀ = 20.1 µM). Importantly, Ru-Cp complexes showed excellent selectivity toward *T. cruzi* parasite (SI > 49) than the mammalian cells.

In comparison to free bioactive ligands, some ruthenium complexes exhibited higher antiparasitic activity. Interestingly, the Ru-Cp complexes are more effective compared to other Ru classical and organometallic complexes and showed selectivity toward the parasites compared to mammalian cells. Further studies of these complexes can be helpful in developing the antiparasitic agents.

14.2.1.4 Ferrocenyl- and cyrhetrenyl-imine complexes

Organic molecules, such as 5-nitrothiophene and 5-nitrofuran derivatives, produce free radicals during bio-reduction of the nitro group to nitro anion radical and exhibit antitrypanosomal activity [21,22]. When 5-nitroheterocyclic moieties are attached to ferrocenyl (Fc) and cyrhetrenyl (Cy) through imine linker then it results in enhancement in their activity. The complexes (**13** and **14**; Fig. 14.6) showed in vitro activity against epimastigote and trypmastigote form. The activity of compounds containing electron-withdrawing cyrhetrenyl group (**14a** and **14b**) was higher than compounds containing electron-donating ferrocenyl group (**13a** and **13b**) against *T. cruzi* epimastigote and their activity potential was almost similar to that of Nifurtimox. Compound **14a** (IC₅₀ = 3.7 ± 0.1 µM) and **14b** (IC₅₀ = 3.7 ± 0.7 µM) showed much higher activity against trypomastigote form than that of Nifurtimox (IC₅₀ = 19.8 µM) and activity of **13a** (IC₅₀ = 12.3 ± 1.7 µM) and **13b** (IC₅₀ = 18.3 ± 0.7 µM) were comparable with Nifurtimox [23].

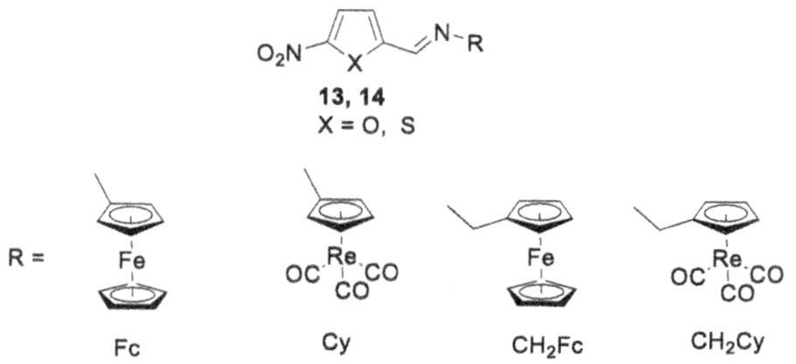

Role of organometallic compounds in neglected tropical diseases Chapter | 14 359

FIGURE 14.6 Molecular structure of furan and thiophene based organometallic compounds.

The electronic communication between organometallic fragment and 5-nitroheterocylic moiety via conjugated imine system is the key factor in antichagasic activity. The cyrhetrenyl fragment enhances the reduction of NO_2 group to NO_2^- radical anion. In comparison to conjugated complexes, nonconjugated analogous of these complexes, that is, presence of alkyl group between the imine and the organometallic moiety, are less active. Although the initial results are based on in vitro studies but the comparison of electronic effect can be helpful and interesting factor in developing new promising anti-*T. cruzi* drugs.

14.2.2 Human African trypanosomiasis

Human African trypanosomiasis, also known as sleeping sickness is a parasitic disease caused by unicellular protozoan parasite species named *Trypanosoma brucei*. Either of the subspecies of *T. brucei*; *T. brucei gambiense* and *T. brucei rhodesiense* can cause infection which is transmitted by the bite of infected tsetse flies. The *T. brucei rhodesiense* and *T. brucei gambiense* affect millions of people in southern/eastern and central/western sub-Saharan Africa, respectively, (Fig. 14.7) [24].

In the early stage of infection, *T. brucei* causes itching, headache, swelling of organs and fever. Later, it disrupts the sleep/wake cycle and causes sleeping sickness. Currently, there are few drugs, such as Sumarin, Pentamidine, Eflornithin, Melarsoprol, and Nifurtimox, available for the treatment of sleeping sickness [25]. However, toxicity and effectiveness depending on stage and type of sleeping sickness is still a problem with use of these drugs.

14.2.2.1 Ferrocenyl and ruthenocenyl complexes of quinoline and benzimidazole

The antimalarial drug ferroquine (FQ) showed in vitro trypanocidal activity against both subspecies of *T. brucei* [26]. In in vivo murine studies, the

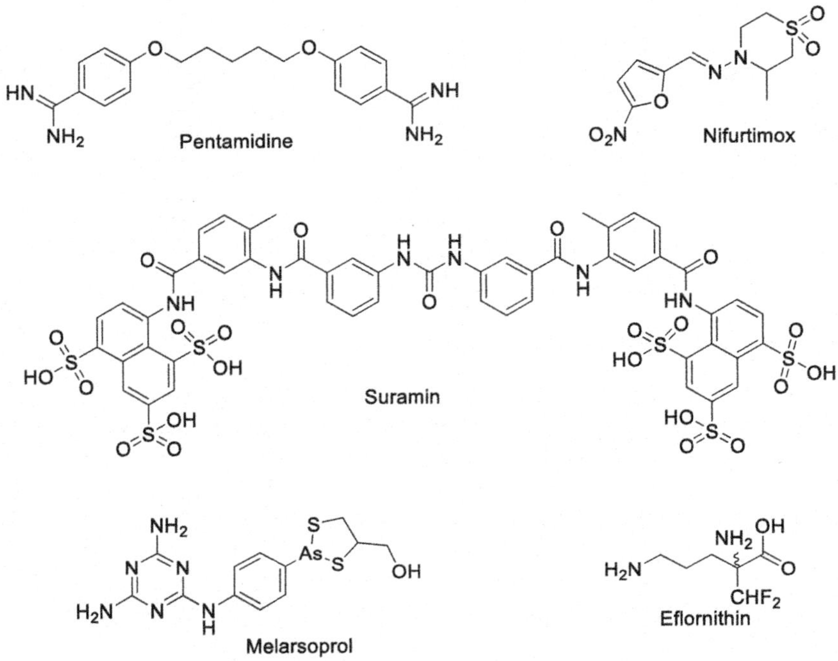

FIGURE 14.7 Some selected drugs for treatment of Human African trypanosomiasis.

ferroquine was found to be nontoxic at low concentrations (50 and 20 μmol/kg). Even though at low concentration there was the delay in appearance of parasites in blood was observed but still it was unable to cure the infected mice. In order to increase the trypanosomal activity of FQ, various organometallic complexes (Fig. 14.8) have been developed [27].

The in vitro studies on *T. brucei gambiense* showed that the attachment of methyl group on nitrogen at C-4 of quinoline ring of compound FQ (**18a**) and RQ (**18b**) results in a decrease in activity (Fig. 14.8). The ferrocenyl complexes of benzimidazole **15a** ($IC_{50} = 10.99$ μM) and **15b** ($IC_{50} = 6.80$ μM) showed higher activity than ferrocenyl-benzimidazole compounds with dimethylaminomethyl substituent **16a** ($IC_{50} > 100$ μM) and **16b** ($IC_{50} = 34.15$ μM). There was no significant change in the activity on attaching two benzimidazole groups to the ferrocene moiety (**17a** and **17b**). The in vitro study of various ferrocenyl and ruthenocenyl complexes of quinoline showed that ruthenocenyl complexes (**18b** and **19b**) were found more selective than ferrocenyl complexes (**18a** and **19a**). The ferrocenyl derivatives of benzimidazole are even less selective than FQ. In vivo studies of some selected compounds (**15b** and **RQ**) on *T. brucei brucei* infected Swiss mice showed no toxicity and were unable to lessen parasitemia or cure the infected mice at 50 and 100 μmol/kg doses.

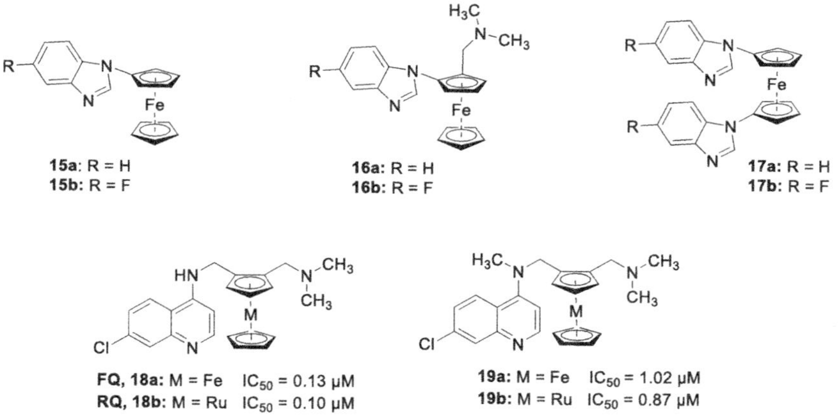

FIGURE 14.8 Ferrocenyl and ruthenocenyl complexes of quinoline and benzimidazole derivatives.

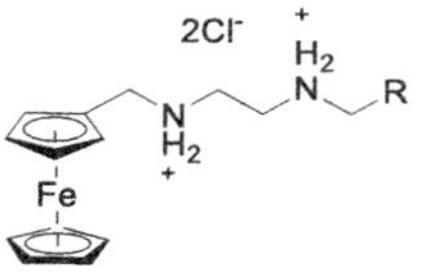

20a: R = C₆H₅
20b: R = 4-OMeC₆H₄
20c: R = 2-OMeC₆H₄
20d: R = 2-OHC₆H₄
20e: R = Fc
20f: R = Py (3Cl-)
20g: R = 2,4-ClC₆H₃
20h: R = 2-BrC₆H₄
20i: R = 4-NO₂C₆H₄
20j: R = 4(R)-7-chloroquinoline

FIGURE 14.9 Structure of various ferrocenyl-diamine complexes.

14.2.2.2 Ferrocenyl-diamine complexes

The ferrocenyl diamines (**20a-20j**) were tested in vitro against *T. brucei* 427, using pentamidine as a reference drug (Fig. 14.9). The complexes showed potent activity against *T. brucei* 427 with IC$_{50}$ values of 0.35–10.42 μM. Cytotoxicity of these compounds was examined on HepG2 cells. Among these ferrocenyl derivatives, some showed significantly higher antitrypansomal activity together with good selectivity indexes (SI = 26.29–63.90) than the pentamidine reference drug (IC$_{50}$ = 6.34 μM; SI = 24.84) and benznidazole (IC$_{50}$ = 207.57 μM; SI = 10.24) [28].

It is thought that generation of cytotoxic radical species is facilitated by the conversion of ferrocene to ferrocenium ion, foiling trypanosomal defense mechanism against oxidative stress and ability of ferrocenium ion to form charge transfer complexes with trypanosomal proteins that may play a role in its trypanocidal action.

14.2.2.3 Ruthenium-thiosemicarbazone complexes

The drug Nifurtimox is mainly used to treat Chagas disease but it also exhibits some effectiveness against human African trypanosomiasis. For the late stage of *T. b. gambiese* HAT, the nifurtimox/eflornithine combined therapy

(NECT) is used as the first-line treatment [29,30]. The NECT consists of taking Nfx pills and receiving 14 intravenous infusions of eflornithine over the course of seven days. A spinal tap is needed to qualify for this treatment. This therapy must be administered in a hospital setting, which is inconvenient for rural patients. It has been reported that organoruthenium complexes (**11a−11c**; Fig. 14.5), containing two nifurtimox-like ligands were found active against *T. b. brucei* 427 (IC_{50} = 2.9 μM, 0.5 μM and 10.6 μM) with a dual mechanism of action [31]. Compounds **11a** and **11b** exhibited higher activity against *T. b. brucei* than compound **11c**. The organoruthenium component in these metal complexes appeared to facilitate the mode of action. However, the only ruthenium-containing starting material does not exhibit any activity against *T. b. brucei*, demonstrating the ligand importance in ruthenium complexes.

14.2.3 Leishmaniasis

Leishmaniasis is a parasitic disease caused by protozoa parasites from more than 20 different species of *Leishmania*. The parasite spread in humans by the bite of female phlebotomine sandflies [32]. According to WHO, around 1 million new cases of Leishmaniasis come annually. There are several different forms of leishmaniasis, the most common forms are Visceral leishmaniasis, Cutaneous leishmaniasis and Mucocutaneous leishmaniasis. Visceral leishmaniasis (VL) is the most fatal form, also known as Kala-azar caused by leishmania donovnia and leishmania infantum. It causes weight loss, irregular fever, anemia, and enlargement of liver and spleen. In 2020 over 90% of new cases of this disease occurred in India, China, Yemen, Brazil, Kenya, Ethiopia, Eritrea, Somalia, Sudan, and South Sudan. Cutaneous leishmaniasis (CL), caused by leishmania major, is the most common type of leishmaniasis. It causes lesions and skin ulcer. In 2020, more than 85% of new cases of this occurred in Afghanistan, Iraq, Pakistan, Syrian Arab Republic, Brazil, Columbia, Peru, Algeria, Libya, and Tunisia. Mucocutaneous leishmaniasis (ML) caused by Leishmania braziliensis, destroy the mucous membranes of mouth, nose and throat. More than 90% of cases of ML are found in Brazil, Bolivia, Ethiopia and Peru [33].

A variety of treatments exists for the three forms of leishmaniasis. The antifungal Amphotericin B is used to treat VL [34]. The two antimony-based compounds, that is, sodium stibogluconate and meglumine antimoniate are also used for the primary treatment of CL and ML (Fig. 14.10) [35,36].

In 2014, the US Food and Drug Administration approved miltefosine, an anticancer drug, for the treatment of leishmaniasis [37]. Miltefosine is the only oral drug available for the treatment of leishmaniasis but its teratogenicity is a major concern and cannot be administered during pregnancy. So, further developments are still required to treat leishmaniasis.

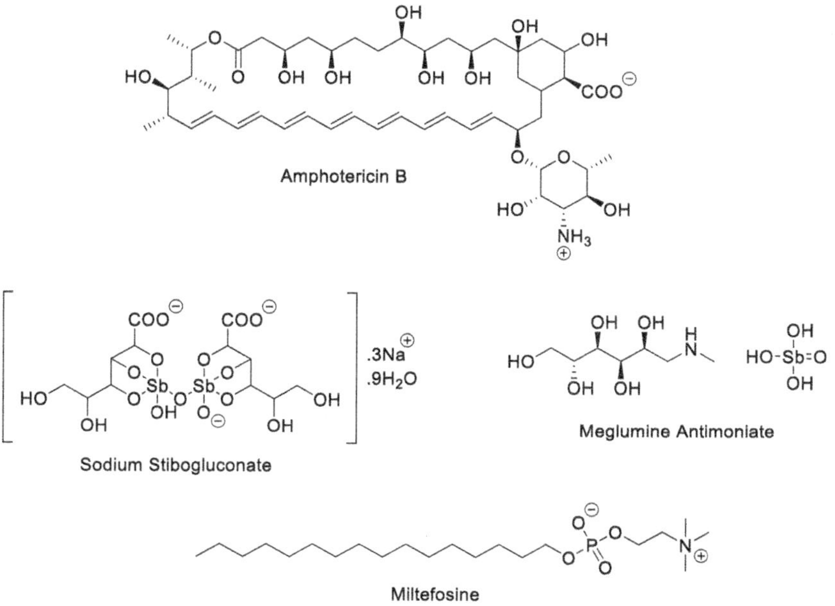

FIGURE 14.10 Chemical structures of some recommended drugs for treatment of leishmaniasis.

14.2.3.1 Ruthenium-azole complexes

The ruthenium complexes (**4–7**; Fig. 14.3) of CZT and KTZ showed potential against *Chagas* disease and can also treat *leishmaniasis* disease. Various ruthenium complexes with ruthenium in center and coordinated with four ligands exhibit in vitro activity against promastigotes form of *L. major* ($IC_{50} = 0.015–9.53$ μM) [14,15]. The activity of organoruthenium complexes containing KZT or CZT is higher than the free KZT or CZT as well as ruthenium complexes lacking KZT or CZT. Complex **7a** ($IC_{50} = 0.015$ μM) was found to be 110 times more effective against *L. major* than free CZT and did not show any toxicity to human osteoblasts. Also, the antileishmanial activity of complex **7a** is retained against *L. major* amstigote ($IC_{70} = 29$ nM). On coordination with ruthenium, the activity of KTZ and CZT was increased, with low toxicity to human osteoblasts and fibroblasts.

14.2.3.2 Iron-based complexes

The antimalarial drug primaquine **21** exhibits activity against visceral leishmaniasis (VL). Due to premature oxidative deamination (Fig. 14.11), it has no applicability in a clinical setting for this disease.

However, the complex of primaquine with ferrocenyl **22** in which ferrocene unit was appended to 8-NH_2 position of primaquine via glycine linker

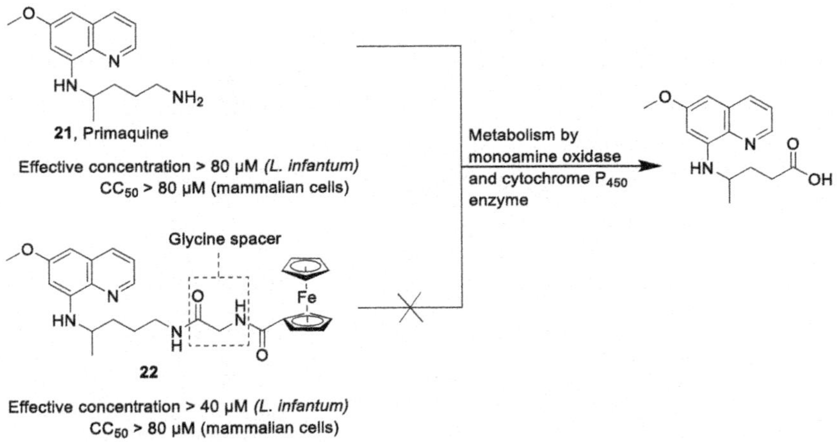

21, Primaquine
Effective concentration > 80 µM *(L. infantum)*
CC$_{50}$ > 80 µM (mammalian cells)

Metabolism by monoamine oxidase and cytochrome P$_{450}$ enzyme

22
Glycine spacer
Effective concentration > 40 µM *(L. infantum)*
CC$_{50}$ > 80 µM (mammalian cells)

FIGURE 14.11 Primaquine and its structural ferrocenyl analog.

showed significant activity against *L. infantum* [38]. The organometal complex (**22**; Fig. 14.11) prevents the metabolism of primaquine and exhibited good in vitro activity against the intracellular amastigoate form. Complex **22** was found less cytotoxic to mammalian macrophages as compared with reference antileishmainal drugs sitamaquine and miltefosine. The other ferrocene organometallics containing different *N*-heterocycles, such as imidazole, quinoline, pyridine, and imidazole derivatives, attached to ferrocene through amide or ester bridging were studied (**23–25**, Fig. 14.12) [39].

In vitro studies on *L. infantum* promastigotes showed that such compounds are less potent than reference drug miltefosine. However, the cytotoxicity test of these compounds on human macrophages confirmed that they have a higher therapeutic index with good selectivity than miltefosine. Further analysis on Leishmania amastigote form showed that these compounds exhibit high activity against intracellular amastigotes (IC$_{50}$ = 5.19–5.73 µM) and higher selectivity (SI = 16–89) as compared with reference drug, miltefosine (IC$_{50}$ = 11.0 µM; SI = 6). The increased antileishmanial activity of these complexes might be due to oxidation of Fe(II) to Fe(III) inside the host cell which is facilitated by the intracellular conditions.

14.2.3.3 Palladium-imine complexes

Palladium is an attractive candidate for drug compounds due to its ability to intercalate with DNA and forms square planar complexes in +2 oxidation state. The compounds of palladium with two types of imine ligands (**26** and **27**; Fig. 14.13) exhibited activity against the amastigote form *L. donovani* [40].

The free imine ligands and their complexes with palladium despite being less active are more selective than the reference drug; pentamidine.

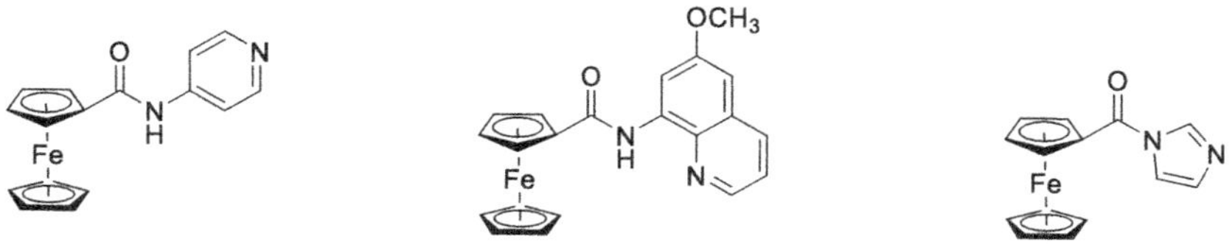

Role of organometallic compounds in neglected tropical diseases Chapter | 14   365

23
IC$_{50}$ = 5.73 μM
TI = 88.5

24
IC$_{50}$ = 5.19 μM
TI = 16.6

25
IC$_{50}$ = 5.69 μM
TI = 56

FIGURE 14.12 Antileishmanial compounds with activity against *L. Infantm*.

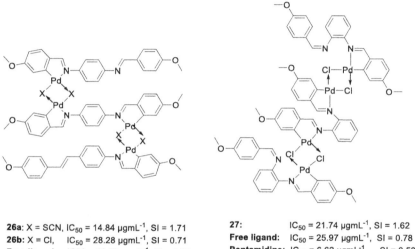

26a: X = SCN, IC$_{50}$ = 14.84 μgmL^{-1}, SI = 1.71
26b: X = Cl,   IC$_{50}$ = 28.28 μgmL^{-1}, SI = 0.71
Free ligand: IC50 = 41.70 μgmL^{-1}, SI = 0.61

27:   IC$_{50}$ = 21.74 μgmL^{-1}, SI = 1.62
Free ligand: IC$_{50}$ = 25.97 μgmL^{-1}, SI = 0.78
Pentamidine: IC$_{50}$ = 6.62 μgmL^{-1},   SI = 0.58

FIGURE 14.13 Structures of cyclometalated organopalladium(II) complexes with activity against amastigote form of *L. donovani*.

14.2.4 Echinococcosis

Echinococcosis is an infectious disease caused by parasites, namely tapeworm of the genus *Echinococcous*. It is transferred to humans through direct contact with infected animals. *Echinococcosis* occurs in 4 forms: cystic echinococcosis (caused by *E. graulosus*), alveolar echinococcosis (caused by *E. multiloculairs*), polycystic echinococcosis (caused by *E. vogeli*), and unicystic echinococcosis (caused by *E. oligarthrus*) [41]. Out of these cystic echinococcosis (CE) and alveolar echinococcosis (AE) are clinically relevant and important form [42]. CE causes hydatid cysts most often in the liver, lungs and less frequently in central nervous system, kidneys and bones. The common symptoms of CE are vomiting, nausea, chest pain, abdominal pain and shortness of breath. While AE causes tumor like lesion in the brain, lungs, liver and other organs, which

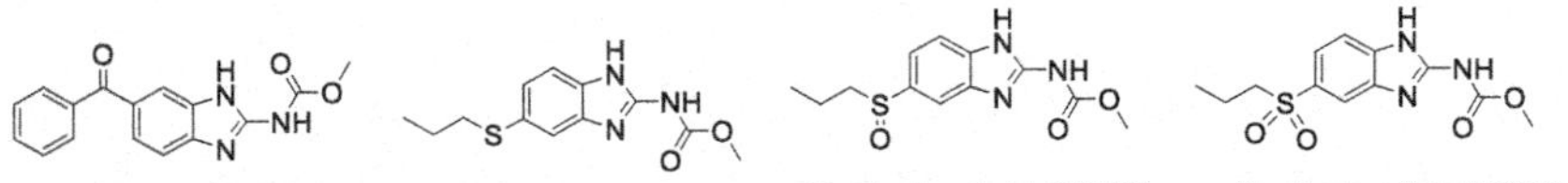

Mebendazole (MBZ) Albendazole (ABZ) Albendazole sulfoxide (ABZSO) Albendazole sulfone (ABZSN)

FIGURE 14.14 Recommended drugs for the treatment of ecchinococoosis.

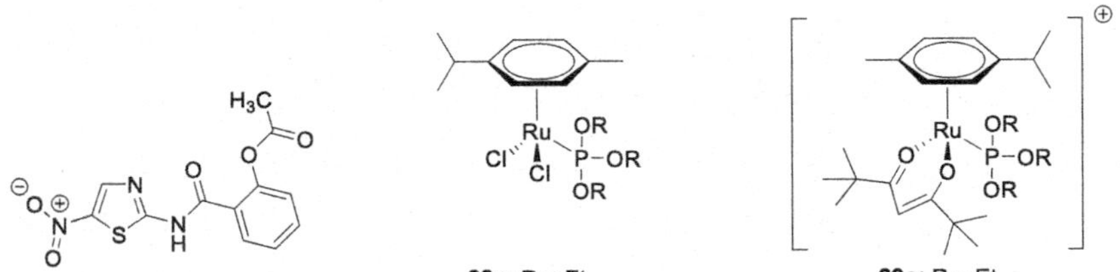

Nitazoxanide

28a: R = Et
28b: R = *i*Pr
28c: R = Ph

29a: R = Et
29b: R = *i*Pr
29c: R = Ph

FIGURE 14.15 Chemical structure of nitazoxanide and η^6-areneruthenium(II)phosphate complexes.

is fatal if left untreated. Cystic echinococcosis is found across the world whereas alveolar echinococcosis is mainly confined to the northern hemisphere.

The treatment of both echinococcosis is often complicated and expensive. The cysts caused by CE can be treated via surgery, PAIR (Puncture, Aspiration, Injection, and Reaspiration) technique or medical treatment, depending upon the characteristics of the cyst. The treatment of alveolar echinococcosis requires chemotherapy with or without surgery.

The benzimidazole compounds like albendazole (ABZ) and mebendazole (MBZ) are mainly used for the medical treatment of echinococcosis and both MBZ and ABZ are parasitostatic (Fig. 14.14) [43]. Any interruption in treatment causes the reappearance of the disease. Also, these drugs may induce teratogenic effects or hepatotoxixcity in some patients.

Ruthenium complexes, η^6-areneruthenium(II)phosphate complexes [Ru(η^6-*p*-cymene)Cl_2{P(OR)$_3$}] **28** and [Ru(η^6-*p*-cymene)(*t*Bu_2acac){P(OR)$_3$}]-[BF_4] **29** where R = Et, *i*Pr and Ph (Fig. 14.15) were tested for their in vitro behavior against *E. multilocularis* metacestodes, vero cells, rat hepatoma cells and human foreskin fibroblasts [44].

It has been reported that the complexes **28a–28c** showed only minor toxicity against E. *multilocularis* metacestodes and were not much effective as the reference drug nitazoxanide. In contrast, complexes **29** showed almost the same metacestodicidal effects as observed for nitazoxanide drug. The in vitro toxicity toward matacestodes was in similar range to that of human fibroblasts (IC_{50} = 1.1–2.9 μM) and vero cells (1.2–8.9 μM), indicates the poor selectivity of these complexes for *E. matacestodes*. However, these complexes could be interesting and helpful for the development of new anthelmintic metal complexes.

14.2.5 Schistosomiasis

Schistosomiasis, also known as bilharzia is an acute and chronic parasitic disease caused by blood flukes of genus *Schistosoma* [45]. Freshwater snails release the infectious form of the parasite in water. The parasite enters into the human body via penetration through human skin when the body comes into contact with contaminated water. The larvae, which grow into adult schistosomes inside the human body, live in the blood vessels. Female worm releases eggs in blood vessels; some of them pass out through urine or feces while some become trapped in body tissues, causing immune reactions and damage to the organs. Five types of parasitic worm: *S. mansoni*, *S. japonicum*, *S. mekongi*, *S. intercalatum*, and *S. haematobium* can cause Schistosomiasis. In 2019 about 236.6 million people suffering from schistosomiasis were in need to have preventive treatment, whereas only around 105.4 million people received preventive treatment [46].

Praziquental (PZQ) is the only drug used for the treatment of all species of schistosomiasis (Fig. 14.16). The PZQ has certain short-term side effects, such as abdominal pain, nausea, and vomiting, but no long-term ones, also safe for usage during pregnancy and in children. Unfortunately, PZQ is less effective against eggs and immature worms [47,48]. On the other hand, artemether, a derivative of the artemisinin drug, is effective against the immature state of *Schistosoma japonicum* and *Schistosoma mansoni* [49]. To prevent schistosomiasis, artemisinin in combination with PZQ is recommended but the long-term use of artemisinin could develop artemisinin resistance to malaria parasites. In the following section, organometallic complexes derived from the existing drugs (PZQ) are discussed.

The modification of the praziquantel drug was done by attaching ferrocenyl group to the nitrogen of the heterocyclic part of PZQ while in another modification the tricarbonylchromium moiety was coordinated to the aromatic ring [50,51]. Unfortunately, the ferrocenyl derivatives (**30** and **31**) showed less in vitro activity against adult *Schistosoma mansoni* (IC_{50} = 25.6−68 μM) compared to praziquental (IC_{50} = 0.1 μM). While two out of four tricarbonylchromium complexes (**32a** and **32d**; Fig. 14.17) showed comparable activity to that of parent drug praziquental. Also, these complexes were nontoxic toward MRC-5 human fibroblast cells, and hence selective toward parasite. However, the promising results were not observed

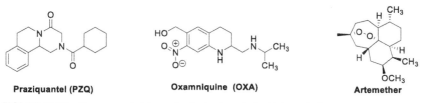

FIGURE 14.16 Recommended drugs for the treatment of schistosomiasis.

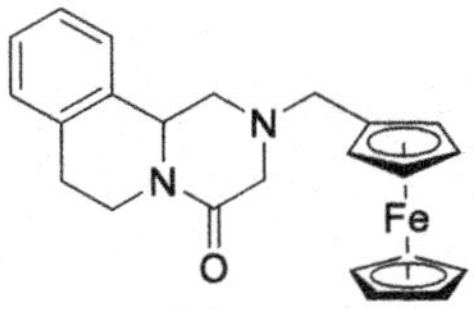

30

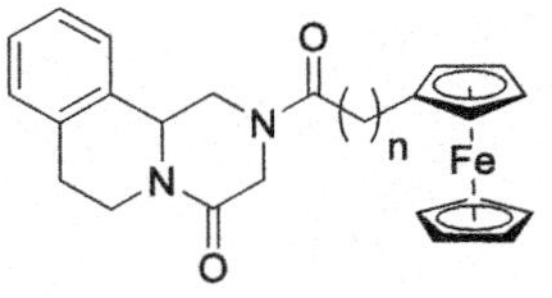

31

31a: n = 1
31b: n = 2
31c: n = 3

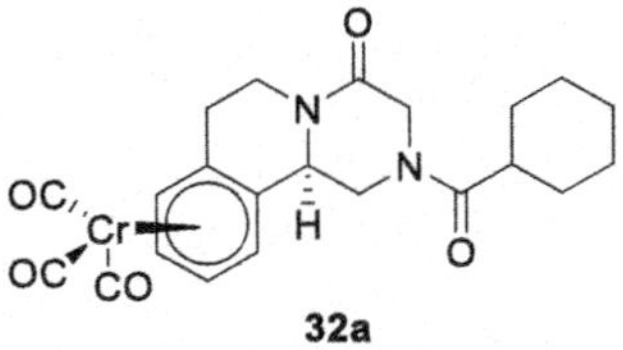

32a
IC_{50} = 0.13 μM

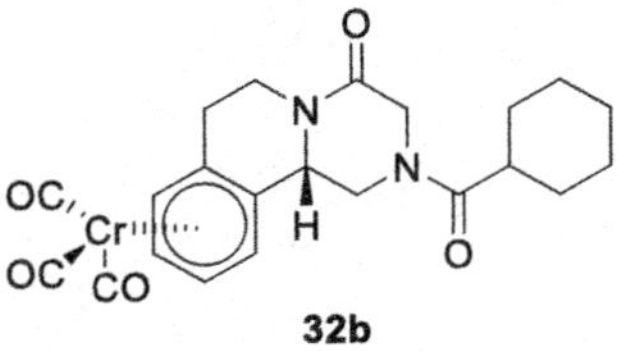

32b
IC_{50} > 66.9 μM

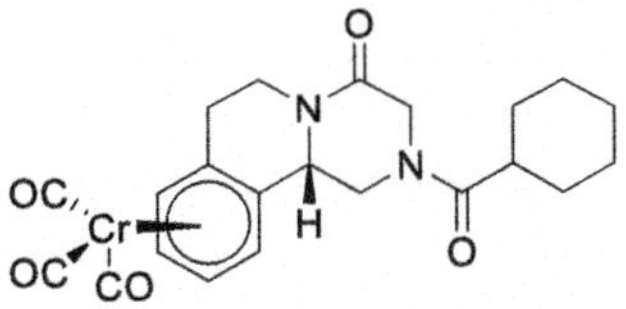

32c
IC_{50} > 66.9 μM

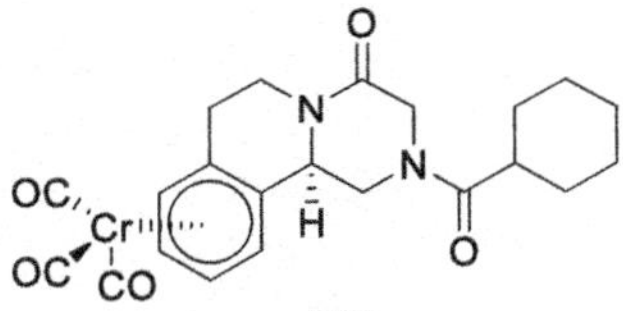

32d
IC_{50} = 0.08 μM

FIGURE 14.17 Ferrocenyl and chromium tricarbonyl derivatives of praziquantel.

in in vivo studies. The single dose of racemic mixture of **32a** and **32c** to *S. mansoni* infected mice results in low worm burden reduction to 24% and 29% which is low as compared to PZQ. Although these organometallic complexes are not very effective against *schistosomiasis* but might be helpful in understanding and developing new drug candidates.

14.2.6 Lymphatic filariasis

Lymphatic filariasis is a parasitic disease, commonly known as elephantiasis. It is caused *Wuchereriabancrofit*, *Brugiamalayi*, and *Brugiatimori* filarial worms. Lymphatic filariasis spreads through the bite of microfilariae infected mosquito. When infected by mosquito bites, the mature parasite larvae penetrate the body through skin and migrate to the lymph vessels where they grow into adults. Lymphatic filariasis infection causes damage to the lymphatic system and swelling/thickening of tissues. According to WHO, 863 million people are threatened by lymphatic filariasis in 47 countries [52]. Therefore preventive chemotherapy is required to stop the spread the lymphatic filariasis. Currently, diethylcarbamazine (DEC), albendazole (ALB), ivermectin (IVM), suramin, levamisole hydrochloride, and moxidectin (MOX) are the drugs used

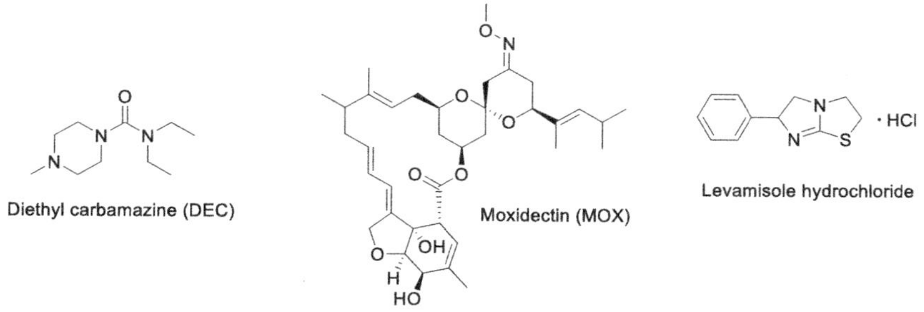

FIGURE 14.18 Structure of some antifilarial drugs.

to treat lymphatic filariasis [53]. Mainly, these drugs are effective for microscopic worm (microfilaremia) circulating in the blood and have limited effect on adult parasite. Therefore further developments are required for the treatment of lymphatic filariasis, specifically, adult parasite. The mono and binuclear complexes [RuH(CO)(PPh$_3$)$_2$(L)]$^+$ (**33**) and [RuH(CO)(PPh$_3$)$_2$(-µ-L) RuH(CO)(PPh$_3$)$_2$]$^{2+}$ (**34**) where L = pyridine-2-carbaldehyde azine (paa), *p*-phenylene-bis(picoline)aldimine (pbp) and *p*-biphenylene-bis(picolene)aldimine (bbp) and a related complex [RuH(CO)(PPh$_3$)$_2$(L)]$^+$ (**35**), L = 2,4,6-tris(2-pyridyl)-1,3,5-triazine (tptz) and 2,3-bis(2-pryridyl)-pyrazine (bppz) have been examined against *Setariacervi* filarial parasite (Fig. 14.18) [54].

These complexes act as inhibitor of topoisomerase II (topo II). The activity of these Ru complexes against filarial parasites is determined by the number and type of coordinated uncoordinated nitrogen atoms in the ligands. Complex **35a** (Fig. 14.19), in which the two nitrogen atoms are coordinated with Ru metal and the remaining four nitrogen remain uncoordinated exhibited higher activity (95% inhibition than other complexes).

14.2.7 Dengue

Dengue fever is a mosquito borne viral disease, caused by the dengue virus (DENV). The dengue virus is a member of the Flaviviridae family. DENV-1, DENV-2, DENV-3, and DENV-4 are 4 distinct serotypes of the dengue virus [55]. The virus transmits through the bite of female mosquitoes *Aedes aegypti* and *Aedes albopitus*. Dengue is found in tropical and subtropical climates. It is estimated that 3.9 billion people in 129 countries are at risk of dengue infection. Despite the risk existing in 129 countries; 70% of the global burden of dengue is in Asia only [56]. Dengue causes high fever with severe headache, nausea, vomiting, muscle and joint pain, swollen glands or rush. The symptoms of dengue appear after 4−7 days of bite of a mosquito. About one in twenty dengue patients have a chance of developing severe dengue. Persistent vomiting, bleeding gums or nose, blood in vomit, severe

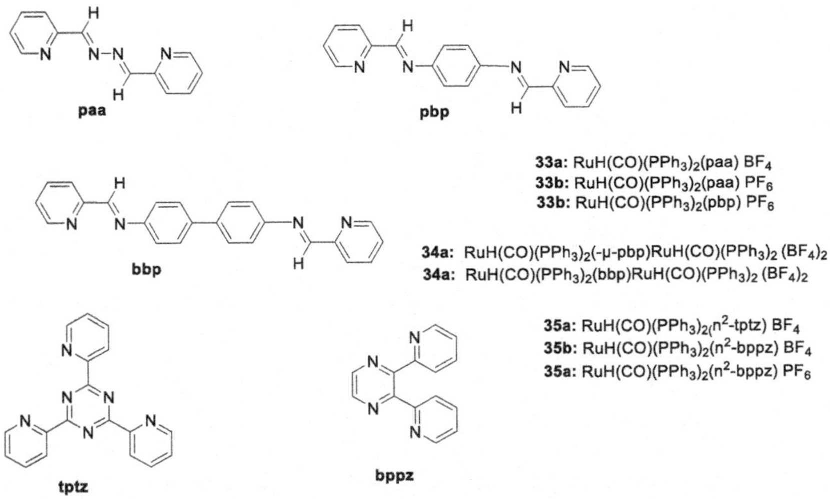

FIGURE 14.19 The paa, pbp, bbp, tptz, and bppz ligands and their ruthenium complexes.

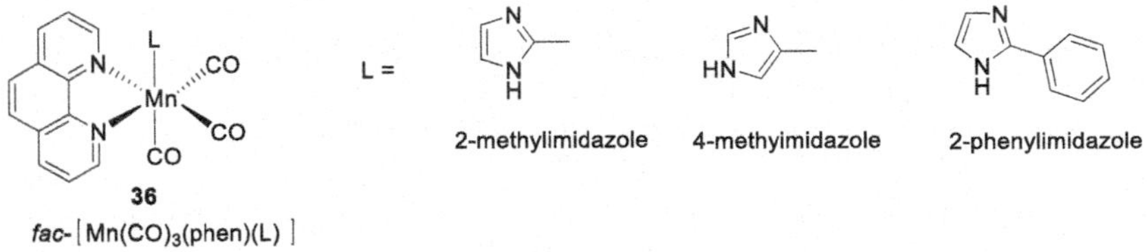

33a: RuH(CO)(PPh$_3$)$_2$(paa) BF$_4$
33b: RuH(CO)(PPh$_3$)$_2$(paa) PF$_6$
33b: RuH(CO)(PPh$_3$)$_2$(pbp) PF$_6$

34a: RuH(CO)(PPh$_3$)$_2$(-μ-pbp)RuH(CO)(PPh$_3$)$_2$ (BF$_4$)$_2$
34a: RuH(CO)(PPh$_3$)$_2$(bbp)RuH(CO)(PPh$_3$)$_2$ (BF$_4$)$_2$

35a: RuH(CO)(PPh$_3$)$_2$(η^2-tptz) BF$_4$
35b: RuH(CO)(PPh$_3$)$_2$(η^2-bppz) BF$_4$
35a: RuH(CO)(PPh$_3$)$_2$(η^2-bppz) PF$_6$

36 fac-[Mn(CO)$_3$(phen)(L)]

L = 2-methylimidazole, 4-methyimidazole, 2-phenylimidazole

FIGURE 14.20 The Mn(I) complexes with different imidazole derivatives.

abdominal pain, rapid breathing, and enlargement of the liver are the common symptoms of severe dengue.

There is no specific vaccine or medication available for the treatment of dengue. Patients with dengue are advised to stay hydrated, get rest, and use painkillers and fever reducer. The use of ibuprofen, aspirin or any other NSAID should be avoided due to the risk of bleeding. Since dengue affects millions of people worldwide, so further developments in dengue treatments are still required.

The organometallic complex fac-[Mn(CO)$_3$(phen)(L)]$^+$ **36** (**20**), wherein L = 2-methylimidazole, 4-methylimidazole, and 2-phenylimidazole and phen = phenanthroline (Fig. 14.20) are found to be effective against *A. Aegypti* larvae and showed mortality up to 90% at 0.33–0.46 g/L concentrations in 4 days [57]. It was also observed that these complexes have ability to interrupt the growth of larvae to pupae. These studies showed that such complexes have great potential to control mosquito proliferation. The high level of P450 in mosquitoes contributes toward a higher insecticidal resistance. In presence of light, the Mn (I) complexes interact strongly with P450. Under the irradiation of light, the Fe

(III) reduces to Fe(II) and Mn(I) converts to Mn(II). The reduction of P450 lowered the level of insecticidal resistance and increased the susceptibility of A. *aegypti* toward insecticides.

14.3 Summary

Undoubtedly, organometallics have made significant progress in the field of medicine. Over the last few decades, scientists focus on either modification of bioactive compounds by attaching an organometallic moiety to it or developing novel organometallic compounds to treat various NTDs. The different organometallic compounds which have shown activity against Chagas disease, Human African trypanosomiasis, Schistosomiasis, Echinococcosis, Leishmaniasis, Lymphatic filariasis and Dengue have been discussed. Indeed, this field is underdeveloped and most of the results discussed in this chapter are based on in vitro studies reported in literature. The in vivo studies and exploration of mechanism of action of various organometallic compounds are still needed. This chapter summarized the different studies particularly related to the development of organometallic compounds used for the treatment of neglected diseases and might be helpful to provide valuable information to the scientific community for the development of new drug candidates in future.

14.4 Problems with solutions

Q1. What are neglected tropical diseases?

Answer: Neglected tropical diseases are a diverse set of diseases that are common in people living in poverty, predominantly in tropical and subtropical areas. The WHO currently prioritized 20 NTDs, namely, Buruli ulcer, Chagas disease, dengue and chikungunya, dracunculiasis (Guinea-worm disease), echinococcosis, foodborne trematodiases, human African trypanosomiasis (sleeping sickness), leishmaniasis, leprosy (Hansen's disease) lymphatic filariasis, mycetoma, chrornoblastomycosis and other deep mycoses, onchocerciasis (river blindness), rabies, scabies and other ectoparasitoses, senistosomiasis, soil-transmitted helminthiases, snakebite envenoming, tenias/cysticercosis, trachoma, and yaws and other endemic treponematoses.

Q2. In which regions of the world the NTDs are a major health problem?

Answer: The NTDs are found in 149 different countries of Africa, Asia, and Latin America. NTDs are especially common in tropical areas and affect over 1 billion people. Most people with neglected tropical disease live in Asia but on a per capita basis Africa is the most affected continent.

Q3. Why NTDs are called neglected?
Answer: NTDs, such as dengue, chagas diseases, leishmaniasis, lymphatic filarisis, are called "neglected" because they affect the world's poorest population and have not received as much attention as other diseases. NTDs tend to thrive among the poorest people who live in rural, urban slum or conflict areas where access to healthcare, clean water, and adequate sanitation is limited. Unfortunately, the pharmaceutical industries either ignored or have less interest in developing novel drugs for NTDs because most patients with such diseases lives in less developed countries with low socio-economic status and they are not able to pay for expensive drugs.

Q.4 List the NTDs that can be controlled or even eliminated through mass drug administration.
Answer: The following five NTDS can be controlled or even eliminated through mass drug administration:

Disease	Drug used for treatment
Lymphatic filariasis	Albendazole, Ivermectin, and DEC
Schistosomiasis	Praziquantel
Onchocerciasis	Ivermectin
Soil-transmitted helminths	Albendazole and Mebendazole
Trachoma	Azithromycin

14.5 Objective type questions

i. What is tropical disease?
   a. A disease confined to the North Africa region.
   b. A disease that occurs solely, or principally, in tropical countries.
   c. A disease that is neglected by western medical research.
   d. An infectious disease that thrives in humidity and extreme climates.
ii. Which of the following is not a tropical disease?
   a. Malaria
   b. Chagas disease
   c. Clonorchiasis
   d. Lymphatic filariasis
iii. In 2014, the US Food and Drug Administration approved miltefosine drug for the treatment of which neglected tropical disease?
   a. leishmaniasis
   b. Dengue
   c. Chagas disease
   d. Human African Trypanosomiasis
iv. Chagas disease is caused by...
   a. Trypanosoma Cruzi
   b. Trypanosoma brucei

c. Filarial worms
d. Leishmania

v. The inorganic complex of CZT which exhibited anti-*T. cruzi* activity is...
   a. $RuBr_2(CZT)_2$
   b. $RuCl_2(CZT)_2$
   c. $RuCl_3(CZT)$
   d. $RhCl_2(CZT)_2$

vi. PAIR (Puncture, Aspiration, Injection, and Reaspiration) technique is used for the treatment of...
   a. Echinococcosis
   b. Schistosomiasis
   c. Leishmaniasis
   d. Dengue

vii. Which drug is used for the treatment of all species of schistisomiasis?
   a. Albendazole
   b. Miltefosine
   c. Praziquental
   d. Clotriamzole

viii. For the treatment of dengue the use of any NSAID should be avoided due to ......
   a. Abdominal pain
   b. Vomiting
   c. Rapid breathing
   d. Risk of bleeding

References

[1] World Health Organization. Neglected tropical diseases. World Health Organization. Available at: https://www.who.int/health-topics/neglected-tropical-diseases; 2022

[2] World Health Organization. Accelerating work to overcome the global impact of neglected tropical diseases: a roadmap for implementation: executive summary. World Health Organization. Available at: https://apps.who.int/iris/handle/10665/70809; 2012.

[3] World Health Organization. Ending the neglect to attain the sustainable development goals: a road map for neglected tropical diseases 2021–2030. World Health Organization. Available at: https://apps.who.int/iris/handle/10665/338565; 2020.

[4] Gasser G, Metzler-Nolte N. The potential of organometallic complexes in medicinal chemistry. Curr Opin Chem Biol 2012;16(1–2):84–91.

[5] Ong YC, Gasser G. Organometallic compounds in drug discovery: past, present and future. Drug Discov Today Technol 2020;37:117–24.

[6] World Health Organization. Neglected tropical diseases. World Health Organization. Available at: https://www.who.int/news-room/fact-sheets/detail/chagas-disease-(american-trypanosomiasis); 2022.

[7] Ferreira LG, Oliva G, Andricopulo AD. From medicinal chemistry to human health: current approaches to drug discovery for cancer and neglected tropical diseases. An Acad Bras Cienc 2018;90(1):645−61.

[8] Patterson S, Wyllie S. Nitro drugs for the treatment of trypanosomatid diseases: past, present, and future prospects. Trends Parasitol 2014;30(6):289−98.

[9] Ong YC, Roy S, Andrews PC, Gasser G. Metal compounds against neglected tropical diseases. Chem Rev 2019;119(2):730−96.

[10] Castro JA, deMecca MM, Bartel LC. Toxic side effects of drugs used to treat chagas' disease (American Trypanosomiasis). Hum Exp Toxico 2006;25:471−9.

[11] Sánchez-Delgado RA, Navarro M, Lazardi K, Atencio R, Capparelli M, Vargas F, et al. Toward a novel metal based chemotherapy against tropical diseases 4. Synthesis and characterization of new metal-clotrimazole complexes and evaluation of their activity against Trypanosoma cruzi. Inorg Chim Acta 1998;275 − 76:528−40.

[12] Navarro M, Lehmann T, Cisneros-Fajardo EJ, Fuentes A, Sánchez-Delgado RA, Silva P, et al. Toward a novel metal-based chemotherapy against tropical diseases: part 5. Synthesis and characterization of new Ru(II) and Ru(III) clotrimazole and ketoconazole complexes and evaluation of their activity against *Trypanosoma cruzi*. Polyhedron 2000;19(22 − 23):2319−25.

[13] Navarro M, Cisneros-Fajardo EJ, Lehmann T, Sanchez-Delgado RA, Atencio R, Silva P, et al. Toward a novel metal-based chemotherapy against tropical diseases. 6. Synthesis and characterization of new copper(II) and gold(I) clotrimazole and ketoconazole complexes and evaluation of their activity against *Trypanosoma cruzi*. Inorg Chem 2001;40(27):6879−84.

[14] Martínez A, Carreon T, Iniguez E, Anzellotti A, Sánchez A, Tyan M, et al. Searching for new chemotherapies for tropical diseases: ruthenium-clotrimazole complexes display high in vitro activity against leishmania major and Trypanosoma cruzi and low toxicity toward normal mammalian cells. J Med Chem 2012;55(8):3867−77.

[15] Iniguez E, Sánchez A, Vasquez MA, Martínez A, Olivas J, Sattler A, et al. Metal-drug synergy: new ruthenium(II) complexes of ketoconazole are highly active against leishmania major and *Trypanosoma cruzi* and nontoxic to human or murine normal cells. J Biol Inorg Chem 2013;18(7):779−90.

[16] Arce ER, Sarniguet C, Moraes TS, Vieites M, Tomaz AI, Medeiros A, et al. A new ruthenium cyclopentadienyl azole compound with activity on tumor cell lines and trypanosomatid parasites. J Coord Chem 2015;68(16):2923−37.

[17] Leite ACL, Moreira DRM, Cardoso MVO, Hernandes MZ, Pereira VRA, Silva RO, et al. Synthesis, cruzain docking, and in vitro studies of aryl-4-oxothiazolylhydrazones against Trypanosoma cruzi. Chem Med Chem 2007;2(9):1339−45.

[18] Donnici CL, Araújo MH, Oliveira HS, Moreira DRM, Pereira VRA, de Assis Souza M, et al. Ruthenium complexes endowed with potent anti-trypanosomacruzi activity: synthesis, biological characterization and structure-activity relationships. Bioorg Med Chem 2009;17(14):5038−43.

[19] Demoro B, Rossi M, Caruso F, Liebowitz D, Olea-Azar C, Kemmerling U, et al. Potential mechanism of the anti-trypanosomal activity of organoruthenium complexes with bioactive thiosemicarbazones. Biol Trace Elem Res 2013;153(1−3):371−81.

[20] Fernández M, Arce ER, Sarniguet C, Morais TS, Tomaz AI, Azar CO, et al. Novel ruthenium(II) cyclopentadienyl thiosemicarbazone compounds with antiproliferative activity on pathogenic trypanosomatid parasites. J Inorg Biochem 2015;153:306−14.

[21] Olea-Azar C, Rigol C, Mendizabal F, Morello A, Maya JD, Moncada C, et al. ESR spin trapping studies of free radicals generated from nitrofuran derivative analogues of nifurtimox by electrochemical and Trypanosoma cruzi reduction. Free Radic Res 2003;37(9):993–1001.

[22] Olea-Azar C, Atria AM, Di Maio R, Seoane G, Cerecetto H. Electron spin resonance and cyclic voltammetry studies of nitrofurane and nitrothiophene analogues of nifurtimox. Spectrosc Lett 1998;31(4):849–57.

[23] Arancibia R, Klahn AH, Buono-Core GE, Contreras D, Barriga G, Olea-Azar C, et al. Organometallic Schiff bases derived from 5-nitrothiophene and 5-nitrofurane: synthesis, crystallographic, electrochemical, ESR and anti *Trypanosoma cruzi* studies. J Organomet Chem 2013;743:49–54.

[24] Brun R, Blum J, Chappuis F, Burri C. Human African trypanosomiasis. Lancet 2010;375:148–59.

[25] Bacchi CJ. Chemotherapy of human African trypanosomiasis. Interdiscip Perspect Infect Dis 2009;2009:195040.

[26] Pomel S, Biot C, Bories C, Loiseau PM. Antiprotozoal activity of ferroquine. Parasitol Res 2013;112(2):665–9.

[27] Pomel S, Dubar F, Forge D, Loiseau PM, Biot C. New heterocyclic compounds: synthesis and antitrypanosomal properties. Bioorg Med Chem 2015;23(16):5168–74.

[28] Krishna ADS, Panda G, Kondapi AK. Mechanism of action of ferrocene derivatives on the catalytic activity of topoisomerase II alpha and beta–distinct mode of action of two derivatives. Arch Biochem Biophys 2005;438(2):206–16.

[29] Yun O, Priotto G, Tong J, Flevaud L, Chappuis F. NECT is next: implementing the new drug combination therapy for *Trypanosoma brucei* gambiense sleeping sickness. PLoS Neglected Trop Dis 2010;4(5):e720.

[30] Kansiime F, Adibaku S, Wamboga C, Idi F, Kato CD, Yamuah L, et al. A multicentre, randomised, non-inferiority clinical trial comparing a nifurtimox-eflornithine combination to standard eflornithinemono therapy for late stage *Trypanosoma brucei* gambiense human African trypanosomiasis in Uganda. Parasit Vectors 2018;11(1):105.

[31] Demoro B, Sarniguet C, Sánchez-Delgado R, Rossi M, Liebowitz D, Caruso F, et al. New organoruthenium complexes with bioactive thiosemicarbazones as co-ligands: potential anti-trypanosomal agents. Dalton Trans 2012;41:1534–43.

[32] Desjeux P. Leishmaniasis: current situation and new perspectives. Comp Immunol Microbiol Infect Dis 2004;27(5):305–18.

[33] World Health Organization. Neglected tropical diseases. World Health Organization. Available at: https://www.who.int/news-room/fact-sheets/detail/leishmaniasis; 2022.

[34] Rodrigo C, Weeratunga P, Fernando SD, Rajapakse S. Amphotericin B for treatment of visceral leishmaniasis: systematic review and meta-analysis of prospective comparative clinical studies including dose-ranging studies. Clin Microbiol Infect 2018;24(6):591–8.

[35] Haldar AK, Sen P, Roy S. Use of antimony in the treatment of Leishmaniasis: current status and future directions. Mol Biol Int 2011;2011:571242.

[36] Yesilova Y, Surucu HA, Ardic N, Aksoy M, Yesilova A, Oghumu S, et al. Meglumine antimoniate is more effective than sodium stibogluconate in the treatment of cutaneous Leishmaniasis. J Dermatol Treat 2016;27(1):83–7.

[37] Dorlo TPC, Balasegaram M, Beijnen JH, de Vries PJ. Miltefosine: a review of its pharmacology and therapeutic efficacy in the treatment of leishmaniasis. J Antimicrob Chemother 2012;67(11):2576–97.

[38] Vale-costa S, Vale N, Matos J, Tomás A, Moreira R, Gomes P, et al. Peptidomimetic and organometallic derivatives of primaquine active against Leishmania infantum. Antimicrob Agents Chemother 2012;56(11):5774−81.

[39] Quintal S, Morais TS, Matos CP, Robalo MP, Piedade MFM, de Brito MJV, et al. Synthesis, structural characterization and leishmanicidal activity evaluation of ferrocenyl N-heterocyclic compounds. J Organomet Chem 2013;745−746:299−311.

[40] Franco LP, de Góis EP, Codonho BS, Pavan ALR, Pereira IO, Marques MJ, et al. Palladium(II) imine ligands cyclometallated complexes with a potential leishmanicidal activity on Leishmania (L.) amazonensis. Med Chem Res 2013;22:1049−56.

[41] Moro P, Schantz PM. Echinococcosis: a review. Int J Infect Dis 2009;13(2):125−33.

[42] Hemphill A, Walker M. Drugs against Echinococcosis. Drug Des Rev−Online 2004;1(4):325−32.

[43] Hemphill A, Stadelmann B, Rufener R, Spiliotis M, Boubaker G, Müller J, et al. Treatment of echinococcosis: albendazole and mebendazole − what else? Parasite 2014;21:70.

[44] Küster T, Lense N, Barna F, Hemphill A, Kindermann MK, Heinicke JW, et al. A new promising application for highly cytotoxic metal compounds: η^6-areneruthenium(II) phosphite complexes for the treatment of Alveolar Echinococcosis. J Med Chem 2012;55(9):4178−88.

[45] Colley DG, Bustinduy AL, Secor EV, King CH. Human schistosomiasis. Lancet. 2014;383:2253−64.

[46] World Health Organization. Neglected tropical diseases. World Health Organization. Available at: https://www.who.int/news-room/fact-sheets/detail/schistosomiasis; 2022.

[47] Vale N, Gouveia MJ, Rinaldi G, Brindley PJ, Gärtner F, da Costa JM. Praziquantel for Schistosomiasis: single-drug metabolism revisited, mode of action, and resistance. Antimicrob Agents Chemother 2017;61(5):e02582 -16.

[48] Levecke B, Vlaminck J, Andriamaro L, Ame S, Belizario V, Degarege A, et al. Evaluation of the therapeutic efficacy of praziquantel against schistosomes in seven countries with ongoing large-scale deworming programs. Int J Parasitol Drugs Drug Resist 2020;14:183−7.

[49] Elmorshedya H, Tanner M, Bergquist RN, Sharaf S, Barakat R. Prophylactic effect of artemether on human schistosomiasis mansoni among Egyptian children: a randomized controlled trial. Acta Tropica 2016;158:52−8.

[50] Patra M, Ingram K, Pierroz V, Ferrari S, Spingler B, Keiser J, et al. Ferrocenyl derivatives of the anthelmintic praziquantel: design, synthesis, and biological evaluation. J Med Chem 2012;55(20):8790−8.

[51] Patra M, Ingram K, Pierroz V, Ferrari S, Spingler B, Gasser RB, et al. [(η^6-Praziquantel) Cr(CO)$_3$] derivatives with remarkable in vitro anti-schistosomal activity. Chem- Eur J 2013;19(7):2232−5.

[52] World Health Organization. Neglected tropical diseases. World Health Organization. Available at: https://www.who.int/news-room/fact-sheets/detail/lymphatic-filariasis; 2022.

[53] Singh PK, Ajay A, Kushwaha S, Tripathi RP, Misra- Bhattacharya S. Towards novel antifilarial drugs: challenges and recent developments. Future Med Chem 2010;2(2):251−83.

[54] Chandra M, Sahay AN, Pandey DS, Tripathi RP, Saxena JK, Reddy VJM, et al. Potential inhibitors of DNA Topoisomerase II: ruthenium(II) poly-pyridyl and pyridyl-azine complexes. J Organome Chem 2004;689(13):2256−67.

[55] Hasan S, Jamdar SF, Alalowi M, Al Beaiji SMA. Dengue virus: a global human threat: review of literature. J Int Soc Prev Community Dent 2016;6(1):1−6.

[56] World Health Organization. Neglected tropical diseases. World Health Organization. Available at: https://www.who.int/news-room/fact-sheets/detail/dengue-and-severe-dengue; 2022.

[57] de Aguiar I, dos Santos ER, Mafud AC, Annies V, Navarro-Silva MA, dos Santos Malta VR, et al. Synthesis and characterization of Mn(I) complexes and their larvicidal activity against Aedes aegypti, vector of dengue fever. Inorg Chem Commun 2017;84:49−55.

Chapter 15

A computational approach toward the role of biomimetic complexes in hydroxylation reactions

Monika[1,*], Oval Yadav[1,*], Manjeet Kumar[1], Ranjan Kumar Mohapatra[2], Vitthalrao Swamirao Kashid[3,4] and Azaj Ansari[1]

[1]*Department of Chemistry, Central University of Haryana, Mahendergarh, Haryana, India,* [2]*Department of Chemistry, Government College of Engineering, Keonjhar, Odisha, India,* [3]*Department of Chemistry, Gaya College of Engineering, Gaya, Bihar, India,* [4]*Department of Humanities and Science (Chemistry), Malla Reddy Engineering College for Women, Hyderabad, Telangana, India*

15.1 Introduction

Metalloenzymes are eminent natural catalysts, hold paramount importance for sustaining diverse life processes and embarked on the origin of mankind [1]. Examples, such as oxygenase, hydroxylase, halogenase, and reductase, catalyze a diverse array of important reactions that are biochemically relevant, such as hydroxylation and oxidation [2–6]. Metalloenzymes have always been a subject of contemporary interest among scientific communities owing to their enhanced efficiency, selectivity, and their ability to operate under mild conditions. Nature has served as a huge source of inspiration in pursuit of seeking knowledge about the functioning of these metalloenzymes. Besides these, the study of metalloenzymes is directed toward new synthetic routes in producing materials that may be crucial for human consumption (Scheme 15.1) [7–14,10,11].

Hydroxylase enzymes are one of the celebrated and widely studied categories among enzymes that are able to perform the most energetically difficult hydroxylation of C–H bond at saturated carbon centers in various organic compounds, yielding highly stereo and regioselective hydroxylated products

* These authors contributed equally to this work.

Recent Advances in Organometallic Chemistry. DOI: https://doi.org/10.1016/B978-0-323-90596-1.00013-4

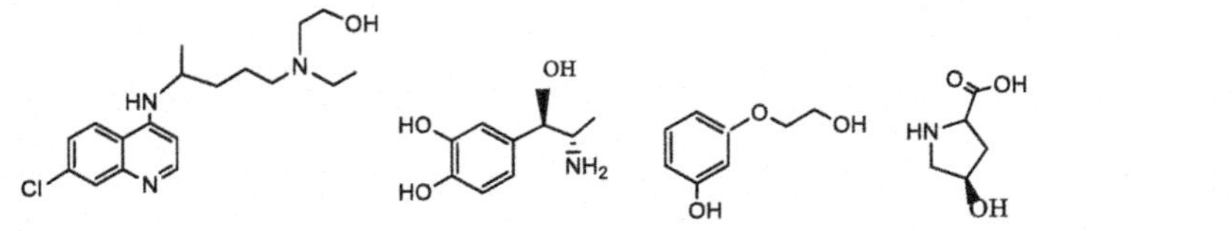

Hydroxycholoroquine (Antimalarial) **Levonordefrin** (Vasoconstrictor) **3HEP** (Antioxidant) **trans-4-hydroxy-L-proline** (Cardioprotector)

SCHEME 15.1 Some hydroxylated products of pharmaceutical and synthetic importance. *Adapted from Refs. [7−12].*

$$\text{R-H} + \text{O}_2 + 2\text{e}^- + 2\text{H}^+ \xrightarrow[\text{(e.g. cytochrome P450)}]{\text{monooxygenase}} \text{R-OH} + \text{H}_2\text{O}$$

SCHEME 15.2 Hydroxylation carried out by by monooxygenase [16]. *Adapted from Meunier B, Brudvig G, Mclain JL, Murahashi SI, Pecoraro VL, Riley D, editors. Biomimetic oxidations catalyzed by transition metal complexes. World Scientific. Imperial College Press; 2000.*

and hence play important roles in essential biological transformation reactions [15]. A few examples of hydroxylase enzymes are methane monoxygenase, AlkB, TauD, Tyrosinase, flavins, alpha-ketoglutarate-dependent hydroxylases and some diiron hydroxylases, cytochromes P450, nitric oxide synthases, prostaglandin H synthases, chloroperoxidases, etc. [15−18]. A typical hydroxylation reaction carried out by monooxygenase is shown below (Scheme 15.2).

Several biochemical relevant reactions are prevalent in nature which involves hydroxylation, such as hydroxylation of methane in methanotrophs, desaturation of fatty acids in plants, DNA and RNA repairs formation of blood vessels, and all these required the controlled oxidation of organic substrates by metal-mediated activation of dioxygen (O_2) [17−19]. Heme enzymes, such as peroxygenases and monooxygenases, carry the hydroxylation of many aromatic compounds efficiently. Flavin-having hydroxylases are less promiscuous but have good regioselectivity. These monooxygenases activate molecular oxygen by transient formation of C(4a)-hydroperoxide by insertion of an oxygen atom into the substrate [20−22]. Its regio and stereoselective properties are used in pharmaceuticals, fine chemicals, flavors and fragrances. In chemical synthesis, flavin-containing hydroxylases are attractive catalysts as their aromatic product shows anticancer and antimicrobial properties [20,23,24]. Metal-mediated selective hydroxylation has emerged as a recent active and popular research area in past years as it yields precursors that find huge utility in the pharmaceutical industry, polymer industry and synthetic utility [21,22].

Methane to methanol conversion has gained supreme importance in the scientific community due to its industrial importance. In nature, this conversion is carried out by soluble methane monooxygenase (sMMO) [25].

The capability of enzymes to catalyze difficult inert reactions with ease and that too with requisite selectivity has been the focal point of interest for scientists. The motivation behind the elucidation of functioning patterns of natural enzymes and the study of fine details of their catalytic cycles, commenced a series of development of biomimetic models. The enzymes cannot withstand the harsh conditions of industrial processes due to their fragile nature and less heat tolerating capacity making them prone to lose their functionalities. In such a scenario, synthetic models come into play. These models assisted in gaining deep insights about the diverse aspects of enzymes, such as functioning patterns, nature of active sites, formation of reactive intermediates, and fine mechanistic details of catalytic reaction. Innumerous efforts have been made by experimentalists as well as theoreticians all over the world to obtain robust and environment benign functional biomimetics of these metalloenzymes which work at optimal reaction conditions, at low activation barrier.

The first biomimetic model for C—H bond hydroxylation was developed by Groves and coworkers in 1979 [26]. This model consisted of an iron catalyst: [Fe^{III}(por)Cl] (por = a porphyrinatodianion) which catalyzed the hydroxylation of cyclohexane and adamantane. Que and coworkers reported a complex [Fe^{II}(TPA)(CH$_3$CN)$_2$]$^{2+}$ (TPA = tris(2-pyridylmethyl)amine), which is capable of performing stereoselective C—H bond hydroxylation. Iron-containing enzymes have a great efficiency to carry out hydroxylation of alkyl C—H bond. Iron is very important in biochemistry due to its very low toxicity as living things are evolved in iron-containing media. Iron-containing compounds carry out several catalytic reactions, such as the activation of dioxygen, hydrogen peroxide, and superoxide [27]. Getting inspired from iron enzymes, chemists developed many synthetic iron-based complexes, one of the popular choices among them is tetradentate N-donor ligands bound to Fe(II) and have two cis labile sites this is the biomimetic model of Rieske dioxygenase. In 2007, White and Costas performed the C—H bond activation using H$_2$O$_2$ as an oxidant. The selectivity for the 3° and 2° sites for intermolecular C—H bond activation is governed by steric, stereoelectronic, and electronic rules.

Besides these, different ligand architecture were also employed for study of alkane hydroxylation, such as N$_4$Py (1,1-di(pyridin-2-yl)-*N*,*N*-bis(pyridin-2ylmethyl)methanamine), BPMEN (*N*,*N*′-dimethyl-*N*,*N*′-bis(2-pyridylmethyl) ethane-1,2-diamine), H$_2$salen, (salen: 2,2′-Ethylenebis(nitrilomethylidene) diphenol, *N*,*N*′-Ethylenebis(salicylimine)) [9].

Manganese has also been used as an analog of iron in these functional biomimetic porphyrin models due to the comparatively higher stability of Mn-oxo species over the iron counterparts. Mn-oxo porphyrins have been detected and characterized spectroscopically and are well established as potent catalysts for alkane hydroxylation.

In the 1980s a series of ruthenium porphyrin-based biomimetic models were reported for hydroxylation of alkanes by various research groups. James and coworkers reported [Ru^{III}(oep)(PPh$_3$)Br] or [Ru^{III}(tmp)(P^{n}Bu$_3$)

Br] + PhIO, Groves and Quinn reported [Ru^{VI}(tmp)O$_2$] + O$_2$ and Hirobe and coworkers reported [Ru^{II}(por)(CO)] or [Ru^{VI}(por)O$_2$]$^+$ 2,6-Cl$_2$pyNO (oep = octaethylporphyrin; tmp = tetramesitylporphyrin and por = porphyrin) [28].

A lot of theoretical work has been devoted toward understanding the biomimetic role of mMMO catalyzing the conversion of CH$_4$ into CH$_3$OH, that is, hydroxylation. Yoshizawa and coworkers carried out density functional theory (DFT) study of the conversion of CH$_4$ into CH$_3$OH catalyzed by active transition metal oxides. Several theoretical models were developed for methane hydroxylation mediated by MMOs [29]. Friesner and Lippard and coworkers developed Fe^{IV}(μ-O)$_2$Fe^{IV} model to gain deep insights about mechanism of hydroxylation reaction. A detailed computational investigation was performed to obtain mechanistic details of methane hydroxylation reaction. DFT calculations were also performed on pMMO inspired by copper based clusters like (**1**, tricopper − cluster), bis(μ-oxo)-Cu(III)Cu(III), (**2**, bicopper cluster), and the bis (μ-oxo)-Cu(II)Cu(III) (**3**, mixed valence dicopper cluster). Out of these three models, model 3 offers easiest pathway for CH$_4$ hydroxylation. Several bis-(μ_3-oxo)-Cu(II)Cu(II)Cu(III) models of compound Q of MMO have been prepared in last few years [30]. Musaev and coworkers exploited the mechanism of hydroxylation reaction of fluorinated hydrocarbons (methane, ethane and methyl fluoride) using two models of compound Q, namely cis-(H$_2$O)-(NH$_2$)Fe (μ-O)$_2$(η^2-HCOO)$_2$Fe(NH$_2$)(H$_2$O) and cis-(HCOO)(Imd)Fe(μ-O)$_2$(η^2-HCOO)$_2$Fe (Imd)(HCOO) (Imd = Imidazole). Taking inspiration from copper based tyrosinase enzymes, a mechanistic study was performed on aromatic hydroxylation with PPN (bis(triphenylphosphine)iminium) ligand based dinuclear Cu$_2$O$_2$ complex using the computational tool. Computed data suggested that the reaction occurred via the formation of bis (μ-oxido) dinuclear complex as the chief intermediate followed by 1,2-H shift transition state. Dihydroxylation of styrene has been also studied by DFT/MM methods by employing osmium catalyst (OsO$_4$(DHQD)$_2$PYDZ); (DHQD)$_2$PYDZ = bis(dihydroquinidine)-3,6-pyridazine to explore all the possible reaction pathway of olefin dihydroxylation where it was found that the selectivity of system was governed by the orientation of the substrate.

Tetraamido macrocyclic ligand (TAML) is one of the potent ligands which stabilize the high-valent metal centers. In the mid-1990s Collins developed the TAML activators, one of them is iron with TAML which mimics peroxidase [31]. The peroxide-bound Fe^{III} complex gets converted into a complex upon addition of mCPBA at room temperature, and this is capable to carry out the C−H bond oxidation [32]. It has been hypothesized that Fe^V = O (first complex) shows lower electrophilicity because of strong σ-donating deprotonated amide N atoms make it unreactive toward aromatic rings and selective for 3-degree aliphatic C−H bond activation [33]. Iron-TAML activators mimic the properties of peroxidase and cytochrome P450 enzymes. With TAML systems, several Fe(IV) complexes are well known, and the TAML Fe(V) = O species is more reactive than the corresponding Fe(IV). Apart from TAML, several other ligands, such as TMC (e.g., 14-TMC = 1,4,8,11-tetramethyl-1,4,8,11-tetraazacyclotetradecane), TPA, BPMEN, and H$_3$buea (tris[(N'-tert-butylureayl)-N-ethyl]-(6-pivalamido-2-pyridylmethyl)aminato), also stabilize the high-valent metal complexes [34,35]. The high-valent iron complexes also play an important role in green catalytic oxidation reactions. The chemistry mimicking these processes is greener than those based upon the biochemically unfamiliar elements and oxidizing agents. Green chemistry replaced the catalyst made up toxic metals for the more abundant and less toxic metals as Mn, Fe, Cu, and Zn. The metal and nature of ligand determine the catalytic reactivity and selectivity of such complexes. Metal peroxo and hydroxo complexes have been synthesized and characterized and their reactivity toward the electrophilic and nucleophilic oxidation reaction is also studied [36]. Several biomimetic metal superoxo complexes have been synthesized and characterized to know about the properties of metal superoxo species involved in enzymatic reactions.

Besides alkane hydroxylation, numerous studies have been devoted to hydroxylation of aromatic systems as it yields products of pharmaceutical importance. Considerable progress has been made in the area of hydroxylation reactions carried out by biomimetic iron complexes. The corresponding synthetic work for hydroxylation of arene substrates has been dedicated toward using a modification of TPA ligand with single R-arene substituents. In one of the studies, the behavior pattern and mechanism of ortho-hydroxylation of aromatic compounds have been explored by DFT study using Fe(II) precursor: [Fe^{II}(TPA)(CH$_3$CN)$_2$]$^{2+}$. Hydroxylation reaction proceeds via the formation of high-valent species, Fe^V = O which triggers the catalytic reaction rather than the Fe^{IV} = O species which is mostly proposed as an oxidant for various oxidation reactions. Although Fe^V = O, which has been spectroscopically characterized and reported as the potential oxidant in the ortho-hydroxylation reaction [21].

A model complex of enzyme tyrosinase was also developed using bis (pyrazolyl)methane ligand. Tyrosinase enzyme bears dicopper(II)peroxide core as its active center and its enzymatic reactions include ortho-hydroxylation of phenols to catechols (and further oxidation to quinones) [37]. To date, a significant contribution has been made in designing ligand systems that can efficiently catalyze phenol hydroxylation reactions. Starting from Cu$_2$O$_2$ complex by Casella et al. to copper (I) with pyridine and/or imine moiety, imine and benzimidazole and imine and pyrazolyl moieties, respectively, by Tuckzek et al., imidazolyl containing tris(2-pyridylmethyl) amine (TMPA) derivative by Karlin and coworkers, tripodal ligand containing imidazolyl moieties by Limberg et al., remarkable efforts have been made in the rational design of new catalysts in the field of catalytic phenol hydroxylation (see Scheme 15.3) [38].

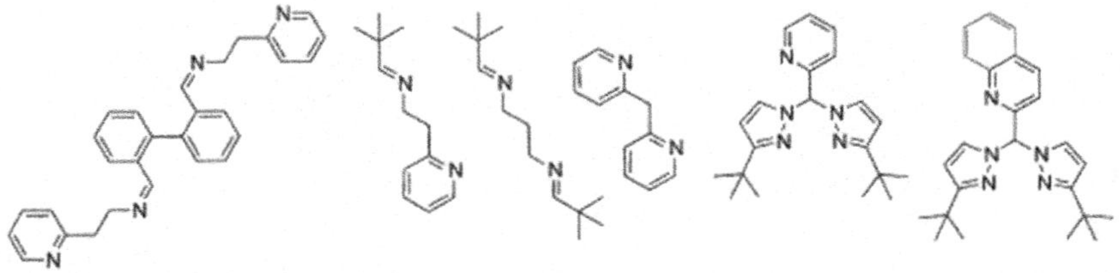

SCHEME 15.3 Pyridine/imine based ligands employed for hydrocarbon hydroxylation. *Adapted from Wilfer C, Liebhuser P, Hoffmann A, Erdmann H, Grossmann O, Runtsch L, et al. Efficient biomimetic hydroxylation catalysis with a bis(pyrazolyl)imidazolylmethane copper peroxide complex. Chem Eur J 2015;21(49):17639–17649.*

15.2 Computational details

Computational chemistry uses mathematical methods to study chemical relevance. It deals with the electronic ground state, excited states of individual atoms and molecules, and also the transition states and intermediates during reaction pathways of chemical transformation. Many approaches, such as Hartree-Fock, semi-empirical methods, DFT, Monte Carlo methods, and coupled cluster methods, are used for theoretical study. Calculations are performed using many programs, such as Jaguar and Gaussian [39,40]. Functionals and basis sets are used during calculations. There are many functionals, such as B3LYP [41], B3LYP-D [42], wB97XD [43], B97D [42], M06-2X [44], TPPSh [45], and MP2 [46], and the basis sets LACVP/LACV3P + */6−31G/6−311 + G*/D95**[17]/TZVP/Def2-TZVPP, etc., comprising the LanL2DZ Los Alamos effective core potential and (14s9p5d)/[9s5p3d] primitive set of Wachters-Hay supplemented with/without function can be used during calculations (geometry optimization/single-point energy) [47−53]. The development of DFT is a very big achievement in computational chemistry. Optimization (locating stationary points of a function), is one of the important methods to solve the majority of quantum chemical problems. The optimized structures are subjected to various theoretical and experimental investigations to study properties of interest. The frequency calculations are computed on the optimized structures to verify the minima on the potential energy surface and also to calculate the free energy corrections and zero-point energy corrections. All local minima have real frequencies, and the transition states have one imaginary frequency for the correct mode. Solvation is also performed for the chemical reactions, the self-consistent reaction field [54] solvation model which is based on Onsager's reaction field theory [55] is generally used with the appropriate solvent.

15.3 Recent developments with examples

Selective hydroxylation of aliphatic and aromatic compounds is of great interest because it forms an important precursor in the pharmaceutical industry [56]. Hydroxylation reactions catalyzed by iron with Fenton's reagent is nonselective due to the formation of hydroxyl radicals [57]. In recent years, many metal complexes with TPA, BPMEN and PyTACN (1-(2-pyridylmethyl)-4,7-dimethyl-1,4,7-triazacyclononane) ligands have reported, they catalyze such reactions efficiently and selectively [58,59]. Reactions catalyzed by nonheme iron oxidants are also important because of their regio and stereo selective oxidation of organic substrates. Nonheme high-valent ferryl-oxo species play an important role in selective C—H bond activation in an efficient manner [60]. The catalytic reactivity of metal-oxo species is greatly influenced by their redox properties, and also the basicity of metal-oxo complexes toward C—H bond activation [61]. Phthalocyanines ligated nonheme complexes have been used in the C—H bond oxidation, with many oxidants as dioxygen, hydrogen peroxide (H_2O_2), m-chloroperbenzoic acid (mCPBA), and iodosylbenzene [62]. Upon the oxidation of inactivated alkanes, which included several substrates, such as natural products, hydroxylation reaction is observed mostly at the 3° C—H bonds with 3°:2° selectivity up to 100:1 [63]. Recently, mononuclear nonheme manganese complexes are reported as efficient catalysts in the catalytic oxidation of hydrocarbons. High-valent Mn(V)-oxo complexes act as active oxidants that afford high regio-, stereo-, and enantioselectivity in the catalytic oxidation reactions. DFT calculations has also been performed with Mn(V)-oxo species, $[(S-PMB)Mn^V(O)(AcO)]^{2+}$ for the conversion of cyclohexane to cyclohexanol [64].

Along with mononuclear, dinuclear species also carries hydroxylation. Dinuclear-μ-nitrido iron species [(TPP)(m-CBA)Fe(IV)(μ-N)Fe(IV)(O)(TPP$^{\bullet+}$)]$^-$ (TPP = meso-tetraphenylporphyrin) shows the catalytic abilities toward hydroxylation. Rajaraman et al. have carried out the hydroxylation of methane with the dinuclear μ-nitrido iron species and the studies show that the barrier for C—H bond activation is 6.4 kcal/mol (using Gaussian/B3LYP-D/Lanl2DZ) indicating that it is efficient catalyst [65]. Dicopper(II) complex, $[Cu_2(\mu-OH)(6-hpa)]^{3+}$, (6-hpa = 1,2-bis[2-[bis(2-pyridylmethyl)- aminomethyl]-6-pyridyl]ethane), generates an oxyl radical of $Cu^{II}O^\bullet$ and catalyzes the selective hydroxylation of benzene. The catalytic performance for the hydroxylation of methane to methanol by the dicopper complex is investigated by using DFT calculations. The activation barrier for C—H bond activation is computed to be 10.2 kcal/mol, which is low enough for reactions taking place under normal conditions. It indicates that the dicopper complex has a precondition to hydroxylate methane to methanol. Experimental verification is required to look in detail at the reactivity of this dicopper complex [66]. Recently, Pedro J. Perez et al. have reported a direct hydroxylation of benzene, by the reaction of Tp^{x}Cu (Tp^x = hydrotrispyrazolylborate ligand) with hydrogen peroxide proceeds via copper-oxyl species of type Tp^{x}Cu-O$^\bullet$. Cu-oxyl species can serve in the design of novel catalysts in the future for the direct functionalization of benzene into phenol [67]. It is reported that hydroxylation is affected by ligand design. Rajaraman et al. studied hydroxylation with two different complexes, that is, {[(BPMEN)Fe^{III}-OOH]$^{2+}$ and [(TPA)Fe^{III}-OOH]$^{2+}$} using the computational tool and found that the orientation of the pyridine ring affects the hydroxylation barrier. The pyridine ring parallel to the Fe = O bond interacts with Fe(d_{yz})-O(p_y) this delocalizes the electron densities to decrease the electrophilic reactivity during the hydroxylation [68]. Designing of equatorial ligands also plays an important role in tuning the reactivity of high-valent metal-oxo species studied. Selective hydroxylation of alkanes, mainly methane, is highly desirable for near future energy sources. It is of great interest in the economy because methanol has enormous potential as an energy carrier and chemical feedstock.

Production of phenol by the direct oxidation of benzene has attracted much attention to overcome the drawbacks of cumene process [67]. Arene oxidation with hydrogen peroxide is known as Fenton chemistry and in this process formation of radical occurs and is capable of oxidizing the arene rings [69]. In recent Karlin, Fukuzumi and coworker [70] have reported that [Cu(tmpa)]$^{2+}$ {tmpa = tris(2-pyridylmethyl)amine} species carry the direct hydroxylation of benzene by generating free HO$_2^{\bullet}$ radical from H$_2$O$_2$, that interacts with the arene. Iron iron oxychloride (FeOCl) is reported as the efficient catalyst for the selective hydroxylation of benzene. Oxidation using H$_2$O$_2$ has gained a significant interest in catalytic oxidative process, because it is green, economical and environment friendly, and water or oxygen is produced as product [71].

15.4 Mechanism involved

Hydroxylation is carried out in both aliphatic and aromatic compounds. Several mechanisms have been proposed for hydroxylation.

15.4.1 C—H bond activation followed by oxygen rebound mechanism

Metal-oxo (or superoxo/peroxo/hydroxo) activates the hydrogen of organic substrates to generate a radical. Thereafter, OH rebound proceeds via the rebound mechanism generating the hydroxylated product or the dissociation of substrate radical is more favorable, that is, produced after hydrogen abstraction from hydrocarbon than the rebound and desaturation process [72], In this, H-atom abstraction/oxygen rebound mechanism, the formal oxidation state of Fe^{IV} is reduced to Fe^{III} and formed the alcohol as a product (Scheme 15.4). Soluble methane monooxygenase enzymes also catalyze the hydroxylation of methane to methanol via hydrogen abstraction followed by an oxygen rebound mechanism as shown in Scheme 15.5.

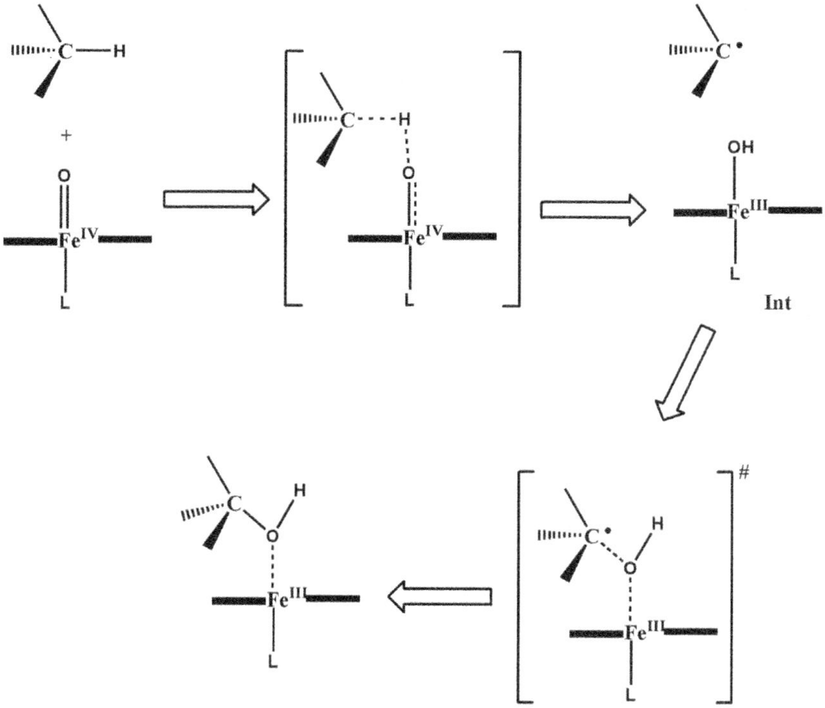

SCHEME 15.4 Schematic representation for reaction pathway for hydroxylation of methane. Adapted from Kumar R, Ansari A, Rajaraman G. Axial vs. equatorial ligand rivalry in controlling the reactivity of iron(IV)-oxo species: single-state vs. two-state reactivity. Chem Eur J 2018;24(26):6818–6827.

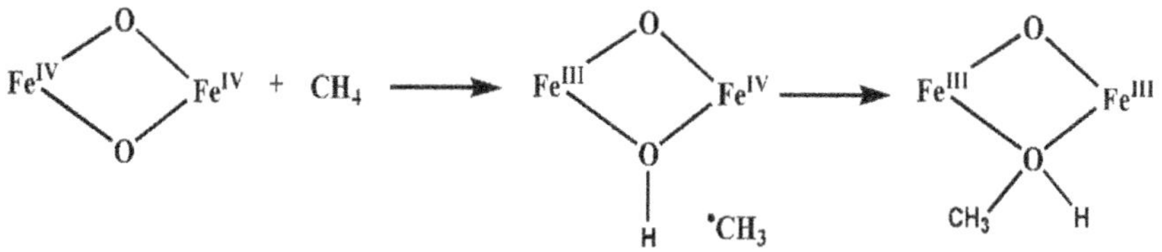

SCHEME 15.5 Rebound mechanism for the hydroxylation of substrate by sMMO enzyme. Adapted from Kumar R, Ansari A, Rajaraman G. Axial vs. equatorial ligand rivalry in controlling the reactivity of iron(IV)-oxo species: single-state vs. two-state reactivity. Chem Eur J 2018;24(26):6818–6827.

Hydroxylation reactions are generally catalyzed by mononuclear and dinuclear heme and nonheme complexes [68,73]. The first example [Fe^{II}(TPA)(CH$_3$CN)$_2$]$^{2+}$ was published by Que and coworkers, and the complex is capable of performing stereo selective C—H bond hydroxylation. Sason Shaik and his colleagues carried out the analysis of methane hydroxylation by a ferry-porphyrin cation radical which is the model species for the enzyme cytochrome P450 using the DFT and predicted that the C−H bond activation occurs at 26.5 kcal/mol (using Gaussian/B3LYP/LACVP) [74]. Wonwoo Nam carried out the hydroxylation of cyclohexane by the complex [(SPMB)Mn^V(O)(OAc)]$^{2+}$, and by DFT calculations it was found that the hydroxylation follows the oxygen rebound mechanism, in which C−H bond activation occurs at 6.8 kcal/mol with cyclohexane and the oxygen rebound step is barrier less (B3LYP/Def2-TZVPP//LACVP) [64]. Rajaraman et al. also carried out hydroxylation of methane with [(LNHC) Fe^{IV}(O)(CH$_3$CN)]$^{2+}$ species (LNHC = 3,9,14,20-tetraaza-1,6,12,17-tetraazoniapenta-cyclohexane cosane-1(23),4,6(26),10,12(25),15,17(24),21-octaene) follow the rebound mechanism, C−H bond activation occurs at 29.4 kcal/mol and OH rebound occurs at 33.1 kcal/mol (using Gaussian/B3LYP-D/LACVP) [73]. de Visser et al. also studied hydroxylation (also oxygen atom transfer) of propene activated by the nonheme biomimetic [Fe^{IV}O(TMCS)]$^+$ complex and DFT calculations (using Jaguar/UB3LYP/LACV3P + *) showed that hydroxylation (15 kcal/mol) is preferred over epoxidation (17.9 kcal/mol). This is the first iron-oxo complex where hydroxylation has a lower barrier than the epoxidation reaction. In this case, also follow hydrogen abstraction followed by the oxygen rebound [75].

15.4.2 Hydrogen abstraction followed by oxygen nonrebound mechanism

Nonheme Mn^{IV}O, Fe^{IV}O, Cr^{IV}O, and Ru^{IV}O species show an alternative oxygen nonrebound mechanism [76,77]. After the hydrogen abstraction by metal(IV)-oxo, metal(III)-hydroxo along with carbon radical is formed, and this radical escapes from the cage then the reaction takes place by the radical species and such mechanism is known as oxygen nonrebound mechanism. In this mechanism, one electron reduction takes place instead of two electron reduction.

Here the retention in stereochemistry of oxygenated product will be low (Scheme 15.6) [64]. Thus it may be stereo-, regio-, and enantioselectivity of C−H bond activation reactions are determined by high-valent metal-oxo intermediates. Nam et al. also observed similar mechanism while using [(Bn-TPEN)-Fe^{IV}O]$^{2+}$ and [(N$_4$Py)Fe^{IV}O]$^{2+}$ during hydroxylation reaction [78].

15.4.3 Direct attack of oxygen atom

Diverse reaction mechanisms were postulated for benzene hydroxylation (Schemes 15.7 and 15.8), and DFT calculations predicted the reaction takes place via the formation of iron-arene-σ-pathway (pathway b in Scheme 15.8). In the aromatic hydroxylation, an electrophilic attack occurs at the benzene ring.

On the basis of experimental and theoretical results carried by Nam and de Visser et al., it has been concluded that mononuclear nonheme iron(IV)-oxo

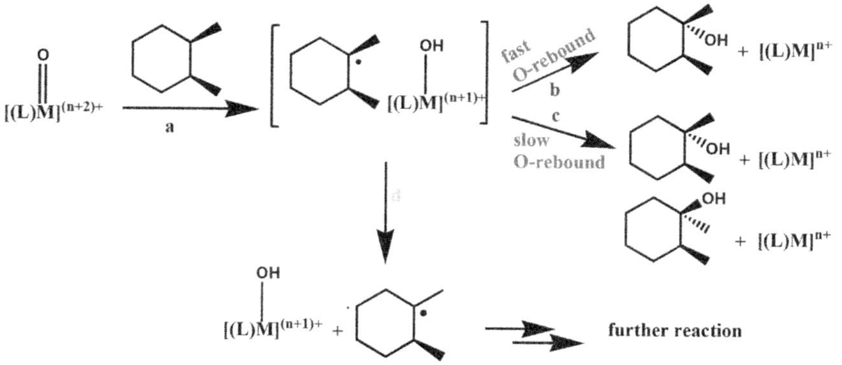

SCHEME 15.6 Oxygen rebound versus oxygen nonrebound mechanism for selective oxidation reactions. *Adapted from Li X.-X, Guo M, Qiu B, Cho K-B, Sun W, Nam W. High-spin Mn(V)-oxo intermediate in non-heme manganese complex-catalyzed alkane hydroxylation reaction: experimental and theoretical approach. Inorg Chem 2019;58(21):14842–14852.*

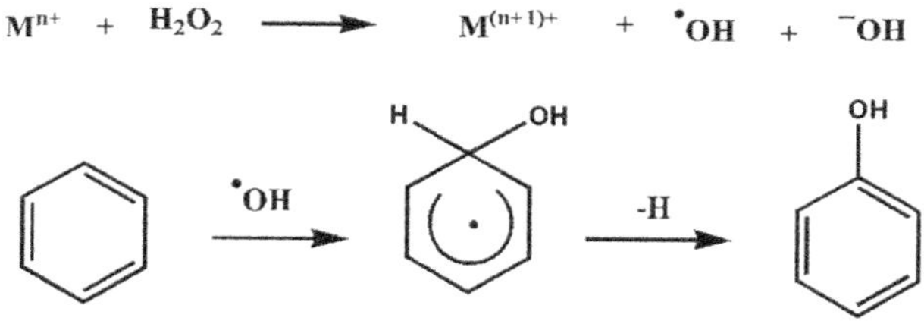

SCHEME 15.7 Top: the metal-mediated catalytic transfer of carbene, nitrene, and oxo/oxyl groups; bottom: the Fenton-like oxidation of benzene with hydrogen peroxide. *Adapted from Luo G, Lv X, Wang X, Yan S, Gao X, Xu J, Ma H, Jiao Y, Li F, Chen J. Direct hydroxylation of benzene to phenol with molecular oxygen over vanadium oxide nanospheres and study of its mechanism. RSC Adv 2015;5(114):94164–94170.*

species carries the aromatic ring oxidation via the initial electrophilic attack on the π-system of the aromatic ring to generate a tetrahedral radical or cationic σ-complex instead of hydrogen atom abstraction [57]. Rajaraman et al. also studied the mechanism for the ortho-hydroxylation of benzoic acid. For which two paths have been proposed (Scheme 15.9); in one pathway C–H bond activation generates the radical followed by OH rebound, in another pathway, electrophilic attack of ferryl-oxo on the benzene ring generates Fe^{III}-radical with the de-aromatization of benzene [21].

In the next step, hydrogen from the sp^3 C is transferred to ferryl-oxo leading in the formation of ortho-hydroxylation. DFT calculations found that the direct (electrophilic) attack is more favorable than the C–H bond activation [71]. Direct hydroxylation of benzene is also reported by the Tp^x Cu (Tp^x = hydrotrispyrazolylborate ligand) with hydrogen peroxide proceeds via copper-oxyl species of type Tp^x Cu-O$^\bullet$.

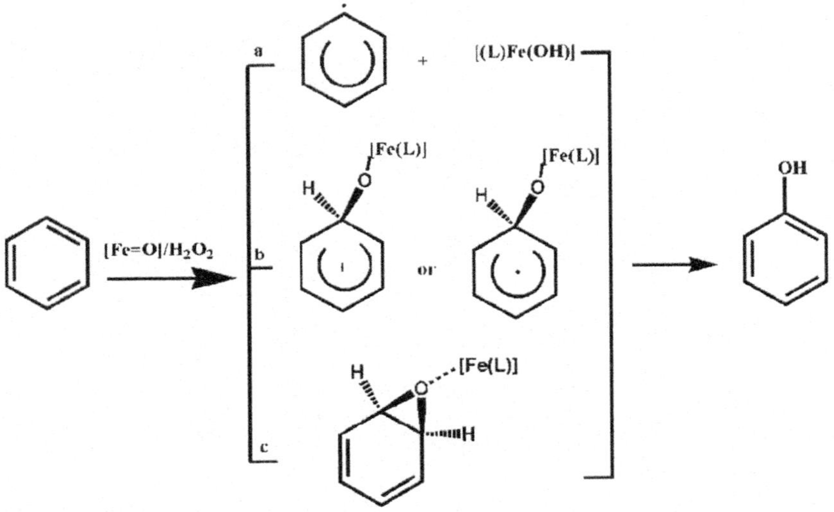

SCHEME 15.8 Diverse mechanistic pathways for the iron-catalyzed hydroxylation of benzene. *Adapted from Luo G, Lv X, Wang X, Yan S, Gao X, Xu J, Ma H, Jiao Y, Li F, Chen J. Direct hydroxylation of benzene to phenol with molecular oxygen over vanadium oxide nanospheres and study of its mechanism. RSC Adv 2015;5(114):94164–94170.*

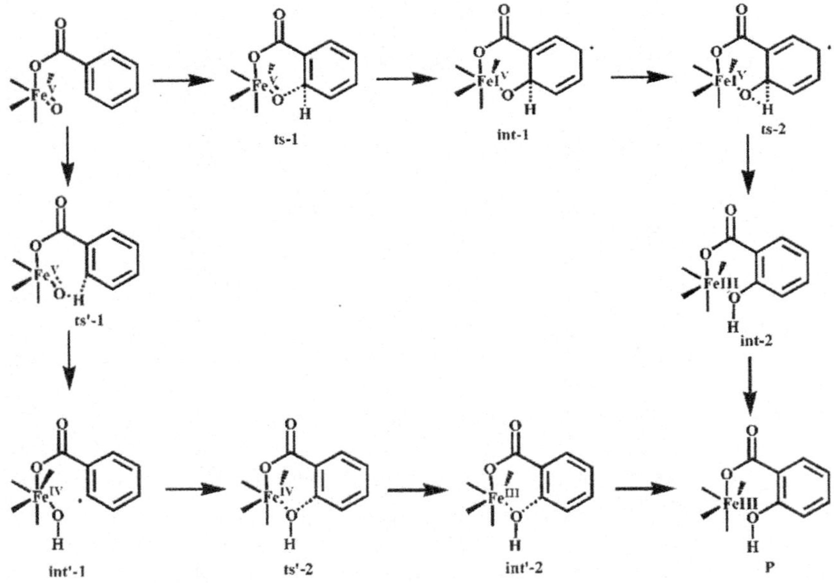

SCHEME 15.9 Adapted mechanism for C–H bond activation and oxygen attack during ortho-hydroxylation by putative Fe^V = O species. *Adapted from Ansari A, Kaushik A, Rajaraman G. Mechanistic insights on the ortho-hydroxylation of aromatic compounds by non-heme iron complex: a computational case study on the comparative oxidative ability of ferric-hydroperoxo and high-valent FeIVO and FeVO intermediates. J Am Chem Soc 2013;135(11):4235–4249.*

15.4.4 Nonradical mechanism

Nonradical, radical, and nonsynchronous concerted mechanisms have been proposed for methane hydroxylation by sMMO, as shown in Schemes 15.10 and 15.11 [79]. In the conversion of methane to methanol by FeO$^+$, the hydroxy intermediate (HO-Fe$^+$-CH$_3$) is energetically more favorable than the one via a methoxy intermediate (H-Fe$^+$-OCH$_3$). This conversion occurs

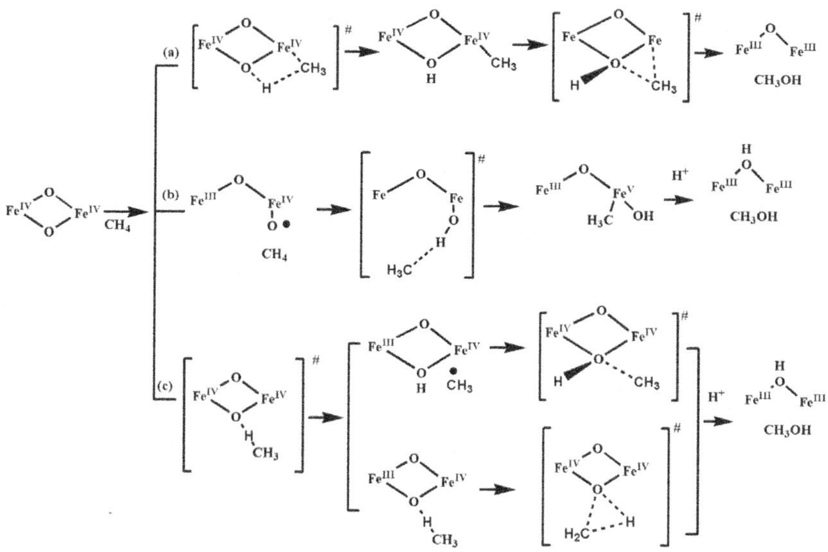

SCHEME 15.10 Theoretical reaction pathways proposed for methane hydroxylation. *Adapted from Huang S-P, Shiota Y, Yoshizawa K. DFT study of the mechanism for methane hydroxylation by soluble methane monooxygenase (sMMO): effects of oxidation state, spin state, and coordination number. Dalton Trans 2013;42(4):1011–1023.*

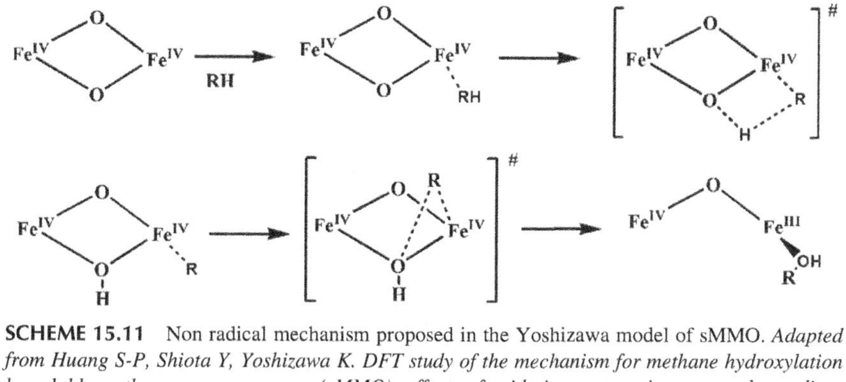

SCHEME 15.11 Non radical mechanism proposed in the Yoshizawa model of sMMO. *Adapted from Huang S-P, Shiota Y, Yoshizawa K. DFT study of the mechanism for methane hydroxylation by soluble methane monooxygenase (sMMO): effects of oxidation state, spin state, and coordination number. Dalton Trans 2013;42(4):1011–1023.*

reasonably well on the basis of the nonradical mechanism (Scheme 15.10). This is also similar to the reactions carried by the FeO$^+$ with methane proposed by Schwarz and coworkers.

15.4.5 Regioselectivity of aliphatic versus aromatic hydroxylation

DFT studies have been also carried out on the aliphatic and aromatic hydroxylation taking the example of ethyl benzene using Fe^{II}-superoxo (Scheme 15.12) and found that barrier height for aliphatic hydroxylation using the [Fe^{II}-OO(TMC)]$^+$ requires 22.1 kcal/mol (TS$_A$) and the benzene hydroxylation requires energy greater than 31.4 kcal/mol (TS$_B$) (Fig. 15.1; using Gaussian/B3LYP/LACVP) indicates the favorability of aliphatic hydroxylation. This suggests that weak aliphatic C—H bond hydroxylation can occur by [Fe^{II}-OO(TMC)]$^+$ species [80].

A similar observation was also found during aliphatic and aromatic hydroxylation with nonheme iron(IV) = O complexes bearing a TMC ligand system with acetonitrile (NCCH$_3$) or chloride (Cl) as an axial ligand: [Fe^{IV} = O(TMC)(L)]$^{n+}$ (L = NCCH$_3$ or Cl) using the ethylbenzene as the substrate [81]. Barrier heights of Fig. 15.2A and B also suggested higher reactivity are found with the aliphatic hydroxylation rather than aromatic [81].

15.4.6 Tunneling of reaction pathway from epoxidation to hydroxylation using cyclohexene

Nam et al. studied the effect of temperature by using [(TMP$^{•+}$)Fe^{IV}(O)(Cl)]$^+$ (TMP = meso-tetramesitylporphyrin dianion) and cyclohexene as the substrate. By the DFT calculations, rate constant of the C = C bond epoxidation and the

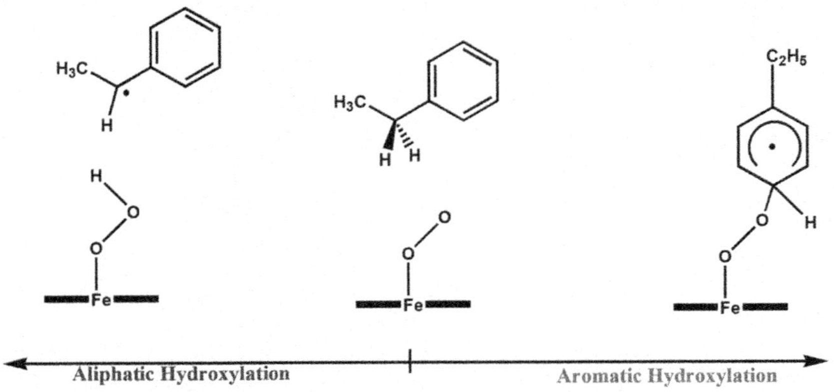

SCHEME 15.12 Aliphatic and aromatic hydroxylation by iron(II)-superoxo species. *Adapted from de Visser SP, Latifi R, Tahsini L, Nam W. The axial ligand effect on aliphatic and aromatic hydroxylation by non-heme iron(IV)-oxo biomimetic complexes. Chem Asian J 2011;6 (2):493–504.*

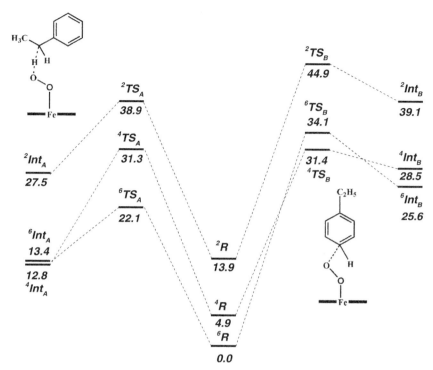

FIGURE 15.1 Energy surface of [(TMC)Fe^{II}OO]$^{+}$ with ethylbenzene (EB) leading to aliphatic hydroxylation (mechanism from the center to the left) or aromatic hydroxylation (mechanism from the center to the right). *From Latifi R, Tahsini L, Nam W, de Visser Sam P. Regioselectivity of aliphatic versus aromatic hydroxylation by a non-heme iron(II)-superoxo complex. Phys Chem Chem Phys 2012;14(7): 2518–2524.*

C–H bond hydroxylation of cyclohexene with the species [(TMP$^{•+}$)Fe^{IV}(O)(Cl)]$^{+}$ has been studied at different temperatures, and it is predicted that the reaction pathway changes to C–H hydroxylation from C=C epoxidation by decreasing the temperature because with decrease in temperature KIE values for the hydroxylation increases. The barrier for epoxidation is found to be 10.4 kcal/mol whereas 13.6 kcal/mol for the hydroxylation indicating epoxidation is favorable. Energy difference is 3.2 kcal/mol which can be achieved, if the conditions are favorable, such as lowering of temperature, in which tunneling effect becomes more important than the other factors which are insensitive toward the change in temperature (B3LYP/LACV3P + */LACVP) [82].

15.4.7 Sigma pathway and pie pathway

The spin surface of reaction pathway is selectively determined by geometric orientation. If the alpha electron from the substrate is transferred to the σ_z

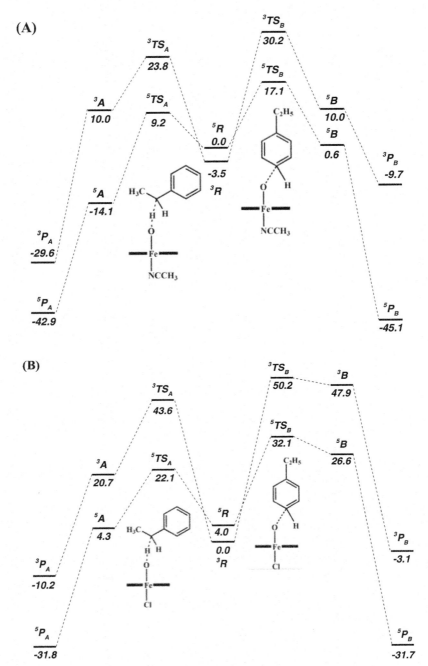

FIGURE 15.2 Energy profile of benzyl and phenyl hydroxylation with ^{3,5}R. Energies obtained with UB3LYP with zero-point, entropic, and thermal corrections at UB3LYP and (A) acetonitrile at axial and (B) chlorine at the axial. *Adapted from de Visser SP, Latifi R, Tahsini L, Nam W. The axial ligand effect on aliphatic and aromatic hydroxylation by non-heme iron(IV)-oxo biomimetic complexes. Chem Asian J 2011;6(2):493–504.*

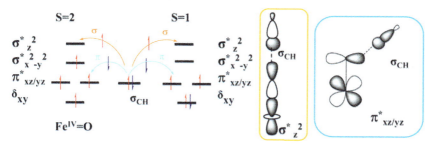

SCHEME 15.13 Schematic summary of the electronic structure changes along the reaction pathway in the quintet and triplet state of mononuclear nonheme iron(IV)-oxo complexes.

[2] orbital of metal then the pathway is known as sigma (σ) pathway and the angle of approach is in the range of 180°. If the beta electron from the substrate is transferred to the $\pi_{yz/xz}$ orbital of metal then this pathway is known as pie (π) pathway and the angle of approach in this pathway is nearer to 120° (Scheme 15.13).

There are two spin surfaces such as quintet and triplet of Fe^{IV} then the triplet surface will be destabilized by the electrostatic repulsion in the π-pathway while it will be aligned along the z-axis in the quintet transition state. When the electron-donating group is present at the axial position then the σ pathway reactivity decreases, whereas it is the opposite in the π-pathway [83].

15.5 Factors affecting hydroxylation

There are several factors that affect C—H bond activation (see below).

15.5.1 Oxidation state (spin state) of metal ion

It is found that the reactivity of complexes toward the C—H bond increases with an increase in the oxidation state. C—H bond activation by Fe^V = O requires less barrier as compared to Fe^{IV} = O, and the Fe^{II}-superoxo hydroxylation barrier is higher than that of Fe^{IV} = O and Fe^V = O indicating that high reactivity increases with an increase in oxidation state [84]. The effect of the oxidation state is studied by Neese et al. by modeling the hexacoordinated complexes of iron-oxo in different oxidation state IV/V/VI with four NH$_3$ ligands at equatorial along with one OH at an axial position to Fe = O moiety. The computational studies on these modeled complexes revealed that reactivity increases with the increase in oxidation states by a decrease in barrier height for C—H activation (16.7 kcal/mol for Fe^{IV} = O, 4.0 kcal/mol for Fe^V = O and barrier less for Fe^{VI} = O). When the oxidation state of metal inceases then the electrophilicity of metal also increases, which in turn enhances the ability of metal to abstract a proton [85].

15.5.2 Nature of ligand

With ligand change the ground state of the species also gets changed. For example, iron(IV) = O species shows both S = 1 and S = 2 spin states as the ground state depending upon the ligands. Nonheme iron(IV)-oxo species, such as the active species of TauD, has a quintet spin ground state well separated from the triplet. Whereas $[Fe^{IV}-(N_4Py)(O)]^{2+}$ species has a triplet spin ground state which is in agreement with the previous reports of Sason Shaik [84,86]. Thus as the ground state changes, the reactivity of species also changes. Nature of the ligand with iron (using $[(BPMEN)Fe^{III}-OOH]^{2+}$ and $[(TPA)Fe^{III}-OOH]^{2+}$) also changes the reactivity toward the hydroxylation reaction (Fig. 15.3) [21,68,87].

15.5.3 Ligand design

From the previous studies, it has been found that ligand architecture also plays an important role in the reactivity of species. Orientation of the

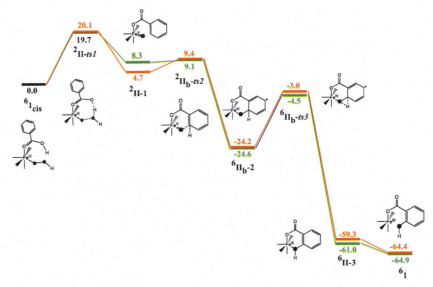

FIGURE 15.3 All energy are reported here including free energy corrections (green color for BPMEN and red for TPA). All energies are in kcal/mol. *Adapted from Ansari A, Rajaraman G. ortho-Hydroxylation of aromatic acids by a non-heme FeV = O species: how important is the ligand design? Phys Chem Chem Phys 2014;16(28):14601–14613.*

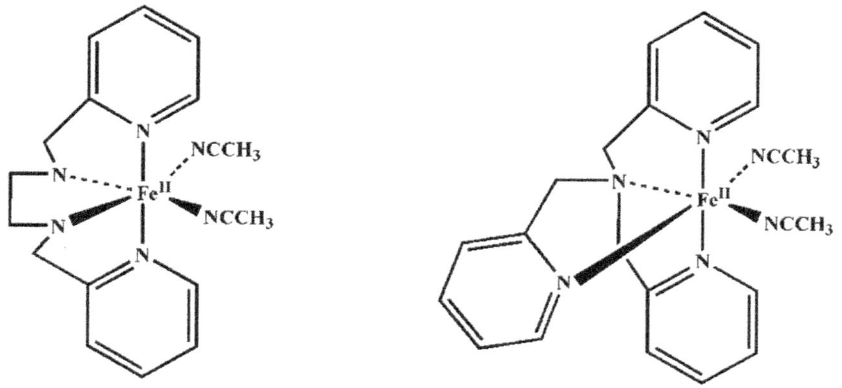

[Fe^II(BPMEN)(CH₃CN)₂]²⁺ **[Fe^II(TPA)(CH₃CN)₂]²⁺**

SCHEME 15.14 Schematic diagram of [Fe^II(BPMEN)(CH₃CN)₂]²⁺ and [Fe^II(TPA)(CH₃CN)₂]²⁺. *Adapted from de Visser, SP. What factors influence the ratio of C-H hydroxylation versus C=C epoxidation by a non-heme Cytochrome P450 biomimetic?. J Am Chem Soc 2006;128(49):15809–15818.*

pyridine ring affects the hydroxylation barrier (Scheme 15.14) while using [(BPMEN)Fe^{III}-OOH]$^{2+}$ and [(TPA)Fe^{III}-OOH]$^{2+}$ [21,68,87]. Rajaraman et al. reported that the pyridine ring parallel to the Fe = O bond interacted with Fe(d_{yz})-O(p_y) and this delocalized the electron densities decreased the electrophilic reactivity. If the pyridine ring in the BPMEN were perpendicular they would enhance the reactivity. Thus the barrier with BPMEN ligand is small (Fig. 15.3) [68]. When the TMC ligand is substituted by TMCS (TMCS = 1-mercaptoethyl-4,8,11-trimethyl-1,4,8,11-tetraza cyclotetradecane) then the hydroxylation of propene is favored over the epoxidation, this is due to the steric hindrance between the protons of the TMCS ligand with atoms of the propene destabilize the epoxidation barriers considerably and make the hydroxylation process favorable. Thus simply by tuning the ligand architecture, one can achieve selectivity with high efficiency [75]. Bond dissociation energies of the C−H bond that is undergoing activation, and the redox potential of high-valent species and the basicity of the terminal oxo unit play an important role in C−H bond activation by metal-oxo complexes [86].

15.5.4 Axial and equatorial ligands

The axial and equatorial ligands affect the reactivity of metal species toward the C−H bond activation. The reactivity toward the C−H bond activation has been studied by using experiments and theory. The reactivity can be varied by changing the nature of the axial ligand. This has been studied in detail for the cytochrome P450 enzymes. It is found that the weaker axial ligands are found to decrease the reactivity toward hydrogen atom transfer (HAT) reactions, for example, the -SH axial ligand has higher reactivity as compared to the imidazole axial ligand. Hydrogen atom abstraction by iron(IV)-oxo becomes reactive as the axial ligand becomes electron donating [88]. To study the effect of axial ligand, Xuri Huang et al. carry the DFT study with nonheme iron(IV)-oxo model systems $[Fe^{IV}(O)(NH_3)_4L]^+$ (where $L = CF_3CO_2^-$, F^-, Cl^-, N_3^-, NCS^-, NC^-, OH^-) and investigated the effects of electron donor ability of axial ligands on the reactivity of the distinct reaction channels, and the results indicates that reactivity towards the σ-pathway decreases as the electron-donating ability of the axial ligand strengthens, while it follows opposite trend in the π-pathway. Fig. 15.2 also reports similar observations.

15.5.5 Single-state reactivity (SSR) versus two-state reactivity (TSR)

When the two surfaces do not cross each other, then the single-state reactivity is followed. When the rate-determining transition structure (TS) arises from the spin state other than the ground spin state of the reactant followed by less gap between these spin states gives rise to two-state reactivity (TSR) i.e. the spin changes from the reactant to the transition state followed by the intermediate as shown in Schemes 15.15 and 15.16.

TSR/SSR is that the reaction proceeds at least on two potential energy surfaces with different spin multiplicity, either they may cross each other or they may have approximate energy.

There are four possibilities, (Scheme 15.16) for single-state versus two-state reactivity. In case A, the spin state of the ground state (at the lowest enegry) of the reactant and the transition state is the same then it follows STR whereas spin state of the transition state and the product is different indicating TSR in case of B. In C, the spin state of the reactant and

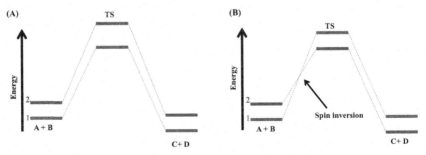

SCHEME 15.15 Systematic representation of (A) single-state reactivity and (B) two-state reactivity.

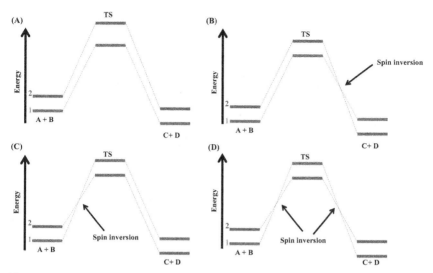

SCHEME 15.16 Various spin state scenarios in chemical reactions. *Data taken from Schröder D, Shaik S, Schwarz H. Two-state reactivity as a new concept in organometallic chemistry. Acc Chem Res 2000;33(3):139–145.*

the transition state is different shows TSR. Similarly in D, the spin state of the transition state changed from the reactant/product it also showed TSR [84].

Example; hydroxylation of cyclohexane catalyzed by $[(N_4Py)Fe^{IV}O]^{2+}$ species can occur at two spin surfaces then it shows two-state reactivity [78,84]. In the reactant, the triplet spin state (4.8 kcal/mol) is found as the ground state. The quintet state (15.8 kcal/mol) has energy higher than the triplet state (16.7 kcal/mol) at the transition state during hydrogen atom abstraction. These observations indicate that the spin state at the reactant and the transition state are different which shows the two-state reactivity (Fig. 15.4).

15.6 Applications in emerging fields

Modeling active site of the enzymatic system has emerged as a rapidly expanding discipline in the field of biomimetic chemistry [6,12,68,81,85]. It has motivated a range of biomimetic studies that have led to the development of contemporary catalytic protocols with increased efficiency and selectivity that have a ubiquitous significant impact on the synthetic catalytic field with pharmaceutical and industrial importance [68,81,85]. Ligand design helps in understating the reaction pattern and clarifying the mechanism. Experimental and theoretical studies have proved that axial ligand (trans to oxo group) has a significant effect on the catalytic activity of heme enzymes bearing high-valent oxo active species. Significant axial ligand effects have been reported

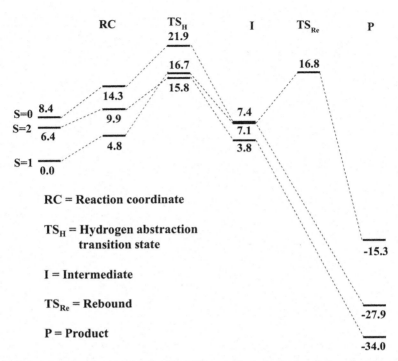

FIGURE 15.4 B3LYP calculated the TSR/MSR scenario during hydroxylation of cyclohexane by [(N₄Py)Fe^{IV}O]$^{2+}$ (B3LYP/LACVP). *Adapted from Kumar D, Hirao H, Que L Jr, Shaik S. Theoretical investigation of C-H hydroxylation by (N4Py)FeIV = O2 + : an oxidant more powerful than P450? J Am Chem Soc 2005;127(22):8026−8027.*

on ethylbenzene hydroxylation where axial acetonitrile ligand leads to phenyl hydroxylation products and an axial chloride anion giving primarily benzyl hydroxylation products. Nam and co-workers reported that DFT calculations on hydroxylation of ethylbenzene catalyzed by [Fe^{IV} = O (Por$^+$)L] (Por = porphyrin; L = NCCH₃ or Cl$^-$) which yields 1-phenylethanol and p-ethylphenol products. The hydroxylation reaction was tested for two different oxidants: one with acetonitrile and the other with chloride (axial) ligands. There was a remarkable effect of the axial ligand on the orbital energies of the oxidant and on the orbitals of the iron-oxo oxidant, as predicted by the DFT study [68]. Thus the fine-tuning axial ligand is a very crucial factor to obtain desired hydroxylated product distributions and this can be important to improve the efficiency of the oxidant and to clarify the relevant reaction mechanism [68,81].

Another interesting example of ligand design toward fine-tuning catalytic transformation reaction is the ortho-hydroxylation of aromatic compounds catalyzed by synthetic nonheme iron-oxo complexes. Makhlynets et al. reported experimental study for the ortho-hydroxylation of aromatic compounds by the catalysts [(BPMEN)Fe^{III}-OOH]$^{2+}$ and [(TPA)Fe^{III}-OOH]$^{2+}$ species, and Rajaraman et al. investigated mechanistic study on the same reaction using DFT calculations [68,87]. Both the species catalyze hydroxylation reaction via the formation of the putant high-valent Fe^V = O species. The difference in ligand architecture has a direct impact on the reactivity of both species. The different reactivity portrayed by these ligand systems is due to the parallel or perpendicular orientation of the pyridine ring(s) to the FeV = O bond. Due to such orientation, mixing of the orbital of the Fe-oxo with pyridine ring (π-$_{Fe(dyz)-O(py)}$)* takes place which leads to a reduction in the electrophilicity of the ferryl oxygen atom [68].

15.7 Future prospective and scope

Recent developments in the field of computational modeling and simulation have resulted in multifold advances in the field of biomimetic hydroxylation [21,68,79]. Computational techniques have helped in developing useful insights about the fine details of the nature of complexes and mechanism of hydroxylation reactions. The integration of other domains and methods are utmost required to explore many unanswered experimental questions related to biomimetic hydroxylation. The successive improvements in the computing facilities will immensely help in modeling complex and also will help in coming generation of scientists to thoroughly understand large data base with precision and within a short time duration, that is, computational approaches will go deeper and deeper [21,29,34,66,68].

Understanding of functioning of the active site of metalloenzymes especially that of dinuclear core is very complex. Many of the fine details are still missing which are yet to be explored. So theoretical study that too with advances in computational tools and techniques is required in the near future. Fine-tuning ligand modeling will enjoy a prominent position in the near future which will enable to obtain robust systems with desired catalytic properties and applications of interest.

One of the challenges particularly faced is the emulation of stereoselective aspects of enzymes. Natural enzymatic systems achieve stereoselectivity with quite ease. Modeling of enzymes has been of key interest to achieve good stereo and regioselectivity through modulating electronic effects of ligand, positioning, and changing effect of polar and/or charged groups, substitution at metal center, etc. The symbiosis of engineering with computational simulations may help in achieving robust and functional catalytic systems to carry out desired transformation reactions. The coming scenario will open new scopes for engineering technology to design bioinspired catalyst systems that aim at focussing on different aspects of biomimetics, such as structure-relation aspects, modulation of primary and secondary coordination sphere, and fine-tuning ligand architecture.

Besides these, the evolving developments and integration of various spectroscopic techniques will also play an important role in the characterization and detection of reactive intermediates. The integration of theoretical tools with sophisticated spectroscopic and kinetic methods may help in gaining deep insights about the unexplained aspects of the functioning of natural enzymatic systems.

The major breakthrough for obtaining a robust and selective catalytic system for hydroxylation of alkane may be chiefly achieved by ligand designing [68,81]. The fine-tuning of properties of ligands, such as modulation of electronic aspects, reverting the steric properties, proper substitution at active core, and stabilizing high-valent core with appropriate ligand environment, are a few of the key points that must be adhered to in the coming future to get desired catalytic activity of the catalyst systems which may be used via integration of computational tools with synthetic techniques.

15.8 Summary

Modeling of metalloenzymes has gained a significant amount of importance over past several decades. The study of hydroxylase enzymes has motivated a range of biomimetic studies and has become a widely studied research domain due to its importance in several disciplines, such as pharmaceuticals, synthetic organic chemistry, and agrochemical industry. Along with experimental study, computational tools have assisted in understanding the functioning of these enzymes. The computational study, especially DFT has helped in understanding the structural-functional characteristics of hydroxylase enzymes, the nature of active sites, the formation of the reactive intermediate, the role of these intermediates toward product formation, and the study of catalytic cycles of hydroxylase enzymes. Ligand designing has proved to be a vital factor in achieving good yields of the desired products. Besides this, chemo selectivity can also be achieved by making the selection of appropriate ligand systems with requisite substituent selections. DFT study has helped immensely to sort out many of the unclear doubts regarding the mechanistic aspects of the catalytic cycle of hydroxylation reaction.

At the end, theoretical studies have a huge potential to impart current research directions in the field of biomimetic chemistry of hydroxylase enzymes and in the coming future will definitely assist in answering many unanswered experimental findings.

Problems with solutions

Q1. Describe the catalytic activity of metalloenzymes.

Answer: Metalloenzymes bearing transition metals, such as Mn, Fe, Co, Zn, and Cu, catalyze a diverse range of chemical reactions. Metalloenzymes, such as methane monoxygenase, cytochromes p450, hemerythrin (Hr), Tyrosinase, and flavins, catalyze important reactions, such as bond breakage and formation, protein transfer, hydrolysis, oxygen flow, O_2 insertion into substrate redox process, and radical reactions, which are biochemical relevant reactions of life process.

Q2. What is the importance of hydroxylase enzyme?

Answer: Hydroxylase enzymes are an important and widely studied class of enzymes that are able to catalyze the hydroxylation of inert C—H bond in various organic compounds at mild conditions and yield highly stereo and regioselective hydroxylated products. These enzymes play important roles in essential biological transformation reactions. For example, methane monoxygenase, flavins, some diiron hydroxylases, cytochromes P450, etc. Hydroxylation reaction plays an important role in several life process reactions, such as hydroxylation of methane in methanotrophs, fatty acid desaturation in plants, genome material repairs, and formation of blood vessels. These reactions involve the activation of dioxygen (O_2) catalyzed by metal-mediated reactions.

Q3. What are biomimetic complexes?

Answer: The synthetic complexes that mimic the natural enzymes are referred as biomimetic complexes. These model systems were developed to understand various aspects of enzymatic systems, such as nature of active sites, their role in forming reactive intermediates, mechanistic details of the catalytic cycle, nature of key intermediates formed during the catalytic cycle, oxygen binding ability, and low activation barrier for the given chemical reaction.

Q4. What is importance of DFT in study of biomimetic hydroxylation reaction?

Answer: DFT study helps to study important properties of reactive intermediates, such as their electronic structures, stability, nature of reactive sites (electrophilic and nucleophilic centers), transition states, activation barrier, and help to figure out mechanistic routes for catalytic reactions. DFT study further helps in elucidating structural-functional aspects of reactive intermediates and their thermodynamic formation energy.

Q5. How does spin density analysis help in understanding a particular mechanism of interest?

Answer: The presence of significant spin density at atoms of intermediate helps in predicting the probable mechanistic pathways. For example, the presence of significant spin density corresponding to the radical character helps in building the possible mechanism to follow the radical formation pathway. This also helps in figure out about reaction proceed through the stepwise or concerted, proton transfer and electron transfer, electron transfer and proton transfer, proton coupled electron transfer, etc., during chemical reaction.

Q6. How does DFT help in understanding the chemo selectivity of hydrocarbon in case of hydroxylation versus epoxidation reaction?

Answer: DFT study helps in studying the geometries of transition states which can also help in determining rate of reaction/process. In case, if steric hindrance is more in the transition state corresponding to the epoxidation reaction than the transition states undergoing hydroxylation which suggests hydroxylation is more dominant as compared to the epoxidation reaction. Thus DFT helps in understanding the chemoselectivity of a given reaction.

Q7. Explain the role of high-valent oxo species.

Answer: High-valent oxo species, such as metal(IV/V)-oxo plays an important role in catalyzing important chemical reactions involving activation of inert bonds, such as C−H/O−H/N−H. These species are reported as reactive intermediates which trigger triggers catalytic reactions, such as hydroxylation, epoxidation, and halogenation.

Q8. What are methane monooxygenase enzymes?

Answer: Methane monooxygenase enzymes are iron-containing enzymes which catalyze hydroxylation of methane. These enzymes are present in aerobic methanotrophs which utilize carbon an energy source. These enzymes serve as very interesting motives for biotechnological applications, synthesis of hydroxylated products, etc.

Q9. What is the two-state reactivity?

Answer: When the reactions take place at surfaces, and during the reaction the spin state of the reactions changes on going from the reactant to the transition state (i.e., the reaction occurs more than one spin surface). The energy gap between the spin surfaces should be less (see Scheme 15.16).

Q10. What do you mean by sigma and pie pathway?

Answer: When the electron entering goes to d_{z^2} orbital then the reaction follows the sigma pathway, and when the entering electron goes to $d_{xz/yz}$ orbital then the reaction follow the pie pathway. In sigma pathway, the angle is near to 180 degrees whereas angle is near to 120 degrees in the case of pie pathway.

Objective type questions

i. Which of the following is an example of hydroxylase enzyme?
   (a) RNR (ribonucleotide)
   (b) MMO(methane monooxygenase)

Role of biomimetic complexes in hydroxylation reactions Chapter | 15   **405**

   (c) Hemocynanin
   (d) Peptidase
   Answer: (b)

ii. Which transition metal is commonly present at active site of enzymes?
   (a) Mn
   (b) Co
   (c) Cu
   (d) Fe
   Answer: (d)

iii. Which of the following reaction(s) is catalyzed by MMO?
   (a) Hydrogenation of propene
   (b) Ethane to ethanol conversion
   (c) Methane to methanol conversion
   (d) Hydrolysis of ethene
   Answer: (c)

iv. The first high-valent metal complex with the following ligands is
   (a) TACN (1,4,7-triazacylononane)
   (b) BPMEN (N,N'-dimethyl-N,N'-bis(2-pyridylmethyl)ethane 1,2diamine)
   (c) TPA (Tris(2-pyridylmethyl)amine)
   (d) TPEN (N,N,N',N'-tetrakis(2-pyridinylmethyl)-1,2-ethanediamine)
   Answer: (c)

v. Which of the following reaction(s) runs parallel to hydroxylation reaction disturbing its chemoselectivity?
   (a) Hydrolysis
   (b) Epoxidation
   (c) Hydrogenation
   (d) Reduction
   Answer: (b)

vi. Which of the following parameter(s) corresponding to a transition state are studied by DFT method?
   (a) Structural parameters
   (b) Spin density
   (c) Activation barrier
   (d) All of the them
   Answer: (d)

vii. Which drug does not contain hydroxyl moiety?
   (a) Morphine
   (b) Levonodefrin

(c) cis-Platin
(d) Hydroxycholoroquine
Answer: (c)

viii. Who developed the first model of C—H bond hydroxylation?
   (a) Groves and coworkers
   (b) Que and coworkers
   (c) Hay and coworkers
   (d) None of the above
   Answer: (a)

ix. What does DFT stand for?
   (a) Density Functional Theory
   (b) Density Force Theory
   (c) Dense Fluid Theory
   (d) None of the above
   Answer: (a)

x. When was the first biomimetic hydroxylation carried out?
   (a) 1987
   (b) 1950
   (c) 1999
   (d) 1979
   Answer: (d)

References

[1] de Montellano PRO. Cytochrome P450: structure, mechanism, and biochemistry. 3rd ed. New York: Kluwer Academic/Plenum Publishers; 2005.
[2] Fishman A, Tao Y, Rui A, Wood TK. Controlling the regiospecific oxidation of aromatics via active site engineering of toluene para-Monooxygenase of Ralstonia pickettii PKO1. J Biol Chem 2005;280(1):506—14.
[3] Sazinsky MH, Bard J, Donato AD, Lippard SJ. Crystal structure of the toluene/o-xylene monooxygenase hydroxylase from Pseudomonas stutzeri OX_1. J Biol Chem 2004;279(29):30600—10.
[4] Murray LJ, Lippard SJ. Substrate trafficking and dioxygen activation in bacterial multicomponent monooxygenases. Acc Chem Res 2007;40(7):466—74.
[5] Fitzpatrick PF. Mechanism of aromatic amino acid hydroxylation. Biochemistry 2003;42(48):14083—91.
[6] Flatmark T, Stevens RC. Structural insight into the aromatic amino acid hydroxylases and their disease-related mutant forms. Chem Rev 1999;99(8):2137—60.
[7] Timmins A, de Visser SP. A comparative review on the catalytic mechanism of non-heme iron hydroxylases and halogenases. Catalysts 2018;8(8):314.
[8] Niwa T, Murayama N, Imagawa Y, Yamazaki H. Regioselective hydroxylation of steroid hormones by human cytochromes P450. Drug Metab Rev 2015;47(2):89—110.

[9] He Y, Gorden JD, Goldsmith CR. Steric modifications tune the regioselectivity of the alkane oxidation catalyzed by non-heme iron complexes. Inorg Chem 2011;50:12651–60.
[10] Hirao H, Kumar D, Jr. LQ, Shaik S. Two-state reactivity in alkane hydroxylation by non-heme iron–oxo complexes. J Am Chem Soc 2006;128(26):8590–606.
[11] Bahramian B, Mirkhani V, Tangestaninejad S, Moghadam M. Catalytic epoxidation of olefins and hydroxylation of alkanes with sodium periodate by water-soluble manganese (III)salen. J Mol Catal A: Chem 2006;244:139–45.
[12] Ma J, Gu Y, Xu P. A roadmap to engineering antiviral natural products synthesis in microbes. Curr Opin Biotechnol 2020;66:140–9.
[13] Dalle KE, Meyer F. Modelling binuclear metallobiosites: insights from pyrazole-supported biomimetic and bioinspired complexes. Eur J Inorg Chem 2015;2015 (21):3391–405.
[14] Sheldon RA, Kochi JK. Metal-catalyzed oxidations of organic compounds in the liquid phase: a mechanistic approach. Adv Catal 1976;25:272–413.
[15] Jensen MP, Lange SJ, Mehn MP, Que EL, Que L. Biomimetic aryl hydroxylation derived from alkyl hydroperoxide at a non-heme iron center. Evidence for an $Fe^{IV}O$ oxidant. J Am Chem Soc 2003;125(8):2113–28.
[16] Biomimetic oxidations catalyzed by transition metal complexes. In: Meunier B, Brudvig G, Mclain JL, Murahashi SI, Pecoraro VL, Riley D, editors. World Scientific. Imperial College Press; 2000.
[17] Costas M, Mehn MP, Jensen MP, Que Jr. L. Dioxygen activation at mononuclear non-heme iron active sites: enzymes, models, and intermediates. Chem Rev 2004;104 (2):939–86.
[18] Hausinger RP. Fe (II)/α-ketoglutarate-dependent hydroxylases and related enzymes. Crit Rev Biochem Mol Biol 2004;39(1):21–68.
[19] Kovaleva EG, Lipscomb JD. Versatility of biological non-heme Fe(II) centers in oxygen activation reactions. Nat Chem Biol 2008;4(3):186–93.
[20] Massey V. Activation of molecular oxygen by flavins and flavoproteins. J Biol Chem 1994;269(36):22459–62.
[21] Ansari A, Kaushik A, Rajaraman G. Mechanistic insights on the ortho-hydroxylation of aromatic compounds by non-heme iron complex: a computational case study on the comparative oxidative ability of ferric-hydroperoxo and high-valent $Fe^{IV}O$ and $Fe^{V}O$ intermediates. J Am Chem Soc 2013;135(11):4235–49.
[22] Lindhorst AC, Drees M, Bonrath W, Schütz J, Netscher T, Kühn FE. Mechanistic insights into the biomimetic catalytic hydroxylation of arenes by a molecular Fe(NHC) complex. J Catal 2017;352:599–605.
[23] Romero E, Castellanos JRG, Gadda G, Fraaije MW, Mattevi A. Same substrate, many reactions: oxygen activation in flavoenzymes. Chem Rev 2018;118(4):1742–69.
[24] Dong J, Fernández-Fueyo E, Hollmann F, Paul CE, Pesic M, Schmidt S, Wang Y, Younes S, Zhang W. Biocatalytic oxidation reactions: a chemist's perspective. Angew Chem Int Ed 2018;57(30):9238–61.
[25] Nguyen HHT, Elliott SJ, Yip JHK, Chan SI. The particulate methane monooxygenase from Methylococcus capsulatus (Bath) is a novel copper-containing three-subunit enzyme: isolation and characterization. J Biol Chem 1998;273(14):7957–66.
[26] Groves JT, Haushalter RC, Nakamura M, Nemo TE, Evans BJ. High-valent iron-porphyrin complexes related to peroxidase and cytochrome P450. J Am Chem Soc 1981;103 (10):2884–6.

[27] Jasniewski AJ, Que Jr. L. Dioxygen activation by non-heme diiron enzymes: diverse dioxygen adducts, high-valent intermediates, and related model complexes. Chem Rev 2018;118(5):2554−92.
[28] Che CM, Huang JS. Metalloporphyrin-based oxidation systems: from biomimetic reactions to application in organic synthesis. Chem Comm 2009;27:3996−4015.
[29] Yoshizawa K, Shiota Y, Yamabe T. Methane-methanol conversion by MnO^+, FeO^+, and CoO^+: a theoretical study of catalytic selectivity. J Am Chem Soc 1998;120(3):564−72.
[30] Mahyuddin MH, Shiota Y, Staykov A, Yoshizawa K. Theoretical overview of methane hydroxylation by copper-oxygen species in enzymatic and zeolitic catalysts. Acc Chem Res 2018;51(10):2382−90.
[31] Collins TJ. TAML oxidant activators: a new approach to the activation of hydrogen peroxide for environmentally significant problems. Acc Chem Res 2002;35(9):782−90.
[32] Ghosh M, Singh KK, Panda C, Weitz A, Hendrich MP, Collins TJ, Dhar BB, Sen GS. Formation of a room temperature stable $Fe^V(O)$ complex: reactivity toward unactivated C-H bonds. J Am Chem Soc 2014;136(27):9524−7.
[33] Kundu S, Thompson JVK, Ryabov AD, Collins TJ. On the reactivity of mononuclear iron (V)oxo complexes. J Am Chem Soc 2011;133(46):18546−9.
[34] Monika Ansari A. Mechanistic insights into the allylic oxidation of aliphatic compounds by tetraamido iron(V) species: a C-H vs. O-H bond activation. New J Chem 2020;44(44):19103−12.
[35] Monika Ansari A. Effect of the ring size of TMC ligands in controlling C−H bond activation by metal-superoxo species. Dalton Trans 2022;51(15):5878−89.
[36] Cho J, Jeon S, Wilson SA, Liu LV, Kang EA, Braymer JJ, Lim MH, Hedman B, Hodgson KO, Valentine JS, Solomon EI, Nam W. Structure and reactivity of a mononuclear non-haem iron(III)-peroxo complex. Nature 2011;478(7370):502−5.
[37] Peterson RL, Himes RA, Kotani H, Suenobu T, Tian L, Siegler MA, Solomon EI, Fukuzumi S, Karlin KD. Cupric superoxo-mediated intermolecular C-H activation chemistry. J Am Chem Soc 2011;133(6):1702−5.
[38] Wilfer C, Liebhuser P, Hoffmann A, Erdmann H, Grossmann O, Runtsch L, Paffenholz E, Schepper R, Dick R, Bauer M, Dürr M, Ivanović-Burmazović I, Herres-Pawlis S. Efficient biomimetic hydroxylation catalysis with a bis(pyrazolyl)imidazolylmethane copperperoxide complex. Chem Eur J 2015;21(49):17639−49.
[39] Bochevarov AD, Harder E, Hughes TF, Greenwood JR, Braden DA, Philipp DM, Rinaldo D, Halls MD, Zhang J, Friesner RA. Jaguar: a high-performance quantum chemistry software program with strengths in life and materials sciences. J Quantum Chem 2013;113(18):2110−42.
[40] Frisch M.J., Trucks G.W., Schlegel H.B., Scuseria G.E., Robb M.A., Cheeseman J.R., Scalmani G., Barone V., Petersson G.A., Nakatsuji H., Li X., Caricato M., Marenich A. V., Bloino J., Janesko B.G., Gomperts R., Mennucci B., Hratchian H.P., Ortiz J.V., Izmaylov A.F., Sonnenberg J.L., Williams-Young D., Ding F., Lipparini F., Egidi F., Goings J., Peng B., Petrone A., Henderson T., Ranasinghe D., Zakrzewski V.G., Gao J., Rega N., Zheng G., Liang W., Hada M., Ehara M., Toyota K., Fukuda R., Hasegawa J., Ishida M., Nakajima T., Honda Y., Kitao O., Nakai H., Vreven T., Throssell K., Montgomery J.A. Jr, Peralta J.E., Ogliaro F., Bearpark M.J., Heyd J.J., Brothers E.N., Kudin K.N., Staroverov V.N., Keith T.A., Kobayashi R., Normand J., Raghavachari K., Rendell A.P., Burant J.C., Iyengar S.S., Tomasi J., Cossi M., Millam J.M., Klene M., Adamo C., Cammi R., Ochterski J.W., Martin R.L., Morokuma K., Farkas O., Foresman J.B., Fox D.J. Wallingford, CT: Gaussian, Inc.; 2009.

[41] Becke AD. Density-functional thermochemistry. III. role exact. Exch J Chem Phys 1993;98(7):5648–52.
[42] Grimme SJ. Semiempirical GGA-type density functional constructed with a long-range dispersion correction. Comput Chem 2006;27(15):1787–99.
[43] Chai JD, Head-Gordon M. Long-range corrected hybrid density functionals with damped atom-atom dispersion corrections. Phys Chem 2008;10(44):6615–20.
[44] Zhao Y, Truhlar DG. The M06 suite of density functionals for main group thermochemistry, thermochemical kinetics, noncovalent interactions, excited states, and transition elements: two new functionals and systematic testing of four M06 functionals and 12 other functionals. Theor Chem Acc 2008;120(1):215–41.
[45] Tao JM, Perdew JP, Staroverov VN, Scuseria GE. Climbing the density functional ladder: nonempirical meta-generalized gradient approximation designed for molecules and solids. Phys Rev Lett 2003;91(14):146401–4.
[46] Møller C, Plesset MS. Note on an approximation treatment for many-electron systems. Phys Rev 1934;46(7):618–22.
[47] Hay PJ, Wadt WR. Ab initio effective core potentials for molecular calculations. Potentials for the transition metal atoms Sc to Hg. J Chem Phys 1985;82(1):270–83.
[48] Wadt WR, Hay PJ. Ab initio effective core potentials for molecular calculations. Potentials for main group elements. J Chem Phys 1985;82(1):284–98.
[49] Krishnan R, Binkley JS, Seeger R, Pople JA. Self-consistent molecular orbital methods. XX. A basis set for correlated wave functions. J Chem Phys 1980;72(1):650–4.
[50] Schaefer A, Horn H, Ahlrichs R. Fully optimized contracted Gaussian-basis sets for atoms Li to Kr. J Chem Phys 1992;97(4):2571–7.
[51] Weigend F, Ahlrichs R. Balanced basis sets of split valence, triple zeta valence and quadruple zeta valence quality for H to Rn: design and assessment of accuracy. Phys Chem Chem Phys 2005;7(18):3297–305.
[52] Dunning TH, Jeffrey HP. Gaussian basis sets for molecular calculations. In: Schaefer HF, editor. *Methods of electronic structure theory. Modern theoretical chemistry.* Boston (MA): Springer; 1977. p. 1–27.
[53] Wachters AJH. Gaussian basis set for molecular wavefunctions containing third-row atoms. J Chem Phys 1970;52(3):1033–6.
[54] Wong MW, Frisch MJ, Wiberg KB. Solvent effects 1 The mediation of electrostatic effects by solvents. J Am Chem Soc 1991;113(13):4776–82.
[55] Onsagar L. Electric moments of molecules in liquids. J Am Chem Soc 1936;58 (8):1486–93.
[56] Punniyamurthy T, Velusamy S, Iqbal J. Recent advances in transition metal catalyzed oxidation of organic substrates with molecular oxygen. Chem Rev 2005;105(6):2329–64.
[57] de Visser SP, Oh K, Han A-R, Nam W. Combined experimental and theoretical study on aromatic hydroxylation by mononuclear non-heme iron(IV)-oxo complexes. Inorg Chem 2007;46(11):4632–41.
[58] i Payeras AM, Ho RYN, Fujita M, Que Jr. L. The reaction of [Fe^{II}(tpa)] with H$_2$O$_2$ in acetonitrile and acetone—distinct intermediates and yet similar catalysis. Chem Eur J 2004;10 (20):4944–53.
[59] Quinonero D, Musaev DG, Morokuma K. Theoretical studies of the complex [(BPMEN) Fe(II)(NCCH$_3$)$_2$]$^{2+}$, precursor of non-heme iron catalysts for olefin epoxidation and cis-dihydroxylation. Inorg Chem 2003;42(25):8449–55.
[60] Gunay A, Theopold KH. C-H bond activations by metal oxo compounds. Chem Rev 2010;110(2):1060–81.

[61] Prokop KA, de Visser SP, Goldberg DP. Unprecedented rate enhancements of hydrogen-atom transfer to a manganese(V)-oxo corrolazine complex. Angew Chem Int Ed 2010;49(30):5091–5.

[62] Hitomi Y, Arakawa K, Funabiki T, Kodera M. An Iron(III)-monoamidate complex catalyst for selective hydroxylation of alkane C-H bonds with hydrogen peroxide. Angew Chem Int Ed 2012;24(14):3504–8.

[63] Chandra B, Singh KK, Gupta SS. Selective photocatalytic hydroxylation and epoxidation reactions by an iron complex using water as the oxygen source. Chem Sci 2017;8(11):7545–51.

[64] Li X-X, Guo M, Qiu B, Cho K-B, Sun W, Nam W. High-spin Mn(V)-oxo intermediate in non-heme manganese complex-catalyzed alkane hydroxylation reaction: experimental and theoretical approach. Inorg Chem 2019;58(21):14842–52.

[65] Ansari A, Ansari M, Singha A, Rajaraman G. Interplay of electronic cooperativity and exchange coupling in regulating the reactivity of diiron(IV)-oxo complexes towards C-H and O-H bond activation. Chem Eur J 2017;23(42):10110–25.

[66] Hori Y, Shiota Y, Tsuji T, Kodera M, Yoshizawa K. Catalytic performance of a dicopper–oxo complex for methane hydroxylation. Inorg Chem 2018;57(1):8–11.

[67] Vilella L, Conde A, Balcells D, Díaz-Requejo MM, Lledós A, Pérez PJ. A competing, dual mechanism for catalytic direct benzene hydroxylation from combined experimental-DFT studies. Chem Sci 2017;5(8):8373–83.

[68] Ansari A, Rajaraman G. ortho-Hydroxylation of aromatic acids by a non-heme Fe^V = O species: how important is the ligand design? Phys Chem Chem Phys 2014;16(28):14601–13.

[69] Shilov AE, Shulpin GB. Activation of C-H bonds by metal complexes. Chem Rev 1997;97(8):2879–932.

[70] Fukuzumi S, Karlin KD. Kinetics and thermodynamics of formation and electron-transfer reactions of $Cu-O_2$ and Cu_2-O_2 complexes. Coord Chem Rev 2013;257(1):187–95.

[71] Conde A, Diaz-Requejo MM, Perez PJ. Direct, copper-catalyzed oxidation of aromatic C-H bonds with hydrogen peroxide under acid-free conditions. Chem Commun 2011;47(28):8154–6.

[72] de Visser SP, Kumar D, Cohen S, Shacham R, Shaik S. A predictive pattern of computed carriers for C-H hydroxylation by compound I of Cytochrome P450. J Am Chem Soc 2004;126(27):8362–3.

[73] Kumar R, Ansari A, Rajaraman G. Axial vs. equatorial ligand rivalry in controlling the reactivity of iron(IV)-oxo species: single-state vs. two-state reactivity. Chem Eur J 2018;24(26):6818–27.

[74] Huang X, Groves1 JT. Beyond ferryl-mediated hydroxylation: 40 years of the rebound mechanism and C–H activation. J Biol Inorg Chem 2017;22(1):185–207.

[75] de Visser SP. What factors influence the ratio of C-H hydroxylation versus C = C epoxidation by a non-heme Cytochrome P450 biomimetic? J Am Chem Soc 2006;128(49):15809–18.

[76] Cho K-B, Hirao H, Shaik S, Nam W. To rebound or dissociate? This is the mechanistic question in C-H hydroxylation by heme and non-heme metal-oxo complexes. Chem Soc Rev 2016;45(5):1197–210.

[77] Yuan Y-C, Bruneau C, Dorcet V, Roisnel T, Gramage-Doria R. Ru-catalyzed selective C-H bond hydroxylation of cyclic imides. J Org Chem 2019;84(4):1898–907.

[78] Kumar D, Hirao H, Que Jr L, Shaik S. Theoretical investigation of C-H hydroxylation by (N4Py)Fe^{IV} = O^{2+}: an oxidant more powerful than P450? J Am Chem Soc 2005;127(22):8026–7.

[79] Huang S-P, Shiota Y, Yoshizawa K. DFT study of the mechanism for methane hydroxylation by soluble methane monooxygenase (sMMO): effects of oxidation state, spin state, and coordination number. Dalton Trans 2013;42(4):1011–23.

[80] Latifi R, Tahsini L, Nam W, de Visser Sam P. Regioselectivity of aliphatic versus aromatic hydroxylation by a non-heme iron(II)-superoxo complex. Phys Chem Chem Phys 2012;14(7):2518–24.

[81] de Visser SP, Latifi R, Tahsini L, Nam W. The axial ligand effect on aliphatic and aromatic hydroxylation by non-heme iron(IV)-oxo biomimetic complexes. Chem Asian J 2011;6(2):493–504.

[82] Gupta R, Li X-X, Cho K-B, Guo M, Lee Y-M, Wang Y, Fukuzumi S, Nam W. Tunneling effect that changes the reaction pathway from epoxidation to hydroxylation in the oxidation of cyclohexene by a compound I model of Cytochrome P450. J Phys Chem Lett 2017;8(7):1557–61.

[83] Tang H, Guan J, Zhang L, Liu H, Huang X. The effect of the axial ligand on distinct reaction tunneling for methane hydroxylation by non-heme iron(iv)−oxo complexes. Phys Chem Chem Phys 2012;14(37):12863–74.

[84] Schröder D, Shaik S, Schwarz H. Two-state reactivity as a new concept in organometallic chemistry. Acc Chem Res 2000;33(3):139–45.

[85] Geng C, Ye S, Neese F. Does a higher metal oxidation state necessarily imply higher reactivity toward H-atom transfer? A computational study of C-H bond oxidation by high-valent iron-oxo and -nitrido complexes. Dalton Trans 2014;43(16):6079–86.

[86] Cho K-B, Hirao H, Shaik S, Nam W. To rebound or dissociate? This is the mechanistic question in C−H hydroxylation by heme and nonheme metal−oxo complexes. Chem Soc Rev 2016;45:1197–210.

[87] Makhlynets OV, Das P, Taktak P, Flook M, Mas-Balleste R, Rybak-Akimova EV, Que Jr. L. Iron-promoted ortho- and/or ipso-hydroxylation of benzoic acids with H_2O_2. Chem Eur J 2009;15:13171–80.

[88] Sastri CV, Lee J, Oh K, Lee YJ, Lee J, Jackson TA, Ray K, Hirao H, Shin W, Halfen JA, Kim J, Que Jr L, Shaik S, Wonwoo Nam W. Axial ligand tuning of a nonheme iron(IV)-oxo unit for hydrogen atom abstraction. Proc Natl Acad Sci USA 2007;104(49):19181–6.

Index

Note: Page numbers followed by "*f*" and "*t*" refer to figures and tables, respectively.

A

Activation by hydrolysis, 269
  Rh^{III} and Ir^{III} Cp complexes, 270–271
  Ru^{II} and Os^{II} half-sandwich complexes, 270
  square-planar Pt^{II} complexes, 269
  Ti^{IV} dichloride complexes, 269
Addition reactions, 14
Agostic interaction, 21
Al_{50}(cyclopentadienyl)$_{10}$, 93, 94*f*
Aliphatic hydroxylation, 393*f*
Alkyl lithium reagents, 3
Alkyne cyclotrimerization, 3–4
American Chemical Society, 1–2
American trypanosomiasis. *See* Chagas disease
Amino alcohol ligands, 102
2-Amino benzoselenazoles, 117
2-Aminobenzothiazoles, 117
Aminophosphines
  cis/trans isomerization, 175
  cyclodiphosphazanes
    description, 175
    macrocyclic, 179
    reactivity of, 177–179
    synthesis of, 175–176
Anti-Alzheimer activity
  AChE inhibitory activity, 263–264
  multitargeted treatments, 263, 263*f*
  selenodihydropyrimidinones, 263, 263*f*
  selenyl-dihydrofurans, 262
Antibacterial activity, 287–288, 295–296, 296*f*
Anticancer properties
  copper complexes, 243–245
  gold complexes, 248
  iron complexes, 241–243
  rhodium and iridium complexes, 248–250
  ruthenium complexes, 245–248
Antifilarial drugs, structure of, 369*f*
Antifungal activity
  ATZ imidazole for, 295*f*
  ferrocenyltriazole and thiazole derivatives for, 301*f*
  ferrocenyl 1,2,4-triazole-containing compounds for, 300*f*
  KTZ- and CTZ-based metal complexes for, 297*f*, 299*f*
  mono-, bis-, and tris-imidazole based organoruthenium complexes for, 294*f*
Antifungal drugs
  structures of imidazole-based, 289*f*
  structures of triazole-based, 290*f*
Antileishmanial compounds, 365*f*
Antimalarial agents
  hybrid organometallic, 317–323
  molecular structures of, 316*f*
Antimalarial complexes
  organoosmium-based, 321
  organoplatinum-based, 322–323
  organorhodium-based, 320–321
Antimalarial compounds, hybrid organometallic, 315–317
Antimalarial drugs
  approach to hybrid, 314–315
  candidates, 315*f*
  examples of, 312–314
  molecular structures of, 312*f*
Antimalarial hybrid complexes, organoiridium based, 319–320
Arachno, 95–96
Arachno boranes, 95–96
Aromatic hydroxylation, 393*f*
Arsenic-based organo-compound, 3
Arsphenamine, 3
Artemisinin, 313
  mechanism of action, 313, 314*f*
  1,2,4-Trioxane scaffold of, 313*f*
4-Arylisoquinolines
  regioselective synthesis of, 128, 128*f*, 129*f*
  synthetic application of, 128–129, 130*f*

414 Index

4-Arylisoquinolones, 127–128, 128f
Aryl-4-oxothiazolylhyrazone (ATZ), structure of, 357f
Aryl-substituted isoquinolines, 119–120, 120f
Asymmetric catalysis, NHCs-TM
 applications, 148–150, 149f
 chiral catalyst, 140
 chiral NHCs, 144–145
 enantiopure pharmaceuticals production, 148–150
 future, 150
 metal-assisted, 148, 149f
 methoxylation and intramolecular cyclization, 1,6-enyne, 145–147, 147f
 transition-metal, 140
Asymmetric hydrogenation, E-aryl alkenes, 144, 145f
Atorvastatin, 82
Azole-based ligands, metallocenyl derivatives with, 298–301
Azole based metal-complexes, use in medicinal chemistry, 292
Azole-based organometallics
 complexes, mechanistic approach of, 291, 301–302
 compounds, advances of, 292–301
Azole derivatives
 based ruthenium organometallic complexes, 293–295
 copper- and zinc-based organometallic complexes with, 298
 copper-, gold-, and platinum-based organometallic complexes with, 296–297
Azole ligands
 manganese based organometallic complexes with, 295–296
 metallocenyl derivatives with, 298–301
Azoles, 287
 as antifungal drugs, 288–291
 combination of, 288
 type compounds, 287–288

B

Band theory, 4–5
Beer-Lambert law, 51
Benzene hydroxylation, 388, 389f, 390f
Benzisoselenazolones, 255–256, 256f, 262
β-Agostic-metal-alkyl compounds, 15f
β-H Elimination reactions, 15
β-Hydride elimination, 63–64, 67

Bilharzia. See Schistosomiasis
Biochemical sensor, 230–231
Biological activities, organoselenium compounds
 anti-Alzheimer activity, 262–264
 cytoprotection, 264–265
 GPx activity, 259–262
 insecticidal activity, 266–268
Biological target, 230–231
Bio-organic chemistry, 155
Bioorganometallic chemistry (BOC), 81–82
 description, 239–240
 mechanisms of action
  activation by hydrolysis, 269–271
  catalytic metallodrugs, 278–280
  ionizing radiation, sonodynamic and thermal activation, 276–278
  photoactivation, 275–276
  redox activation, 271–275
 naturally occurring moieties, 239–240, 240f
 organoselenium compounds
  biological activities of, 259–268
  as chemotherapeutic agents, 250–259
  platinum metal based therapeutics to biocompatible less toxic nonplatinum complexes, 241–250
Bioorganometallics, 239–241. See also Bioorganometallic chemistry (BOC)
Bio-probes for cellular imaging, 79–81
BNCT. See Boron neutron capture therapy (BNCT)
BOC. See Bioorganometallic chemistry (BOC)
Bonding
 covalent, 32–34
 dative bonds, 36, 37f
 delocalized bond, polynuclear systems, 35, 36f
 electron deficient/polycentric localized, 34–35, 35f
 hydrogen, 36–38
 ionic, 32
 M-C bonds, 32
 modes, sensing
  concerted track based, 222–223
  ligand substitution, 219
  oxidative addition, 219–221
Borane
 arachno, 95–96
 closo, 94–96, 95f
 clusters, 94–95, 95f

Index   415

hypho, 94–95
klado, 94–95
nido, 95–96
reactions of, 99
Boron hydrides
　categories, 96t
　diborane, 93–94
Boron neutron capture therapy (BNCT), 108
BPMEN ligand, 396f
Bromo substituted diselenide compounds, 264–265

C

Cadet's fuming arsenical liquid, 2
Cages and clusters
　Al$_{50}$(cyclopentadienyl)$_{10}$, 93, 94f
　applications, 103–108
　carbonyl clusters, 96–97
　diborane, 93–94
　mechanism, 100–101
　molecular boranes (BnHn), 94–95
　noble metals, 108–109
　noncovalent interactions, 93
　preparation methods, 97–99
　principles, 97
　recent developments, 101–108
Carbometallation, 63–64
Carbon-carbon bond formation, 115–117, 117f, 134–136
Carbon dioxide fixation, 85–86
Carbonylation, 63–64
Catalysis
　homogeneous, 155–157
　organometallic complexes used in, 17, 18f
　silicon and germanium organometallic compounds, 205–210
　tridentate chelating pincer complexes, 169–172
Catalytic metallodrugs
　biomacromolecules, degradation and cleavage of, 280
　NADH and thiols, oxidation of, 279
　reduction enroute transfer hydrogenation, 278–279
Cationic organometallic hydrides, 37, 38f
Cativa process, 17
CCS. See CO$_2$ capturing and storage (CCS)
Cellular imaging, 79–81
3-Center-2-electron interaction, 15
C-H activation
　iron-catalyzed, 118–119

preparation method and mechanisms, 117–133
principle of, 116–117
rhodium-catalyzed, 119–133
ruthenium-catalyzed, 117–118
Chagas disease
　chemical structures of selected drugs, 355f
　ferrocenyl- and cyrhetrenylimine complexes, 358–359
　ruthenium-azole complexes, 355–356
　ruthenium-hydrazone complexes, 356
　ruthenium-thiosemicarbazone complexes, 357–358
C-H bond activation
　adapted mechanism for, 390f
　pincer ligands, 158, 158f, 159f
Chemical sensors, 217–218, 225–228
Chemopreventive compounds
　ISCs, 258–259
　ITCs, 258–259
　organofluorine ISC analogs, SFN, 259
Chemotherapeutics
　1,2-benzisoselenazole-3[2H]-one derivatives, 255–256, 256f
　indole chalcone compounds, 257
　inorganic selenium compounds, 251–252
　isoselenocyanate compounds, 258–259
　organic selenium compounds, 252
　organoselenium compounds, 250–259
　Sec, 253
　seleninic acids, 254–255, 255f
　selenium-bearing 4-anilinoquinazoline compounds, 256–257
　selenophene-based derivatives, 255, 255f
　selol, 253–254, 254f
Chirality, 5–6
Chiral N-heterocyclic carbenes (NHCs)
　applications, 148–150
　asymmetric hydrosilylation reaction, 144, 145f
　asymmetric transition-metal catalysis, 140
　bicyclic triazolium, 142–143
　bottelable, 139–140, 151
　chiral inclusion, 141–142, 141f, 150
　C2 symmetry, 144–145, 146f
　cyclometalated, 142–143
　donor qualities, 140, 150
　enantioselective reaction, 140, 151
　mechanism, 143–144, 144f
　nitrogen atoms, presence of
　　nucleophilicity, 141
　　resonance condition, 141f

416 Index

Chiral N-heterocyclic carbenes (NHCs) (*Continued*)
  and phosphines, comparative structure, 140*f*
  preparation methods, 142–143
  principles, nitrogen atoms, presence of, 141–142
  recent developments, 144–148
  symmetry, 150
Chiral tridentate chelating pincer complexes, 172–174
Cisdiamminedichloroplatinum(II), 241
cis-Dihydrobenzimidazo[2,1-a]isoquinolines, 130–131, 131*f*
Cisplatin, 81–82, 241–242, 246–248, 269–270, 276–277
Closo boranes, 94–96, 95*f*
Clotriamzole (CTZ), structures of, 356*f*
Clusters, 94–95, 95*f*
Cluster valence electrons (CVEs), 96–97. *See also* Cages and clusters
CMCs. *See* Coordination molecular cages (CMCs)
CMD. *See* Concerted metalation deprotonation (CMD)
CMOS FETs. *See* Complementary metal oxide semiconductor field effect transistors (CMOS FETs)
CO_2 capturing and storage (CCS), 85
Coenzyme B12, 17
Complementary metal oxide semiconductor field effect transistors (CMOS FETs), 198–199
Complex (multicenter) organometallic compounds, 12
Concerted metalation deprotonation (CMD), 124
Cooperative catalysis, 148
Coordination compounds, 181
  phosphorus-based, 155–157
  pincer ligands, 157, 161–169
Coordination molecular cages (CMCs), 93
Copper(II)-borondipyrromethene conjugates, 244–245
Copper complexes, anticancer properties, 243–245
Copper II cyclen, 243
Coupling reactions, 9–10
  C-C bond formation, 115–117, 117*f*, 134–136
  C-H activation
    iron-catalyzed, 118–119
    preparation method and mechanisms, 117–133

  principle of, 116–117
  rhodium-catalyzed, 119–133
  ruthenium-catalyzed, 117–118
  cross-coupling, 10
  homocoupling, 10
Covalent organometallic compounds, 12
σ-bonding, 12
Cross-coupling reactions, 10
CVEs. *See* Cluster valence electrons (CVEs)
Cyclized isocoumarinoselenazoles, 117
Cycloaddition reaction
  maleimide with 2-aryl-1H-benzo [d] imidazoles, 130–131, 131*f*
  Rh-catalyzed, 131–132, 132*f*
Cyclodiphosphazanes, 181, 183
  cis/trans isomerization, 175, 183
  description, 175
  macrocyclic, 179
  reactivity of, 177–179
  synthesis of, 175–176
Cyclometalated organopalladium(II) complexes, structures of, 365*f*
Cyclopentadiene (Cp) compounds, preparation of, 14
Cytoprotection
  bromo substituted diselenide compounds, 264–265
  organoselenium antioxidant compounds, 264
Cytotoxicity
  ferroquine, 274
  ISCs, 258
  Noyori-type Ru^{II} arene complexes, 278–279
  Se containing compounds, routes of, 252*f*

D

Dative bonds, 36, 37*f*
Decaborane, deprotonation of, 101, 101*f*
Decarbonylation, 15, 63–64
Decoupling, 40
Delocalized bond, polynuclear systems, 35, 36*f*
Dengue virus (DENV), 369–370
Density functional theory (DFT), 382
DEPT. *See* Distortionless enhancement by polarization transfer (DEPT)
Diarylalkynes, 118
Diborane, 93–94
1,2-Dichloroisobutane (DCIB), 118
Diels alder synthesis, 79*f*

3,4-Dihydropyrimido[1,6-a]indol-1(2*H*)-ones, 124, 125*f*
Dihydroquinazolin-4(1*H*)-one
   oxidant-free synthesis of, 130, 131*f*
   Rh(III)-catalyzed synthesis of, 129, 130*f*
Dimethylmercury, 2−3
Dinitrosyl iron complex (DNIC), 242−243
Distortionless enhancement by polarization transfer (DEPT), 41−42
DNIC. *See* Dinitrosyl iron complex (DNIC)
DSSC. *See* Dye-sensitized solar cell (DSSC)
Dye-sensitized solar cell (DSSC), 83−85

E

EAN rule. *See* Effective atomic number (EAN) rule
E-aryl alkenes, 144, 145*f*
Ebselen, 255−256
Ecchinococoosis, drugs for the treatment of, 366*f*
EDA. *See* Energy decomposition analysis (EDA)
Effective atomic number (EAN) rule, 7
Electro-catalysis, 19−20
Electron-deficient bond, 34−35, 35*f*
Electron nuclear double resonance (ENDOR), 47−49, 48*f*, 49*f*
Electron paramagnetic resonance (EPR), 44−47
18-Electron rule, 7−8, 20, 96−97, 111
Electron spin echo envelope modulation, 49−50
Elephantiasis. *See* Lymphatic filariasis
Emerging fields, applications in
   of dye sensitized solar cell, 344−345
   of medicine, 343−344
ENDOR. *See* Electron nuclear double resonance (ENDOR)
Energy decomposition analysis (EDA), 36
Enzyme inhibition, 82−83
EPR. *See* Electron paramagnetic resonance (EPR)
Ethaselen, 255−256
Ethyl mercury chloride, 196−197
EXAFS. *See* X-ray absorption fine structure spectrum (EXAFS)

F

Ferrocene, 241−242, 274
Ferrocenyl-diamine complexes, structure of, 361*f*

Ferrociphenols, 242, 242*f*
Ferroquine, 274
   antimalarial drug, 17−18, 18*f*
Fluxional behavior, 9*f*, 23
Fluxionality, 9
Friedal-Crafts reaction, 100
Furan derivatives, tri-substituted, 126, 127*f*

G

Gas sensor, 228−229
Germanium. *See also* Silicon and germanium organometallic compounds
   2-D structural representations, 198*f*
   electronic configuration, 197*f*, 197*t*
   films, PRAM devices, 198−199
   Gr-IV semiconductors, 198−199
   high costs, 197−199
   physical properties, 197*t*
   porous, 200*f*
Germylenes, 209, 210*f*
Gilman's reagent, transition metal in, 22
(+)-Gliocladin C, synthesis of, 79, 79*f*
Global warming, 83, 85−86
Glutathione peroxidase (GPX) activity
   benzisoselenazolones, 262
   2,7-dialkoxy-substituted naphthalene-1,8-peridiselenides, 259
   pH-sensitive organoselenium compounds, 259−261
Gold complexes, anticancer properties, 248
Green chemistry, 115−116
Grignard reagent, 3, 18, 63
   preparation of, 13, 21

H

Half-sandwich ruthenium (II) complexes, anticancer activity of, 246−248
Halide-free organometallic complexes, 71−72, 73*f*
Hapticity, 7−8, 8*f*, 195
Heck coupling reaction, 22
Heck cross-coupling reactions, 170−171, 182−183
Heck reaction, 17
Heptacity, 111
Heteronuclear experiments
   DEPT, 42
   incredible natural abundance double quantum transfer, 42−44
   INEPT, 41−42
   J modulation test, 41

Homocoupling reactions, 10
Homogenous/heterogeneous catalytic applications, 155
Homonuclear decoupling, 40
Human African trypanosomiasis
  drugs for treatment of, 360f
    ferrocenyl and ruthenocenyl complexes of quinoline and benzimidazole, 359–360
    ferrocenyl-diamine complexes, 361
    ruthenium-thiosemicarbazone complexes, 361–362
Hund's rule, 80–81
Hybrid drug approach, 314
Hybrid organometallic compounds
  with antiplasmodial activity, 318f
  ferrocene-quinoline containing, 318, 318f
  future prospective and scope, 323–324
  molecular structure of ruthenocene containing, 319f
  structures of trioxaferroquine and trioxaferrocene, 317f
Hybrid organometallic drug, trioxaferroquine containing, 317–318
Hydrido nickel complexes, 71, 72f
Hydrocarbon hydroxylation, pyridine/imine based ligands, 384f
Hydroformylation, olefins, 3–4
Hydrogenation
  of alkenes, 196–197
  E-aryl alkenes, 144, 145f
  ketones, 73–74, 74f
  nitroarenes, 74–75, 75f
Hydrogen atom transfer (HAT), 397–398
Hydrogen bonding
  categories, 36–37
  cationic organometallic hydrides, 37, 38f
  hydride ligands, 38
  transition elements, 37–38
Hydrometallation, 63–64
Hydrosilylation
  metal-catalyzed, 200, 201f
  Si-C bonding, 200
Hydrosilylation reaction, acetophenone, 144, 145f
Hydroxylase enzymes, 379–380
Hydroxylated products of pharmaceutical and synthetic, 380f
Hydroxylation
  applications in emerging fields, 399–401
  benzyl and phenyl, 394f
  B3LYP calculated TSR/MSR scenario during, 400f

computational details, 384
factors affecting, 395–399
  axial and equatorial ligands, 397–398
  ligand design, 396–397
  nature of ligand, 396
  oxidation state (spin state) of metal ion, 395–396
  single-state reactivity vs. two-state reactivity, 398–399, 398f
mechanism involved, 386–395
  C-H bond activation followed by oxygen rebound, 386–388
  direct attack of oxygen atom, 388–390
  hydrogen abstraction followed by oxygen non-rebound, 388
  nonradical, 391–392
  regioselectivity of aliphatic vs. aromatic hydroxylation, 392, 392f
  sigma pathway and pie pathway, 393–395, 395f
  using cyclohexene, 392–393
of methane, 387f
by monooxygenase, 380f
overview, 379–383
prospective and scope, 401–402
recent developments with examples, 384–386
of substrate by sMMO enzyme, 387f
Hypho, 94–95

I

Imatinib, 82
Imidazole derivatives, Mn(I) complexes with, 370f
Indazole derivatives, 124–126, 125f, 126f
Indenamines/aminoindanes
  synthesis of, 122, 122f
  trifluoromethyl substituted, 120–122, 121f
Indole chalcone compounds, 257
Indolones
  dihydropyrimidoindolones, 124
  isochromenoindolones synthesis, 122–124, 123f
  pyrimidine fused, 124, 124f
Indomethacin, 244
INEPT. See Insensitive nuclei enhanced by polarization transfer (INEPT)
Infrared (IR) spectroscopy, 50–51, 51f
Inorganic complex of CZT, structures of, 356f
Inorganic selenium compounds, 251–252, 253f

Index   419

Insensitive nuclei enhanced by polarization transfer (INEPT), 41–42
Insertion, migratory, 66–67
Ionic bond, 32, 33f
Ionic organometallic compounds, 12
Ionizing radiation, 276–278
Iridium-chloroquine complexes, chemical structures of, 320f
Iridium complexes
  catalytic activity of, 75–76, 76f, 77f
  half sandwich, 75–76
  synthesis, 76–77, 77f
Iron-catalyzed C-H activation
  alkyne and Grignard reagent, annulation reaction of, 118–119, 120f
  1,2,3,4-tetraphenylnaphthalene, 118, 119f
Iron complexes, 76, 77f
  anticancer properties, 241–243
IR spectroscopy. *See* Infrared (IR) spectroscopy
ISCs. *See* Isoselenocyanates (ISCs)
Isochromenoindolones, synthesis of, 122–124, 123f
Isocoumarinoselenazoles, 117–118, 119f
Isoquinolines
  aryl-substituted, 119–120, 120f
  Rh-catalyzed synthesis of, 120, 121f
Isoselenocyanates (ISCs), 117, 258–259
Isothiocyanates (ITCs), 258–259
ITCs. *See* Isothiocyanates (ITCs)

J
J Modulation test, 41

K
Ketoconazole (KTZ), structures of, 356f
Klado, 94–95

L
Leishmaniasis
  chemical structures of drugs for treatment, 363f
  iron-based complexes, 363–364
  palladium-imine complexes, 364
  ruthenium-azole complexes, 363
Lewis base cleavage reaction, 100
Ligands. *See also* Pincer ligands
  bidentate, 156
  halide functionalized, 159
  lithiation of, 159
  monoanionic terdentate, 160
  phosphorus-based, 156–157
  and their ruthenium complexes, 370f
  tridentate chelating pincer, 160–161, 160f
  tridentate meridional, 157

M
Magic bullet, curing of syphilis, 3
Malaria
  affects, 310–311
  life cycle of, 311–312
  methylene blue and quinine, molecular structures of, 310f
  mosquito-borne infectious disease, 310
  overview, 309–311
Malaria parasites
  biological evolution of, 309
  life cycle of, 311f
  resistance of, 324
MB. *See* Methylene blue (MB)
Mechanisms
  Heck reaction, 17
  oxidative addition, 16–17
  simplified catalytic, coupling reactions, 15–16, 16f
Medicine, 17–18
Metal-based carbonyl compounds, 3
Metal boranes, classification of, 93–94, 96f, 96t
Metal carbonyl, preparation of, 14
Metal-hydrogen/metal-halogen exchange reaction, 63–64
Metalladithioacetals, 207–208
Metallathiazolidines, 207–208, 209f
Metallocenes, 8, 8f
Metallocenyl derivatives, with azole-based ligands, 298–301
Metallo-drug action
  catalytic metallodrugs
    biomacromolecules, degradation and cleavage of, 280
    NADH and thiols, oxidation of, 279
    reduction enroute transfer hydrogenation, 278–279
  photoactivation, 275–276
  ruthenium-based complexes, 246–247
Metalloenzymes, 379
Metalloporphyrins, 277–278
Metal-metal bond, 6, 23, 112
Metaloenediynes, 277–278
Metal organic polyhedrals (MOPs), 93

Metathesis, 14
  metal halides with alkylating agents, 63−64
Methane hydroxylation, 391f
Methylene blue (MB), 105−106
Methyl orange (MO), 105−106
Methylselenic acid, 254−255, 255f
Methylselenocysteine (MSC), 254−255
Migratory insertion, 66−67
M-M bonds
  in Ir$_4$(CO)$_{12}$, 112
  quadrupole bonds, 112
Mn(I) based organometallic complex, 67−68, 68f
MO. *See* Methyl orange (MO)
Mode of bonding, 11f, 12
Monsanto acetic acid process, 17
MOPs. *See* Metal organic polyhedrals (MOPs)
MSC. *See* Methylselenocysteine (MSC)
Multitargeted treatments for Alzheimer's disease, 263, 263f

N

N-alkylation, 117
Natural orbital for chemical valance (NOCV), 36
N-bonding, 12
Neglected tropical diseases (NTDs), 292, 353−354
  diseases, 354−371
    Chagas disease, 354−359
    dengue, 369−371
    echinococcosis, 365−366
    human African trypanosomiasis, 359−362
    leishmaniasis, 362−364
    lymphatic filariasis, 368−369
    schistosomiasis, 367−368
NHCs. *See* Chiral N-heterocyclic carbenes (NHCs)
NHCs-TM. *See* N-heterocyclic carbene transition metal (NHCs-TM)
N-heterocyclic carbene transition metal (NHCs-TM). *See also* Asymmetric catalysis, NHCs-TM
  applications, 148−150
  future, 150
  mechanism, 143−144, 144f
  preparation methods, 142−143
  recent developments, 144−148
Nido, 95−96
Nifurtimox/eflornithine combined therapy (NECT), 361−362
Nitazoxanide, chemical structure of, 366f
N-methylated ebselenamine, antioxidant property, 264
NMR. *See* Nuclear magnetic resonance (NMR)
Noble-gas rule. *See* 18-Electron rule
NOCV. *See* Natural orbital for chemical valance (NOCV)
NOE. *See* Nuclear overhauser enhancement (NOE)
Nonplatinum anticancer compounds
  copper complexes, 243−245
  gold complexes, 248
  half-sandwich ruthenium (II) complexes, 246−248
  iron complexes, 241−243
  rhodium and iridium complexes, 248−250
  ruthenium complexes, 245−248
N-substituted 2-amino isocoumarinoselenazoles, 117, 118f
Nuclear magnetic resonance (NMR)
  decoupling difference, 40
  heteronuclear experiments, 41−44
  NOE, 40−41
Nuclear overhauser enhancement (NOE), 40−41

O

Olefins, hydroformylation, 3−4
Optical sensor, 219, 229−230
Organic selenium compounds, 252, 253f
Organochlorosilanes, 2−3
Organometallic catalysis
  applications
    bio-organometallic chemistry, 81−82
    bio-probes for cellular imaging, 79−81
    enzyme inhibition, 82−83
    photoredox catalysis, 78−79
    solar cells, 83−86
  fundamental principles, 62
  mechanism, 65−67
  oxidative addition, 65−67
  preparation methods
    metals with chemical compounds, reaction of, 63−64
    metals with organic halides, reaction of, 62−63
  recent developments, 67−77
Organometallic complexes
  with azole derivatives, 296−298
  azole derivatives based ruthenium, 293−295

with azole ligands, 295–296
of CTZ and KTZ, structures of ruthenium-based, 357f
ferrocenyl and ruthenocenyl complexes of quinoline and benzimidazole derivatives, 361f
representative structures of platinum-based, 323f
ruthenium-cyclopentadienyl-based, 295f
structures of ruthenium-thiosemicarbazone, 358f
Organometallic compounds
as antimalarial candidates, 315–317
applications in emerging fields, 17–20
bonding, 32–38
fashion, 1
model (band theory), metal-alkene complexes, 4–5
chirality concept, 5–6
classifications, 10–13
bonding pattern based, 10
metal type based, 12
coupling reactions, 9–10
fluxionality, 9
fuel additive, used as, 23, 218f
hapticity, 7–8, 8f
metallocenes, 8, 8f
metal-metal bond, 6, 23, 38–39
molecular structure of furan and thiophene based, 359f
multicenter bonding, 21
prefix organo, 218
preparation methods, 13–15
problems with solutions, 371–372
representative structures of the ruthenium- and osmium-based, 322f
sandwich complexes, 4–5, 8–10, 9f
selectivity, 217–218
silicon-based, 3
spectral characteristics, 39
summary, 371
tautomerism, 5–6
transition metal-based, 2
Organometallics
bonding, 32–38
definition, 31
M-C bonds, 195, 199
organophosphorus compounds
phosphorus-based ligands, 156
pincer complexes, classification of, 160–181
preparation methods, 157–160

principles, 156–157
phosphorus-based compounds, 155–157
pincer ligands, 157, 161–169
silicon and germanium compounds
bonding through carbon, 202–205
catalysis, 205–210
direct synthesis, 200
future, 210–211
hydrosilylation, 200
silenes, 200
silicones, silyl ethers, silanols, siloxanes, and siloxides, 201–202
spectral characteristics, 39
Organophosphorus compounds
phosphorus-based ligands, 156
pincer complexes, classification of, 160–181
preparation methods, 157–160
principles, 156–157
Organoruthenium aminoquinoline-trioxane hybrid, 318–319
Organoselenium compounds
biological activities of, 259–268
anti-Alzheimer activity, 262–264
cytoprotection, 264–265
GPx activity, 259–262
insecticidal activity, 266–268
as chemotherapeutic agents, 250–259
1,2-benzisoselenazole-3[2H]-one derivatives, 255–256, 256f
indole chalcone compounds, 257
inorganic selenium compounds, 251–252, 253f
isoselenocyanate compounds, 258–259
organic selenium compounds, 252, 253f
seleninic acids, 254–255, 255f
selenium-bearing 4-anilinoquinazoline compounds, 256–257
selenocysteine, 253
selenophene-based derivatives, 255, 255f
selol, 253–254, 254f
Organosilicon compounds
bonding through carbon, 202–205
hydrosilation, 200
silanols, 202
silenes, 200
silicones, 201, 202f
siloxanes, 202
siloxides, 202
silyl ethers, 201–202
synthesis, 200

422 Index

Oxidative additions, 15, 63−64
  bimolecular nucleophilic substitution, 337
  concerted mechanism, 65
  concerted track, 336−337
  description, 65
  hydride elimination, 67
  ionic-based mechanism, 337
  migratory insertion, 66−67
  radical-based mechanism, 338
  reductive elimination, 65−66
  requirements, 65
  SN^2 mechanism, 65
Oxo-process, 3−4
Oxygen rebound *vs.* oxygen nonrebound mechanism, 389*f*
Oxymetallation, 63−64

P

PCP pincer complexes
  carbon monoxide, activation of, 162, 162*f*
  alkane metathesis, iridium pincer, 163, 164*f*
  CO insertion
    nickel pincer complex, 165−166, 166*f*
    platinum pincer complex, 165, 166*f*
  dihydride and tetrahydride complexes, 161*f*, 162, 162*f*
  norbornene, hydrogenation of, 163, 163*f*
  binuclear, 165, 165*f*
  dehydroaromatization reaction, 163, 164*f*
  dehydrochlorination, rhodium, 161−162, 161*f*
  dihydride iridium(III) complexes, 163*f*
  rhodium(II) and iridium complexes, 161−162
  tridentate-chelating pincer ligands, 161*f*
  olefin group, replacement of, 166, 166*f*
  palladium(II) and platinum (II), 163−165, 165*f*
Penta-aurated indolium compound, 36*f*
1-Pentene hydroformylation, 88
Phase-change random access memory (PRAM) devices, 198−199
Phosphacyclophanes, 179−182, 180*f*
Phosphorus-based ligands
  bite angle, 156
  classification of, 156−157, 156*f*
  cone angle, 156
  ligating ability, 156
  steric demand, 156
  trivalent, 156−157

Phosphorus compounds
  classification, 181
  coordination/organometallic chemistry, 155
  phosphorus-based ligands, 156−157
Photoactivation, 275−276
Photoredox catalysis
  C-H functionalization, 78
  cycloaddition reactions, 79
  (+)-gliocladin C, synthesis of, 79, 79*f*
  hexa-coordinated metal complexes, 78
  Pschorr reaction, 78
Picolinamides, 72−73, 73*f*
Pincer ligands
  classification, 182
    aminophosphines, 174−179
    chiral tridentate chelating pincer complexes, 172−174
    PCP pincer complexes, 161−166
    phosphacyclophanes, 179−181
    PNP pincer complexes, 166−169
    transition metal chemistry, 161−169
    tridentate chelating pincer complexes, catalysis, 169−172
  description, 157
  donor sites, 182
  example of, 157*f*
  synthesis, complex
    C-H activation, 158, 158*f*, 159*f*
    oxidative addition, 159
    trans-cyclometallation, 160, 160*f*
    trans-metallation, 159, 159*f*
PNP pincer complexes
  β-aminoethyl complex, synthesis of, 168−169, 169*f*
  cationic palladium, 169, 169*f*
  C-C coupling reaction, 167−168, 168*f*
  CO reaction, iron complexes, 167−168, 168*f*
  dearomatization, pyridine rings, 168−169, 168*f*
  iron containing, 166−167, 167*f*
  silver(I) complex, 169, 170*f*
  symmetrical, 167*f*, 183
Polycentric localized bonds, 34−35, 35*f*
Polymer, 158
Polynuclear organometallic complexes
  applications in emerging fields, 343−345
  common methods of preparation, 332−334
    metal fragments condensation, 334
    oxidative addition reactions, 334
    substitution of ligand on metal carbonyls, 333

template method of preparation, 333–334
fundamental principles, 332
future prospective and scope, 345
mechanism involved, 335–343
  hydrometalation/β-hydride elimination, 339–340
  ligand substitution, 335–336
  migratory insertion reaction, 339
  oxidative addition, 336–338
  recent developments with examples, 340–343
  reductive elimination, 338–339
  transmetalation, 340
Polypyridinic complexes, 83–84
Porphyrins, 277–278
PRAM devices. *See* Phase-change random access memory (PRAM) devices
Primaquine, 364*f*
Protein kinase, 82–83
Prussian blue, 61–62
Pschorr reaction, 78
Pyrimidine fused indolones, 124, 124*f*

R

Raman spectroscopy (RS)
  alkyl-chloro aluminum compounds, 52–53, 53*f*
  description, 52
  Li organo-chloro-aluminum complexes, 53, 54*f*
Red blood cells (RBCs), 311
Redox activation
  ligand oxidation, 274–275
  ligand reduction, 273
  metal oxidation, 274
  metal reduction, 271–272
Reduction enroute transfer hydrogenation, 278–279
Reductive carbonylation, metal oxides, 15, 16*f*, 63–64
Reductive elimination, 65–66
Reversible addition-fragmentation chain-transfer (RAFT) polymerization approach, 19
Rh(I)COD, structural representation of, 321*f*
Rh^{III} and Ir^{III} Cp complexes, 270–271
Rhodium and iridium complexes, anticancer properties, 248–250
Rhodium-catalyzed C-H activation
  4-arylisoquinolines, 128–129, 128*f*, 129*f*, 130*f*

C(vinyl)-H activation, 127, 127*f*
  indenamines/aminoindanes, 122, 122*f*
  oxidation and reduction reactions, 133
  pyrimidine fused indolones, 124, 124*f*
  tri-substituted furan derivative, 126, 127*f*
4-arylisoquinolones, 127–128, 128*f*
  aryl-substituted isoquinolines, 119–120, 120*f*
  isoquinolines, 120, 121*f*
  [4 + 2]-cycloaddition method, 130–132, 131*f*, 132*f*
  3,4-dihydropyrimido[1,6-a]indol-1(2H)-ones, 124, 125*f*
[4 + 1] spiro annulations
  asymmetric spiro annulation reaction, 132–133, 133*f*
  dihydroquinazolin-4(1*H*)-one, 129–130, 130*f*, 131*f*, 132, 133*f*
  indazole derivatives, 124–126, 125*f*, 126*f*
  isochromenoindolones, 122–124, 123*f*
  trifluoromethyl substituted indenamines/aminoindanes, 120–122, 121*f*
RS. *See* Raman spectroscopy (RS)
Ru-ATZ complex, structure of, 357*f*
Ru(II) complexes, half sandwich, 68, 69*f*, 70–71, 71*f*, 74–75
Ru^{II} and Os^{II} half-sandwich complexes, 270
Ru(n^6-p-cymene)-(PTA)Cl$_2$ (RAPTA-C), 246–247, 246*f*
Ruthenium-catalyzed C-H activation
  isocoumarinoselenazoles, 117–118, 119*f*
  N-substituted 2-amino isocoumarinoselenazoles, synthesis of, 117, 118*f*
Ruthenium complexes
  anticancer properties, 245–248
  half-sandwich, anticancer activity, 246–248
  Ru(II)-arene complexes, 246, 246*f*
Ruthenium hydrido carbonyl complexes, 73–74
Ruthenium polypyridyl complexes, 83–84

S

Sandwich complexes, 4–5, 8–10, 9*f*
σ-bonded organometallic compounds, 33–34, 34*f*, 35*f*
Schistosomiasis
  drugs for treatment of, 367*f*
  praziquantel drug, 368*f*
Sec. *See* Selenocysteine (Sec)

Selectivity, 217–218
Selenazoles
  2-amino benzoselenazoles, 117
  isocoumarinoselenazoles, 118, 119f
  N-substituted 2-amino
    isocoumarinoselenazoles, 117, 118f
Seleninic acids, 254–255, 255f
Selenium-bearing 4-anilinoquinazoline
  compounds, 256–257
Selenocysteine (Sec), 253
Selenodihydropyrimidinones, Alzheimer's
  disease, 263, 263f
Selenoethers, 266
Selenophene, 268
Selenophene-based derivatives, 255, 255f
Selenyl-dihydrofurans as anti-Alzheimer
  agents, 262
Selol, 253–254, 254f
Sensing
  applications
    biochemical sensor, 230–231
    chemical sensors, 217–218, 225–228
    gas sensor, 228–229
    optical sensor, 219, 229–230
  future, 231
  mechanism
    concerted track based, 222–223
    ligand substitution, 219
    oxidative addition, 219–221
  preparation methods, 219
  principles, 218–219
  recent developments, 223–225
Sensors, 19
SFN. See Sulforaphane (SFN)
Sigma antibonding, 219
Silanols, 202
Silenes, 200
Silicon. See also Silicon and germanium
  organometallic compounds
  2-D structural representations, 198f
  electronic configuration, 197f, 197t
  oligomers and polymers, 196
  ordered surfaces, 198–199, 199f
  physical properties, 197t
  porous, 200f
  silicon-based rubbers, 196–197
  silicon carbide, 196
Silicon and germanium organometallic
  compounds
  bonding through carbon, 202–205
  catalysis, 205–210
  direct synthesis, 200

future, 210–211
germanium-carbon bond, 208, 210f
germylenes, 209, 210f
hydrosilylation, 200
silanols, 202
silenes, 200
silicones, 201, 202f
siloxanes, 202
siloxides, 202
silyl ethers, 201–202
Silicon-based organometallic compounds, 3
Silicones, 201, 202f
Siloxanes, 202
Siloxides, 202
Silver NHC catalysts, 144–145, 146f
Silyl ethers, 201–202
Single-state vs. two-state reactivity, 398–399,
  399f
single walled carbon nanotube (SWNTs),
  33–34
Sleeping sickness. See Human African
  trypanosomiasis
Solar cells
  carbon dioxide fixation, 85–86
  DSSC, 83–85
  ruthenium-based complexes,
    photosensitizer, 84f, 84t
  ruthenium polypyridyl complexes,
    83–84
Soluble methane monooxygenase (sMMO),
  380
Sonodynamic therapy, 277–278
Sonogashira coupling reaction, 16–17
Sonoluminescence, 277–278
Spectral characteristics, 39
  electron spin echo envelope modulation,
    49–50
  ENDOR, 47–49
  EPR, 44–47
  IR, 50–51
  NMR, 39–44
  RS, 52–53
  UV-Vis, 51
  XAS, 53–55
Spiro annulations
  asymmetric, 132–133, 133f
  O-pivaloyl oximes with a-diazo compounds,
    132, 133f
Square-planar PtII complexes, 269
SSR. See Synchrotron stereotactic
  radiotherapy (SSR)
Staurosporine, 82–83

Sulforaphane (SFN), 259
SWNTs. *See* single walled carbon nanotube (SWNTs)
Synchrotron stereotactic radiotherapy (SSR), 277
Synthesis, silicon and germanium organometallic compounds, 200
Synthetic organometallic complexes, 331

T

Tautomerism, 5–6
Tetraamido macrocyclic ligand (TAML), 382–383
Tetraethylsilane, preparation, 200
1,2,3,4-Tetraphenylnaphthalene, synthesis of, 118, 119*f*
Ti^{IV} dichloride complexes, 269
Titanocene Y, 17–18, 19*f*
Trans-alkylation approach, 3
Transition metal chemistry, pincer ligands
  PCP pincer complexes, 161–166
  PNP pincer complexes, 166–169
Transition metal NHC complexes. *See* N-heterocyclic carbene transition metal (NHCs-TM)
Transition structure (TS), 398
Transmetallation, 13–14, 21
Tridentate chelating pincer complexes
  chiral, organic transformation, silver(I) complex, 173*f*
  Heck cross-coupling reactions, 170–171
  hydrophosphination reaction, 172
  organic transformation, silver(I) complex, 172–174
    examples, 173*f*
    palladium complexes, 173
    stereochemical centers, 172
    transcyclometallation, 174, 174*f*
  oxidative addition on aryl halide, 174, 174*f*
Trifluoromethyl substituted indenamines/aminoindanes, 120–122, 121*f*

Tubulin polymerization inhibition, 256–257
Two-state reactivity (TSR), 398

U

Ultraviolet (UV) spectroscopy, 39, 51
Ultraviolet-visible (UV-Vis) spectroscopy, 51, 52*f*
US Food and Drug Administration, 362
UV-Vis spectroscopy. *See* Ultraviolet-visible (UV-Vis) spectroscopy

V

Vaska's complex, 4–5, 65
Vitamin B$_{12}$, 17

W

Weel inhibitor, 248–250
Wilkinson's catalyst, 15, 23, 88, 196–197
World Health Organisation (WHO), 353–354

X

XAS. *See* X-ray absorption spectroscopy (XAS)
X-ray absorption fine structure spectrum (EXAFS), 53–54, 54*f*
X-ray absorption near edge spectrum (XANES), 53–54, 54*f*, 55*f*
X-ray absorption spectroscopy (XAS), 53–55, 54*f*, 55*f*

Y

Yoshizawa model of sMMO, 391*f*

Z

Zeise's salt, 2, 61–62, 195–196
Ziegler-Natta polymerization process, 19
Z-selenoenynes, 268

Printed and bound by CPI Group (UK) Ltd, Croydon, CR0 4YY
21/06/2024
01013803-0006